DRAFTING

Paul Wallach

Glencoe Publishing Co., Inc.
Encino, California
Collier Macmillan Publishers
London

Glencoe Publishing Co., Inc.
17337 Ventura Boulevard
Encino, California 91316
Collier Macmillan Canada, Ltd.

Library of Congress Catalog Card Number: 80-84394

ISBN 0-02-829600-1

1 2 3 4 5 6 7 8 9   85 84 83 82 81

Contents

Preface

Drafting is a form of graphic communication. It expresses ideas and conveys specific information by means of geometric shapes, lines, and dimensions. These elements comprise the language of drafting, a language that is uniform and standardized throughout the world.

The drafting language is expressed in one of two forms: the U.S. Customary system and the metric system. In this book, the U.S. Customary measurement is the primary learning and teaching tool. It is emphasized because, at present, this system continues to be the preferred standard system of measure in the United States.

At the same time, the metric system is introduced to students who may not be familiar with it. Measurement equivalents in both systems are explained, and the technique of converting dimensions from one system to the other is described. The fundamentals of drafting are taught in customary and metric terms, and the chapter exercises are planned to give students practice in solving problems in either system.

Purpose

The main purposes of *Drafting* are to develop the drafting student's ability to:

1. Visualize objects in three dimensions
2. Prepare working drawings
3. Read working drawing blueprints
4. Think through drafting problems clearly and completely
5. Draw clearly and accurately
6. Obtain substantial practice in drafting techniques

Organization

To a certain extent, the organization of *Drafting* was dictated by the subject matter. The chapters follow the sequence of subjects that many drafting instructors normally use in teaching the drafting course. In addition, each chapter is a self-contained unit. Chapters can be presented in the order found in the book, or they can be arranged in an order that suits the instructor's preference.

Drafting is divided into six parts. Each part represents an important but fundamentally separate aspect of drafting.

Part I, "Introduction to Industry," introduces the student, in a general way, to the subject of drafting and how it relates to industry. Major career areas in drafting are identified and described in a comprehensive overview.

Part II, "Drafting Conventions," is a short survey of the applications of engineering drawing. This gives the student the opportunity to quickly grasp the major conventions of technical drafting. Among the topics covered are aeronautical, aerospace, and ship drafting, cartography, and technical illustration.

Part III, "Preparation for Drafting," presents the basic techniques, skills, and knowledge required by drafting students before they start to draw technical drawings. The subjects covered in this part include freehand sketching, lettering, drafting equipment, line conventions, drafting scales, introduction to metrics, sheet formats, and geometric construction.

Part IV, "Drafting Concepts," surveys the different types of technical drawings. The chapters in this part discuss pictorials and isometric drawings, orthographic projection, dimensioning, tolerancing, surface finishes, sections, revolutions, perspective, and auxiliary view.

Part V, "Industrial Drafting," covers the specialized applications of drafting used in industry. Examples of actual industrial working drawings illustrate the instructions and demonstrate the high level of drawing skill that is essential. In this section, the student is introduced to technical illustration, treads, gearing, cams, electronics, pipe drafting, design/working drawings, the design process, architectural drafting, and computer graphics.

Part VI, "Reproduction," presents a look at reproduction methods and the uses of reproduced engineering drawings. Microfilming is covered in this section.

The Appendixes include glossaries of technical terms, abbreviations, engineering reference tables, and conversion tables.

The Program

The *Drafting* program consists of this textbook, a problems book, and an instructor's guide. The problems book contains more than 400 drawing problems. It is an excellent skill-building workbook and, in addition, it helps students become familiar with the types of problems that are found in typical work situations.

All exercises in the problems book are given in both U.S. Customary and metric measure. Thus, the teacher has a choice of method and the student gains experience in both measurement systems.

The instructor's guide provides a complete overview of the program. It contains chapter objectives, general teaching suggestions, additional activities, discussion questions, and a set of transparency masters. The transparency masters aid the teacher in visually presenting complicated concepts in an effective way. The masters may be used to make overhead transparencies, or they can be duplicated and distributed to the students as assignments or for reference.

Acknowledgment

The author wishes to acknowledge and to thank the many students and teachers who contributed to the development of this text. In particular, we wish to express our appreciation to Professor Cortland C. Doan of California State University, Los Angeles; Mr. Andrew J. Oliver of Newport Harbor High School, Newport Beach, California; and Mr. Fred Baer, District Supervisor of Industrial Education of the Los Angeles City Unified School District for their help and suggestions.

Special appreciation is expressed to the following for their aid in the preparation of the specific chapters:

Dean Chowenhill (Positional Tolerancing, Gearing, Cams, Pipe Drafting)

Dave Duncan (Technical Illustration)

Helen Hunter (Introduction to Industry)

Mike Robbins (Electronics)

Irv Simon (The Design Process)

Jack Brown (Computer Graphics)

Paul I. Wallach

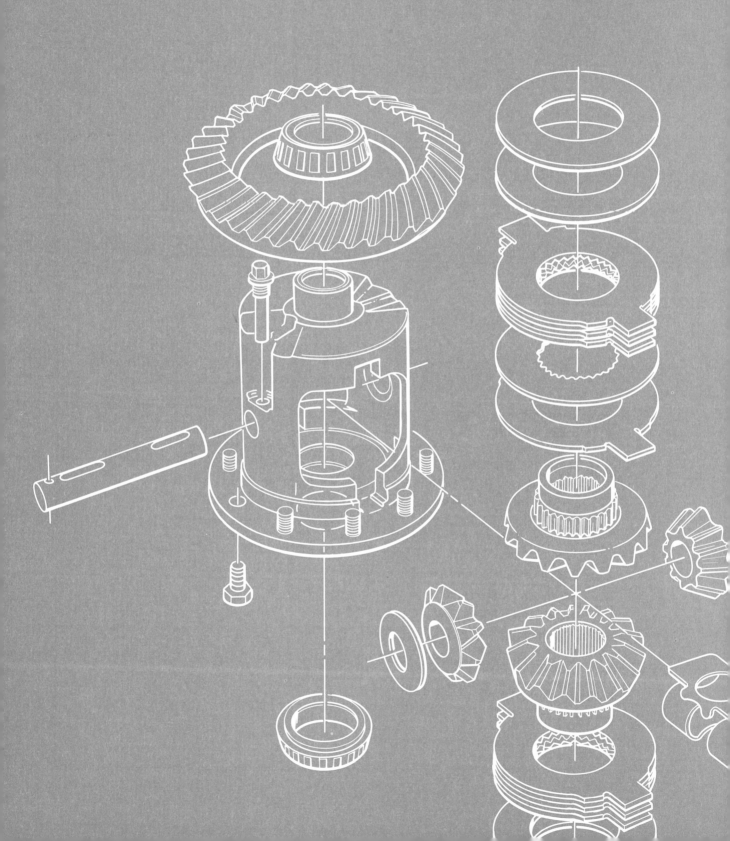

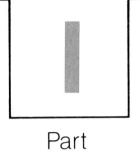

Part

Introduction

We are living in an age of machinery. Our industrial complex has a profound effect on our lives and how we live them. Engineers, designers, and draftspersons are continually developing and improving machinery.

Part I of this text is an introduction to our amazing industrial complex, and the part you, as a student of drafting, can play in it.

1

Chapter

Introduction to Industry

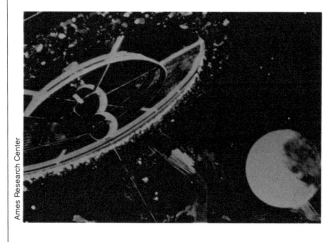

Ames Research Center

'Twas brillig, and the slithy toves
　　Did gyre and gimble in the wabe;
All mimsy were the borogoves,
　　And the mome raths outgrabe.
　　　　　　　　　"Jabberwocky" by Lewis Carroll

Gibberish may be used in poetic fantasy, but we avoid it when we want to understand each other. Clear, precise language allows us to communicate effectively. Communication includes speaking, writing, and reading. The more exact our expression, the more likely our meaning will be understood.

Drafting is graphic communication—a visual expression of information and ideas. Drafting communicates thoughts and specific information. Therefore, it can actually be considered a language. And it is an international language. Drafting techniques allow technical data to be recorded on paper so that a blueprint reader anywhere in the world can translate the information precisely into a specific product.

In this book the language of drafting measurement is expressed in two ways—the U.S. Customary System and the metric system. At present, the United States is in a transitional period from one system of measurement to another. The U.S. Customary System is still the prevalent method of measuring in the United States. But conversion to the metric system has already begun, as you probably have noticed from the products that you buy every day.

FIG. 1-2 A vast number of precise drawings were necessary to reach this stage of the space shuttle's development.

FIG. 1-3 The drafting process (General Electric Co.) (Teledyne Post)

When the United States has completed its conversion to the metric system, all the major industrial nations of the world will be using one common language of measure.

Graphic drawings have undergone many changes since people first expressed their perceptions of the world on cave walls and on the surface of the earth. Today, there are many specific types of drawings. Each conveys a distinct message involving technical information. Many types of drawings used in the technical graphic language will be covered in the chapters that follow.

In the field of drafting there are two primary functions of a drawing. A drawing may be designed to provide a general view of a subject, as a perspective. Or it may be designed to convey information for accurate production, as a working drawing.

The Space Shuttle is depicted in different types of drawings. The Viking Mission to Mars also re-

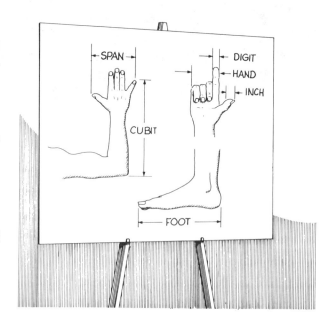

FIG. 1-4 Origin of the U.S. Customary System

FIG. 1-5 Evolution of Drafting (Compucircuit)

Before Grading

Rotation= 20   Elevation= 20   Elevation multiplier= 1.20

After Grading

EXTERIOR ARRANGEMENTS
747, 747B, 747C, 747F

69 800 mm

59 640 mm   22 170 mm

11 020 mm

19 330 mm

68 600 mm

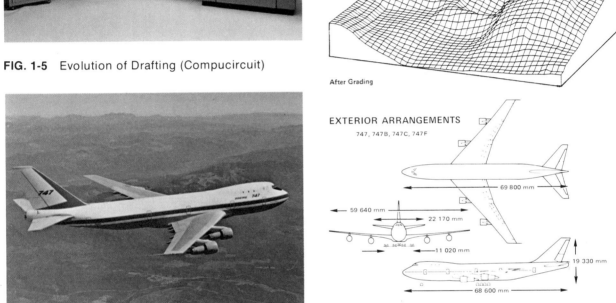

FIG. 1-6 The object and the working drawing (Boeing Commercial Airplane Co.)

Reaction Control Subsystem (RCS)

1 FORWARD RCS MODULE, 2 AFT RCS PODS
38 MAIN THRUSTERS (14 FORWARD, 12 PER AFT POD)
THRUST LEVEL = 3,870 NEWTONS (870 LBI)
6 VERNIER THRUSTERS (2 FORWARD, 4 AFT)
THRUST LEVEL = 111.2 NEWTONS (25 LBI)
BIPROPELLANTS: NITROGEN TETROXIDE
MONOMETHYLHYDRAZINE

MAIN THRUSTERS
(12 PER AFT POD)
3,870 NEWTONS
(870 LBI) THRUSTERS
(12 PER POD)

OMS HELIUM TANK

RCS HELIUM TANKS

OMS PROPELLANT
TANKS

FIXED
THRUSTERS

RCS
PROPELLANT
TANKS

VERNIER
THRUSTERS
(2 PER AFT POD)
111.2 NEWTONS
(25 LBI)

VERNIER
THRUSTER   FORWARD RCS MODULE

AFT PROPULSION SUBSYSTEM
(OMS/RCS POD)

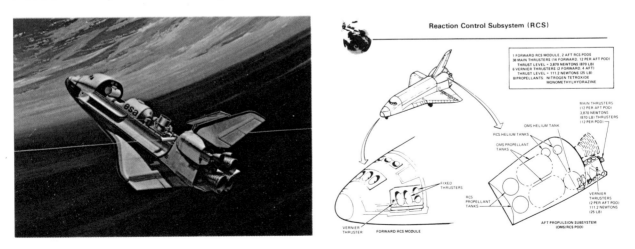

FIG. 1-7 Drawings needed for the Space Shuttle (Rockwell International)

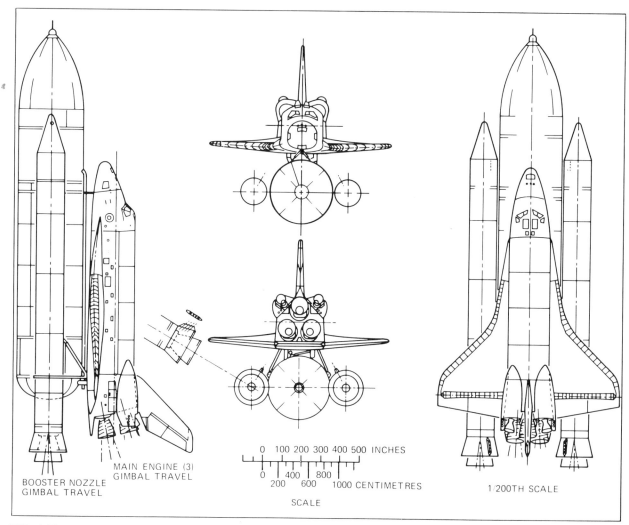

MAIN ENGINE (3)
GIMBAL TRAVEL

BOOSTER NOZZLE
GIMBAL TRAVEL

0 100 200 300 400 500 INCHES

0   400   800
200   600   1000 CENTIMETRES

SCALE

1/200TH SCALE

FIG. 1-7(cont.)   Drawings needed for the Space Shuttle (Rockwell International)

quired a vast number of drawings for the design of its numerous components.

Working drawings are needed by a variety of people inside and outside of industry. For an example within an industry, imagine that you are an industrial engineer. You have thought of a unique method of improving a conveyor system which involves heavy equipment.

Your method requires the installation of newly designed machinery. With the help of a designer you roughly, but carefully, sketch your plan on paper. You note material and size specification details. The sketch and notations are then taken to skilled draftspersons who must be able to read your drawing. After explaining the new concept to the draftspersons, you leave it to these professionals to draw a detailed and accurate graphic representation of the new equipment to be manufactured. After many checks and some changes, you must read the completed drawings carefully and give them your OK. The drawings then proceed to the manufacturer who must produce the newly designed pieces of equipment. Everything must fit together and run smoothly in your new system. Therefore, it is important that the manufacturer be precise. You won't be able to supervise every aspect of production. Everyone must rely on the drawings that you and the draftspersons brought to them. Each

FIG. 1-8 A conveyor system in the steel industry (Westinghouse Electric Corp., Marine Division, Sunnyvale, CA)

FIG. 1-9 Assembling steel components (Evans Products Co.)

screw must be mated by size and location to an internal threaded part. Each part must fit at a certain position.

By accurately following the working and assembly drawings, the manufacturer will produce exactly what you need.

With this new equipment now on hand, your company will dismantle the old conveyor system and install your system. How will the installers know what to put where and how the new system will operate? The drawings are in demand again. They will assure accurate compliance with your scheme so that the operation will function as you have planned.

You've done a terrific job! The new conveyor system has increased production by 30%. It has reduced costs significantly. After a few months, however, there's a snag in the system. It's not major, but it requires adjustment and repair. Bring out the drawings again. The service department will have to be able to read the drawings to determine where the problem may be, how to get to it, and how to fix it in the most efficient manner.

In a few years some of these parts may need replacing. The company will have to see that the manufacturer has the drawing specifications. This is

so that the equipment can be accurately and efficiently replaced.

Meanwhile, even the company's sales representatives will have copies of your drawings. When they present your new conveyor system to other interested industries, they too will have to be able to read the drawings and understand and explain all that they symbolize. Your technical drawings have certainly done a lot of communicating.

Because so many elements of industry rely on technical drawings, they must be produced efficiently. The most important factors in executing a drawing are accuracy and speed. A drawing that is completed accurately and quickly will make it possible to manufacture a well-made product at a reasonable cost.

The development of efficient drafting methods has helped people achieve rapid advancements in every industry from deep sea exploration to outer space probes. The draftsperson has a vital role in promoting progress and efficiency within industry. Technology and scientific information become more and more complex as people's knowledge seems to expand geometrically. This complexity requires close work among many professions. Each profession contributes its expertise toward solving

FIG. 1-10 Deep sea exploration (Lockheed Missiles and Space Co.)

FIG. 1-11 Outer space exploration (JPL)

modern problems. Although the purposes of industries vary greatly, they all rely on the coordinated efforts of skilled personnel. They all require the same combination of professional skills. The nature of each industry determines the specialization required of each profession, but the basic skills used are similar.

The following are brief descriptions of some professions needed most often in areas of technical production.

DRAFTING PROFESSIONS

Drafting is a profession that requires the accurate execution of planning, organizing, and drawing before a drawing is considered complete.

Because of the complexity of industry today, there are many areas of responsibility even within the drafting field. Each serves a vital need in the speedy production of accurate drawings. In most situations, one draftsperson cannot be responsible for all phases of all drawings and still be efficient. A well-qualified senior draftsperson usually has spent much time learning all phases of the profession.

Tracer

A draftsperson may begin as a tracer who traces or copies drawings another draftsperson may have

FIG. 1-12 A modern injection-molded part (General Electric Co.)

made. The copying process helps an individual develop a greater understanding of drafting techniques and better skills.

Junior Detailer

A junior detailer will, upon instruction, make limited simple detail drawings of work prepared by a tracer. These drawings will graphically illustrate the specifics of a design.

Senior Detailer

Senior detailers may work from layouts and the designs of engineers. They may transform the engineer's sketches or verbal instructions into a drawing and then elaborate on the specifics. A senior detailer has more responsibility, experience, and expertise than a junior detailer.

Checker

A checker must be a draftsperson with much experience and knowledge. In the stages of producing a drawing it is the checker's responsibility to review the final drawings for errors. If an error is discovered, this individual sees that it is corrected before the drawing is accepted.

Junior Designer

A junior designer participates in the innovation, adaptation, and coordination of ideas. Following preliminary sketches, junior designers apply their technical knowledge to help develop a working drawing.

Senior Designer

A senior designer must be an experienced draftsperson. A designer, using imagination, creativity, and technical knowledge, finds solutions to problems. These solutions may be in the form of a product or the application of an idea. A senior designer must have a knowledge of math, strength of materials, and potential costs.

Chief Draftsperson

A chief draftsperson is a person of experience, re-

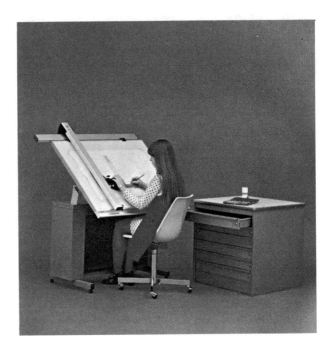

FIG 1-13 Draftsperson at work (Teledyne Post)

sponsibility and imagination who often works from freehand sketches of engineers. This individual is responsible for accurate preparation of working drawings. The senior draftsperson is not only concerned with expression of the product, but also contributes to its design. The development of detail and design requires experience, technical knowledge, and independent judgment.

RELATED NONDRAFTING PROFESSIONS

Industrial Designer

The industrial designers are concerned with the appearance and utility of a product. They work closely with the engineering, marketing, and sales divisions. In so doing, the designer tries to develop a functional and aesthetic product for the company. Designers must synthesize their knowledge of art, materials, and business procedures.

Tool Designer

Tool designers originate designs for cutting tools, using their machine shop experience and formal education. These tools are used to produce machine tools. They are also used to make fixtures to

FIG. 1-14 The output of the industrial designer (General Electric Co.)

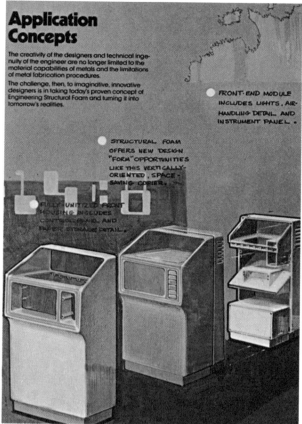

FIG. 1-15 Designer concept drawings (General Electric Co.)

hold the work which is to be machined. Tool designing calls for a knowledge of machines, manufacturing processes, mathematics, sketching, and drawing.

Architect

The architect, like the engineer, applies knowledge of technology and materials toward a design or a solution to a problem. Architectural drafting involves the planning of buildings, their function, location and materials. The architects are concerned with the artistic aspect of a building as well as its function. They design and detail the use of space to serve the specific needs and desires of the people involved. The structure may be a planned community, a single-family house, or a high-rise office building.

FIG. 1-16 The tool designer and his products (International Packings Corp.)

FIG. 1-17 The output of the architect (Georgia-Pacific Corp.)

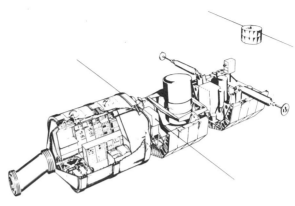

FIG. 1-18 European space agency's Spacelab (Rockwell International)

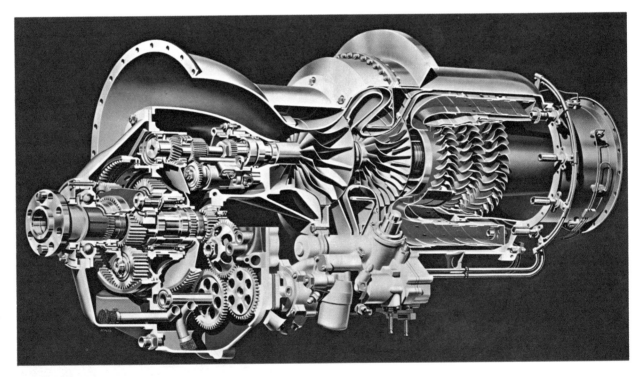

FIG. 1-19 Realistic product renderings (Garrett Corp.)

Technical Illustrator

A technical illustrator draws three-dimensional representations of a product or technical object. The drawings may be line drawing or realistic renderings. The illustrator applies an understanding of descriptive geometry and mechanical drawing techniques. This understanding is used to produce a finished illustration. This illustration may be used in catalogs or other publications to explain or describe a product.

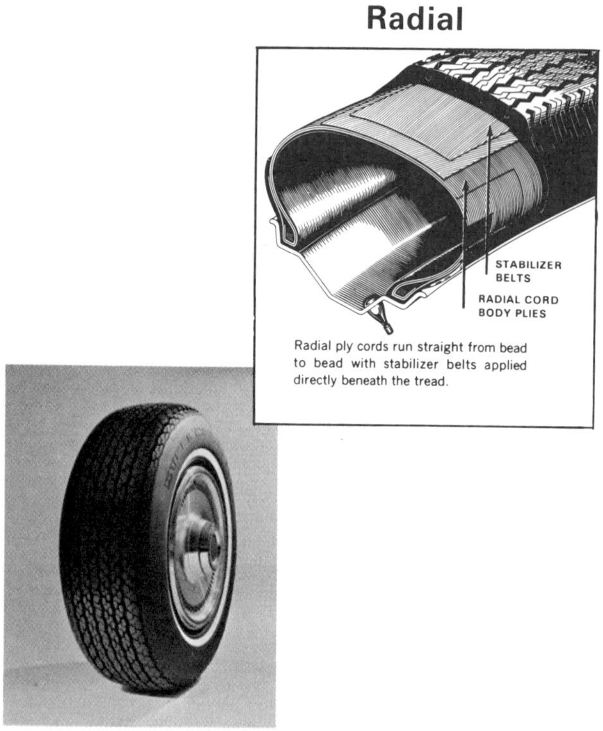

Radial

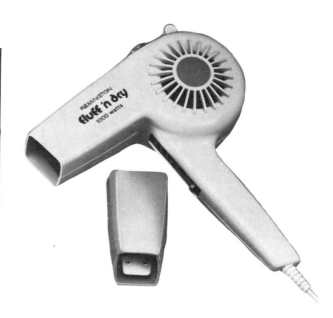

FIG. 1-20   Product technical illustration (Firestone Tire & Rubber Co.)

FIG. 1-21   Commercial art in advertising (Sperry Rand Corp.)

Commercial Artist

A commercial artist is more concerned with the visual appeal than the accuracy of a technical illustration. The artist prepares illustrations for magazines, posters, and books, utilizing a variety of media such as oils, water colors, or charcoal. Commercial artists use their talent and creativity to develop the most effective presentation of a design.

Cartographer

A cartographer designs and produces maps. Cartographers combine drafting and artistic ability. They must also draw from a large background of knowledge and a large source of data. These sources supply the original information on which maps are based. Several different types of maps are produced, each serving a specific need. Maps may

FIG 1-22   A cartographer making maps. (Clark E. Smith)

FIG. 1-23   The drafting class (Teledyne Post)

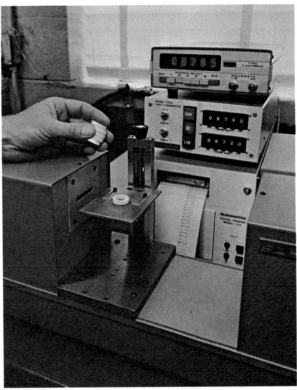

FIG. 1-24   An engineer must make many critical computations (International Packings Corp.)

range from aerial topographies to small-scale representations.

Drafting Instructor

Drafting teachers are needed at many levels of junior and senior high schools, vocational schools, junior colleges, and universities. A drafting instructor must have completed college and must also have actual industrial experience. The combined background of work in industry and formal education permits the instructor to prepare students to become engineers, designers, and drafting technicians.

Technician

A technician may be employed in a field such as engineering, drafting, design, or architecture. A technician is a semiprofessional assistant who works under the supervision of an engineer, architect, or other professional. Being a technician requires some special education and ability to use scientific and mathematical theory. Technicians may use special equipment and tools to calculate, operate, and experiment. They will perhaps assist in the development of models.

ENGINEERING

Engineering is a broad term. The engineering field is generally concerned with the application of technological and scientific principles toward solving problems determined by human needs. There are more than twenty-five engineering specialties recognized by professional societies. Although there are large areas of common knowledge among engineers, most choose to specialize in one branch of the field. Even within a particular branch, however, engineers may apply their knowledge to other fields

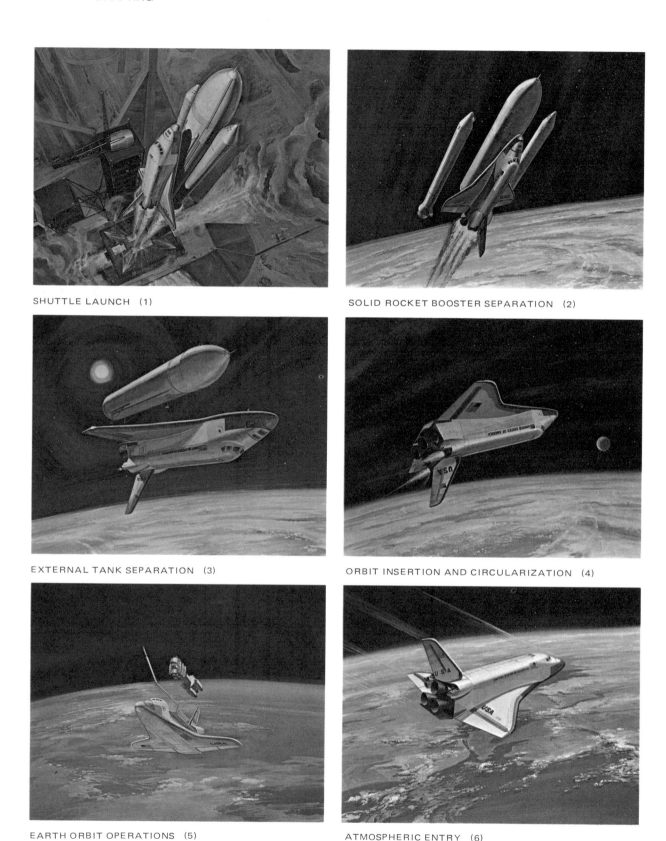

SHUTTLE LAUNCH (1)

SOLID ROCKET BOOSTER SEPARATION (2)

EXTERNAL TANK SEPARATION (3)

ORBIT INSERTION AND CIRCULARIZATION (4)

EARTH ORBIT OPERATIONS (5)

ATMOSPHERIC ENTRY (6)

FIG. 1-25 The exploration of space (Rockwell International, Space Division)

LANDING (7)

MAINTENANCE AND REFURBISHMENT (8)

FIG. 1-25 (cont.) The exploration of space (Rockwell International, Space Division)

SHUTTLE CHARACTERISTICS
(values are approximate)

Length
   System   184 ft [56 m]
   Orbiter   122 ft [37 m]
Height
   System   76 ft [23 m]
   Orbiter   57 ft [17 m]
Wingspan
   Orbiter   78 ft [24 m]
Weight
   Gross lift-off
      4,400,000 lb [2,000,000 kg]
   Orbiter landing
      187,000 lb [85,000 kg]
Thrust
   Solid rocket boosters (2)
      2,650,000 lb [11,800,000 N] of thrust each
   Orbiter main engines (3)
      470,000 lb [2,100,000 N] of thrust each
Cargo Bay
   Dimensions
      60 ft [18 m] long,
      15 ft　[5 m] in diameter
   Accommodations
      Unmanned spacecraft to fully equipped
      scientific laboratories.

such as medicine or space systems. Engineering technology has brought us to manned journeys to the moon, mining of the seas, and precise monitoring of heart patients in hospitals.

Engineering Divisions and Subdivisions

Structural
   Highways
   Bridges
   Buildings
Transportation
   Automotive design and assembly
   Railroad systems
   Nuclear submarines
   Aerospace industry
Industrial production
   Manufacturing
   Petroleum
   Energy
Space exploration
   Fuels
   Control systems
   Communication
   Sanitation
   Guidance systems
   Measurement systems
Communication
   Satellite
   Radio/TV
   Cable

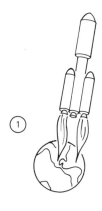

Viking mission sequence: Lift-off from Cape Canaveral (1) aboard Titan III boosters and Centaur rocket.

Boosters separate from Centaur, second stage ignites (2) and puts rocket into "parking orbit" around earth. Launch vehicle then circles earth once or twice so NASA can ensure all equipment is functioning properly.

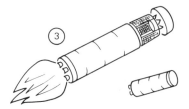

Centaur burn (3) enables Viking to leave earth orbit and head towards Mars. Shortly thereafter, Viking separates from Centaur.

(4) and jettisons bioshield cap (a device that protects Viking from inadvertently taking earthly contaminants to Mars).

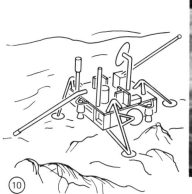

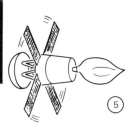

Following space flight of 305 to 360 days, Viking fires rocket for insertion to Mars orbit (5).

Lander's terminal descent rockets activate at 1200 m (9) and finally, five to ten minutes following initial entry of Mars' atmosphere, touchdown.

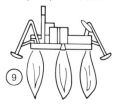

Once orbit is achieved, Lander separates from Orbiter (6).

Parachute deploys at 6000 m (8) and protective heat shield is jettisoned.

Lander deorbits and enters Mars atmosphere at 25 000 m (7).

FIG. 1-26 The Viking Explorer on Mars (JPL)

FIG. 1-27   Electronic Products (Texas Instruments, Inc.)

FIG. 1-28   The auto is an integral part of American industry (Ford Motor Co.)

Sanitation
   Solid and liquid waste
   Water treatment
   Air pollution
Defense
   Weaponry
   Aeronautics
   Naval systems

ELECTRONICS INDUSTRY

Many new applications of electronic principles have developed since World War II. Until that time, electronics was chiefly concerned with radio re-. ceivers and broadcasting equipment. Throughout the 1950s there was a boom in the industry as a result of TV and computer expansion. Between 1950 and 1968 the combined factory sales of the electronics industry rose from $2.7 billion to $24 billion. Today, this industry plays an important role in our daily lives. It develops, produces, and maintains a variety of laborsaving devices, precise measuring tools, communication equipment, and data processing systems.

   The four major market areas for electronics include the consumer, industry, government, and component manufacturers. The consumer products we are most familiar with are TV, radio, stereo, phonograph, and hearing aids. Industry depends on research and development by engineers, scientists, and technicians. This research and development provides testing and measuring systems, processing and control equipment, TV and radio broadcasting equipment, and therapeutic X-ray machines. The government relies on industry to provide the nation's missiles, spacecraft systems, communication equipment, and defense products. Component manufacturers utilize semiconductors, transformers, relays, and switches. For additional information, contact:

   Electronic Industries Association
   2001 Eye Street, N.W.
   Washington, D.C. 20006

AUTOMOTIVE INDUSTRY

The auto has become an integral part of our lives. Because of this, the auto industry has become a major influence on the economic stability of this nation. For example, the decisions of the auto industry also greatly influence the economics of its suppliers of glass, rubber, and electronic components.

   The automotive industry is concerned with the production of all forms of vehicles used for road transportation. These include cars, trucks, taxicabs, fire engines, buses, and ambulances. There are basically four stages of auto production. There are:

FIG. 1-29 Industry is setting new trends in public transportation (Rohr Industries, Inc.)

designing, engineering, testing, and manufacturing.

Designing   The interior and exterior designs of today's automobiles are selected by executives in consultation with stylists. Stylists sketch and model design alternatives. They keep aesthetics, safety features, and function in mind.

Engineering   The cost, performance, and legal requirements of a vehicle must also be considered as the whole vehicle is planned. In addition to styling, engineers are responsible for the car's engine and chassis.

Testing   Once a total design is engineered and approved, it is tested before being put into production. The parts which constitute the whole vehicle must be tested at each stage from design to final assembly. They must be tested for function, durability, and ease of assembly and repair.

Manufacturing   The auto industry relies on a broad variety of skills. The production of a vehicle involves hundreds of operations before it can be assembled. For example, the draftsperson is expected to supply drawings at all phases of production from tooling to testing. In total, more than 27,000

FIG. 1-30 Trucking is a vital part of American industry (White Motor Corp.)

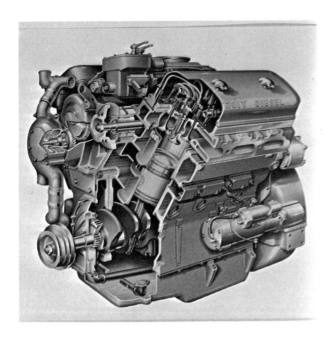

FIG. 1-31 The diesel engine is the main source of energy for the trucking industry (Detroit Diesel Allison)

FIG. 1-32 A drawing being used in construction (Nutone Div., Scoville)

drawings are needed to build an automobile. For additional information, contact:

    Motor Vehicle Manufacturers
      Association of the U.S., Inc.
    320 New Center Building
    Detroit, Michigan 48202

    International Union, United
      Automobile, Aerospace, and
      Agricultural Implement Workers
      of America
    8000 East Jefferson Avenue
    Detroit, Michigan 48214

CONSTRUCTION INDUSTRY

Construction is one of the oldest forms of work. It began when cave dwellers left their natural habitat and started building crude shelters.

The construction industry is concerned with the building, maintenance, and repair of nonmobile structures. The structures may range from small homes, schools, and hospitals to factories and skyscrapers. Other types of construction include highways, heavy construction, and utility structures.

Construction as an industry has a significant influence on the nation's economy. It provides millions of jobs. In addition, this industry requires a volume of materials greater than that of any other. Construction utilizes lumber, cement, and stone. It also utilizes metals such as steel, copper, and aluminum. Construction also relies on the manufacture of machines, tools, and heavy equipment such as cranes and large earthmovers. Other economic influences are seen in the vehicles needed to transport these masses of materials.

Generally there are two types of construction. These types are private and public. *Private* includes residential, business, farm, and utilities structures. *Public* includes public housing projects, military facilities, highways, and conservation projects. For additional information, contact:

AFL-CIO
Building and Construction Trades Department
815 16th Street, N.W.
Washington, D.C. 20006

Associated General Contractors
   of America, Inc.
1957 I Street, N.W.
Washington, D.C. 20006

FIG. 1-33   A few of the products of the aerospace industry (TRW, Cessna Aircraft Company, Boeing Commercial Airplane Co., Garrett Corp.)

AEROSPACE INDUSTRY

Aerospace usually refers to applied research, development, and production related to space and aeronautics. It includes passenger planes, launch vehicles, satellites, landing modules, missiles, rockets, and spacecraft.

Within this industry there is a great interdependence among many technical disciplines from biology to electronics. Almost every field of science and engineering is represented. There are small companies which make components and large companies which assemble the parts.

Accuracy in this industry is vital. For example, NASA requires a 99.9999 percent reliability. Even this will result in a few defective components. Rigid specifications are demanded because of the high cost of failure. For additional information, contact:

American Institute of Aeronautics and
  Astronautics, Inc.
1290 Avenue of the Americas
New York, N.Y. 10019

Aerospace Industries Association
1725 DeSales Street, N.W.
Washington, D.C. 20036

STEEL INDUSTRY

The American steel industry had its start in 1864 in Wyandotte, Michigan. William Kelly, an American, and Sir Henry Bessemer of England devised a process to produce steel by the ton instead of by the pound. The Bessemer process, used for many years in the U.S., has given way to more modern methods. An abundance of the raw materials (iron ore, coal, and limestone) ensures the stability of the industry. Steel is the most used basic metal of our time. It is used in products ranging from safety pins to skyscrapers. This wide use assures the future of the steel industry in the U.S. Steel alloys are used heavily by other industries. Notable among these are auto production, construction, and containerization. The container field relies on the use of tin-plated steel. Most of the world's tin cans are produced in the U.S. A tin can is 98 percent to 99.5 percent steel. Most steel producing plants in the U.S. are located in the northern and eastern parts of the country. For additional information, contact:

American Iron and Steel Institute
1000 16th Street, N.W.
Washington, D.C. 20036

United Steelworkers of America
1500 Commonwealth Building
Pittsburgh, Pennsylvania 15222

FIG. 1-34 Manufacturing steel products (International Packings Corp., Cleveland Twist Drill Co., Bob McCullough for Inland Steel Co.)

FIG. 1-35   Railroad cars (Pullman-Standard)

RAILROAD INDUSTRY

The first locomotive to pull a train of cars on an American railroad was the steam-powered "Best Friend of Charleston" in 1830. Since then the railroad industry has mushroomed. It now includes several hundred privately owned railroad companies servicing passengers and commercial interests throughout the country. Rail transportation is used by industry and business for transportation of supplies and products. Commercial users take advantage of the flexibility, reliability, and relative low cost of the rail system. There are many challenging positions available within the railroad industry. Specialists seek better lubricating materials, brake equipment, riding quality, and cargo handling methods. Also coming into wider use is the application of electronic computers, data processing, and microwave communication. For further information, contact:

Association of American Railroads
American Railroads Building
1920 L Street, N.W.
Washington, D.C. 20036

United Transportation Union
15401 Detroit Avenue
Lakewood, Ohio 44107

DATA PROCESSING

As our society's collective knowledge increases, we need better, more efficient methods of storing the information we gather. Data processing systems can do just that. They process data so it can be efficiently stored, retrieved, and used.

Data processing systems have revolutionized every aspect of business. Once data processing was used for accounting purposes only. Today's computers are also used in the technical sciences and the social sciences. They help experts calculate population and city growth patterns so we may plan for the future. Computers can help us find solutions to the social problems of pollution, overcrowding, and transportation.

In fact, the computer industry may some day become larger than the auto industry. Today almost every family in the U.S. owns at least one auto. Although we may not all own a computer in tomorrow's world, we will be using the computers of the future for most of our business and communication.

Already we see the use of data processing systems in communications, hospital monitoring, space vehicle guidance, newspaper typesetting, machine

FIG. 1-36   Electronic hardware in the data processing industry (Texas Instruments, Inc.)

tool operation, and many other enterprises. For additional information, contact:

Data Processing Management Assn.
505 Busse Highway
Park Ridge, Illinois 60068

PETROLEUM INDUSTRY

Petroleum is a fossil fuel formed by the decay of living matter. It is used as fuel in our homes and machinery. It also supplies the raw material for synthetic rubber, plastics, and many chemicals.

Petroleum production begins with exploration. Exploration crews study surface and subsurface characteristics of the earth. They do this to try to find more reliable and efficient methods of locating potential oil deposits before a site is drilled. The process requires detailed planning. It also requires specialized equipment. When oil or gas is discovered, equipment must be installed to regulate its flow to the surface. The oil or gas will eventually be transported to a refinery. This refinery may be thousands of miles from the oil field. For additional information, contact:

Society of Petroleum Engineers of AIME
6200 North Central Expressway
Dallas, Texas 75206

American Geological Institute
2201 M Street, N.W.
Washington, D.C. 20037

FIG. 1-37 Modernized oil well pump (Union Oil Company of California)

MINING INDUSTRY

The mining industry is concerned with the location and removal of valuable raw minerals from the ground. These resources are converted into materials that keep our society functioning. Coal, for example, is used as a fuel. It is also used as an ingredient in both steel and cosmetics.

Early man used minerals in their raw form. Technology gave people the ability to change the molecular structure of these minerals to make them stronger, harder, or more pliable. For instance, one of the most widely mined minerals today is bauxite. From it we produce aluminum.

The process of mining involves several functions. These functions are prospecting and exploring; underground and surface development; refining; and marketing. The method of locating a

FIG. 1-38 The mining industry (Clark E. Smith)

deposit depends largely on the type of mineral sought. Once a deposit has been located, it must be determined which type of mine is most appropriate —surface or subsurface. Coal for example, can be mined either on the surface or underground. Accordingly, the mine must be carefully designed and constructed. The mining process begins with drilling, blasting, and bulldozing as necessary. Determination and execution of methods require intelligent planning with patience and imagination. For additional information, contact:

United Mine Workers of America
900 15th Street, N.W.
Washington, D.C. 20005

PHOTOGRAPHIC INDUSTRY

The word "photograph" literally means, "to write with light." This is an accurate description of the process. Photography as we know it began with an experimental discovery. This discovery was by Dr. Johann Schultze of eighteenth century Germany. His find led to the work of the French painter Daguerre in 1839—the first photographer. Daguerre perfected the process of using plates coated with silver iodide inside a small box. Mercury vapor was used to develop the plates. The daguerreotype was very popular. It was George Eastman, however, who brought photography into widespread use. His invention was a simple camera and flexible roll film.

Today there are many kinds of film and equipment, and photography may have both artistic and technical usage. A commercial photographer is one who takes pictures of merchandise and fashions. These pictures may be used in selling or advertising. An industrial photographer may help a company improve its image or its product by taking pictures of workers on the job and equipment and machinery in operation. This photographer may also use photography for copying, enlarging, reducing, or microfilming drawings. Photographic specialties include aerial, educational, and scientific photography. For additional information, contact:

Professional Photographers of America, Inc.
1090 Executive Way
Oak Leaf Commons
Des Plaines, Illinois 60018

American Newspaper Guild
1126 16th Street, N.W.
Washington, D.C. 20036

FIG. 1-39   The use of paper (Mike Thorn)

PAPER INDUSTRY

Ancient Egyptians made their writing material from papyrus, a type of reed. From "papyrus" comes our word "paper." Wood-based paper, as we use it today, was invented about 100 A.D. in China. Paper was made laboriously by hand until the Industrial Revolution. Mass production of paper from wood pulp was made possible in the eighteenth and nineteenth centuries. Machines were invented that crushed, ground, and pressed the pulp.

Today, the versatility of paper is as great as our dependency on it. Paper can be made tissue thin or very thick. It is used for newsprint or cardboard packaging. And it can be processed for books, business forms, computer cards, checks, tickets, and greeting cards. Paper is adapted for use in all industries. There are many types of paper, vellum, and film used for drafting.

In the United States, there are well over 7000 different paper products. These are used at a rate of over 600 pounds per person per year. For additional information, contact:

American Paper Institute
260 Madison Avenue
New York, N.Y. 10016

Paper Industry Management Assn.
2570 Devon Avenue
Des Plaines, Illinois 60018

PACKAGING INDUSTRY

The packaging industry is concerned with the wrapping, boxing, and bottling of products. It is one of the largest and fastest growing industries in this country. Package research concerns itself with several factors. For example, how will the product look on the shelf if it is to be retailed? Researchers seek new color combinations, shapes, and designs of packages. They even seek new materials such as plastic rather than paper. A change in a product or introducing a new product calls for innovative packaging.

A combination of skills—engineering, advertising, and chemistry—is needed to solve such problems as: What will attract the consumer to this product? How is the product best preserved? What shape is best for shipping and storing? How will the product be used by the consumer? There are many opportunities for designers and draftspersons in this field. For additional information, contact:

Better Packaging Advisory Council
331 Madison Avenue
New York, N.Y. 10017

Package Designers Council
299 Madison Avenue
New York, N.Y. 10017

FOOD PROCESSING INDUSTRY

One challenge facing the world today is that of providing food for an ever increasing population of hungry people. The food processing industry is concerned with every aspect of delivering food to the consumer.

Food is no longer processed primarily in the home. Much of the food we eat today is processed by industrial firms. It is then sold at the supermarket. The procedure includes growing, harvesting, handling, processing, packaging, transporting, and finally marketing. The future of the food industry is open-ended. In a short period of time we have gone from experimenting with frozen foods to preparing foods suitable for space travel.

The industry seeks to improve the quality and durability of existing products. It also seeks to create new food items. Foods must not only be tasty, they must also meet certain nutritional and health standards. It is desirable for food products to be attractive and capable of maintaining freshness and quality as they go from the processing plant to the carrier to the market and into your home.

We have already seen the successful development of methods such as freezing, dehydrating (which brings us such products as instant coffee, instant potatoes, and powdered milk), and freeze-dried products. The process of the future will be irradiation. This will keep meats, vegetables, fruits, and even seafoods fresh. These foods will be stored on shelves, without freezing, for years. Because of irradiation, they will still be fresh when they are finally cooked. For additional information, contact:

The Institute of Food Technologists
Suite 2120
221 North LaSalle Street
Chicago, Illinois 60601

For additional information about the drafting professions in general, contact:

The Assn. of Collegiate
Schools of Architecture, Inc.
1735 New York Avenue, N.W.
Washington, D.C. 20036

American Institute for Design and Drafting
3119 Price Road
Bartlesville, Oklahoma 74003

American Federation of Technical Engineers
1126 16th Street, N.W.
Washington, D.C. 20036

Engineers Joint Council
345 East 47th Street
New York, N.Y. 10017

National Society of Professional Engineers
2029 K Street, N.W.
Washington, D.C. 20006

Part

Introduction to Drafting

As the largest and most efficient industrial nation in the world, the United States has dominated world trade and therefore our measuring system has become widely accepted. However, with the emergence of the European Common Market, the United States, along with countries all over the world, has been encouraged to adopt the international metric system—Système International (SI).

Part II of this text is an introduction to the metric system, its history and its use.

2

Chapter

Drafting
Conventions

INTRODUCTION

The term *graphics* means expressing ideas by means
of lines and letters. Graphic communication began
when people started to use drawings to communicate with each other and to record and pass on ideas
so they would not be forgotten. The earliest forms
were a kind of picture writing, such as Egyptian
hieroglyphics. An example today would be an engineering drawing, which uses the graphic language of
pictures, lines, and letters to communicate technical
ideas.

The elements of technical drawings have become standardized out of necessity. If each industry,
each company, and each draftsperson made up a
separate method of drawing an object and used its
own drafting language, communication through
technical drawings would be nonexistent. It would
be as though thousands of people were speaking in
languages that no one else understood.

Communication through technical illustrations
has been made possible through the use of conventions. Conventions are very important in drafting. A
convention is a symbol, written rule, or a standard
drawing procedure that is universally used and understood in an engineering drawing. By using uniform drafting conventions, a drawing can be inter-
preted and understood anywhere, even though language differences may exist.

The American Society of Mechanical Engineers
(ASME) has many thousands of members from all

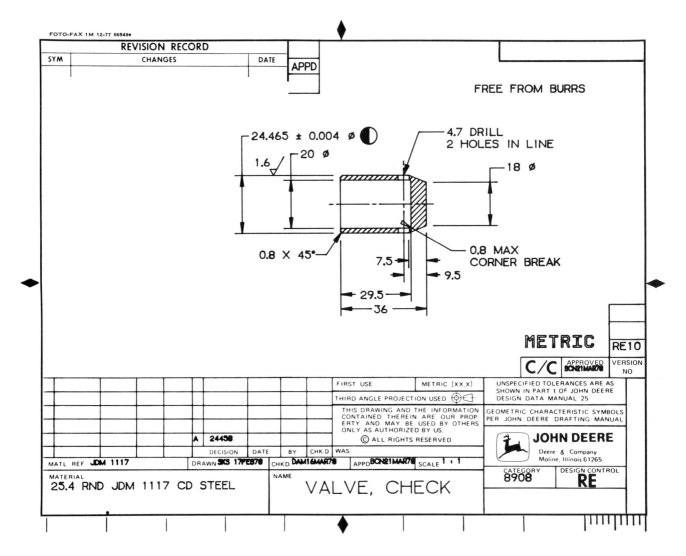

FIG. 2-2

areas of industry, education, business, and government. Its major aim is to advance the profession of mechanical engineering. Its work in developing manuals of standards for drafting is invaluable. These standards apply to written rules and to the principles of engineering drawing. They are revised as necessary to keep up with technical advances and changes. To be aware of all these drafting conventions the draftsperson should have a library of standard drafting conventions. The American National Standard Institute (ANSI) publications are the best source. Information on the ANSI publications can be obtained by writing to:

The American Society of Mechanical Engineers
United Engineering Center
345 East 47th Street
New York, NY 10017

Each segment of an industry that plays a part in making a product must work with engineering drawings. A *drawing* is a set of instructions that tells a person how to make a product. It must be clear, accurate, complete, and current. Being a good draftsperson means more than just drawing well. One must also be creative, and be knowledgeable in the technical field.

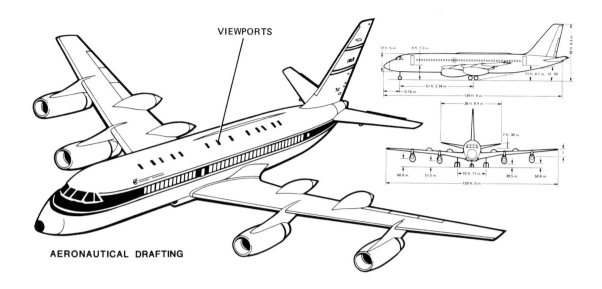

VIEWPORTS

AERONAUTICAL DRAFTING

FIG. 2-3

A draftsperson may work in one specialized area of industry or do general work in many different departments. He or she will usually prepare working drawings of the product from sketches or from design layouts prepared by the designers. It is the draftsperson's responsibility to make finished working drawings, using the standard drafting conventions.

INDUSTRIAL DRAFTING

In industry, it is possible to specialize in one of the many different facets in engineering drawing. Some of these areas are described below.

Machine Drafting involves the design, development, and drawing of all mechanical devices from very small machine parts (Fig. 2-2) to the largest type of machinery. This is the broadest area of engineering drawing. A few of its specific applications are: automotive design, tools, appliances, machinery, engines, and so forth. The draftsperson who wants to advance in this area should have a background in math, design, physics, and engineering drawing.

Aeronautical Drafting is concerned with the design, development and drawing of all types of air-

borne vehicles. These vehicles include everything from balloons to gliders to the most sophisticated type of jet aircraft (Fig. 2-3). To specialize in aeronautical drafting, one must learn about aerodynamics, propulsion systems, and instrumentations systems.

Aerospace Drafting is related to the design, development and drawing of all types of space vehicles. This includes the smallest satellite to very complicated space vehicles, as shown with the Viking Mission in Figure 2-4. The draftsperson should have some knowledge in math, propulsions systems, instrumentation, and astrophysics.

Ship Drafting concerns the design and drawing of all types of watercraft (Fig. 2-5), from the smallest boat to the largest aircraft carrier to all types of undersea crafts. Ship drafting requires knowledge of math, design, lofting, and gas, diesel, and nuclear propulsion engines.

Electrical/Electronic Drafting is the design, development and drawing of the products and the parts of products that use electricity as a power source (Fig. 2-6). The products vary from a simple lamp to an extremely complex computerized system. For this drafting, knowledge of electrical and electronic

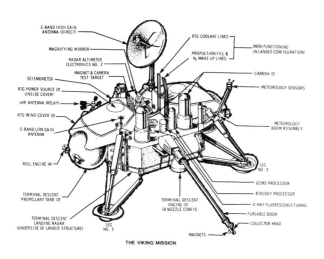

THE VIKING MISSION

FIG. 2-4

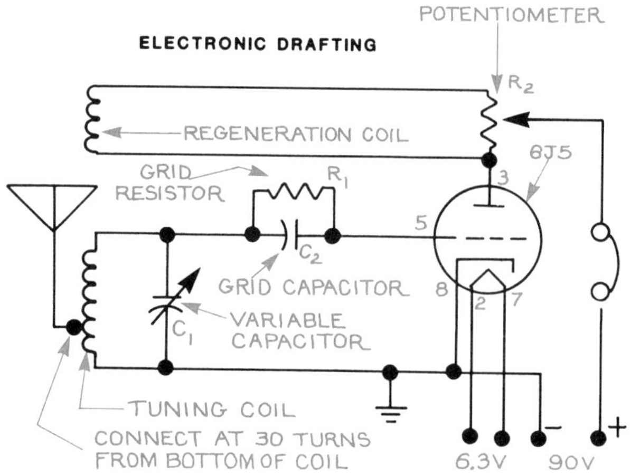

ELECTRONIC DRAFTING

FIG. 2-6

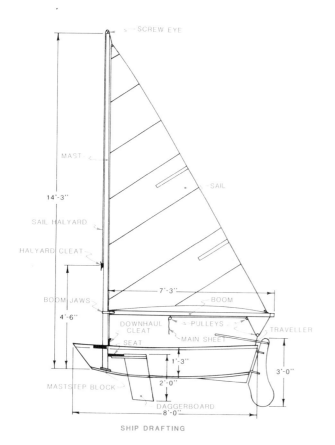

SHIP DRAFTING

FIG. 2-5

theory, math, physics, schematic symbols, and electronic drafting is important.

Cartography concerns the layout and drawing of maps. A map can be a simple street map or a complex contour map (Fig. 2-7). In cartography, one must be familiar with map symbols and cartography drafting techniques. Some background in civil engineering is also helpful.

Technical Illustration applies to the production of various types of drawings, usually in ink, that are used for instructional purposes and in publications (Fig. 2-8). These drawings are usually finished drawings. The subject might be a simple logo or a very complex, expanded view of a mechanical object. Good artistic ability and drafting knowledge are important in technical illustration.

Architectural Drafting involves the working drawings of buildings and structures (Fig. 2-9). The building can be a simple cabin or a huge skyscraper. Background in architectural design and drafting are needed.

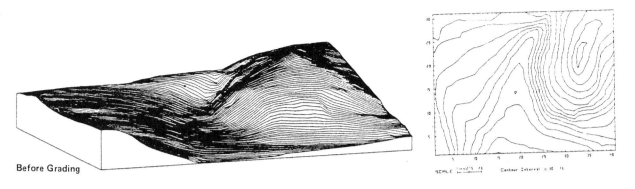

Before Grading

Before Grading

After Grading

After Grading

FIG. 2-7

CARTOGRAPHY

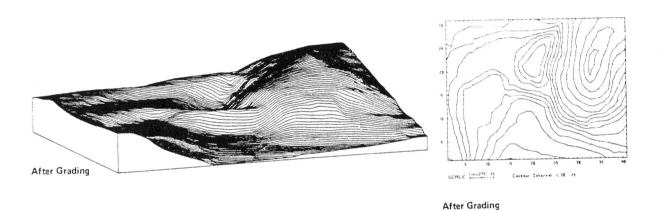

TECHNICAL ILLUSTRATION

VIEWING PORT

ANGLE OF ATTACK
MECHANISM

AIR CIRCULATION DUCT
ENCLOSING AIR CHAMBER

MODEL SUPPORT

ARC HEATER

WATER COOLED
8° CONICAL NOZZLE

HIGH PRESSURE
WATER MANIFOLD

FIG. 2-8

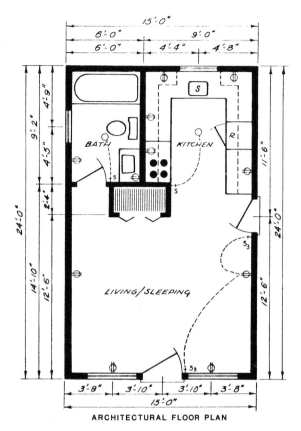

ARCHITECTURAL FLOOR PLAN

FIG. 2-9

WORKING DRAWINGS

The working drawing is a basic tool in making products. It comes from the joint effort of the engineer, the designer, and the draftsperson. Working drawings must contain the information needed by the manufacturer, the technician, and the craftsperson who make a product. Every detail must be clear and understandable.

Selecting the right type of working drawing is important. Several styles are available. A brief overview of the major types of engineering drawings are discussed in this chapter. Later chapters will cover each type of engineering drawing in greater depth. Figures 2–10 through 2–13 are examples of working drawings.

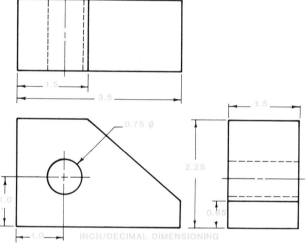

INCH/DECIMAL DIMENSIONING

FIG. 2-11

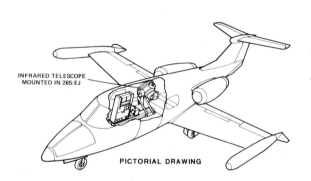

INFRARED TELESCOPE
MOUNTED IN 265-EJ

PICTORIAL DRAWING

FIG. 2-10

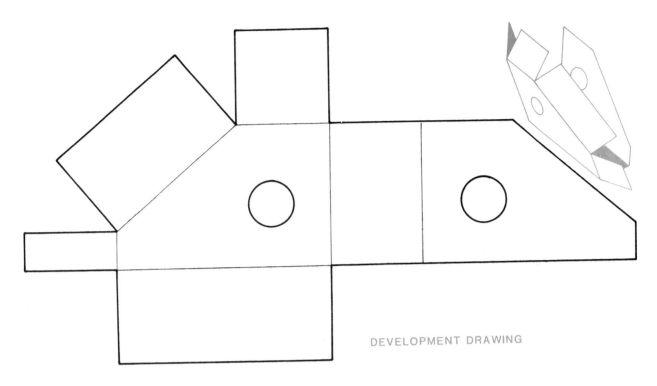

DEVELOPMENT DRAWING

FIG. 2-12

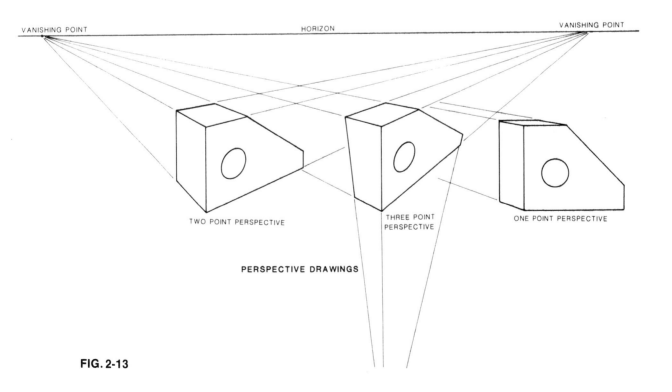

VANISHING POINT   HORIZON   VANISHING POINT

TWO POINT PERSPECTIVE   THREE POINT PERSPECTIVE   ONE POINT PERSPECTIVE

PERSPECTIVE DRAWINGS

FIG. 2-13

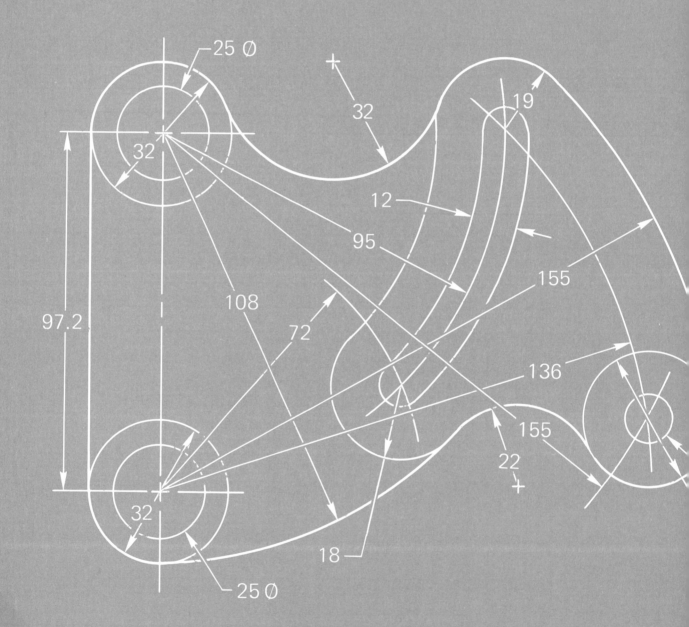

Part

Preparation for Drafting

All draftspersons must have the ability to communicate technical information quickly and accurately through the proper preparation of technical drawings. Lettering, linework, and drawing layouts are only a few of the basic skills a prospective employer will evaluate before hiring a new draftsperson.

Part III of this text is an introduction to the basic skills, required for proper lettering, linework, and drawing layouts.

Freehand Sketching

Clark E. Smith

INTRODUCTION

The freehand sketch (Fig. 3-2) was one of the first forms of communication. Any image drawn on any medium is considered a sketch. The freehand sketch is the simplest form of drawing. It is a convenient and fast way to put an idea into visual form. The freehand sketch is a graphic language used to assist verbal language. Try to describe the object in Figure 3-3. Now describe the object, using Figure 3-4 as an aid. Which method of communication is more successful?

The purposes of sketching and instrument drawing are the same. They present information and communicate ideas with less chance for error.

The sketch is the best way to show an idea immediately. A good idea can be lost if one waits until an instrument drawing can be made. A freehand sketch:

1. Speeds development of a new product

2. Helps organize ideas

3. Helps formulate ideas

4. Helps record ideas

5. Helps in designing

6. Helps with design changes

7. Helps in laying out a drawing

8. Helps find a graphic solution to many problems

9. Saves the time and expense of instrument drawings

FIG. 3-2   The earliest forms of freehand sketching

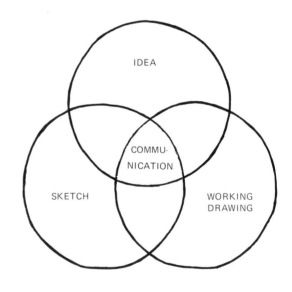

FIG. 3-4

FIG. 3-3   (NASA)

As you study this chapter, you should develop skills to freehand sketch all types of mechanical drawings. Also, you should become familiar with the pencils, drawing media, and erasers needed for freehand drawing.

FIG. 3-5   The sketch in communication

THE DRAFTSPERSON AND THE FREEHAND SKETCH

The main purpose of the sketch is to help the draftsperson communicate with himself and others (Fig. 3-5). The freehand sketch serves designers, engineers, draftspersons, and technicians.

The sketch speeds teaching and learning mechanical drafting techniques and procedures. The draftsperson can use sketching to select the best type of drawing to describe the item. With the sketch, decisions can be made as to what is good or poor, what should be changed, and what should be developed further.

THE FREEHAND SKETCH

The freehand sketch aids planning and clear thinking. It may be discarded or filed for further referral. Freehand sketches can be any type of pictorial or working drawing (Fig. 3-6). However, the most commonly used types are the isometric pictorial and the orthographic working drawing.

Sketches should be drawn quickly (the main advantage over an instrument drawing) but neatly. A sloppy and careless sketch is wasted effort (Fig. 3-7).

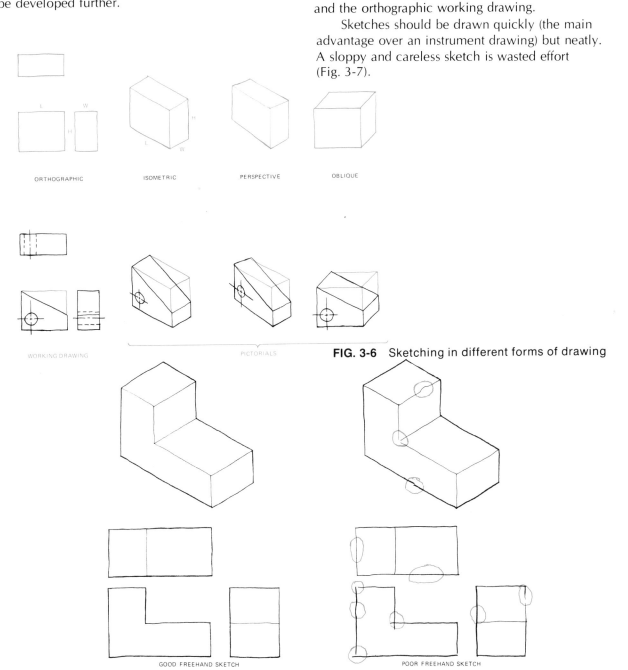

ORTHOGRAPHIC   ISOMETRIC   PERSPECTIVE   OBLIQUE

WORKING DRAWING   PICTORIALS

FIG. 3-6 Sketching in different forms of drawing

GOOD FREEHAND SKETCH   POOR FREEHAND SKETCH

FIG. 3-7 Good and bad examples of freehand sketches

FIG. 3-8 Estimating proportions (Courtesy of Litton Systems, AERO Products Division)

Do not scale a sketch. Scaling will only slow down the sketching process. The proportions of the item being sketched should be estimated as closely as your eye and sketching ability allow (Fig. 34-8). It is important to develop a good sense of proportion. This can be achieved by practice.

EQUIPMENT

One of the advantages of freehand sketching is the simplicity of the required equipment. The only equipment you will need are pencils, erasers, and a drawing medium.

Pencils Use a soft pencil. H, F, HB, or #2 writing pencils are recommended.

Erasers Any regular or art gum eraser will suffice. Do not use an ink eraser or an old hard eraser that will damage the paper.

Drawing medium Any surface that can be drawn on can be used for a sketch. An opaque or transparent paper (for printing copies) is recommended.

The sketching medium can be clear or lined. Lined paper used for sketching is called graph paper. The lines can be square or isometric (Fig. 3-9). It is best to use lines that are spaced either ⅛ inch or ¼ inch apart. Drafting instruments can be used if they reduce sketching time. For example, using a circle template may be faster than laying out a large freehand circle.

PREPARATIONS FOR MAKING FINISHED FREEHAND SKETCHES

There are a number of ways to make a sketch. You must decide which sketching style is the most comfortable for you, and which style will give you the best results.

No two people will make identical sketches. Each person's sketching style will be as individual as his handwriting. It is important to get plenty of practice.

WELL-PROPORTIONED SKETCH

POORLY PROPORTIONED SKETCH

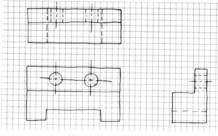

SQUARED GRAPH PAPER

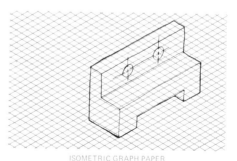

ISOMETRIC GRAPH PAPER

FIG. 3-9 Graph paper used in sketching

Consider the following exercises. Practice them and decide which method of freehand sketching is the most comfortable for you. Then start developing your skills.

1. Hold the pencil in a comfortable position.

2. The pencil is drawn, not pushed (Fig. 3-10).

3. Select the proper degree of softness for your pencil.

4. Practice sharpening the pencil point (Fig. 3-11).

5. Practice line conventions with your pencil. Two line weights, a heavy dark line and a thinner light line, will suffice for most freehand sketching.

6. Move your paper around until the particular line can be comfortably drawn.

FIG. 3-10   Pull your pencil. Do not push it.

FIG. 3-12   Sketching horizontal lines

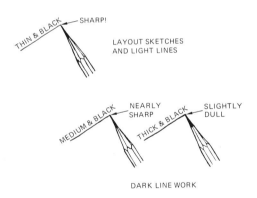

FIG. 3-11   Pencil points for sketching

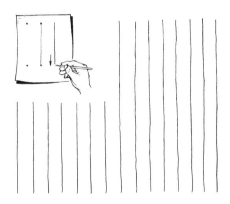

FIG. 3-13   Sketching vertical lines

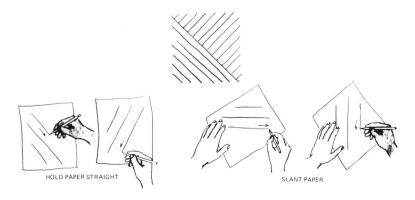

FIG. 3-14   Two methods of sketching slanted lines

7. Practice on different types of paper.
8. Practice horizontal lines (Fig. 3-12).
9. Practice vertical lines (Fig. 3-13).
10. Practice slanted lines (Fig. 3-14).
11. Practice angles (Fig. 3-15).
12. Practice circles (Fig. 3-16).
13. Practice isometric circles (Fig. 3-17).
14. Practice arcs (Fig. 3-18).
15. Practice ellipses (Fig. 3-19).
16. Practice rounds (Fig. 3-20).
17. Practice fillets (Fig. 3-21).

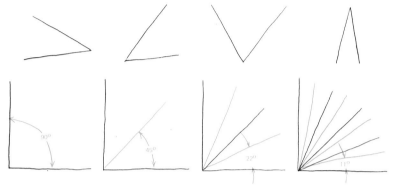

FIG. 3-15 Estimating and sketching angles

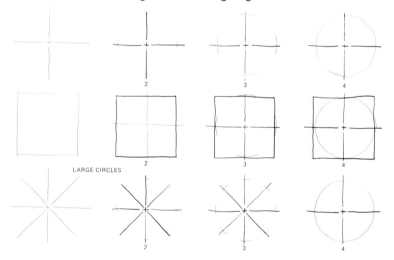

FIG. 3-16 Steps for sketching circles

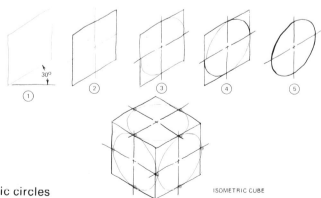

FIG. 3-17 Steps for sketching isometric circles

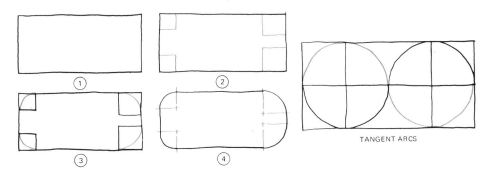

FIG. 3-18   Steps for sketching arcs

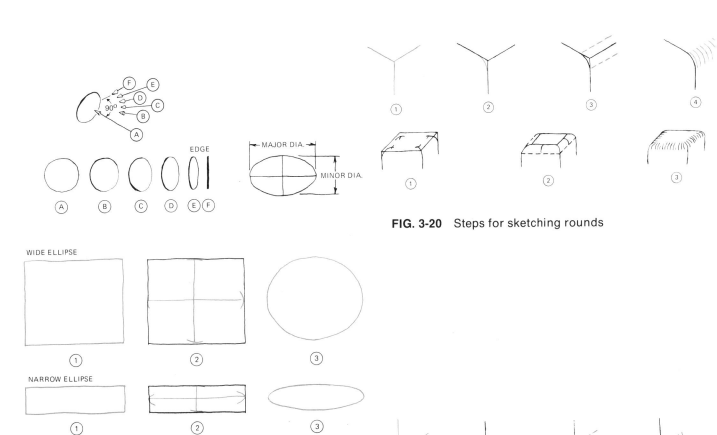

FIG. 3-20   Steps for sketching rounds

FIG. 3-19   Steps for sketching ellipses

FIG. 3-21   Steps for sketching fillets

After your skills and techniques start to develop, you will be ready to practice sketching. The usual method of sketching is:

1. Start the sketch with very light lines that can be changed easily.
2. Block in the views. Concentrate on the length, width, and height so the proportions will look close.
3. Sketch in the outline.
4. Sketch in the inside details.
5. Clean up the sketch.
6. Darken the sketch.

Figure 3-22 follows these steps for an orthographic sketch. Figure 3-23 follows the same steps for an isometric sketch. Follow the same steps regardless of the shape of the item or the type of sketch (Fig. 3-24).

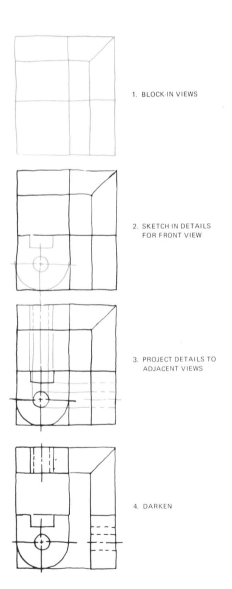

1. BLOCK-IN VIEWS

2. SKETCH IN DETAILS FOR FRONT VIEW

3. PROJECT DETAILS TO ADJACENT VIEWS

4. DARKEN

FIG. 3-22 Steps for developing an orthographic sketch

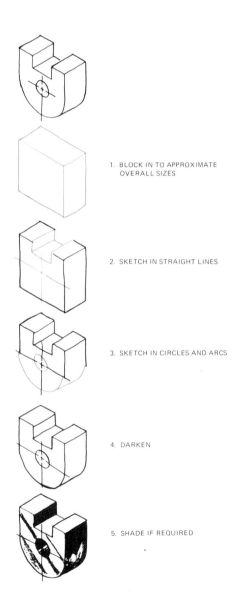

1. BLOCK-IN TO APPROXIMATE OVERALL SIZES

2. SKETCH IN STRAIGHT LINES

3. SKETCH IN CIRCLES AND ARCS

4. DARKEN

5. SHADE IF REQUIRED

FIG. 3-23 Steps for developing a pictorial sketch

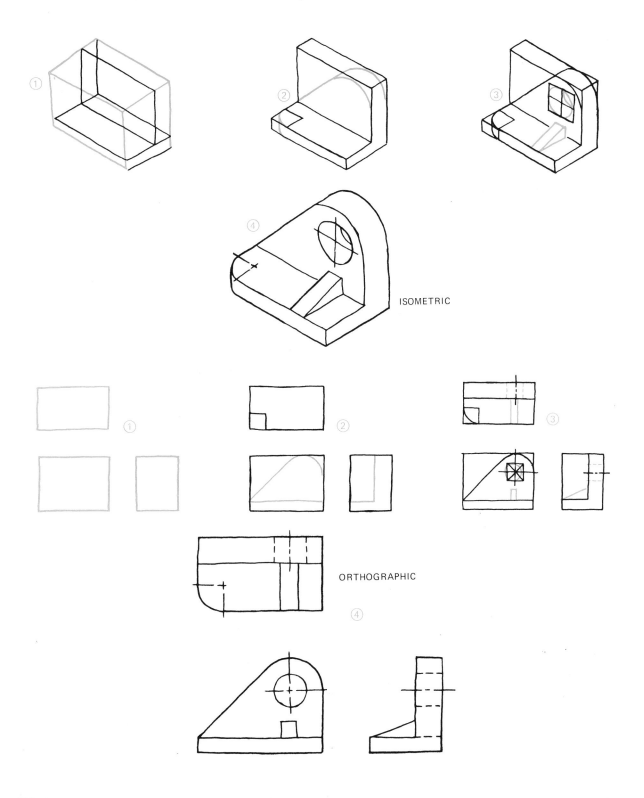

ISOMETRIC

ORTHOGRAPHIC

FIG. 3-24 Steps for developing isometric and orthographic sketches

PROBLEMS

As you sketch the following problems, try to develop speed, neatness, and good proportions.

Follow the instructions for each freehand sketching problem.

SKETCH THE ONE VIEW AND THE ISOMETRIC VIEW.
DRAW ALL OBJECTS ⅛" THICK.

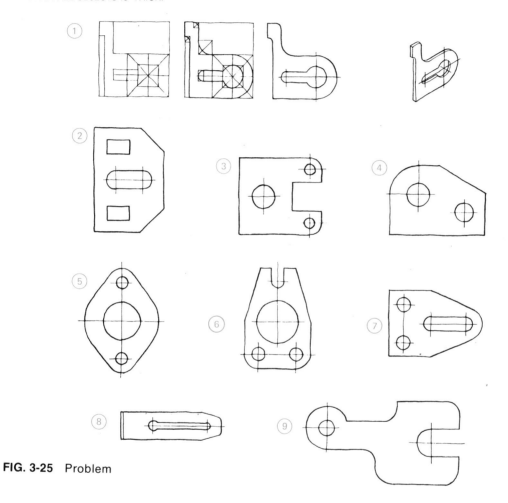

FIG. 3-25 Problem

DRAW A COMPLETED ORTHOGRAPHIC AND ISOMETRIC SKETCH FOR EACH FIGURE.

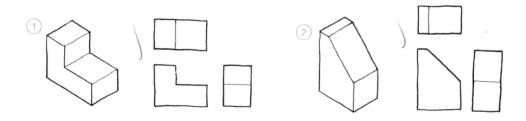

FIG. 3-26 Problem

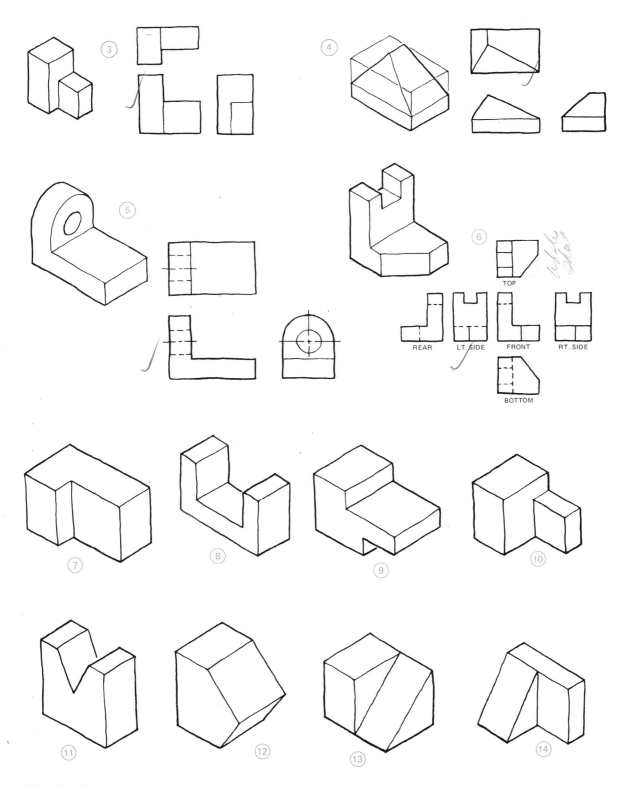

TOP

REAR   LT. SIDE   FRONT   RT. SIDE

BOTTOM

FIG. 3-26 Problem (Cont.)

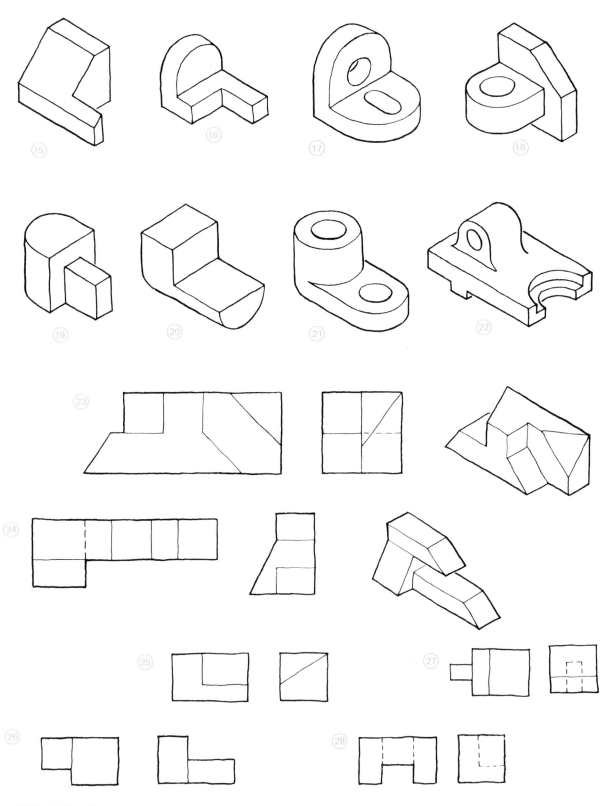

FIG. 34-26 (Cont.)

Lettering

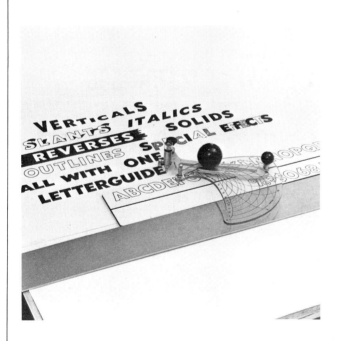

have to be added to communicate with the fabricator. The lettering must be uniform, accurate, sharp, dark, and easy to read. This will eliminate many costly production errors.

A draftsperson's lettering is usually an indication of his drafting ability, and it may be very important when he or she is applying for a drafting position. The general feeling is that good lettering goes with good drawing. As you read this chapter, you should learn to:

1. Print clearly and quickly with vertical and inclined single-stroke Gothic letters.
2. Print clearly and quickly with microfont and architectural style letters.
3. Use proper letter heights, proportions, and spacings.
4. Use the different lettering aids and devices.

SINGLE-STROKE GOTHIC

The American National Standards Institute (ANSI) recommends the use of the single-stroke Gothic style of lettering (Fig. 4-2) as it is easier to read and letter than other styles of lettering (Fig. 4-3). The parts of the Gothic letters are simple and all line widths are the same. The Gothic letter can be lettered in capital letters (uppercase) or lowercase

INTRODUCTION

An engineering drawing shows the exact shape of an object. Dimensions, notes, and specifications

ABCDEFGHIJK
LMNOPQRS
TUVWXYZ&
1234567890

(Fig. 4-4). The uppercase letter is easier to read than the lowercase letter. Therefore, most technical drawings require the use of uppercase Gothic letters. Such areas as civil engineering, mapping, and graphics use lowercase letters.

Single-stroke Gothic letters may be lettered vertically (Figures 4-2 and 4-4), or they may be lettered at a 68 degree incline (Fig. 4-5). Both styles have advantages and disadvantages:

Vertical Gothic

1. Easy to read
2. Difficult to keep strokes vertical
3. Slightest variation is noticeable

Inclined Gothic

1. Not as clear as vertical lettering
2. Needs more practice to learn the slant and curve of the inclined parts of letters

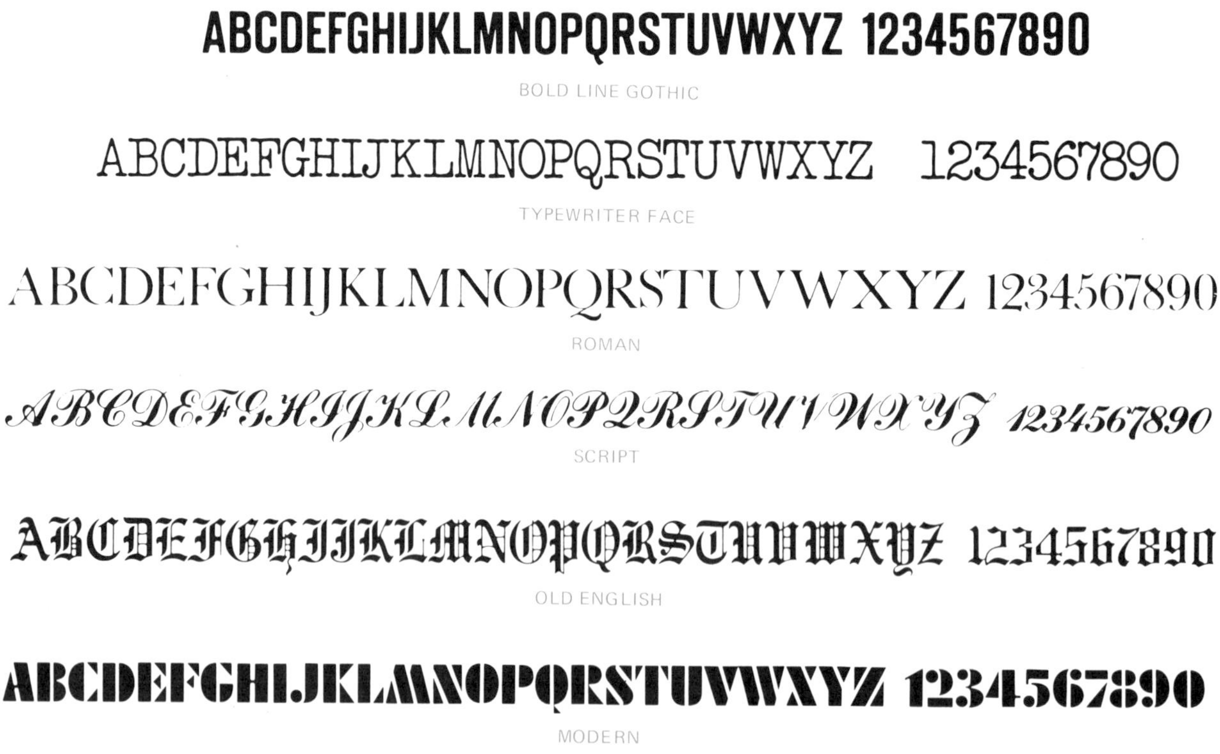

FIG. 4-2   Single-stroke Gothic lettering

ABCDEFGHIJKLMNOPQRSTUVWXYZ 1234567890
BOLD LINE GOTHIC

ABCDEFGHIJKLMNOPQRSTUVWXYZ   1234567890
TYPEWRITER FACE

ABCDEFGHIJKLMNOPQRSTUVWXYZ 1234567890
ROMAN

ABCDEFGHIJKLMNOPQRSTUVWXYZ 1234567890
SCRIPT

ABCDEFGHIJKLMNOPQRSTUVWXYZ 1234567890
OLD ENGLISH

ABCDEFGHIJKLMNOPQRSTUVWXYZ 1234567890
MODERN

FIG. 4-3   Other complex styles of lettering

3. Imperfections not so noticeable
4. Faster to letter

The incline of letters is at a ratio of 5:2, or 68 degrees (Fig. 4-6). Usually the choice of vertical or inclined lettering is specified by the employer. Mixing the two styles on a drawing is not good practice.

abcdefghijklmnopqrstuvwxyz

FIG. 4-4   Lowercase Gothic lettering

ABCDEFGHIJK
LMNOPQ
RSTUVWXYZ&
1234567890
abcdefghijklmno
pqrstuvwxyz

FIG. 4-5   Inclined uppercase and lowercase Gothic lettering

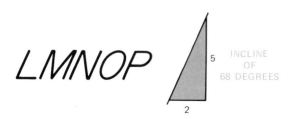

LMNOP

INCLINE
OF
68 DEGREES

5

2

FIG. 4-6   The proper angle for inclined lettering

MICROFONT

With microfilming, it turned out that reducing a large drawing down to a 35 mm size and blowing it back up to full size again makes several letters difficult to read. To eliminate this problem, ANSI adopted the "microfont" type (Fig. 4-7). It is recommended that draftspersons learn this system in addition to vertical and inclined Gothic.

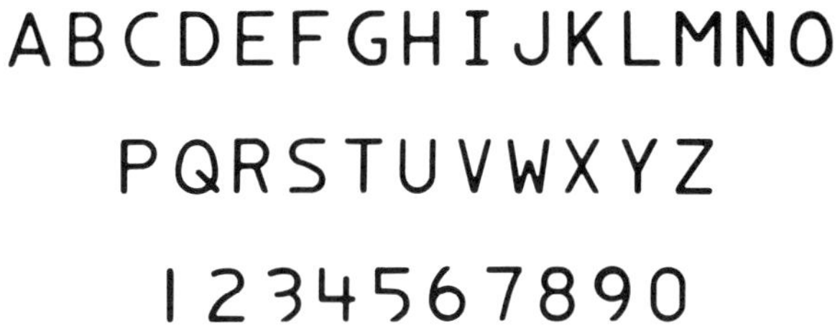

ABCDEFGHIJKLMNO

PQRSTUVWXYZ

1234567890

FIG. 4-7   Microfont lettering

ARCHITECTURAL STYLE

Architectural lettering does not have to conform to engineering standards (Fig. 4-8).

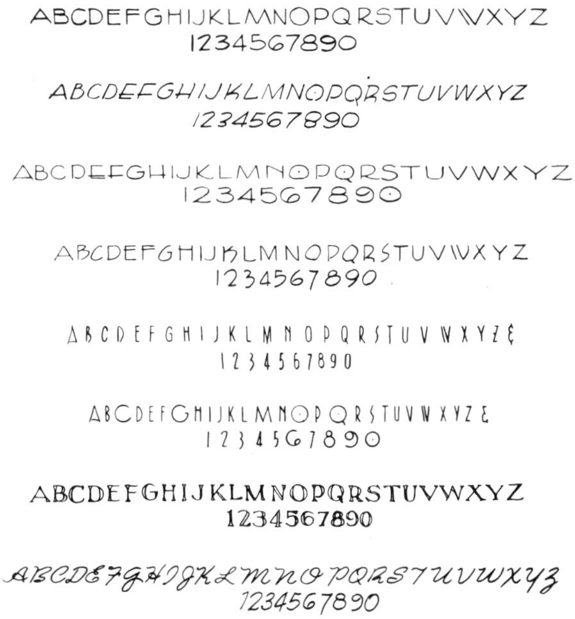

ABCDEFGHIJKLMNOPQRSTUVWXYZ
1234567890

ABCDEFGHIJKLMNOPQRSTUVWXYZ
1234567890

ABCDEFGHIJKLMNOPQRSTUVWXYZ
1234567890

ABCDEFGHIJKLMNOPQRSTUVWXYZ
1234567890

ABCDEFGHIJKLMNOPQRSTUVWXYZ¢
1234567890

ABCDEFGHIJKLMNOPQRSTUVWXYZ¢
1234567890

ABCDEFGHIJKLMNOPQRSTUVWXYZ
1234567890

ABCDEFGHIJKLMNOPQRSTUVWXYZ
1234567890

FIG. 4-8   Architectural lettering styles

ALWAYS USE TWO GUIDE LINES FOR CAPITAL LETTERS.

FIG. 4-9   Guide lines for uppercase letters

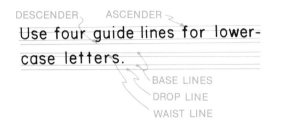

DESCENDER   ASCENDER

Use four guide lines for lower-case letters.

BASE LINES
DROP LINE
WAIST LINE

FIG. 4-10   Guide lines for lowercase letters

VERTICAL GUIDE LINES ARE RANDOMLY SPACED.

INCLINED GUIDE LINES ARE RANDOMLY SPACED.

FIG. 4-11   Vertical and inclined guide lines

GUIDE LINES

Before starting to letter the draftsperson must draw guide lines so that all letters will be uniform in height. Guide lines are light, sharp lines drawn with a hard, sharp pencil lead—usually a 4H or 6H. They should be very light so they cannot be seen on a print or from a short distance, yet dark enough so the draftsperson can use them to make all letters of uniform height. Uppercase lettering needs two guide lines (Fig. 4-9). Lowercase lettering may use four guide lines (Fig. 4-10). Only five lowercase letters have descenders, so draftspersons often leave off the fourth guide line.

Letter Heights

The standard height for most lettering is ⅛ inch. For larger drawings and those that will be microfilmed, a height of ³/₁₆ inch to ¼ inch is recommended. For special notes and title block information see Chapter 9.

Types of Guide Lines

There are three types of guide lines used to make lettering uniform: horizontal, vertical, and inclined (Fig. 4-11). Vertical and inclined guide lines can be randomly spaced to aid in directing the lettering. Horizontal guide lines must be accurately laid out to correct heights. Guide lines should be measured and then drawn in with a T-square or drafting machine.

Mechanical aids are useful also because they reduce drawing time (Fig. 4-12). Guide line devices have countersunk holes in which a pencil with a sharp, hard lead is inserted at the size of guide lines needed. The instrument is drawn across the top of a straight-edge with the pencil. Guide lines for short notes may be drawn quickly with a guide line template as shown in Figure 4-13. Vertical and inclined

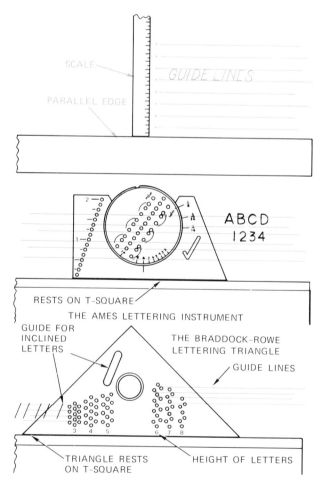

SCALE
GUIDE LINES
PARALLEL EDGE

ABCD
1234

RESTS ON T-SQUARE
THE AMES LETTERING INSTRUMENT

GUIDE FOR INCLINED LETTERS
THE BRADDOCK-ROWE LETTERING TRIANGLE
GUIDE LINES

TRIANGLE RESTS ON T-SQUARE
HEIGHT OF LETTERS

FIG. 4-12   Methods of drawing guide lines

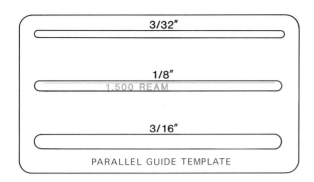

FIG. 4-13   Guide line template

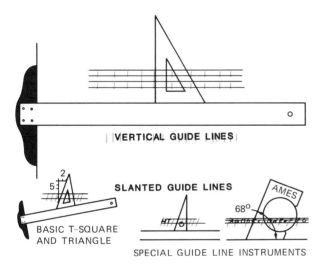

FIG. 4-14   Using guide line instruments

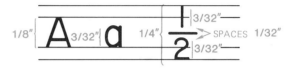

FIG. 4-15   Typical proportions for uppercase and lowercase letters

FIG. 4-16   Line spacing

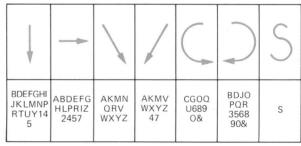

FIG. 4-17   Proper stroke directions

guide lines may be drawn using the procedures shown in Figure 4-14.

Typical proportions for lowercase lettering and fractions are shown in Figure 4-15.

LINE SPACING

The ANSI recommends that the distance between lines of lettering should be one half to one time the height of the letters. Detached notes should be separated by a minimum of two line heights. (Fig. 4-16). When lettering several lines of notes, keep the left side of the notes flush and the right side fairly even.

LETTER PROPORTIONS

After becoming familiar with the different types of lettering, learning to form and space the letters for the best possible composition of letters, words, and sentences is important for the draftsperson. One to four separate strokes are needed to form each Gothic letter properly. A combination of the strokes in Figure 4-17 will form all the Gothic letters and

FIG. 4-18 Strokes for Gothic letters

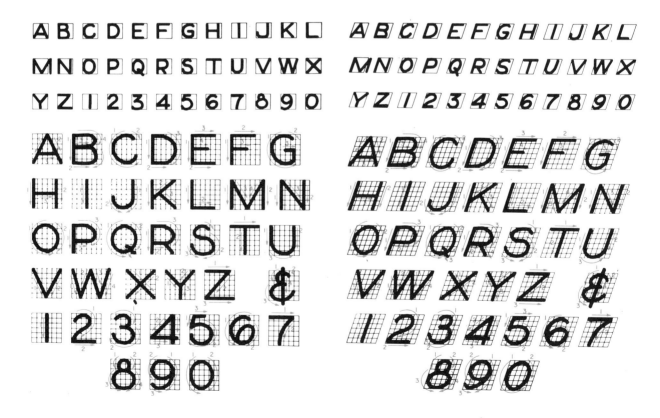

FIG. 4-19 Proportions of vertical Gothic letters

FIG. 4-20 Proportions of inclined Gothic letters

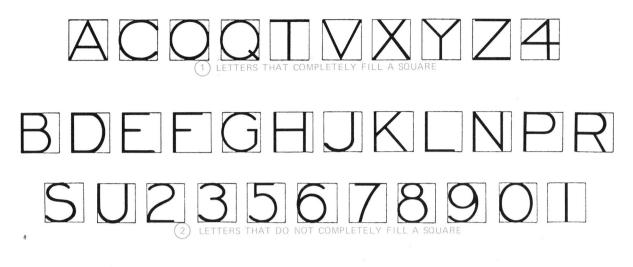

(1) LETTERS THAT COMPLETELY FILL A SQUARE

(2) LETTERS THAT DO NOT COMPLETELY FILL A SQUARE

(3) LETTERS THAT ARE WIDER THAN A SQUARE

FIG. 4-21   Letters grouped by their proportions

EFHILT	7AKMNVWXYZ4	COGJQU BDPRS235689
HORI-ZONTAL VERTICAL	SLANTED HORIZONTAL VERTICAL	CURVED HORIZONTAL VERTICAL

FIG. 4-22   Letters grouped by stroke styles

AOVWXZY

FIG. 4-23   These letters balance around an inclined center line.

numbers. The strokes for vertical Gothic letters are shown in Figure 4-18. Better examples of the widths of vertical and inclined letters are shown in Figure 4-19 and 4-20. There are variations in the approved methods of strokes for Gothic letters. After some practice, the easiest strokes may be selected. Further study breaks the letters into similar divisions as shown in Figure 4-21.

Figure 4-22 divides the letters into characteristic groups of strokes. Letters may vary in width to give them more pleasing proportions. For

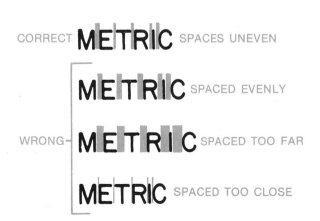

CORRECT **METRIC** SPACES UNEVEN

METRIC SPACED EVENLY

WRONG **METRIC** SPACED TOO FAR

METRIC SPACED TOO CLOSE

FIG. 4-24   Proper letter spacing

example, the tops of the letters B, E, K, S, X, Z are slightly smaller than their bottoms. The rules governing proportions of uppercase Gothic letters are the same for vertical and inclined letters.

In most instances inclined lettering is faster than vertical lettering, except perhaps for the letters A, O, V, W, X, Y, Z (Fig. 4-23). With practice these letters will become uniform. Note how these letters balance around an inclined center line. Always start with the vertical strokes first when forming the letters, and build the letters around these strokes.

SPACING LETTERS, WORDS, AND SENTENCES

Once the draftsperson has become proficient with the letters themselves, he or she must learn the "feel" of spacing the letters in words. The general rule of spacing letters in words is to space them so they will look equal. Because of optical illusions and irregularly shaped letters, the spacing has to be done by eye. Letter spacing should be pleasing to the eye, not too crowded (a minimum space of 1.5 mm), and not spaced too far apart (Fig. 4-24). Proper spacing can be accomplished by concentrating on the areas between letters. These areas

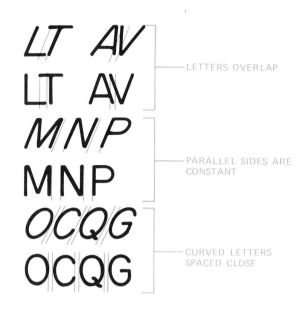

LETTERS OVERLAP

PARALLEL SIDES ARE CONSTANT

CURVED LETTERS SPACED CLOSE

FIG. 4-26   Special rules for spacing

FIG. 4-27   Spacing between words

should appear almost equal, but smaller than the letter; they may be very irregular depending on the shape of the adjacent sides of the letters (Fig. 4-25). Some general rules to follow are shown in Figure 4-26.

The next step in the progression of lettering is spacing between words. This spacing should be equal to the height of the letters (Fig. 4-27). The spacing between sentences is always twice the height of the letters (Fig. 4-28).

LETTERING RULES

There are a few rules to follow when freehand lettering with a drafting pencil:

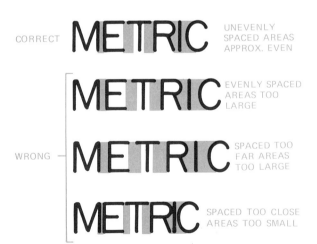

CORRECT — UNEVENLY SPACED AREAS APPROX. EVEN

EVENLY SPACED AREAS TOO LARGE

WRONG — SPACED TOO FAR AREAS TOO LARGE

SPACED TOO CLOSE AREAS TOO SMALL

FIG. 4-25   Spacing by equal areas

LETTER NEATLY. ▮SHARP CLEAR
LETTERS ARE CRITICAL. ▮IT WILL
REDUCE ERRORS IN COMMUNICA-
TION. ▮IT ALSO KEEPS COSTS DOWN.

SPACE EQUALS TWICE THE LETTER HEIGHT

FIG. 4-28   Spacing between sentences

1. Use a pencil with a soft lead for lettering (HB, F, H). A lead that is too soft will smear; one that is too hard will cut the paper and will not reproduce well.
2. Keep a sharp point on the pencil. Rotating the pencil while lettering helps keep the point sharp.
3. Keep pencils and instruments clean.
4. Rest your hand on a piece of paper to protect other pencil lines from smearing.
5. Always pull the pencil for the stroke. Never push it.
6. All lettering strokes are the same width. Never repeat a stroke.
7. The drawing surface under the drawing paper should be firm but not hard. Use a sheet of drawing paper or a commercial board cover.
8. Sit up comfortably. Do not bend over your lettering.

OTHER LETTERING PROCEDURES

Most lettering is done freehand today. However, there are other acceptable procedures for lettering a drawing, such as:

1. **Lettering templates** (Fig. 4-29) come in different heights and styles. A pencil is traced around the inside of the letters. They are slow but neat, and good for titles and other places where a small number of perfectly formed letters are needed.
2. A **typewriter** (Fig. 4-30) can be used to type directly on the drawing. A long or open car-

FIG. 4-29   Lettering templates

FIG. 4-30   Using a typewriter for lettering (Casey Gorman)

riage is required. A black, carbon-typewriter ribbon will give a very black, dense line that reproduces well. The typewriter can also be used to type on adhesive-backed paper or film, which is then attached directly to the drawing. Another type of typewriter, called the **Gritzner typewriter,** is placed on the drawing board and types directly on the drawing while moving across it (Fig. 4-31).

FIG. 4-31   The Gritzner drafting board typewriter (Gritzner Graphics)

FIG. 4-32   A lettering guide (Casey Gorman)

3. **Lettering devices** such as the Leroy and Wrico (Fig. 4-32) use a template, follower, and marker.

4. **Transfer letters** can be purchased in many different sizes and styles (Fig. 4-33). The letter is rubbed onto the drawing.

5. A **Varityper** will produce large or small lettered strips that can be attached to the drawing. Varityper also makes a typewriter that will type directly onto the drawing.

6. **Computer-generated lettering** can be done while the computer is generating drawings (Fig. 4-34).

THE LEFT-HANDER

A set of lettering rules will not be helpful to most left-handed persons. Since their technique of holding a pencil and positioning their drawing varies so

FIG. 4-33 Transfer letters

much, they must work out their own lettering system. It is the finished lettering that counts. Figure 4-35 shows letter strokes that might aid some left-handed persons.

LETTERING IS IMPORTANT

Freehand lettering is a fast way to letter a drawing as each line and curve need not be perfect. If the letters are uniform in height, proportion, slant, vertical position, line weight, and spacing in words and

ABC
DEF

FIG. 4-34 Computer-generated lettering, enlarged

FIG. 4-35 Suggested strokes for left-handed draftspersons

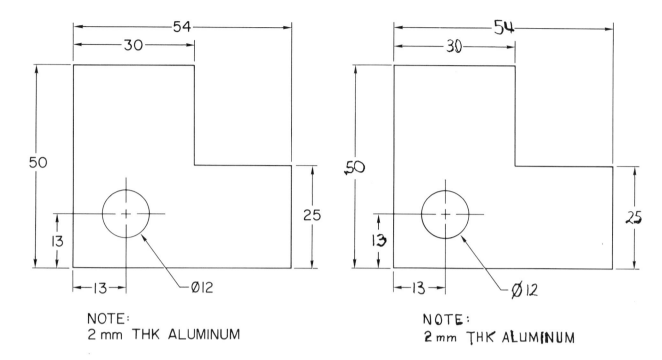

FIG. 4-36 A comparison of lettering qualities

sentences, you will have excellent lettering technique. To achieve this skill requires continual practice, careful observation, determination, and patience. Beginning level draftspersons are usually not hired if their lettering quality and speed are poor. Many people feel that lettering reflects the quality of a drawing and of the draftsperson. People reading a poorly lettered engineering drawing will not have much faith in the drawing or the draftsperson. What feelings do you get from Figure 4-36? The drawings are identical except for the lettering. A good measure of the drawings in Figure 4-36

would be to microfilm then, blow them back up to full size, then see if the lettering is still legible.

EXERCISES

1. Draw up a sheet with ⅛ inch guide lines, and practice vertical lettering.
2. Draw up a sheet with ⅛ inch guide lines, and practice inclined lettering.
3. Draw up a sheet with ⅛ inch guide lines, and practice microfont lettering.

5

Chapter

Drafting Equipment

Jacqueline Marsall

INTRODUCTION

Twenty thousand years ago prehistoric people were scratching drawings on walls with sticks and bones (Fig. 5-2). The pyramids were designed on papyrus by scribing a line, then tracing over the mark with a reed pen and a colored liquid. The Greeks and Romans wrote in wax with a sharp stylus, and later used a piece of metal for scribing lines on a surface. Leonardo da Vinci drew with charcoal on paper, and then went over the lines with a quill pen and ink. Graphite was discovered around 1400 and was used for crude drawings. In the 1600s the graphite was encased in wood holders. Later, clay was added to the graphite to stop it from smearing.

As drawings became more technical, more accurate drawing instruments were developed and improved upon. Drawing became an accurate communication medium for conveying manufacturing and construction ideas. The advancement of technology is related to the accuracy of the available instruments, tools, and products (Fig. 5-3).

As you read this chapter, you should learn how to:

1. Care for and use instruments.
2. Test your T-squares and triangles for accuracy.
3. Use the T-square to draw horizontal lines.
4. Draw vertical and inclined lines using the T-square and triangles.
5. Select the proper grades of pencils used in drawing.

FIG. 5-2   Drawing in prehistoric times

6. Sharpen your pencils correctly.
7. Hold the pencil correctly to produce good line quality.
8. Square up and fasten paper to the drawing board.
9. Make simple corrections on drawings.
10. Draw circles and arcs.
11. Sharpen and adjust a compass lead.
12. Select the proper media on which to make different types of drawings.

MODERN DRAFTING EQUIPMENT

The different types of drawing instruments and equipment used in today's drafting room are quite numerous (Fig. 5-4). They may be divided into three groups:
Beginning student
drawing paper

FIG. 5-3   Drawing in today's world (Teledyne Post)

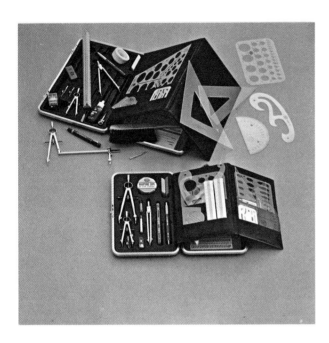

FIG. 5-4 Modern drafting instrument kits (Teledyne Post)

tracing paper
drafting pencils
erasers
erasing shield
drafting boards
drafting tape
pencil sharpeners
T-squares
triangles
scales
compasses
dividers
protractor
templates (circle, isometric)
irregular curve
dusting brush
Advanced student.
drafting machine
drawing surface cover
parallel slide
vellum
adjustable triangle
flexible curve
templates, as needed
lettering templates
guide line template

mechanical pencils
technical pens
ink
transfer letters
shading tones

Professional draftsperson:
track drafting machine
drawing table
electric eraser
drawing film
lofting curves
lettering set
drop bow compass
proportional dividers
drafting lamp
perspective board
pantograph
templates, as needed
computer
duplicator

Drawing media (Fig. 5-5) are available in rolls and sheets (see Chapter 9 for sizes). There are many types of drawing media available. A few are listed below:

1. *Drawing paper* comes in white, buff, and green.

2. *Tracing paper* is thin, inexpensive, and translucent.

3. *Vellum* is heavier than tracing paper, treated with chemical additives, costlier, and translucent.

4. *Polyester film* is an indestructible plastic film, transparent, and very expensive.

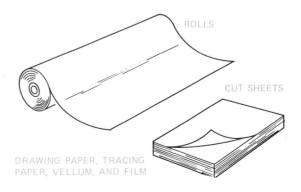

FIG. 5-5 Drafting media

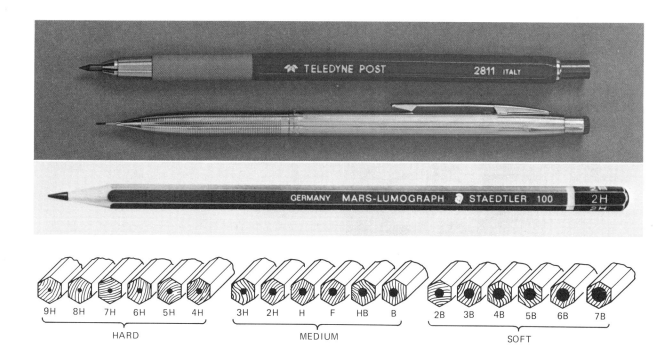

FIG. 5-6 Drafting pencils (J. S. Staedtler, Inc. (Teledyne Post)

5. *Graph paper* is squared drawing paper or vellum with light, nonreproducible lines, available in all sizes

Drafting pencils (Fig. 5-6) or leads are divided into three groups: hard, medium, and soft.

Hard pencils range from 9H to 4H. They leave a very sharp, but light line, and will not smear easily. They are used for first draft outlines.

Medium pencils range from 3H to B. They are used for all finished line work in engineering drawings. Experience is needed to select the best pencil.

Soft pencils range from 2B'to 7B. These pencils are too soft to use on engineering drawings. They are used in renderings and general art work.

A special *plastic lead pencil* is used for drawing on polyester film.

The *lead holder* grips individual drafting leads.

The *ultra-thin lead mechanical pencil* needs no sharpening. It can hold various drafting leads of 0.5 mm diameter.

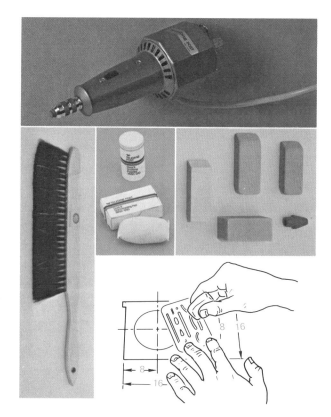

FIG. 5-7 Drawing cleaning equipment (Teledyne Post)

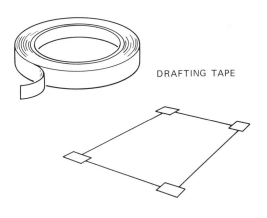

FIG. 5-8 Drafting tape

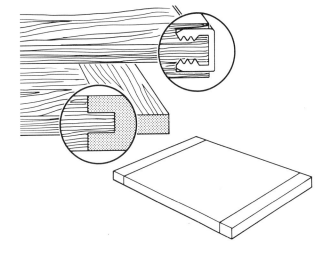

FIG. 5-9 Drafting boards

Cleaners (Fig. 5-7) play a large part in the production of quality drawings. Some cleaning aids are:

1. Basic *white eraser* for all-purpose uses.

2. *Artgum eraser* for erasing light lines and smudges.

3. *Kneaded eraser* for picking up loose graphite that may cause smudging. It is soft and pliable.

4. *Dry cleaner* for cleaning light smudges. It is a powder in a small cloth bag that may be sprinkled on the drawing while doing layout work. The instruments will ride on the powder and keep the drawing clean.

5. *Erasing shield* for protecting parts of line work that are not to be erased.

6. *Electric eraser* for fast and efficient erasing. It does little damage to the surface of the drawing media.

7. *Dusting brush* for removing foreign matter from the drawing. It is cleaner to use a brush than to blow or wipe off foreign matter by hand.

Drafting tape (Fig. 5-8) is the approved method of attaching a drawing to a drawing surface. The tape will not damage the surface.

Drafting boards (Fig. 5-9) come in a variety of sizes. A good size for the beginning student is

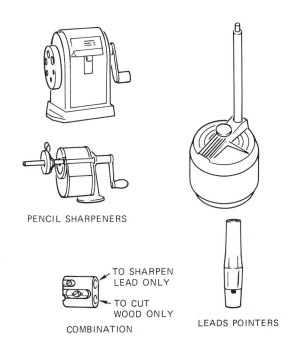

PENCIL SHARPENERS

TO SHARPEN LEAD ONLY

TO CUT WOOD ONLY

COMBINATION

LEADS POINTERS

FIG. 5-10 Pencil sharpeners

20 inches by 26 inches. Larger boards are available for larger size drawings. The board allows the drawing to be easily stored, freeing the desk for other work.

Pencil sharpeners (Fig. 5-10) are available in many styles. There are two types of mechanical pencil sharpeners for the wood-encased drafting

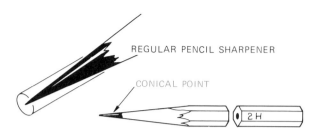

REGULAR PENCIL SHARPENER

CONICAL POINT

2 H

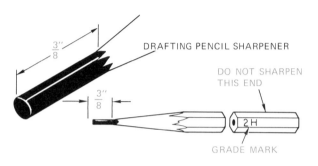

$\frac{3''}{8}$

DRAFTING PENCIL SHARPENER

DO NOT SHARPEN
THIS END

$\frac{3''}{8}$

2 H

GRADE MARK

FIG. 5-11   Methods of sharpening pencils

pencil. One sharpener will sharpen with a long smooth sharp point. The drafting sharpener will remove only the wood. Another operation called pointing the lead is then necessary. A lead pointer will sharpen the exposed end of the drafting lead. Figure 5-11 shows the proper way to sharpen a drafting pencil.

T-squares (Fig. 5-12) come in a variety of sizes to match drafting board sizes. They are used to draw horizontal lines and as a guide for other drafting instruments. The **parallel slide** is a straight edge that is permanently attached to the drawing board. It can be moved quickly and will not slip.

Triangles (Fig. 5-13) are used with the T-square to draw angular and vertical lines.

Scales (Fig. 5-14) come in many different shapes and lengths. Chapter 7 explains the uses of metric scales.

Compasses (Fig. 5-15) come in a variety of shapes and sizes. Using the compass and sharpening the compass lead are shown in Figure 5-16.

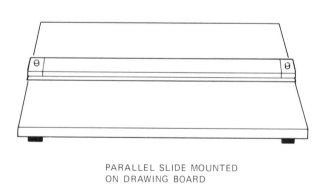

PARALLEL SLIDE MOUNTED
ON DRAWING BOARD

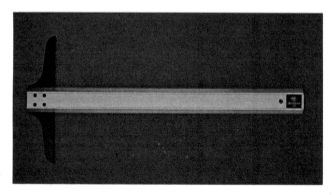

FIG. 5-12   T-square   (Teledyne Post)

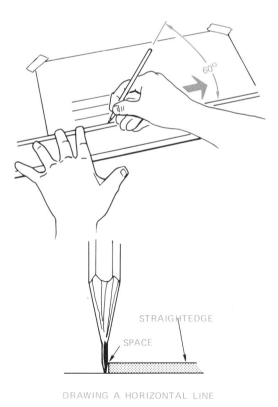

60°

STRAIGHTEDGE

SPACE

DRAWING A HORIZONTAL LINE

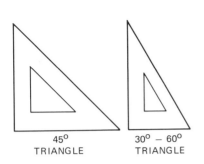

45°
TRIANGLE

30° – 60°
TRIANGLE

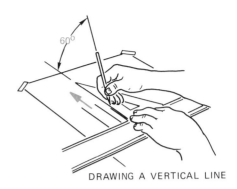

DRAWING A VERTICAL LINE

ALL ANGLES WITH
RESPECT TO HORIZONTAL

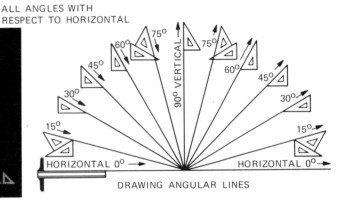

DRAWING ANGULAR LINES

FIG. 5-13 Triangles (Teledyne Post)

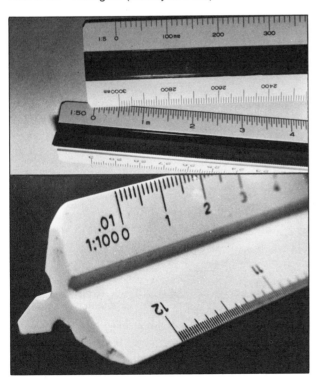

FIG. 5-14 Scales (Clark E. Smith) (Blundell Harling)

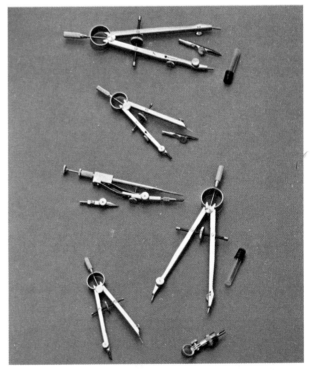

FIG. 5-15 Compasses (Teledyne Post)

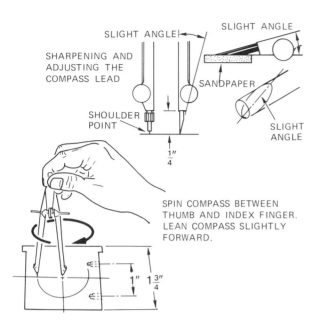

FIG. 5-16 Use of the compass

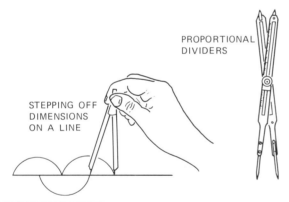

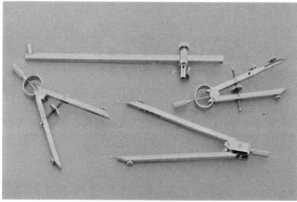

FIG. 5-17 Dividers (Teledyne Post)

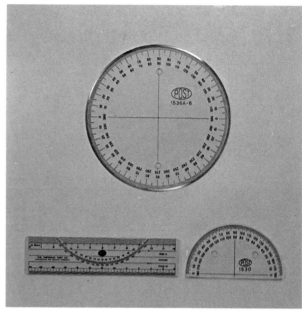

FIG. 5-18 Protractors (Teledyne Post)

Dividers (Fig. 5-17) are used to transfer measurements and divide lines. A proportional divider can be adjusted to different ratios on the opposite end.

Protractors (Fig. 5-18) are used to measure angles.

Templates (Fig. 5-19) are used to reduce drawing time. A tremendous variety of templates are available. Some of the types of templates are:

1. Circles and ellipses
2. Architectural
3. Nut and bolt
4. Lettering
5. Electrical
6. Guide line
7. Math and scientific symbols
8. Piping

Irregular curves (Fig. 5-20) are used to darken irregular lines. A good deal of practice is needed to darken irregular lines smoothly. Irregular curves come in a variety of shapes and forms. A *flexible curve* can be used for long irregular curves. For many long, smooth curves (aircraft and ship drafting) a set of lofting curves can be used. *Lofting curves* are very long to allow for full size drawings.

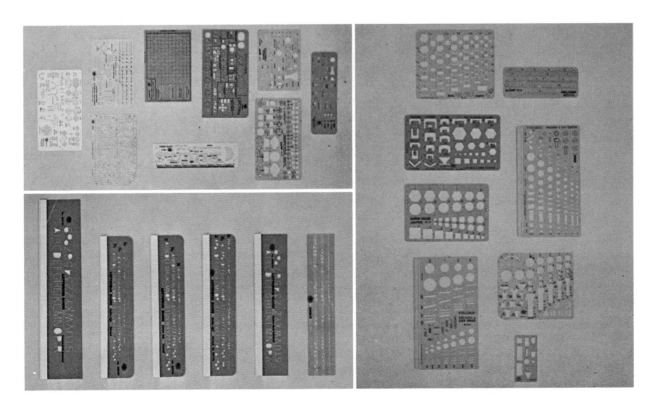

FIG. 5-19 Templates (Teledyne Post, RapiDesign, Inc., Olson Mfg. Co.)

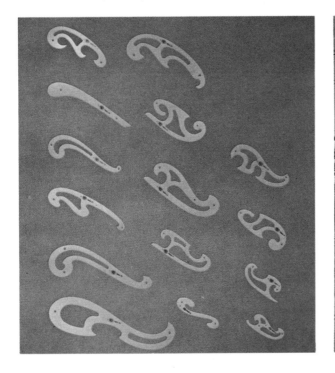

FIG. 5-20 Irregular curves (Teledyne Post)

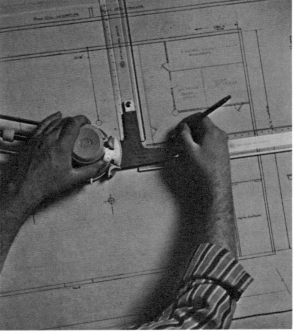

FIG. 5-21 Drafting machines (Teledyne Post)

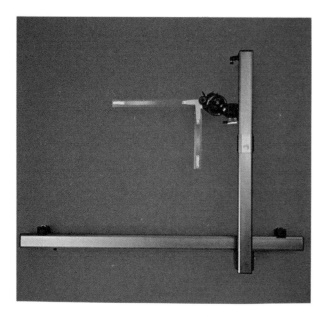

FIG. 5-22   Track drafting machine (Teledyne Post)

FIG. 5-23   Drafting tables (Teledyne Post)

Drafting machines (Fig. 5-21) are used to speed up drawing time. They take the place of T-squares, triangles, scales, and protractors. The arms and scales come in various lengths to accommodate any size drawing board.

Track drafting machines (Fig. 5-22) allow smoother and faster movement than the standard drafting machine. Various sizes of tracks and scales will fit any drafting board and drawing.

Drafting tables (Fig. 5-23) are the draftsperson's work station. Most students will work on drafting boards placed on the table. The professional draftsperson will work directly on the table surface. Tables come in many different sizes to handle any size drawings.

Drawing **surface covers** (Fig. 5-24) provide a smoother drawing surface. A sheet of drawing paper will make an inexpensive, temporary cover. Laminated vinyl is a permanent, almost perfect drawing surface. There is no glare, it cleans easily, and small holes seal themselves. Vinyl covers are easily attached with two-sided tape.

Adjustable triangles (Fig. 5-25) allow different angles to be drawn quickly in conjunction with the T-square.

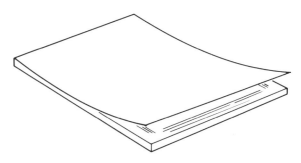

FIG. 5-24   Drawing surface cover

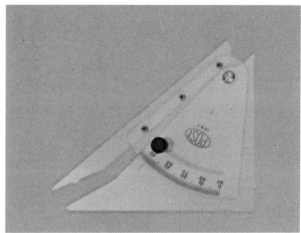

FIG. 5-25   Adjustable triangle (Teledyne Post)

FIG. 5-26 Technical pens (Teledyne Post)

Technical pens (Fig. 5-26) are used for ink drawings. Each pen has its own line width. The most common ink used is black, opaque India ink.

Transfer letters, symbols, and shading tones (Fig. 5-27) are used by the technical illustrator for lettering heads and shading drawings (see Chapter 23). An unlimited number of letter styles, sizes, and variations of shading are available.

FIG. 5-27 Transfer symbols (Teledyne Post)

Lettering sets (Fig. 5-28) are mechanical methods for producing precise lettering. The Leroy and Unitech sets are most common.

FIG. 5-28 Lettering sets (Teledyne Post)

Drafting lamps (Fig. 5-29) are necessary if *drafting room* lighting is not adequate.

Perspective boards (Fig. 5-30) are used to make perspective drawings more rapidly.

Pantographs (Fig. 5-31) are used to copy and change the scale of an original drawing.

Computer drafting (Fig. 5-32) is now being used in the larger drafting rooms. See Chapter 33 for detailed information on computer drafting.

Duplicators (Fig. 5-33) are any machines that can copy a drawing. See Chapter 35 for different methods of reproduction.

FIG. 5-29 Drafting lamp (Teledyne Post)

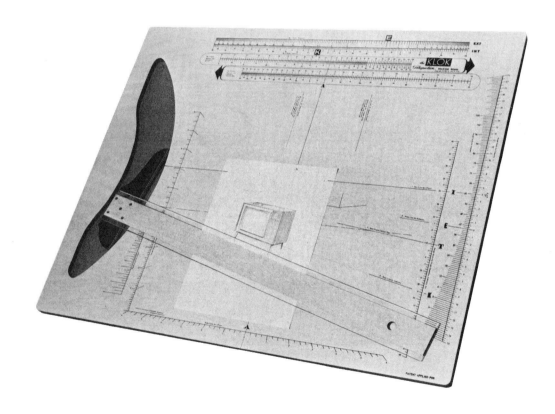

FIG 5-30 Perspective drawing board (Fiberesin, Oconomowoc, Wi.)

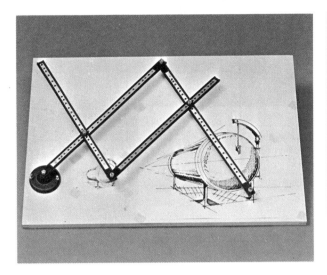

FIG. 5-31 Pantograph (Teledyne Post)

FIG. 5-33 Duplicators (Teledyne Post)

FIG. 5-32 Computer drafting machine (Texas Instruments, Inc.)

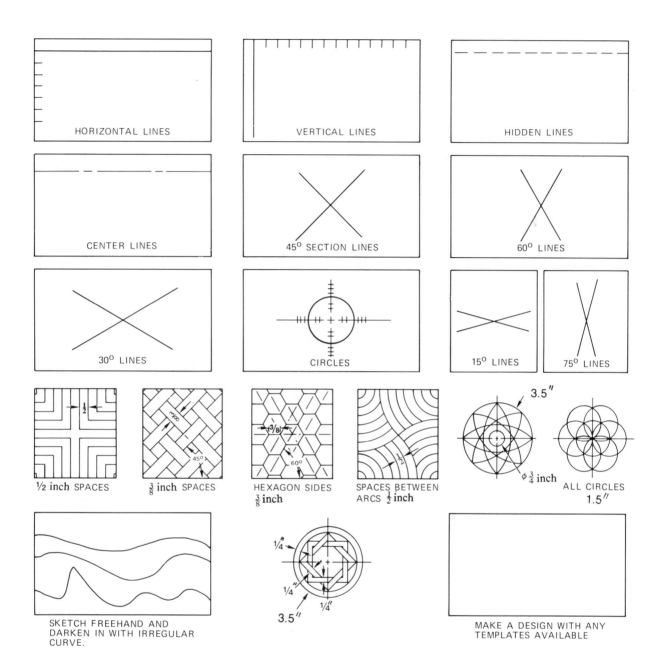

FIG. 5-34 Problems

PROBLEMS

Draw the following exercises with your drafting instruments. Divide a piece of drawing paper into 3-inch squares. Fill in each square.

6

Line Conventions

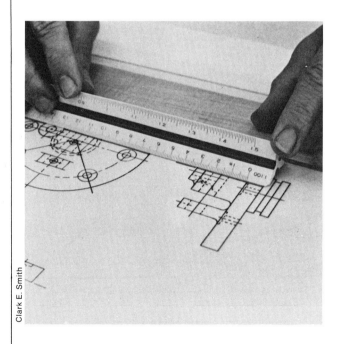

Clark E. Smith

INTRODUCTION

The different types of lines or line symbols used by the draftsperson represent a graphical alphabet, which is referred to as line conventions. Each line has its own specific meaning, and the American National Standards Institute (ANSI) has standardized line conventions for engineering drawings in its publication Y14.2, as shown in Figure 6-2. As you study this chapter, you should learn to recognize and draw all of the line conventions.

LINE WIDTHS

ANSI recommends two line widths (Fig. 6-3). The width of the thick line is between 0.030 inch and 0.038 inch, the average width being 0.035 inch. The width of the thin line is between 0.015 inch and 0.022 inch, the average width being 0.019 inch.

The widths of thick and thin lines on an engineering drawing will vary with the size and type of drawing. Large drawings, or drawings to be reduced, should have thick line widths of 0.038 inch and thin line widths of 0.022 inch. Small engineering drawings should have thick line widths of 0.030 inch and thin line widths of 0.015 inch. Average size drawings should have thick line widths of 0.035 inch and thin line widths of 0.019 inch (Fig. 6-4).

The critical feature of line conventions in engineering drawing is line consistency. Once the line width is determined it must be consistent throughout the completed drawing (Fig. 6-5).

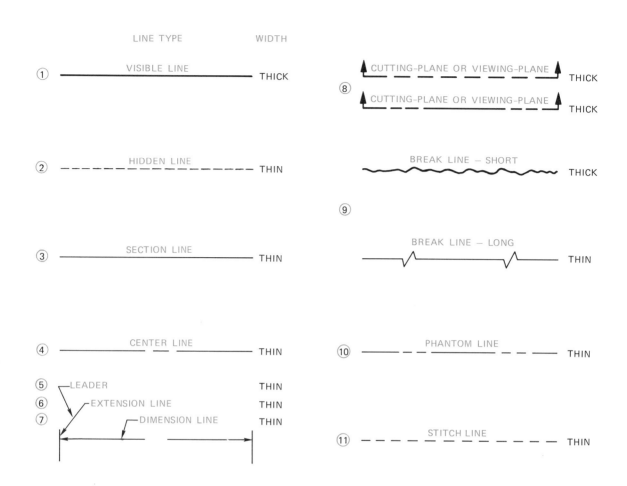

FIG. 6-2   ANSI Y14.2 line conventions

The lengths of large and small dashes and their spaces will also vary slightly with the size of the drawing (Fig. 6-6). The use of correct line conventions will produce a clear and easily read drawing.

DRAFTING PENCIL

The recommended rafting pencils for lines are the F or H for thick lines and the 2H for thin lines. All lines on a drawing should be dense, black, clean-cut, and uniform in width. Line work should never be gray, fuzzy, or inconsistent (Fig. 6-7). Following these standards will ensure good reproduction and microfilming qualities (see Chapter 35).

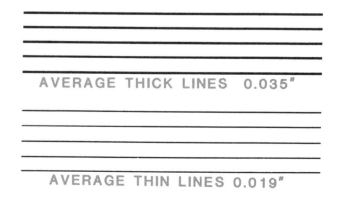

FIG. 6-3   True size line widths

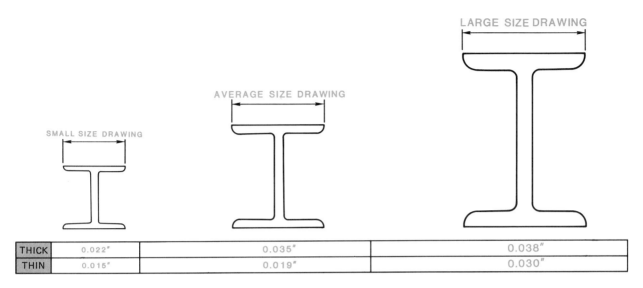

FIG. 6-4   Line widths proportional to drawing size

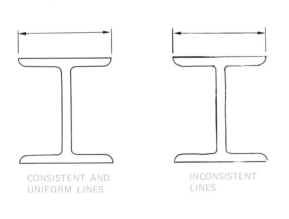

FIG. 6-5   Line consistency

DRAWING LINE CONVENTIONS

Your first attempts at drawing line conventions on an engineering drawing should be made with a scale. Measure the lengths of different types of dashes until you can closely estimate their proper length.

Construction lines and guide lines (Fig. 6-8) are not part of a finished drawing. They should be very light or not be seen at all. These lines are about 0.2 mm wide and are drawn with a 4H or 2H pencil.

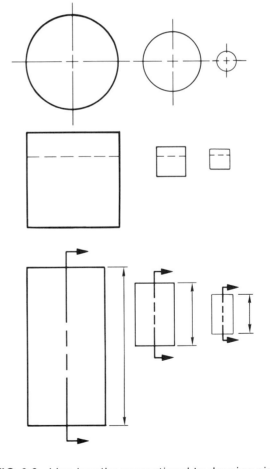

FIG. 6-6   Line lengths proportional to drawing size

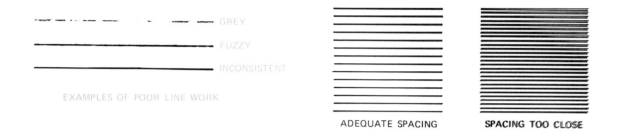

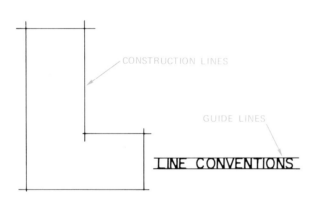

FIG. 6-7   Examples of poor line work

FIG. 6-9   Line spacing

FIG. 6-8   Construction lines and guide lines

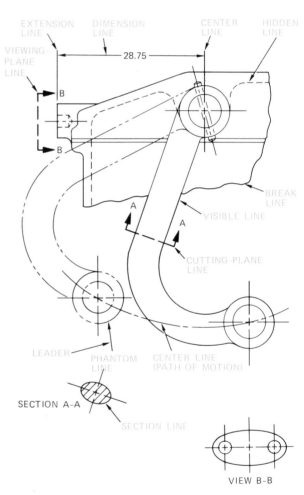

The spacing between parallel lines should be wide enough to prevent "fill-in" during reproduction or microfilming. Spaces between parallel lines should be a minimum of $1/16$ inch (Fig. 6-9).

Examples of all the line conventions used in engineering drawing are shown in Figure 6-10.

A fully detailed chart listing all line conventions is shown in Figure 6-11. Because of photoreduction, many lines are reduced in size, but the difference between thick and thin lines remains constant.

FIG. 6-10   Applications of lines

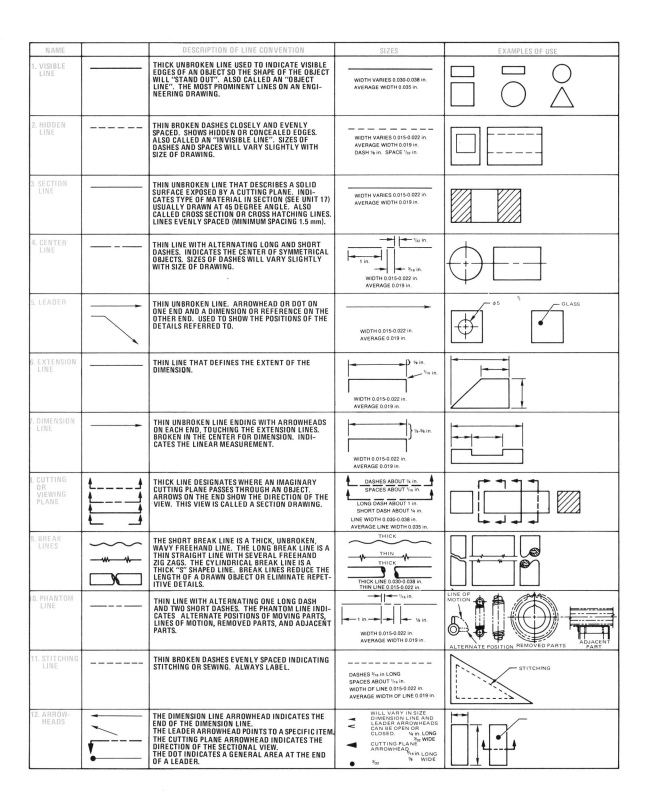

NAME		DESCRIPTION OF LINE CONVENTION	SIZES	EXAMPLES OF USE
1. VISIBLE LINE		THICK UNBROKEN LINE USED TO INDICATE VISIBLE EDGES OF AN OBJECT SO THE SHAPE OF THE OBJECT WILL "STAND OUT". ALSO CALLED AN "OBJECT LINE". THE MOST PROMINENT LINES ON AN ENGINEERING DRAWING.	WIDTH VARIES 0.030-0.038 in. AVERAGE WIDTH 0.035 in.	
2. HIDDEN LINE		THIN BROKEN DASHES CLOSELY AND EVENLY SPACED. SHOWS HIDDEN OR CONCEALED EDGES. ALSO CALLED AN "INVISIBLE LINE". SIZES OF DASHES AND SPACES WILL VARY SLIGHTLY WITH SIZE OF DRAWING.	WIDTH VARIES 0.015-0.022 in. AVERAGE WIDTH 0.019 in. DASH 1/8 in. SPACE 1/32 in.	
3. SECTION LINE		THIN UNBROKEN LINE THAT DESCRIBES A SOLID SURFACE EXPOSED BY A CUTTING PLANE. INDICATES TYPE OF MATERIAL IN SECTION (SEE UNIT 17) USUALLY DRAWN AT 45 DEGREE ANGLE. ALSO CALLED CROSS SECTION OR CROSS HATCHING LINES. LINES EVENLY SPACED (MINIMUM SPACING 1.5 mm).	WIDTH VARIES 0.015-0.022 in. AVERAGE WIDTH 0.019 in.	
4. CENTER LINE		THIN LINE WITH ALTERNATING LONG AND SHORT DASHES. INDICATES THE CENTER OF SYMMETRICAL OBJECTS. SIZES OF DASHES WILL VARY SLIGHTLY WITH SIZE OF DRAWING.	1/32 in. 1 in. 3/16 in. WIDTH 0.015-0.022 in. AVERAGE 0.019 in.	
5. LEADER		THIN UNBROKEN LINE. ARROWHEAD OR DOT ON ONE END AND A DIMENSION OR REFERENCE ON THE OTHER END. USED TO SHOW THE POSITIONS OF THE DETAILS REFERRED TO.	WIDTH 0.015-0.022 in. AVERAGE 0.019 in.	φ5 GLASS
6. EXTENSION LINE		THIN LINE THAT DEFINES THE EXTENT OF THE DIMENSION.	1/8 in. 1/16 in. WIDTH 0.015-0.022 in. AVERAGE 0.019 in.	
7. DIMENSION LINE		THIN UNBROKEN LINE ENDING WITH ARROWHEADS ON EACH END, TOUCHING THE EXTENSION LINES. BROKEN IN THE CENTER FOR DIMENSION. INDICATES THE LINEAR MEASUREMENT.	1/4-5/8 in. WIDTH 0.015-0.022 in. AVERAGE 0.019 in.	
8. CUTTING OR VIEWING PLANE		THICK LINE DESIGNATES WHERE AN IMAGINARY CUTTING PLANE PASSES THROUGH AN OBJECT. ARROWS ON THE END SHOW THE DIRECTION OF THE VIEW. THIS VIEW IS CALLED A SECTION DRAWING.	DASHES ABOUT 1/4 in. SPACES ABOUT 1/16 in. LONG DASH ABOUT 1 in. SHORT DASH ABOUT 1/4 in. LINE WIDTH 0.030-0.038 in. AVERAGE LINE WIDTH 0.035 in.	
9. BREAK LINES		THE SHORT BREAK LINE IS A THICK, UNBROKEN, WAVY FREEHAND LINE. THE LONG BREAK LINE IS A THIN STRAIGHT LINE WITH SEVERAL FREEHAND ZIG ZAGS. THE CYLINDRICAL BREAK LINE IS A THICK "S" SHAPED LINE. BREAK LINES REDUCE THE LENGTH OF A DRAWN OBJECT OR ELIMINATE REPETITIVE DETAILS.	THICK / THIN / THICK THICK LINE 0.030-0.038 in. THIN LINE 0.015-0.022 in.	
10. PHANTOM LINE		THIN LINE WITH ALTERNATING ONE LONG DASH AND TWO SHORT DASHES. THE PHANTOM LINE INDICATES ALTERNATE POSITIONS OF MOVING PARTS, LINES OF MOTION, REMOVED PARTS, AND ADJACENT PARTS.	1/16 in. 1 in. 1/8 in. WIDTH 0.015-0.022 in. AVERAGE WIDTH 0.019 in.	LINE OF MOTION ALTERNATE POSITION REMOVED PARTS ADJACENT PART
11. STITCHING LINE		THIN BROKEN DASHES EVENLY SPACED INDICATING STITCHING OR SEWING. ALWAYS LABEL.	DASHES 3/16 in LONG SPACES ABOUT 1/16 in. WIDTH OF LINE 0.015-0.022 in. AVERAGE WIDTH OF LINE 0.019 in.	STITCHING
12. ARROWHEADS		THE DIMENSION LINE ARROWHEAD INDICATES THE END OF THE DIMENSION LINE. THE LEADER ARROWHEAD POINTS TO A SPECIFIC ITEM. THE CUTTING PLANE ARROWHEAD INDICATES THE DIRECTION OF THE SECTIONAL VIEW. THE DOT INDICATES A GENERAL AREA AT THE END OF A LEADER.	WILL VARY IN SIZE DIMENSION LINE AND LEADER ARROWHEADS CAN BE OPEN OR CLOSED. 1/4 in. LONG 3/32 WIDE CUTTING PLANE ARROWHEAD 3/16 in. LONG 1/8 WIDE 3/32	

FIG. 6-11 Line conventions

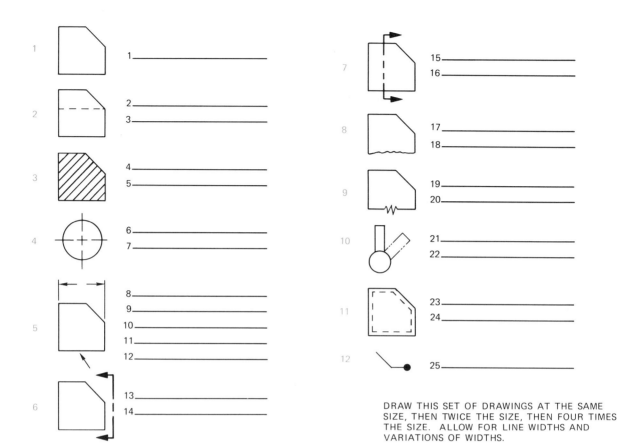

FIG. 6-12   Problems

DRAW THIS SET OF DRAWINGS AT THE SAME
SIZE, THEN TWICE THE SIZE, THEN FOUR TIMES
THE SIZE. ALLOW FOR LINE WIDTHS AND
VARIATIONS OF WIDTHS.

PROBLEM

Name each line convention used in Figure 6-12.

7

Chapter

Drafting Scales

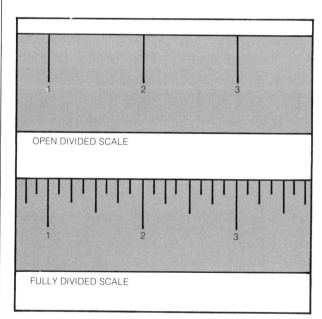

OPEN DIVIDED SCALE

FULLY DIVIDED SCALE

FIG. 7-1

INTRODUCTION

Drafting is a science of accurate measurement. The line work and the measurements on the working drawing for a product must be precise. The drawing must accurately communicate all the information needed by the product's fabricator or maker. As a draftsperson you must learn to draw accurately. To do this, you must have a thorough knowledge of drafting scales and their correct uses.

The term *scale* has two distinct meanings in drafting. It is (1) the actual measuring instrument, which is the ruler, and (2) the size to which the object is drawn. The scale size can be drawn in one of three ways: full size, larger than the object, or smaller than the object. The selection of the drawing scale depends on the size of the drawing format as well as the size of the object. Further information on scale drawings is presented later in this chapter.

The scale, or ruler, is the draftperson's most often used drafting instrument. It is divided into measuring units such as inches or millimetres. The divisions on the drafting scale are either open divided or fully divided. (See Fig. 7-1.) An *open divided* scale gives only the main measuring units on the side of the scale. On the opposite side of the open divided scale, there is also another scale that is either twice the size or half the size of the primary scale. These overlapping scales can double the scale capacity of a measuring instrument.

The *fully divided* scale has all the subdivisions

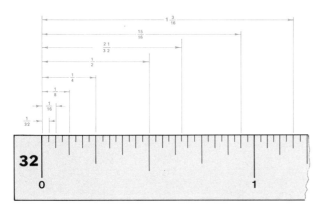

FIG. 7-2   Units for the mechanical engineer's drafting scale (full size)

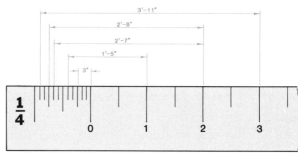

FIG. 7-4   Units for the most common architect's scale (¼″ = 1′-0″)

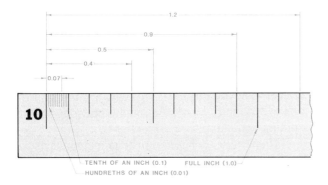

FIG. 7-3   Units for the civil engineer's drafting scale (1:1)

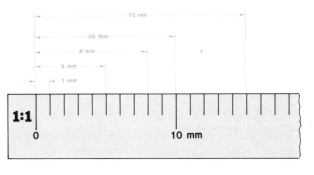

FIG. 7-5   Metric drafting scale (full size)

for each unit marked along the full length of the scale. This scale is easier and faster to read than the open divided scale.

TYPES OF MEASURING SYSTEMS

Because no system of measure is universal, the draftsperson must learn more than one system. Each system of measure has its own scale. The scales used for working drawings in these different systems are explained below.

Mechanical Engineer's Drafting Scale

The mechanical engineer's scale (Fig. 7-2) uses the inch and fractional segments of an inch (i.e., ¼, ½)

as its subdivisions. Its accuracy and quality vary. This scale can be purchased with the inch units divided into ⅛'s, $^{1}/_{16}$'s, $^{1}/_{32}$'s, or $^{1}/_{64}$'s units. The ⅛'s scale is the least accurate, and the $^{1}/_{64}$'s scale is the most accurate. The most commonly used scale is the $^{1}/_{16}$'s scale.

Civil Engineer's Drafting Scale

The civil engineer's scale (Fig. 7-3) is divided into inches and the decimal parts of an inch. On the full-size scale, the inch is divided into ten parts, each part being one-tenth of an inch (0.01). To rule hundredths with this scale you must estimate. But with practice you can learn to measure a hundredth of an inch (0.01).

Thus architectural drawings are reduced in scale. The most common architect's scale is ¼″ = 1′-0″. With this scale, each ¼-inch unit equals one foot. (See Fig. 7-4.) Marked units to the right of the "0" represent feet. Each unit to the left of the "0" represents one inch. Note how a measure of two feet eight inches is indicated. Two units to the right of the "0" equals two feet (2′) and eight units to the left equals eight inches (8″).

Metric Drafting Scale

The metric system of measurement is being used more and more in the United States. The unit of metric measurement for engineering drawings is the millimetre (mm). All measurements, no matter how small or how large, are given only in millimetres. As Figure 7-5 shows, the metric scale is divided into 10-mm units. These are further subdivided into millimetres.

The universal use of the millimetre scale is expected to simplify measuring and to help reduce errors in making and interpreting drawings.

SCALE SHAPES

Drafting scales have triangular and flat shapes (Fig. 7-6). These shapes are: single bevel, alternative double bevel, double bevel, quadruple bevel, and triangular. The scale shape a draftsperson chooses is usually a matter of personal preference.

Each shape has certain advantages. A flat scale, such as the single bevel, is easy to read and handle. But its usefulness is limited by the number of scales on the beveled edge. The advantage of the triangular-shaped scale is the many different scales on one ruler. The typical triangular scale has eleven different scales.

SCALED DRAWINGS

Ideally, drawings should be done in full-size scale. However, this is seldom possible. One reason is that the drawing paper format will not permit it. If the drawing of an object is too large to fit the paper format, then the drawing must be scaled down. For example, the airplane in Figure 7-7 has a 33′-10″ wing span. It would not be drawn full size on drawing paper. Conversely, a full-size drawing of a small object may not be clear enough. To be read cor-

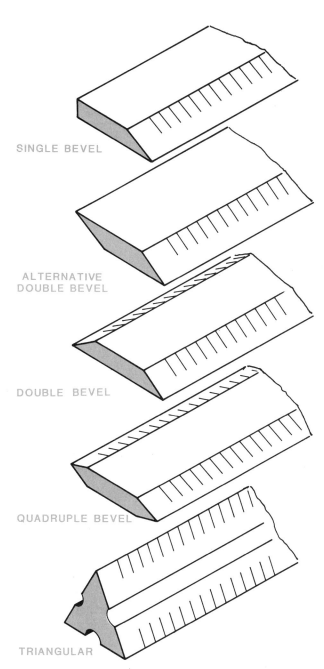

SINGLE BEVEL

ALTERNATIVE DOUBLE BEVEL

DOUBLE BEVEL

QUADRUPLE BEVEL

TRIANGULAR

FIG. 7-6 Shapes of drafting scales

Architect's Drafting Scale

Rarely is any part of an architectural plan drawn in full size. Buildings and sections of buildings are normally too large for the full-size drawing format.

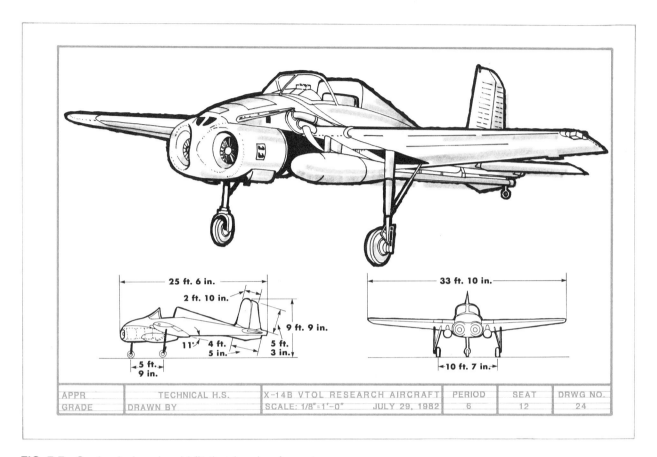

The illustration includes the following labels:

25 ft. 6 in.
2 ft. 10 in.
9 ft. 9 in.
11°   4 ft. 5 in.   5 ft. 3 in.
5 ft. 9 in.
33 ft. 10 in.
10 ft. 7 in.

APPR	TECHNICAL H.S.	X-14B VTOL RESEARCH AIRCRAFT	PERIOD	SEAT	DRWG NO.
GRADE	DRAWN BY	SCALE: 1/8"=1'-0" JULY 29, 1982	6	12	24

FIG. 7-7   Scale choice should fit the drawing format

rectly, the drawing will have to be scaled up as in Figure 7-8. While the full-size drawing in the top half of Figure 7-8 is too small to read, the same object, at an enlarged scale of 5:1, is clear and readable in larger format.

CHOOSING A SCALE

The choice of a size or scale depends on the size of the drawing format and the size of the final drawing. Some commonly used scales for drafting follow.

Figure 7-9 shows six types of scales for the mechanical engineer's scale (ruler). Basic units of these scales are the inch and the fractional parts of an inch. This system of measurement is the standard method in some industries and many schools teaching drafting. A sample drawing with the mechanical engineer's scale is shown in Figure 7-10.

Figure 7-11 illustrates the different engineer's scales. This system of measurement is used by the

majority of American industries. The basic units of these scales are the inch and the decimal parts of an inch. The basic inch unit can be divided into any number of subdivisions that are multiples of 10, such as 20, 50, 80, etc. Also, the scale ratio is flexible. One inch can equal 1000 ft or 0.10 in. (See Fig. 7-12).

The various types of architect's scales are seen in Figure 7-13. Their basic units are feet and inches. Architect's scales are used in the construction industry.

An example of an architect's scale in a drawing is seen in Figure 7-14. All architect's scales are the open divided style. For foot measurements read the scale to the left. For inch measurements, read to the right. Add the number of inches to the number of feet for the total measurement.

The recommended scale selections for architectural drawings are listed in Figure 7-15. Metric drafting scales (Fig. 7-16) have the millimetre as the basic

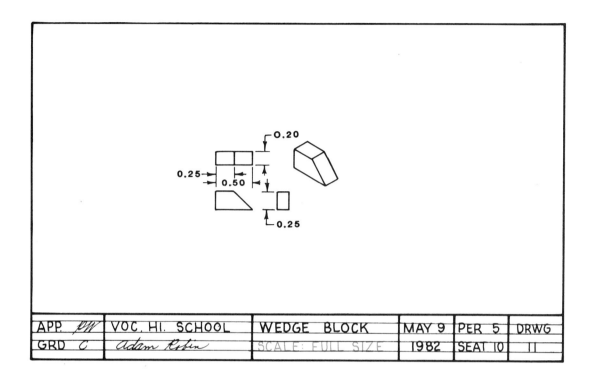

APP. _PW_	VOC. HI. SCHOOL	WEDGE BLOCK	MAY 9	PER 5	DRWG
GRD _C_	_Adam Robin_	SCALE: FULL SIZE	1982	SEAT 10	11

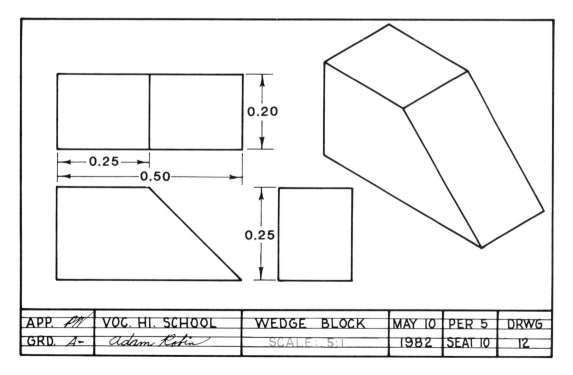

APP. _PW_	VOC. HI. SCHOOL	WEDGE BLOCK	MAY 10	PER 5	DRWG
GRD. _A-_	_Adam Robin_	SCALE: 5:1	1982	SEAT 10	12

FIG. 7-8 The object may be drawn larger or smaller than the real object

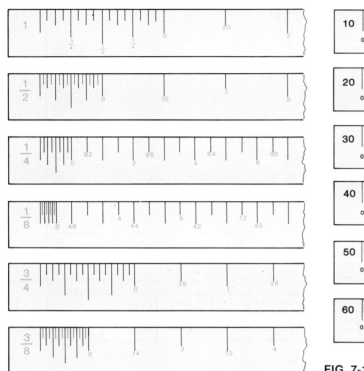

FIG. 7-9　Six types of mechanical engineer's drafting scales

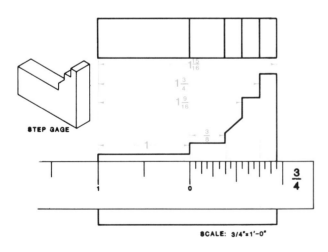

FIG. 7-10　Measuring with a mechanical engineer's drafting scale

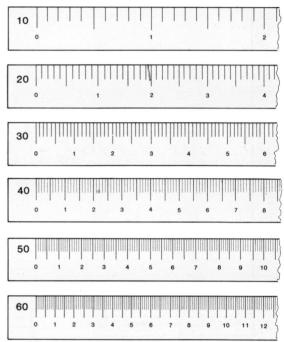

FIG. 7-11　Examples of civil engineer's drafting scales

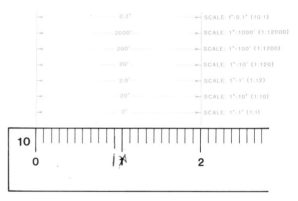

FIG. 7-12　Scales in ratios of 1″ = 0.1″ to 1″ = 1000′, based on the civil engineer's scale

unit of measurement. All the scale drawings are given by ratios. The ratio 1:5 means the drawing is five times smaller than the actual object. The ratio 5:1 means the drawing is five times larger than the actual object. For further study of the metric system of measure, see Chapter 8.

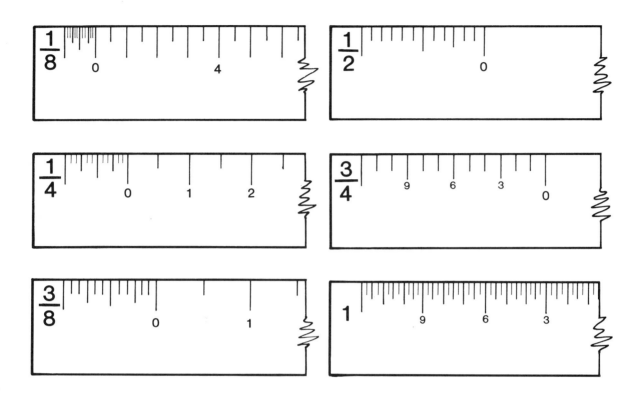

FIG. 7-13 Architect's drafting scales

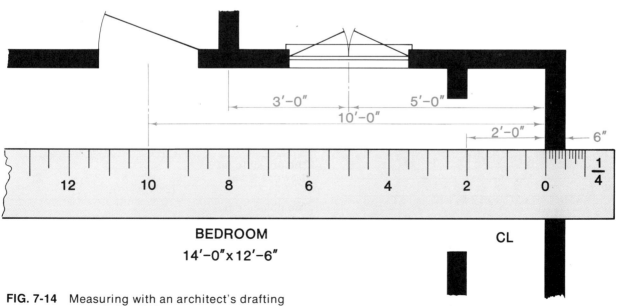

FIG. 7-14 Measuring with an architect's drafting
scale (¼″ = 1′0″)

Scale Selections for Architectural Drawings

Type of Drawing	Architect's Scale	Civil Engineer's Scale	Metric Scale
Plat plan	Not used	1″ = 50′	1:500
Large plot plan	Not used	1″ = 20′	1:200
Small plot plan	⅛″ = 1′-0″	1″ = 10′	1:100
Floor plan	¼″ = 1′-0″	Not used	1:50
Architectural detail drawings	½″ = 1′-0″ ¾″ = 1′-0″ 1″ = 1′-0″	Not used	1:20 1:10 1:5

FIG. 7-15 Recommended scale selections for architectural scales

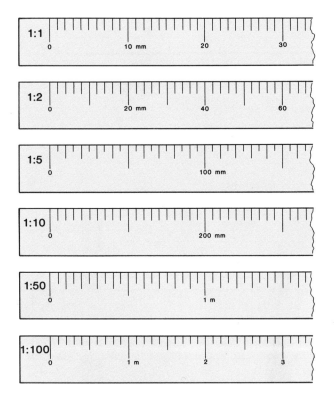

FIG. 7-16 Metric drafting scales

In industry, at least three types of measuring systems are commonly seen in engineering drawings. They are the mechanical engineer's scale, the civil engineer's scale, and the metric scale. The scale selection of preferred sizes and ratios for most engineering drawings are described in Figure 7-17.

PROBLEMS

Complete the following exercises as indicated: Figure 7-18 in mechanical engineer's scale; Figure 7-19 in civil engineer's scale; Figure 7-20 in architect's scale; and Figure 7-21 in metric scale. Measure each line to the right of the starting line at the far left on the drawing. Use the scale indicated. Write the answers on a separate piece of notebook paper. Correct all measurements and remeasure all incorrect answers. It is important for you to find your errors and correct them. Otherwise, you will make these measurement errors on your drawings.

Scale Selection for Engineering Drawings

Type of Drawing	Mechanical Engineer's Scale	Civil Engineer's Scale	Metric Scale
The object to be drawn is slightly smaller than the drawing format.	Full size	1:1	1:1
The object to be drawn is slightly larger than the drawing format.	½″ = 1″	1:2	1:2
The object to be drawn is a gooddeal larger than the drawing format.	⅛″ = 1″	1″ = 10″ 1:10	1:10
The object to be drawn is small.	1″ = ½″ (2:1)	2:1	2:1
The object to be drawn is very small.	1″ = ⅛″ (8:1)	10:1	10:1

FIG. 7-17 Scale selection for engineering drawings

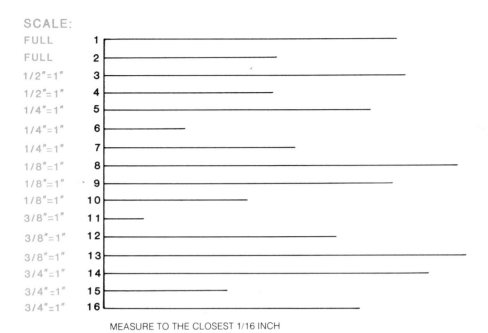

MEASURE TO THE CLOSEST 1/16 INCH

FIG. 7-18 Problem

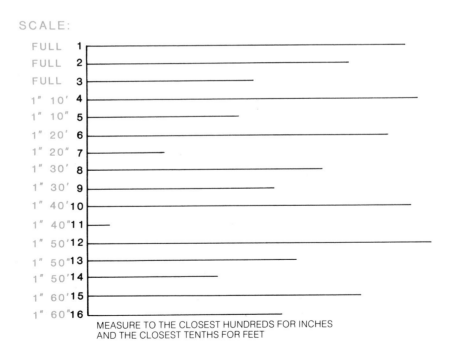

MEASURE TO THE CLOSEST HUNDREDS FOR INCHES
AND THE CLOSEST TENTHS FOR FEET

FIG. 7-19 Problem

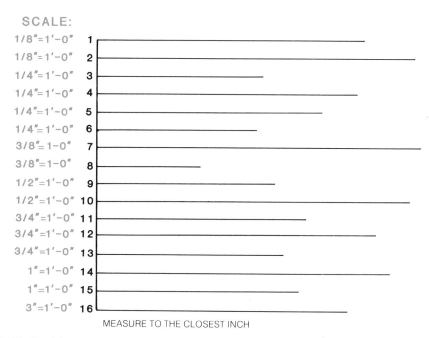

FIG. 7-20 Problem

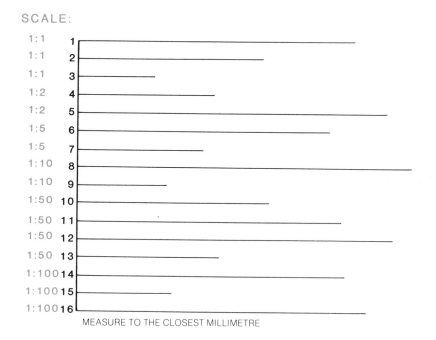

FIG. 7-21 Problem

Introduction to Metrics

Jacqueline Marsall

are uniform, a drawing can be understood anywhere even though language differences may exist. An excellent reference that will help the draftsperson with metric uniformity is the international standard ISO R-1000. When metric standards are not available or not suitable for a particular design, it may be necessary to use an American standard. Examples are steel plates, wires, and lumber sizes. It will be a while before ISO adopts all the metric standards needed to establish a complete international set of standards. The United States is starting to adopt the available ISO standards, and eventually every nation will be designing and manufacturing all parts using the metric system. This section will introduce and describe some of the more important conventions. As you study them, you should recognize how they will affect such areas of drafting as design, drawing formats, working drawings, pictorial drawings, dimensioning, symbol changes, and spelling.

INTRODUCTION

Conventions for metric drafting are important. A convention is a symbol or a standard which is uniformly used and understood. If drafting conventions

DESIGN

One of the primary principles of metric design is to use "first choice sizes." The table in Figure 8-2 gives preferred sizes for component dimensions.

First Choice Sizes (mm)

		Rising by 5		Rising by 10		Rising by 20	Rising by 50		Continuing	
1	5	20	55	80	150	200	300	650	800	3000
1.2	6	25	60	90	160	220	350	700	900	4000
1.6	8	30	65	100	170	240	400	750	1000	
2	10	35	70	110	180	260	450	800	1200	
2.5	12	40	75	120	190	280	500		1500	
3	16	45	80	130	200	300	550		2000	
4	20	50		140			600		2500	

FIG. 8-2   Selection of "first-choice sizes" helps standardize metric designing

This will eliminate many irregular-sized products, making interchangeability more likely. A further study of preferred sizes called Renard Series is required in advanced engineering design.

DRAWING FORMAT

Format means the general shape, size, and makeup of a drawing sheet. This includes the paper size, border lines, divisions, title block, and any other standard information needed on most drawings. Drawing formats are available in several metric sizes. The size is chosen by the draftsperson to suit each particular drawing. Standard formats make reproduction, microfilming, filing, and mailing easier. The information on a drawing format may vary from one company to the next, but in general there is uniformity.

WORKING DRAWINGS

One of the most important things to remember in drafting is that people must read, understand, and use the drawings you make. A good working drawing has the correct views, accurate dimensions from the correct points, and easily understood notes so an item can be manufactured from them. The selection of the type of drawing and the use of conventions are important.

PICTORIAL DRAWINGS

Pictorial drawings are used in instruction books and repair manuals. Assembly or "exploded" view drawings may be drawn in a pictorial manner (Fig. 8-3). Most isometric and pictorial drawings use no

dimensions, as they are most often used for assembly, repair, or sales. A single complex part can often be clarified by using an isometric drawing with dimensions and notes for its fabrication.

ORTHOGRAPHIC PROJECTION

The most common drawing projection convention in American engineering drawings is called third-angle orthographic projection. Almost all assembly and detail drawings are made using this convention (see Figure 8-4).

To show clearly that a drawing uses third-angle orthographic projection, a symbol is shown near the title block on the drawing format (Fig. 8-5). This is done because some countries still use what is called first-angle projection (Fig. 8-6).

DIMENSIONING

Proper dimensioning on a drawing is just as important as a clear picture of the part itself. Dimensions are used to manufacture a part, to set tools, to check a part for accuracy, to arrange parts or equipment in a space, and for many other uses. Not

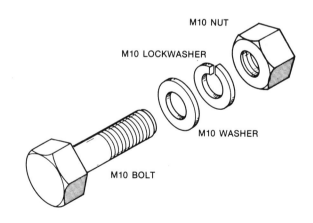

FIG. 8-3   Machine parts in pictorial

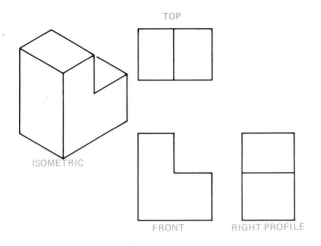

FIG. 8-4 Third-angle orthographic projection

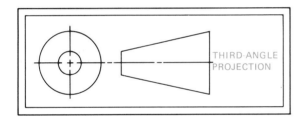

FIG. 8-5 Drawing symbol denoting third-angle projection

the metric system has the advantage of using only one unit of measurement, and no decimal point except when the millimetre is split. Whenever possible, round off all sizes to the closest whole millimetre. Use decimal fractions instead of ratio fractions (for example, use 25.5 mm rather than 25½mm).

Angles on a drawing are shown in degrees (°) and the decimal parts of a degree; for example, 35.85°.

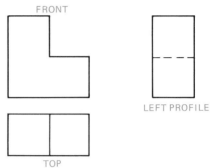

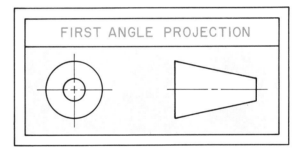

FIG. 8-6 First-angle projection and drawing symbol

only should dimensions be correct, they should be placed for ease of measuring in making the part. Conventions have been developed for dimensioning that suit most cases. Every part should be examined, however, for the best placement of dimensions. The block that was used to illustrate orthographic projection could be dimensioned as shown in Figure 8-7. For simple objects, two projected views will satisfactorily describe its form. For an example of isometric dimensioning see Figure 8-8.

Metric dimension conventions use whole numbers and decimals. Mechanical equipment, however large, requires the use of millimetres for even the overall dimensions. A large diesel engine could have an overall length of 15 285 mm, thus

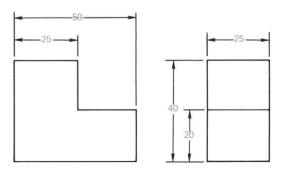

FIG. 8-7 Dimensioned orthographic drawing

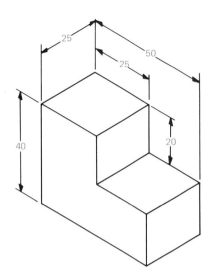

FIG. 8-8   Dimensioned isometric drawing

mm	INCHES
3	0.118
6	0.234
15	0.590
30	1.181
150	5.905

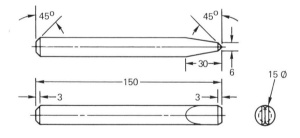

FIG. 8-9   The proper use of inches on a metric drawing

Do not dual-dimension a drawing with millimetres and inches. If it is necessary to show inch dimensions, list the inch dimensions and their metric values in table form on one side of the drawing paper (Fig. 8-9).

The metre is generally used to dimension homes, factories, land, or large areas on maps.

METRIC SYMBOLS

Metric symbols and their use on engineering drawings must conform to the following rules:

1. Leave a space between the number and the symbol (right: 75 mm; wrong: 75mm).

2. Do not leave a space between the prefix and the symbol (right: 25 cm, 30 °C; wrong: 25 c m, 30°C, 30° C).

3. Keep measurements in one unit (right: 12.5 m or 12 500 mm; wrong: 12 m and 500 mm).

4. Do not use a period after a symbol unless the symbol is at the end of a sentence.

5. Use the symbol $\varnothing$ for diameter. The symbol may precede or follow the dimension.

6. Symbols for square area and cubic volume are written with a superscript (5 mm^2, 15 cm^3). Do not spell out the words square or cubic when a symbol is used. Square mm or cubic cm is wrong.

7. The symbol for "per" is an oblique slash as in kilometres per second (km/s).

METRIC SPELLING

The SI spelling used in this textbook will be "metre and litre." There are some people in the U.S. who prefer the spelling to be "meter and liter." Most English speaking countries use the "re" spellings.

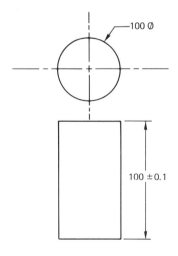

FIG. 8-10   Dimension tolerance convention

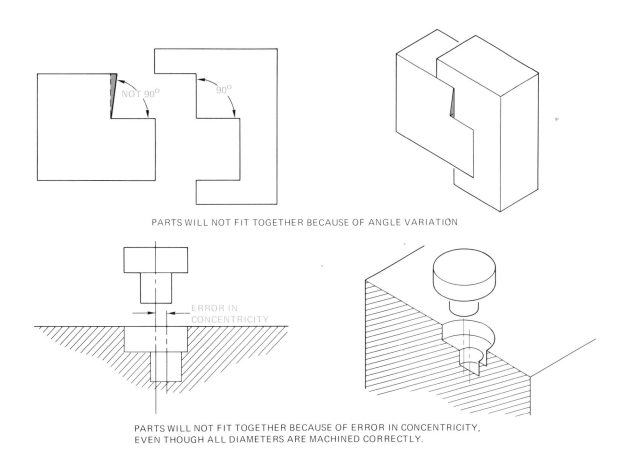

PARTS WILL NOT FIT TOGETHER BECAUSE OF ANGLE VARIATION

ERROR IN CONCENTRICITY

PARTS WILL NOT FIT TOGETHER BECAUSE OF ERROR IN CONCENTRICITY, EVEN THOUGH ALL DIAMETERS ARE MACHINED CORRECTLY.

FIG. 8-11 Geometric tolerance errors

The full spelling of plural units is with an "s" (25 millimetres). The symbols of plural units do not use the "s" (25 mm).

The decimal fraction of a number less than one is singular (0.5 millimetre, 0.5 mm).

All of the metric unit names are not capitalized. The common metric symbols that use capital letters for metric unit names are ampere (A) and kelvin (K).

Most metric prefixes are not capitalized. The only metric prefixes that use capital letters for prefix symbols are exa (E): 10^{18}, peta (P): 10^{15}, tera (T): 10^{12}, giga (G): 10^{9}, and mega (M): 10^{6}.

When using the symbol for one litre (1 ℓ), use a script "ℓ" to avoid errors. Note: It is possible that the symbol for litre will be changed to a capital "L".

DECIMAL POINTS AND SPACING

The decimal marker will be the period (decimal point) on line level (25.750). Some countries use the comma for a decimal marker, or a point in the center of the line (25,750 or 25·750). The latter methods are not used because of the confusion they may cause.

The grouping of numbers is done with a space, not a comma. *A group of four numbers may be written as 1234 or 1 234.* Numbers with five or more numbers are always spaced as 12 345 or 12 345 678 or 9.123 4.

Metric numbers are always expressed with a minimum of one figure to the left of the decimal point (1.50 or 0.50).

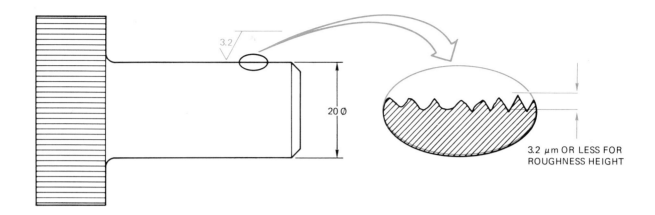

FIG. 8-12   The symbol convention for surface roughness

3.2

20 ∅

3.2 µm OR LESS FOR
ROUGHNESS HEIGHT

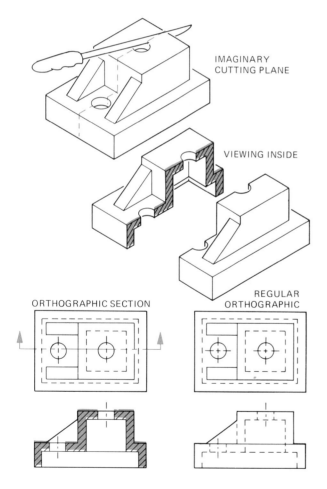

IMAGINARY
CUTTING PLANE

VIEWING INSIDE

ORTHOGRAPHIC SECTION

REGULAR
ORTHOGRAPHIC

FIG. 8-13   The use of cross sections

TOLERANCES

It is not necessary (or possible) to make parts to absolutely exact dimensions. Even though some parts are made that are accurate to 0.000 3 mm (the thickness of a hair divided 250 times), this kind of accuracy is required very seldom, and much larger tolerances are normally used. On a drawing, a tolerance indicates the maximum variation allowed from the stated dimension. In Figure 8-10 a dimension is shown like this: 100 ± 0.1 This means that this dimension can vary 0.1 mm either way from the 100 mm size. Stated another way, the part can be as small as 99.9 mm and as large as 100.1 mm and still be acceptable. These are called dimensional tolerances, and the conventions are well-standardized.

Another type of tolerance required in drafting is called geometric tolerancing. This deals with the shape of a part, and involves the accuracy of angles, flatness of surfaces, roundness of holes, concentricity, and other things that are called "geometry" rather than size or dimension. Two examples are seen in Figure 8-11.

SURFACE FINISHES

Sandpaper is rough, and glass is smooth, and both surfaces serve a distinct purpose. The same is true with machined or cast parts. Bearing or sliding surfaces must be smooth and geometrically accurate. The sides of a drilled clearance hole or the surface

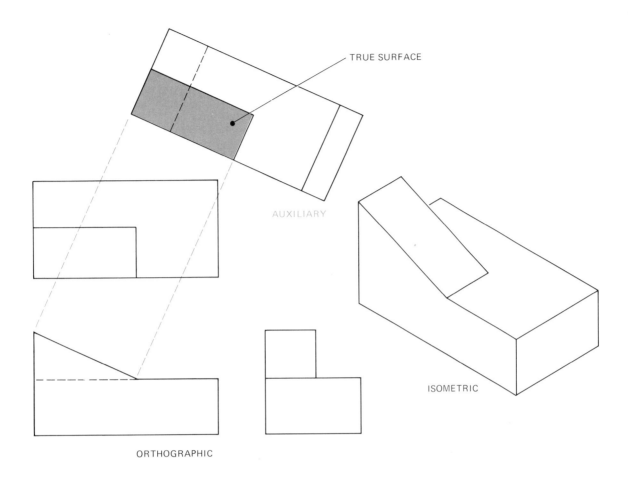

TRUE SURFACE

AUXILIARY

ISOMETRIC

ORTHOGRAPHIC

FIG. 8-14 An auxiliary view used to show a surface accurately

of a protective housing need not be particularly smooth or geometrically accurate. The required smoothness of surfaces is called out on the drawing of the part using a conventional standard. The average roughness height of the surface is expressed in micrometres.

A micrometre (μm) is one millionth of a metre (0.000 001 m).

SECTIONS

A cross section is used to give a clearer picture of a part to be manufactured. Sections show the interior details of a part. No particular metric conventions apply here because sections deal with the "picture" part of drafting only. See Figure 8-13 for an example of a section view.

AUXILIARY VIEWS

Views that are not projected at right angles to the main surfaces of an orthographic drawing are called auxiliary views. These views are used to give a true picture of some angular surface or boss (Fig. 8-14). Several auxiliary views may be required to clearly show the detail required for manufacturing complex parts.

The topics of fasteners, welding, gears, and cams, and methods of indicating these on the drawing using metric conventions are convered in separate chapters.

In summary, the effect of the metric system in engineering drawing will be minimal. Most of the existing conventions will remain as before (Fig. 8-15). The few changes made will be discussed and illustrated in the following chapters.

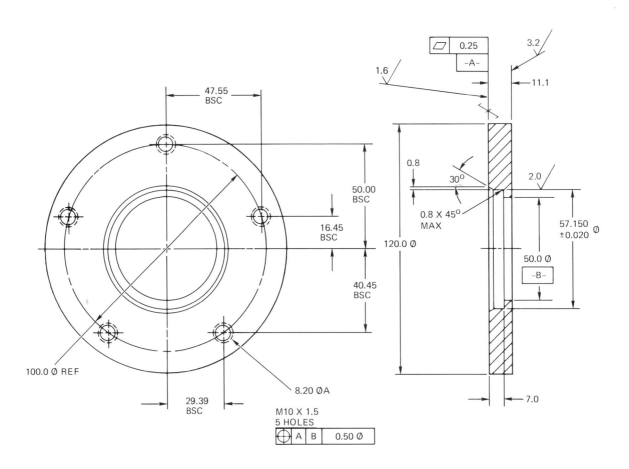

FIG. 8-15   Metric conventions in use on drawings

9
Chapter

Sheet Formats

INTRODUCTION

The detailed information on sheet formats that is presented in this chapter must be divided into two categories: formats used by schools and formats used by industry. Styles and sizes of format will vary according to the needs of the particular school or industry. The important thing to remember is that the sheet must include all of the information needed to build the part being drawn. Formats must be standardized so that all drawings will present their information in an identical manner, making handling, mailing, and storage of the drawings more efficient. As you study this chapter, you should learn how to draw borders and title blocks, and how to fold a drawing. You should also gain a knowledge of paper sizes, drawing zones, micro-filming marks, photoreduction marking, title information, revision records, and material lists.

PAPER SIZES—AMERICAN STANDARD

The first decision the draftsperson must make is the size of the drawing format. The amount of space required for all the parts of the drawing and the necessary notations must be taken into consideration before selecting the smallest size that will accommodate all of the information.

The American standard paper sheet sizes are shown in Figures 9-2 and 9-3. The basic sheet size measures 9 by 12 inches; other sizes are multiples

**American Standard Sheet Sizes
(inches)**

Size Designation	Width	Length	Width	Length
A	9	12	8.5	11
B	12	18	11	17
C	18	24	17	22
D	24	36	22	34
E	36	48	34	44

FIG. 9-2   American standard drawing sheet sizes

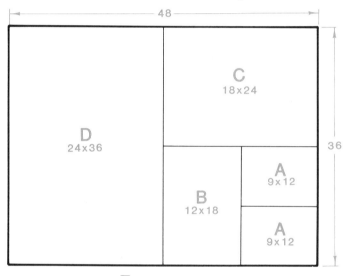

FIG. 9-3   Dimensions of ''A'' to ''E'' paper sizes. All sheets are multiples of 12 inches.

of this size. This sheet is also known as ''A'' paper. Most industries prefer the 9 by 12 size. In doing drawings, it is recommended that you use one size of sheet. Both handling and storage are made easier.

Paper comes in roll form, too. The standard widths of roll paper are 34, 36, 40, 42, 44, 48, and 54 inches. The length of roll paper should be in multiples of 12 inches. When a drawing format calls for paper larger than the ''E'' size (36 by 48 inches), use roll paper. However, always try to scale the drawing to fit the ''E'' format, or something smaller, if possible.

The disadvantages of roll drawings are:

1.   They are difficult to draw.

2.   They are difficult to store.

3.   Preprinted format is not possible.

METRIC PAPER SIZES

The metric ISO paper sizes used in engineering drawing are called the ''A'' series. The largest size in this series is A0. The dimensions for A0 paper are 841 mm ×1 189 mm. This is equal to an area of one square metre. All of the smaller sheet sizes are obtained by dividing the long side of the paper by two (Fig. 9-4). ''A'' paper sizes are shown in Figure 9-5.

Industry will commonly use sizes A4 through A0. Students will usually use sizes A4 and A3. Rolled paper is used for very large drawings. The width of the roll should be a minimum of 841 mm. The length of the drawing should exceed the longest length of available cut-to-size paper.

BORDERS

A general rule for borders is to have the edge used for binding wider than the other three borders. The border used for binding is usually 1 inch to 1¼ inches wide. On ''A'' size drawings the other three borders are usually ¼ inch wide. On ''B,'' ''C,'' ''D,'' and ''E'' drawings the remaining three borders are usually ½ inch (Fig. 9-6).

On roll drawings the right-hand margin is a protective margin and is normally 4 to 8 inches wide (Fig. 9-7).

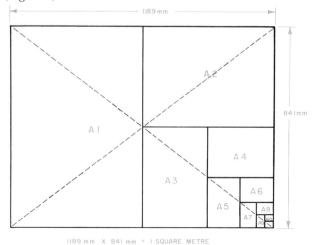

FIG. 9-4   Metric paper sizes are in a ratio of 1:$\sqrt{2}$

ISO Metric Sizes	Millimetres
2A	1189 × 1682
A0	841 × 1189
A1	594 × 841
A2	420 × 594
A3	297 × 420
A4	210 × 297
A5	148 × 210
A6	105 × 148
A7	74 × 105
A8	52 × 74
A9	37 × 52
A10	26 × 37

FIG. 9-5 Dimensions for metric paper

TITLE BLOCKS

Each drafting room or company will have its own type of title block. The important thing is for the title block to contain the correct information and be consistent with the title block format in all drawings. Typical title blocks for drafting students are shown in Figure 9-8. The parts of a student's title block are:

1. Approval of instructor
2. Grade evaluation by instructor
3. Name of school
4. Name of draftsperson
5. Title of drawing
6. Scale of drawing
7. Date the drawing was completed
8. Class period of student
9. Seat number of student
10. Drawing number

Title blocks used in industry are more complex than those used by schools, as they require much more information about the part being drawn. The size, style, and amount of information in the title block will vary with different industries. Figure 9-9 shows an industrial title block with the most common information.

1. Company's name.
2. Title—The name of the part.
3. Model—This identifies the whole layout, or the model for which the part is being designed.
4. Angle of Projection—Specifies first or third angle projection (see Chapter 12).
5. First Used On—The model or designation of machine the drawing part was first used on.
6. Do Not Scale—Because of slight changes in paper size during drawing reproductions, prints should not be measured.
7. Tolerances—A list of machining allowances for parts, related to the number of decimal places in dimensions.
8. Material Specification—This information gives the name, type, and analysis of the material.

FIG. 9-6 Border widths for "A" and "B" paper

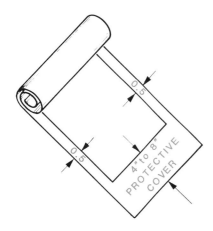

FIG. 9-7   Border dimensions for roll sheets

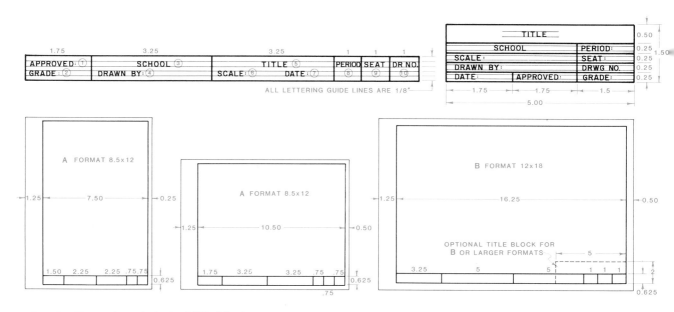

FIG. 9-8   Examples of student title blocks

9.   Angles—The degrees of error allowed in angle tolerances.

10.   Surface Quality—The smoothness of surface finish measured in micrometres.

11.   Draftsperson—Signature of the draftsperson.

12.   Checker—Signature of the checker.

13.   Engineer—Signature of the project engineer.

14.   Manufacturer—Signature approval from the manufacturer.

15.   Authority—A work order number, experiment number, or signature of the person responsible for the project.

16.   Features—Description of special design features of the part.

17.   Drawing Size—Designation of the drawing sheet size.

18.   Finish—Type of surface coating or applied finish.

19.   Reference—The source of any special information related to the part.

METRIC CORP OF AM ①		DIVISION	TITLE ②		
TOLERANCES			MODEL ③		ANGLE OF PROJECTION ④
TWO PLACE DIMENSIONS ±0.XX ONE PLACE DIMENSIONS ±0.XX ⑦			FIRST USED ON ⑤		DO NOT SCALE ⑥
			MATERIAL ⑧		
ANGLE OF TOLERANCE ±0.XX° ⑨	SURFACE QUALITY 0.XX √⎯ ⑩		FEATURES ⑯		
DRAFTSMAN ⑪ *Juan Lopez*		DATE	DRW SIZE ⑰	FINISH ⑱	
CHECKER ⑫		DATE	REFERENCES ⑲		
ENGINEER ⑬		DATE	NEXT HIGHER ASSEMBLY ⑳		
MANUFACTURER ⑭		DATE	DATE ㉑	SCALE ㉒	SHEET ___ OF_ ㉓
AUTHORITY ⑮		DATE	PART NO. ㉔ 1 2 3 4 5 6 7 8 9		

FIG. 9-9   Example of an industrial title block

20. Next Higher Assembly—Where the part will next be assembled.
21. Drawing Date—The date that the drawing was completed.
22. Scale—The proportion of the drawing scale.
23. Sheet Number—Numbering sequence of multiple drawings (1 of 3, 2 of 3, 3 of 3).
24. Part Number—The identifying number of the part is placed in the title block and in the small boxes showing the microfilm frames.

Title blocks can be added to drawings in several ways. They can be drawn in by the draftsperson, preprinted, rubber stamped, or attached to the paper as a transparent transfer. The size of the title block will vary with the paper size and the amount of information required.

A general rule for letter sizes within a title block is to have names and important identifying features ¼ inch high, with the rest of the information ⅛ inch high.

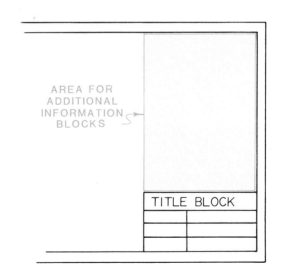

FIG. 9-10   Space for additional information

A 2	202	ADD CHAMF PART 71131	DEC 1, 82	178623	Bruce John	Adams	PW
A 1	201	ADD LOCK-PIN #71131	NOV 18, 82	178623	Bret Wallach	Robins	PW
REV ①	ECO ②	DESCRIPTION ③	DATE ④	AUTHORITY ⑤	DRFT ⑥	CHK ⑦	APP ⑧
REVISION RECORD							

FIG. 9-11   A revision record block

MATERIAL LIST

AMOUNT	PART NO.	STOCK SIZE	MATERIAL

FIG. 9-12   Material list block

ADDITIONAL BLOCKS

Additional information can be inserted above the standard title block (Fig. 9-10). The information most often added consists of revision records and material lists.

MATERIAL LIST

AMT.	PART NUMBER	STOCK NUMBER	MATERIAL
15	56399	28.75 X 14.35	BRASS

C							
B							
A	7811	BEARING	3-9-78	Waled	R.L.B	P.L.	R.H.
REV.	ECO	DESCRIP.	DATE	AUTH.	DRFT.	CHK.	APP.

FIG. 9-13   Positioning of additional blocks

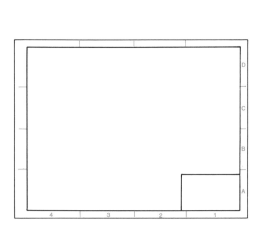

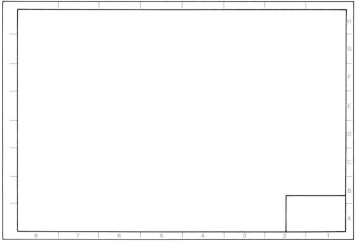

FIG. 9-14   Zone markings for large prints

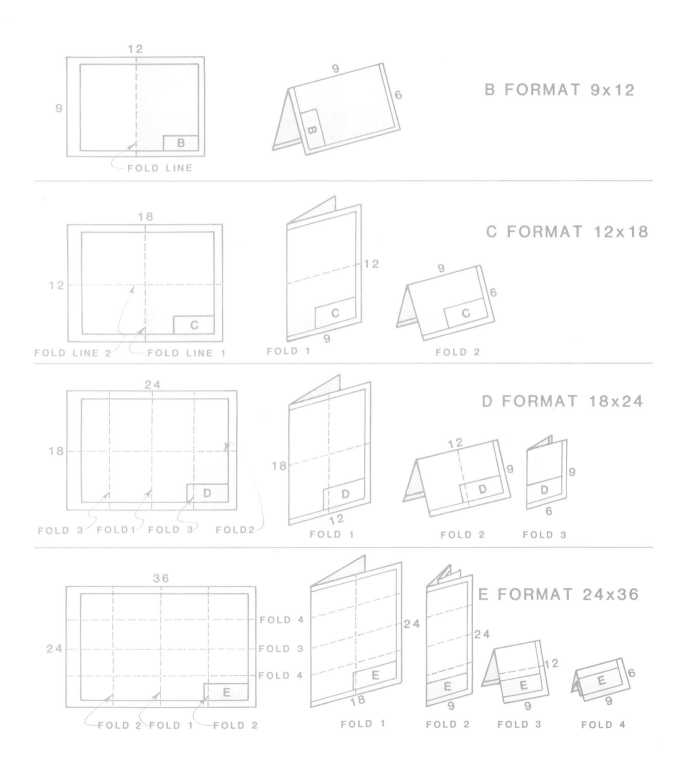

FIG. 9-15 Steps in folding large prints

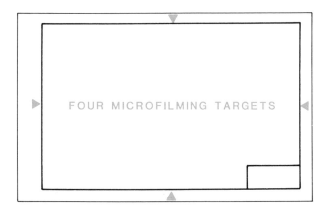

FIG. 9-16   Microfilming targets

REVISION RECORD

A revision record must contain a detailed record of all engineering changes (Fig. 9-11). The required information is:

1. Revision Number—The revisions are numbered in sequence.
2. ECO—The number of the engineering change order.
3. Description—A brief description of the engineering change.

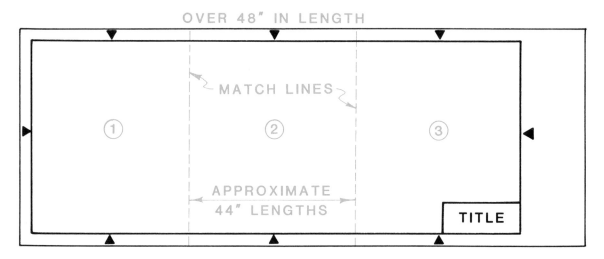

FIG. 9-17   Dividing a roll sheet into three microfilming areas

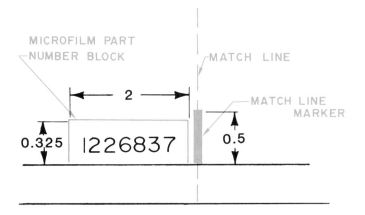

FIG. 9-18   Microfilm match line and marker

4. Date—The date when the engineering change was completed.

5. Authority—The change order number or the name of the individual authorizing the revision.

6. Draftsperson—The signature of the draftsperson making the change.

7. Checker—The signature of the checker.

8. Approval—The final approval.

MATERIAL LISTS

The material list is a compilation of all the pieces that are involved in that one particular drawing (Fig. 9-12). Information in the material list may include:

1. Amount—The number of identical pieces required.

2. Part Number—The identifying number of each part.

3. Stock Size—The overall dimensions of the ordered finished piece.

4. Material—The type of material from which the piece is made.

Additional blocks of information can be inserted from the top of the drawing down, or from the bottom up. Figure 9-13 gives an example of how additional title blocks can be read.

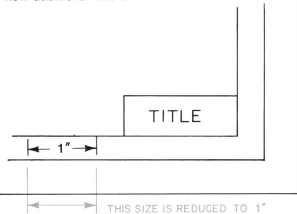

FIG. 9-19 Full-scale mark for reproduction reference

ZONES

Large drawings are divided into imaginary square zones to help the reader locate particular parts of the drawing. These zones are identified by letters along the vertical margin, and numbers along the horizontal margin. "C" size paper, and larger formats, may have these zones, as indicated in Figure 9-14.

FOLD MARKS

All prints should be folded to a 6- × 9-inch filing size, with the title block on the front face. This will facilitate filing and handling the prints. The last fold is toward the longer edge of the drawing, so other prints being filed in the same drawer will not be pushed between folds (Fig. 9-15). Original tracings should not be folded. They should be stored flat or rolled in cylindrical containers.

MICROFILMING MARKS

Diamonds or arrowheads are placed in the center of all margins to serve as centering guides for microfilming (Fig. 9-16). If a drawing is more than 48 inches in length, it can be divided into sections that constitute microfilm frames (Fig. 9-17). The frames are set apart by imaginary match lines (Fig. 9-18). These lines are used to line up and match adjacent prints made from the microfilm (see Chapter 34). The drawing's identifying number is placed in a small box on the border, which is also used to mark off the matching lines. This identification box should be 2 inches long.

PHOTOREDUCTION MARKING

If the drawing is to be reduced to a specific size, the placement of a graphic scale will help the print maker reduce the drawing to the exact size required (Fig. 9-19).

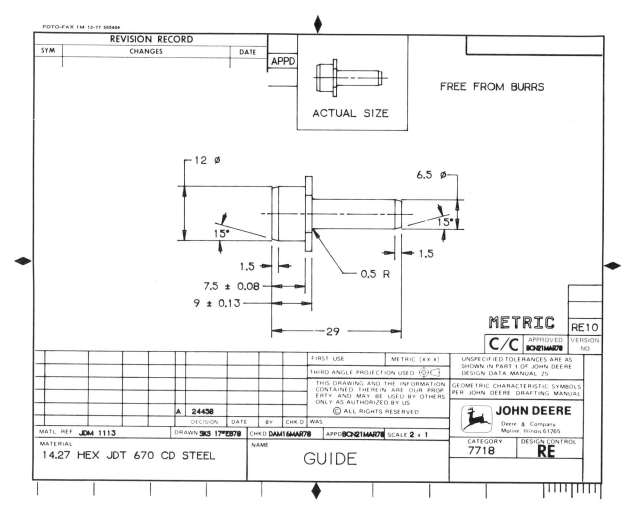

FIG. 9-20   Format for "A" size industrial drawing

PROBLEMS

1. Design a title block for the paper size used in your classroom.

2. Design a new title block for an industry of your choice. Use Fig. 9-20 for reference.

3. Add additional title blocks to problems 1 and 2.

Geometric Constructions

INTRODUCTION

The term *geometry* comes from a Greek word that means the science of earth measure. It deals with the properties, measurements, and relationships of all shapes made from straight and curved lines. Geometric shapes can be observed in nature as well as in handcrafted objects (Fig. 10-2). As you study this chapter, you should become familiar with the definition of plane geometry, develop the skills necessary to make geometric constructions, and develop the skills with drafting instruments required for geometric constructions.

GEOMETRIC FORMS

Geometry is a graphic form of mathematics. Many problems in math use graphical solutions for solid, surface, line, angular, and curved figures. Learning the terminology in Figure 10-3 will be helpful to all draftspersons and mathematicians. All geometric forms fit into one of two categories: plane two-dimensional geometric forms (Fig. 10-4), and three-dimensional forms(Fig. 10-5). Most of the geometric constructions in this chapter will deal with two-dimensional forms.

Many geometric constructions are done with a straight-edge and compass. However, more elaborate drafting instruments will ensure greater accuracy and speed and offer more options to solutions.

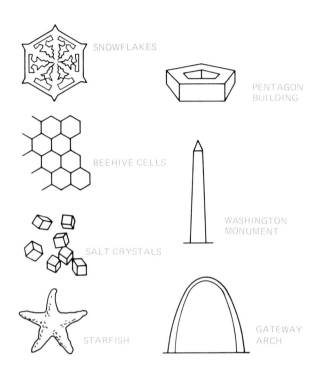

SNOWFLAKES

PENTAGON BUILDING

BEEHIVE CELLS

WASHINGTON MONUMENT

SALT CRYSTALS

STARFISH

GATEWAY ARCH

FIG. 10-2 Natural and handcrafted geometric shapes

All drawings are made of geometric forms (Fig. 10-6). The principles of geometry will be used often to make drawings and solve drafting problems. A graphical solution must be done accurately. The best accuracy is obtained with a sharp, hard lead pencil. A thick line will reduce accuracy.

GEOMETRIC CONSTRUCTION

There are many types of geometric forms that can be constructed in different ways, but the most common ways are shown here. Whenever possible use the simplest solution. Similarly, use templates of geometric forms to speed drawings (see Chapter 5).

Geometric constructions may be divided into the following segments:

1. **Lines** (Fig. 10-7)
2. **Angles** (Fig. 10-8)
3. **Triangles** (Fig. 10-9)
4. **Squares** (Fig. 10-10)
5. **Hexagons** (Fig. 10-11)
6. **Pentagons** (Fig. 10-12)
7. **Octagons** (Fig. 10-13)
8. **Circles** (Fig. 10-14)
9. **Arcs** (Fig. 10-15)
10. **Ellipses** (Fig. 10-16)
11. **Curves** (Fig. 10-17)

PROBLEMS

1. Draw a perpendicular to a straight line using a straight-edge and compass.

2. Draw a perpendicular to a straight line using a T-square and triangle. (Note: A drafting machine or parallel slide may be used in place of a T-square.)

3. Bisect a 5-inch line.

4. Divide a 6-inch line into five equal parts.

5. Draw three parallel lines using a T-square and triangle.

6. Draw two parallel lines using a compass and straight-edge.

7. Trisect a 7-inch line using a T-square and 30-60 degree triangle.

8. Draw a line tangent to a 4-inch diameter circle, using a straight-edge and compass.

9. Draw a line tangent to a 3.5-inch diameter circle, using a T-square and triangle.

10. Bisect a 60 degree angle using a straight-edge and compass.

11. Divide a 60 degree angle into five equal parts using a straight-edge and compass.

12. Copy a 30 degree angle using a straight-edge and compass.

13. Draw an equilateral triangle with sides of 3.75 inches using a straight-edge and compass.

14. Draw an equilateral triangle with sides of 4.25 inches using a T-square and a 30-60 degree triangle.

15. Draw a right triangle with sides of 4.5 inches and 3.5 inches using a T-square and triangle.

16. Draw a right triangle with a side of 7 inches and a hypotenuse of 8.5 inches using a straight-edge and compass.

17. Draw a right triangle with sides of 3 inches,

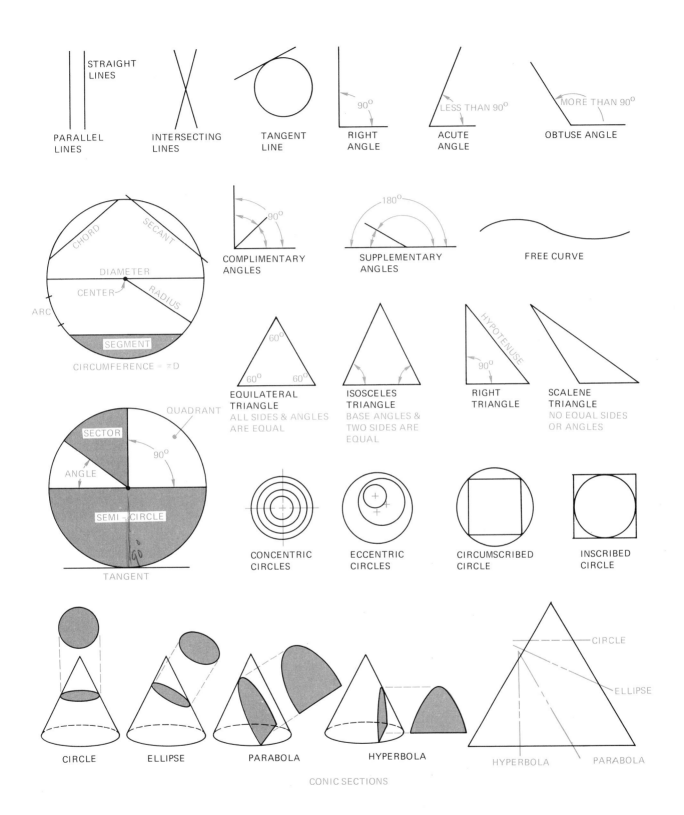

FIG. 10-3 Geometric terminology

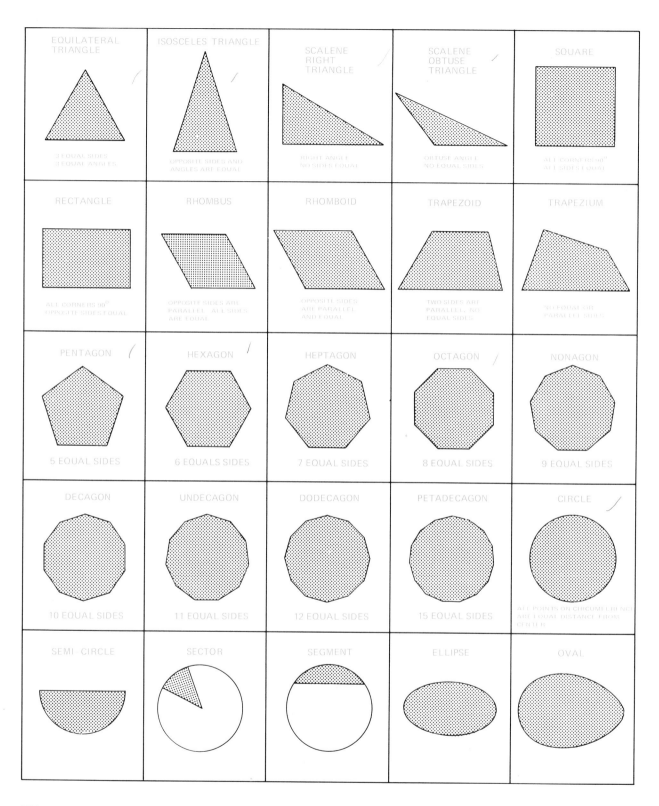

FIG. 10-4 Two-dimensional geometric forms

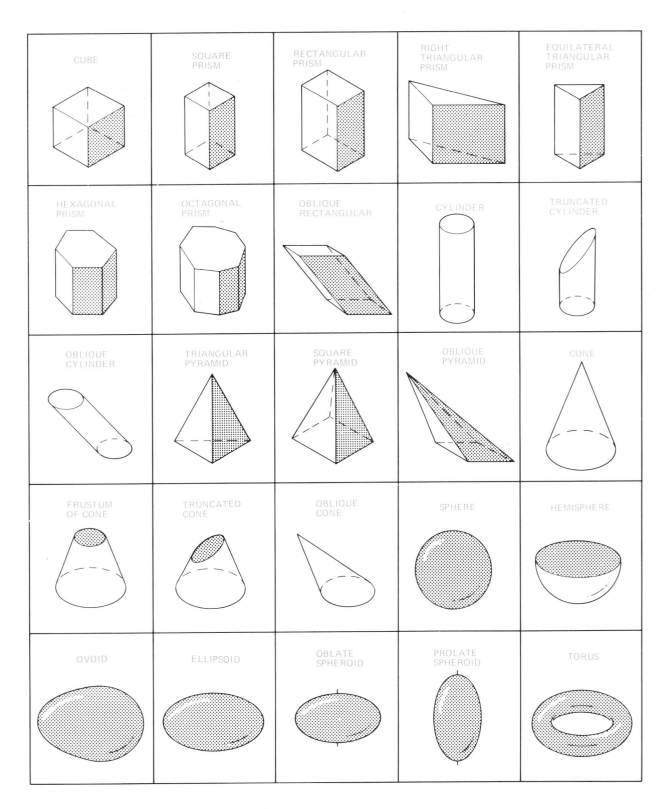

FIG. 10-5 Three-dimensional geometric forms

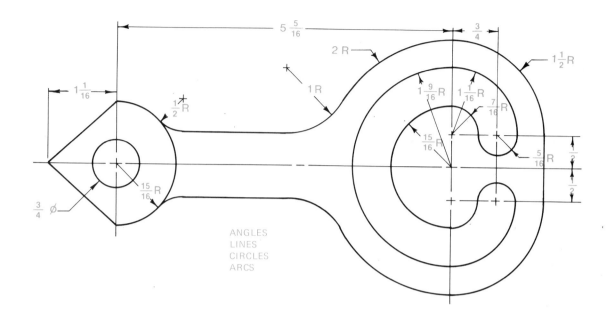

FIG. 10-6   Geometric forms in a drawing

5 inches, and 7 inches using a straight-edge and compass.

18.   Draw an isosceles triangle with a base of 3 inches and sides of 4 inches using a straight-edge and compass.

19.   Draw a square with sides of 2 inches using a T-square and 45 degree triangle.

20.   Draw a square with a diagonal of 2.5 inches using a compass, T-square, and 45 degree triangle.

21.   Draw a square with sides of 2.75 inches using a compass, T-square, and 45 degree triangle.

22.   Draw a hexagon with a distance of 3 inches across flats using a compass, T-square, and 30-60 degree triangle.

23.   Draw a hexagon with a distance of 3.5 inches across corners using a straight-edge and compass.

24.   Draw a hexagon with a distance of 2.5 inches across corners using a T-square and 30-60 degree triangle.

25.   Draw a hexagon with a distance of 3.25 inches across corners using a straight-edge and compass.

26.   Draw a pentagon with a distance of 3 inches across corners using a straight-edge and compass.

27.   Draw an octagon with a distance of 4 inches across flats using a T-square and 45 degree triangle.

28.   Draw an octagon with a distance of 4 inches across corners using a T-square and 45 degree triangle.

29.   Draw an octagon with a distance of 4.5 inches across flats using a straight-edge and compass.

30.   Divide a circle with a diameter of 8 inches in eight equal parts using a straight-edge and compass.

31.   Draw a tangent arc with a radius of 1.5 inches in an angle of 60 degrees using a straight-edge and compass.

32.   Draw a tangent arc with a radius of 2 inches in an angle of 60 degrees using a circle template.

33.   Draw an ellipse with a major axis of 3 inches and a minor axis of 2 inches using a straight-edge, compass, string, and pins.

34.   Draw an ellipse with a major axis of 4 inches and minor axis of 3 inches using a T-square,

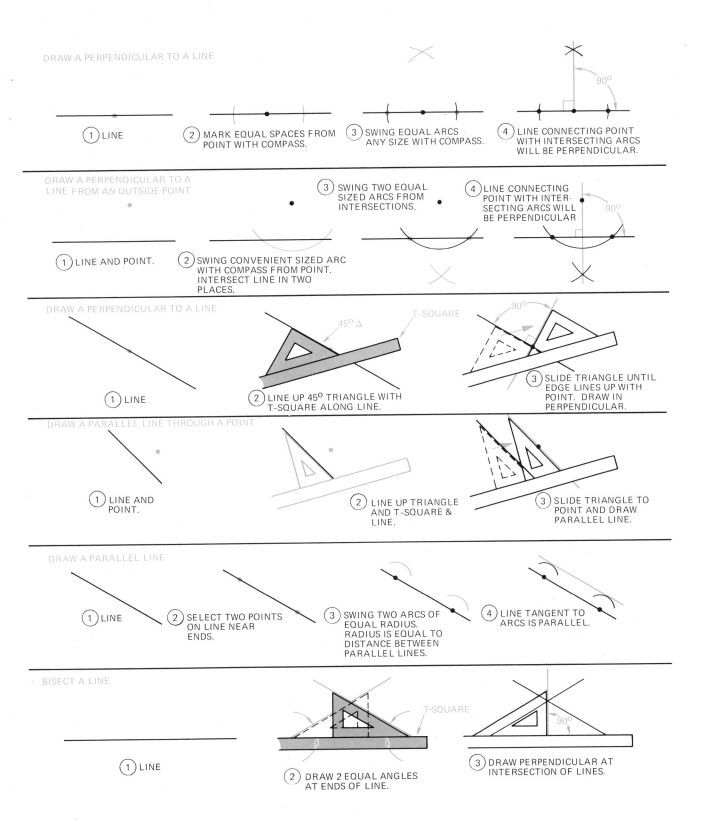

FIG. 10-7 Line constructions

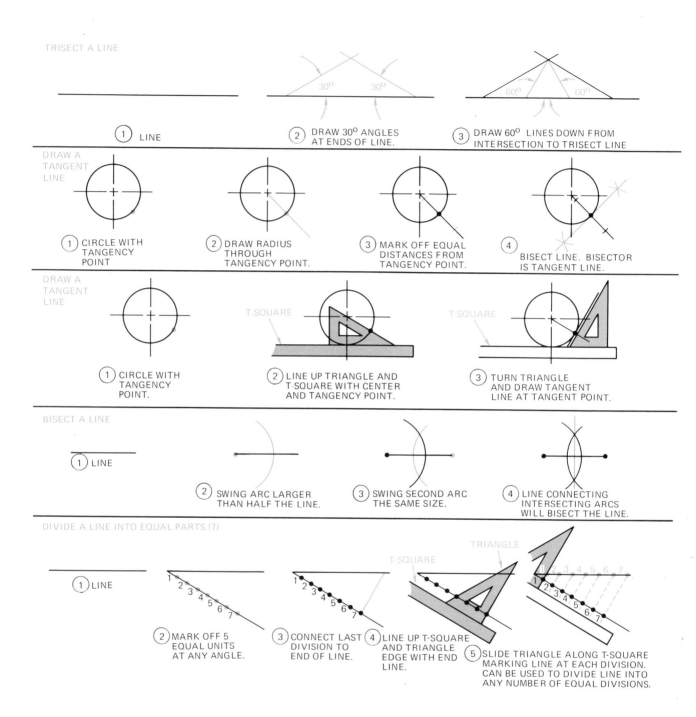

TRISECT A LINE

1 LINE

2 DRAW 30° ANGLES AT ENDS OF LINE.

3 DRAW 60° LINES DOWN FROM INTERSECTION TO TRISECT LINE

DRAW A TANGENT LINE

1 CIRCLE WITH TANGENCY POINT

2 DRAW RADIUS THROUGH TANGENCY POINT.

3 MARK OFF EQUAL DISTANCES FROM TANGENCY POINT.

4 BISECT LINE. BISECTOR IS TANGENT LINE.

DRAW A TANGENT LINE

1 CIRCLE WITH TANGENCY POINT.

2 LINE UP TRIANGLE AND T-SQUARE WITH CENTER AND TANGENCY POINT.

3 TURN TRIANGLE AND DRAW TANGENT LINE AT TANGENT POINT.

BISECT A LINE

1 LINE

2 SWING ARC LARGER THAN HALF THE LINE.

3 SWING SECOND ARC THE SAME SIZE.

4 LINE CONNECTING INTERSECTING ARCS WILL BISECT THE LINE.

DIVIDE A LINE INTO EQUAL PARTS (7)

1 LINE

2 MARK OFF 5 EQUAL UNITS AT ANY ANGLE.

3 CONNECT LAST DIVISION TO END OF LINE.

4 LINE UP T-SQUARE AND TRIANGLE EDGE WITH END LINE.

5 SLIDE TRIANGLE ALONG T-SQUARE MARKING LINE AT EACH DIVISION. CAN BE USED TO DIVIDE LINE INTO ANY NUMBER OF EQUAL DIVISIONS.

FIG. 10-7   (continued)

BISECT AN ANGLE

ANGLE

DRAW CONVENIENT
SIZED ARC, INTERSEC-
TING SIDES OF ANGLE.

DRAW CONVENIENT
SIZE ARC FROM
ONE SIDE.

DRAW SAME SIZE
ARC FROM OTHER
SIDE.

CONNECT INTERSEC-
TING ARCS TO VERTEX
OF TRIANGLE FOR
BISECTOR.

COPY AN ANGLE

ANGLE

DRAW CONVENIENT
SIZE ARC INTER-
SECTING ANGLE.

DRAW ONE SIDE
OF ANGLE.

DUPLICATE ARC.

STEP OFF DISTANCE
OF ARC IN FIGURE 2.

CONNECT END OF
LINE TO INTER-
SECTIONS FOR
SECOND SIDE OF
ANGLE.

DIVIDE AN ANGLE

ANGLE

DRAW CONVENIENT
SIZE ARC.

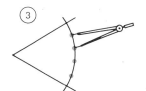

DIVIDE ARC INTO
DESIRED DIVISIONS.

DRAW DIVISIONS.

FIG. 10-8 Angle constructions

triangle, compass, and irregular curve.

35. Draw an ellipse with a major axis of 60 mm and a minor axis of 40 mm using a straight-edge and an irregular curve.

36. Draw an ellipse with a major axis of 75 mm and a minor axis of 50 mm using a trammel.

37. Draw an involute curve for a circle with a diameter of 70 mm using a straight-edge, compass, and an irregular curve. Use half the circumference for the arc.

38. Draw a spiral of Archimedes with a diameter of 70 mm using a straight-edge, compass, and an irregular curve.

39. Draw a helix for a circle with a diameter of 75 mm using a straight-edge, compass, and an irregular curve.

40. Draw a parabolic curve in a right angle with legs of 80 mm and 90 mm using a straight-edge.

41. Draw a circle through three points (Fig. 10-18).

42. Draw tangents to two circles (Fig. 10-19).

43. Draw an ogee curve (Fig. 10-20).

44. Draw tangent curves (Fig. 10-21).

45. Draw geometric Figures 10-22 and 10-23.

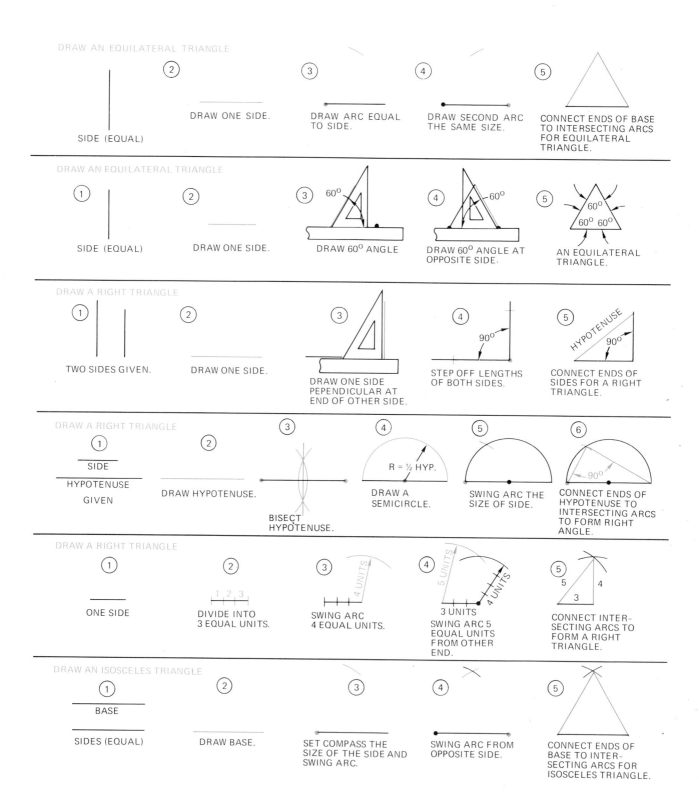

DRAW AN EQUILATERAL TRIANGLE

SIDE (EQUAL)

② DRAW ONE SIDE.

③ DRAW ARC EQUAL TO SIDE.

④ DRAW SECOND ARC THE SAME SIZE.

⑤ CONNECT ENDS OF BASE TO INTERSECTING ARCS FOR EQUILATERAL TRIANGLE.

DRAW AN EQUILATERAL TRIANGLE

① SIDE (EQUAL)

② DRAW ONE SIDE.

③ DRAW 60° ANGLE

④ DRAW 60° ANGLE AT OPPOSITE SIDE.

⑤ AN EQUILATERAL TRIANGLE.

DRAW A RIGHT TRIANGLE

① TWO SIDES GIVEN.

② DRAW ONE SIDE.

③ DRAW ONE SIDE PEPENDICULAR AT END OF OTHER SIDE.

④ STEP OFF LENGTHS OF BOTH SIDES.

⑤ CONNECT ENDS OF SIDES FOR A RIGHT TRIANGLE.

DRAW A RIGHT TRIANGLE

① SIDE / HYPOTENUSE GIVEN

② DRAW HYPOTENUSE.

③ BISECT HYPOTENUSE.

④ DRAW A SEMICIRCLE. R = ½ HYP.

⑤ SWING ARC THE SIZE OF SIDE.

⑥ CONNECT ENDS OF HYPOTENUSE TO INTERSECTING ARCS TO FORM RIGHT ANGLE.

DRAW A RIGHT TRIANGLE

① ONE SIDE

② DIVIDE INTO 3 EQUAL UNITS.

③ SWING ARC 4 EQUAL UNITS.

④ SWING ARC 5 EQUAL UNITS FROM OTHER END.

⑤ CONNECT INTER-SECTING ARCS TO FORM A RIGHT TRIANGLE.

DRAW AN ISOSCELES TRIANGLE

① BASE / SIDES (EQUAL)

② DRAW BASE.

③ SET COMPASS THE SIZE OF THE SIDE AND SWING ARC.

④ SWING ARC FROM OPPOSITE SIDE.

⑤ CONNECT ENDS OF BASE TO INTER-SECTING ARCS FOR ISOSCELES TRIANGLE.

FIG. 10-9 Triangle constructions

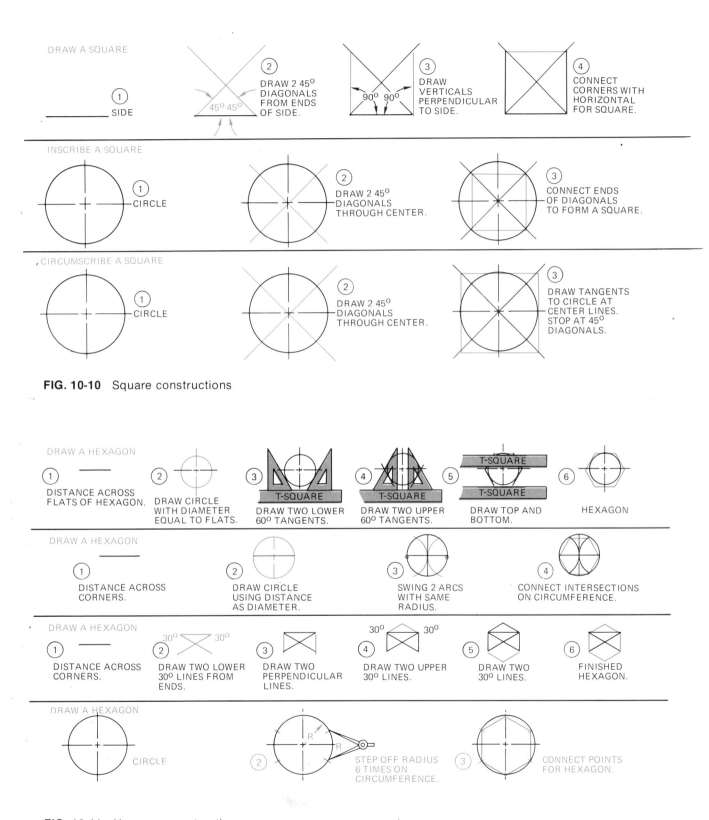

FIG. 10-10   Square constructions

FIG. 10-11   Hexagon constructions

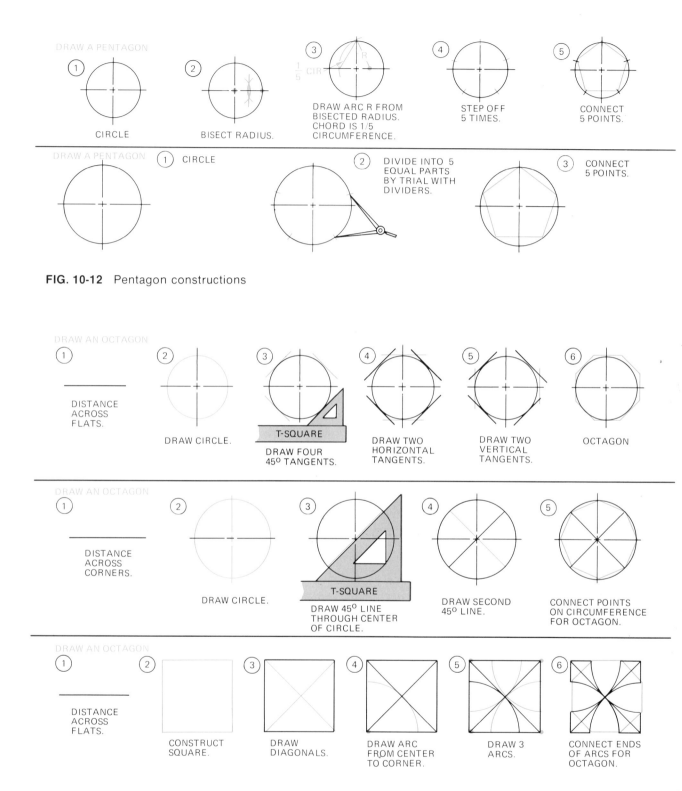

DRAW A PENTAGON

① CIRCLE

② BISECT RADIUS.

③ ⅕ CIR. DRAW ARC R FROM BISECTED RADIUS. CHORD IS 1/5 CIRCUMFERENCE.

④ STEP OFF 5 TIMES.

⑤ CONNECT 5 POINTS.

DRAW A PENTAGON

① CIRCLE

② DIVIDE INTO 5 EQUAL PARTS BY TRIAL WITH DIVIDERS.

③ CONNECT 5 POINTS.

FIG. 10-12   Pentagon constructions

DRAW AN OCTAGON

① DISTANCE ACROSS FLATS.

② DRAW CIRCLE.

③ T-SQUARE DRAW FOUR 45° TANGENTS.

④ DRAW TWO HORIZONTAL TANGENTS.

⑤ DRAW TWO VERTICAL TANGENTS.

⑥ OCTAGON

DRAW AN OCTAGON

① DISTANCE ACROSS CORNERS.

② DRAW CIRCLE.

③ T-SQUARE DRAW 45° LINE THROUGH CENTER OF CIRCLE.

④ DRAW SECOND 45° LINE.

⑤ CONNECT POINTS ON CIRCUMFERENCE FOR OCTAGON.

DRAW AN OCTAGON

① DISTANCE ACROSS FLATS.

② CONSTRUCT SQUARE.

③ DRAW DIAGONALS.

④ DRAW ARC FROM CENTER TO CORNER.

⑤ DRAW 3 ARCS.

⑥ CONNECT ENDS OF ARCS FOR OCTAGON.

FIG. 10-13   Octagon constructions

DIVIDE A CIRCLE INTO EQUAL PARTS

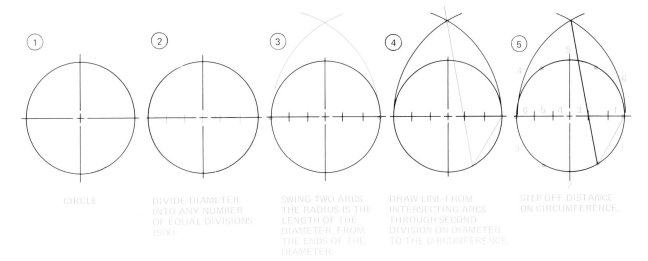

①	②	③	④	⑤
CIRCLE	DIVIDE DIAMETER INTO ANY NUMBER OF EQUAL DIVISIONS (SIX).	SWING TWO ARCS. THE RADIUS IS THE LENGTH OF THE DIAMETER, FROM THE ENDS OF THE DIAMETER.	DRAW LINE FROM INTERSECTING ARCS THROUGH SECOND DIVISION ON DIAMETER TO THE CIRCUMFERENCE.	STEP OFF DISTANCE ON CIRCUMFERENCE.

DRAW A CIRCLE THROUGH 3 POINTS

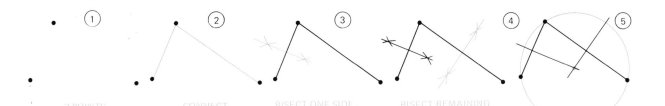

①	②	③	④	⑤
3 POINTS.	CONNECT POINTS.	BISECT ONE SIDE.	BISECT REMAINING SIDE.	INTERSECTING BISECTOR IS CENTER OF CIRCLE. DRAW CIRCLE.

FIG. 10-14 Circle constructions

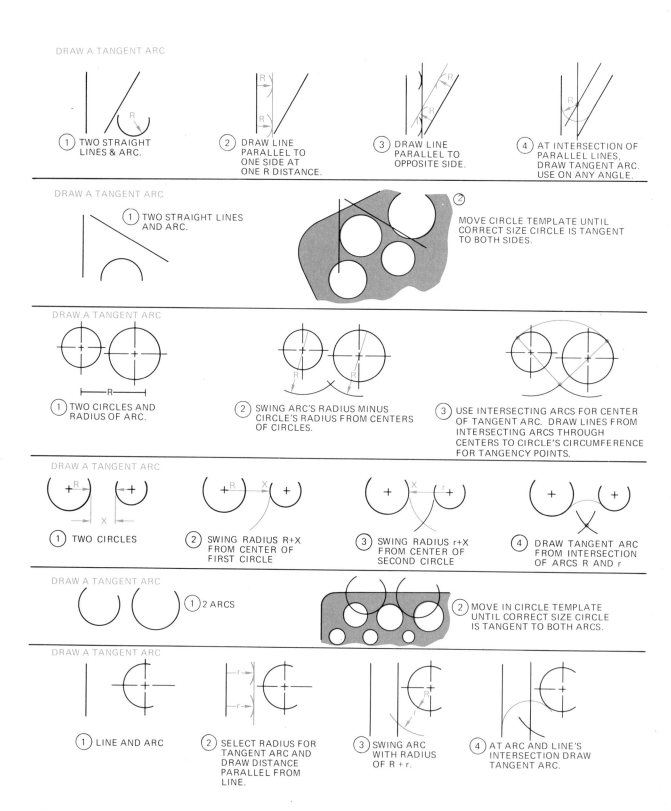

FIG. 10-15 Arc constructions

DRAW AN ELLIPSE

| MAJOR AND MINOR AXES. | SWING RADIUS HALF THE MAJOR AXIS. | INSERT 3 PINS. | TIE STRING SNUGLY. | REMOVE TOP PIN AND INSERT PENCIL. | KEEP STRING TIGHT AND SWING PENCIL. |

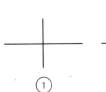   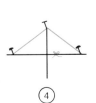 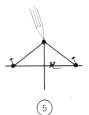 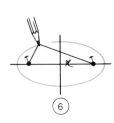

① ② ③ ④ ⑤ ⑥

DRAW AN ELLIPSE

| MAJOR AND MINOR AXES. | DRAW CONCENTRIC CIRCLES WITH MAJOR AND MINOR AXIS FOR DIAMETERS. | DRAW CONVENIENT DIAGONAL. | DRAW HORIZONTAL AND VERTICAL LINES TO PLOT POINTS. | REPEAT FOR ADEQUATE PLOTTING POINTS. | DARKEN WITH IRREGULAR CURVE. |

 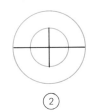 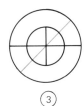 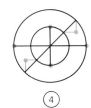  

① ② ③ ④ ⑤ ⑥

DRAW AN ELLIPSE

| MAJOR AND MINOR AXES. | BOX IN. | DIVIDE INTO EQUAL PARTS. | REPEAT IN EACH QUADRANT. |

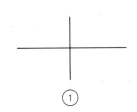 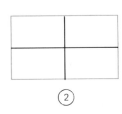 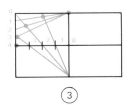 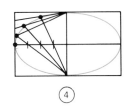

① ② ③ ④

DRAW AN ELLIPSE

| MAJOR AND MINOR AXES. | MAKE A TRAMMEL FROM PAPER. | "X" REMAINS ON MAJOR AXIS. "Y" REMAINS ON MINOR AXIS. PLOT ELLIPSE POINT. | PLOT SUFFICENT POINTS | DARKEN WITH IRREGULAR CURVE. |

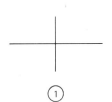 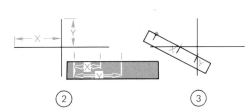 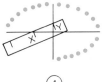 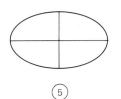

① ② ③ ④ ⑤

FIG. 10-16 Ellipse constructions

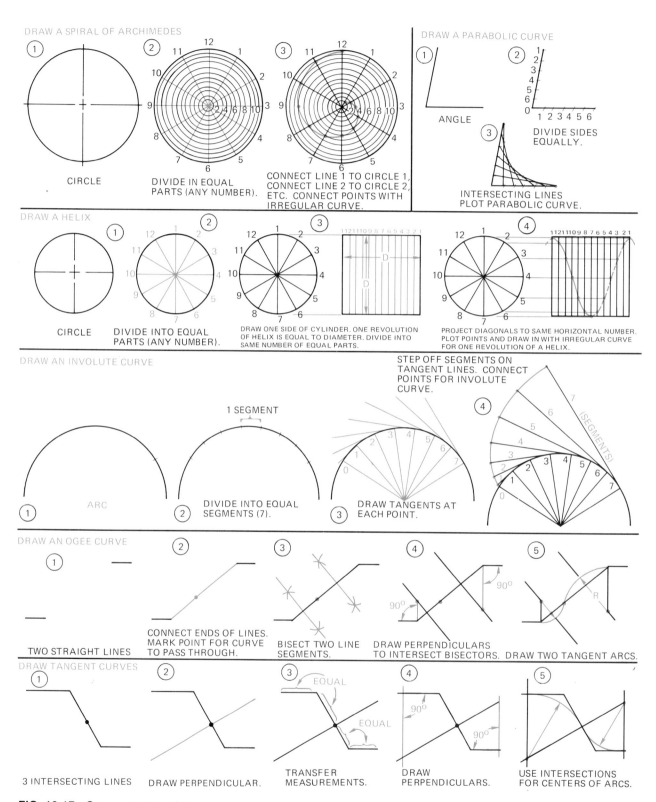

FIG. 10-17 Curve constructions

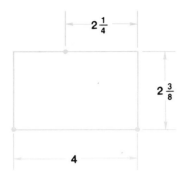

FIG. 10-18 Draw a circle through three points

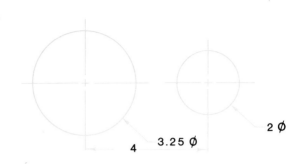

FIG. 10-19 Draw tangents to two circles

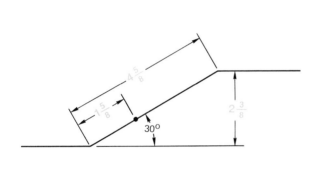

FIG. 10-20 Draw an ogee curve

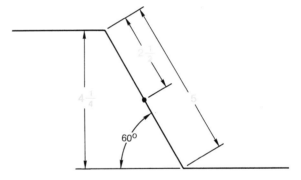

FIG. 10-21 Draw tangent curves

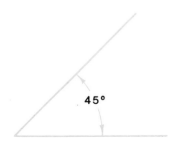

FIG. 10-22 Draw a tangent circle inside the angle with a radius of 1.5 inches

MAJOR AXIS EQUALS 5"
MINOR AXIS EQUALS 3"

FIG. 10-23 Draw an ellipse

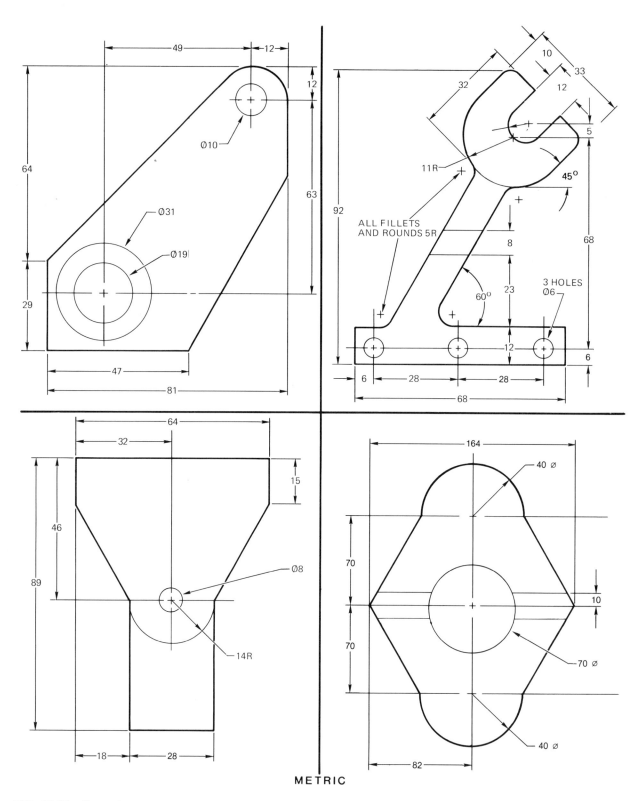

METRIC

FIG. 10-24   Draw these geometric figures

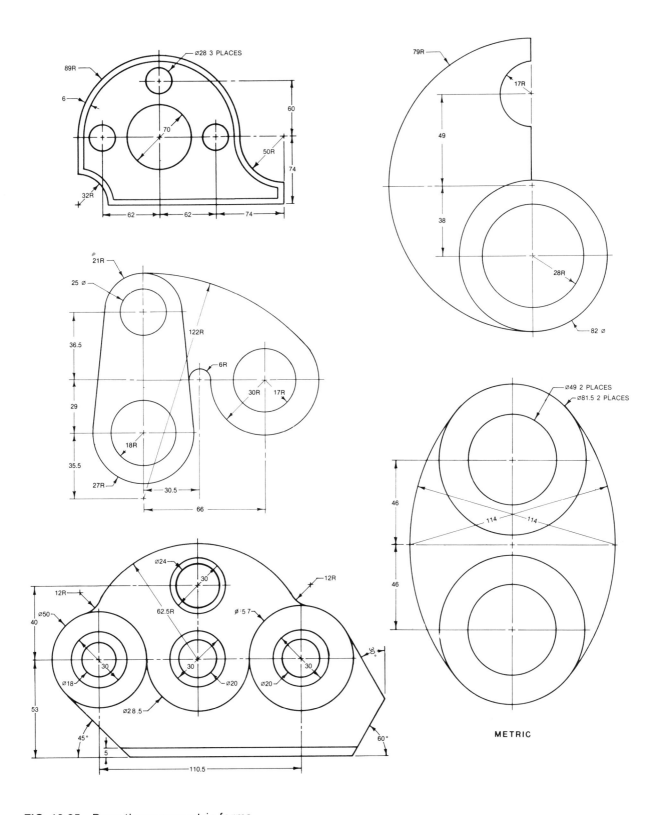

FIG. 10-25   Draw these geometric forms

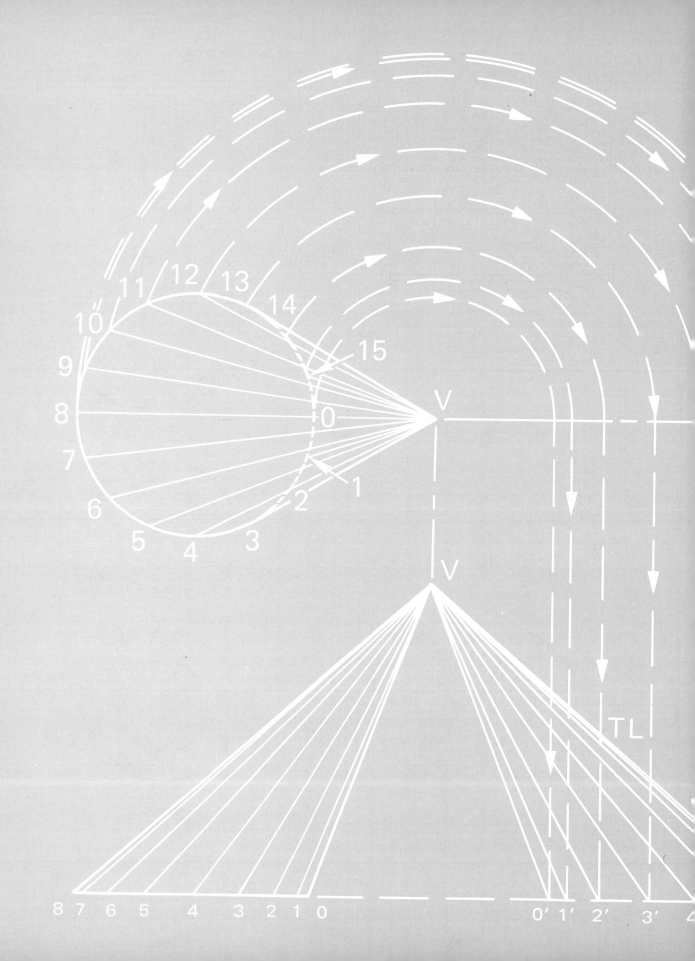

Part

Drafting Concepts

The Chinese philosopher who said "one picture is worth ten thousand words" was probably trying to make something. Words alone cannot communicate the information necessary to manufacture an item. Drafting is a standardized language of lines, symbols, and drawing concepts that lead to universal understanding of technical drawings.

Part IV of this text is an introduction to the major concepts of drafting. Understanding principles presented in this section will prepare the draftsperson to produce finished mechanical drawings that are needed to manufacture any item.

11
Chapter

Isometric / Pictorials

Clark E. Smith

INTRODUCTION

Pictorial drawing is the simplest type of drawing to understand. The isometric drawing is the simplest and fastest type of pictorial drawing to draw. As

you study this chapter, you should learn to understand the following types of pictorial drawings: isometric, dimetric, trimetric, cavalier, cabinet, two point perspective, and one point perspective. You should also develop the skill to make isometric drawings.

PICTORIAL

A single drawing that shows more than one side of an object is called a pictorial drawing. Pictorial drawings are divided into *axonometric, oblique,* and *perspective* drawings (Fig. 11-2). The fastest and most efficient pictorial drawing is an *isometric* drawing. All the lines in an isometric drawing of a rectangular object will be vertical or 30 degree angular lines. The axes will form 120 degree angular lines (Fig. 11-3).

ISOMETRIC DRAWING

All objects, regardless of their shape and size, have three basic dimensions: length, width, and height. From these basic dimensions an isometric drawing can be laid out on the *isometric axes* (Fig. 11-4).

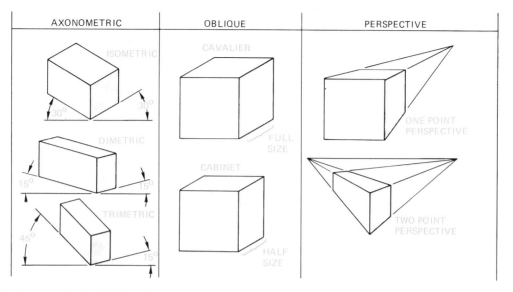

FIG. 11-2 Types of pictorial drawings

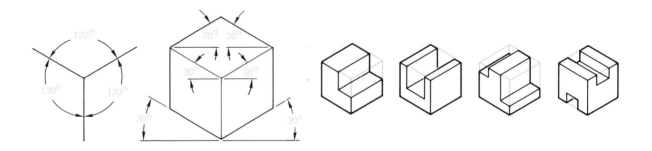

FIG. 11-3 Isometric axes

FIG. 11-5 Parts drawn within the isometric block

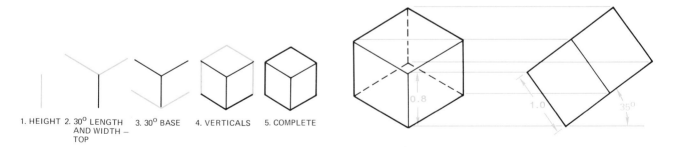

1. HEIGHT 2. 30° LENGTH AND WIDTH — TOP 3. 30° BASE 4. VERTICALS 5. COMPLETE

FIG. 11-4 Steps in drawing an isometric block

FIG. 11-6 Side view shows the actual angle of an isometric drawing

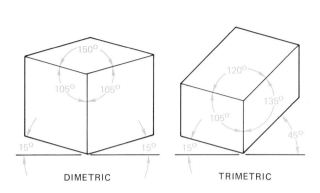

FIG. 11-7 Dimetric and trimetric axonometric drawings

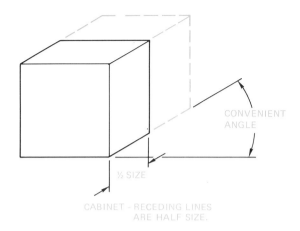

FIG. 11-9 The cabinet oblique

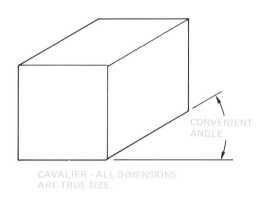

FIG. 11-8 The cavalier oblique

This blocking-in is the starting point for all isometric drawings. Once the overall isometric form is blocked in with the correct dimensions, the actual shape of the object can be drawn with less chance of error (Fig. 11-5).

In an isometric drawing, all vertical and 30 degree lines are drawn true size for convenience and speed. Because the top surface is usually seen in reality, the top, front corner should be slanted forward. This forward slant is slightly over 35 degrees (Fig. 11-6). This is why isometric templates are set at 35 degrees.

Axonometric Drawings

The isometric drawing is one form of axonometric drawings. The other axonometric drawings are the *dimetric* and *trimetric* drawings (Fig. 11-7). Dimetric and trimetric drawings are more difficult to draw than isometric drawings, therefore the isometric drawing is used most often.

Oblique Drawings

The two oblique drawings are called cavalier and cabinet drawings. Oblique means inclined. The *cavalier* (Fig. 11-8) has its front face drawn true size and shape. The receding lines are drawn at any convenient angle between 15 and 45 degrees. All dimensions are drawn to true size. A cavalier drawing will look distorted in shape. To correct this distortion, a cabinet drawing can be used (Fig. 11-9).

The procedure for making a *cabinet* drawing is the same as for the cavalier. However, the receding lines are drawn one-half actual size. Note how the cube in Figure 11-9 now looks like a cube even though the side dimensions are one-half size.

Perspective

The last type of pictorial drawing you will be concerned with is perspective drawing, illustrated in Figure 11-2. Chapter 22 explains how to make perspective drawings.

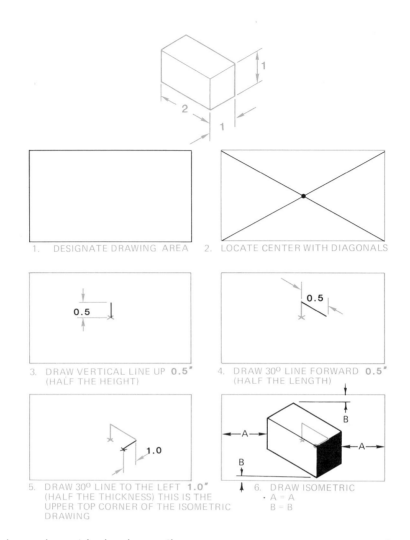

FIG. 11-10 Centering an isometric drawing on the paper

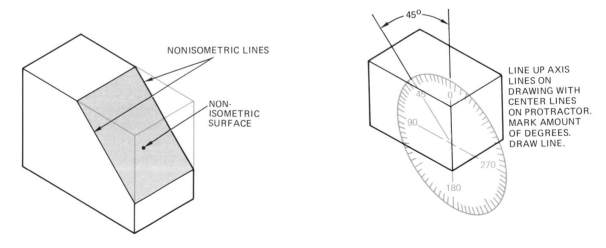

FIG. 11-11 Nonisometric lines and surfaces

FIG. 11-12 Use of an isometric protractor

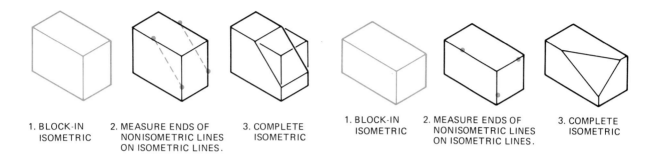

1. BLOCK-IN
   ISOMETRIC

2. MEASURE ENDS OF
   NONISOMETRIC LINES
   ON ISOMETRIC LINES.

3. COMPLETE
   ISOMETRIC

1. BLOCK-IN
   ISOMETRIC

2. MEASURE ENDS OF
   NONISOMETRIC LINES
   ON ISOMETRIC LINES.

3. COMPLETE
   ISOMETRIC

FIG. 11-13   Locating nonisometric lines

FIG. 11-14   An oblique nonisometric surface

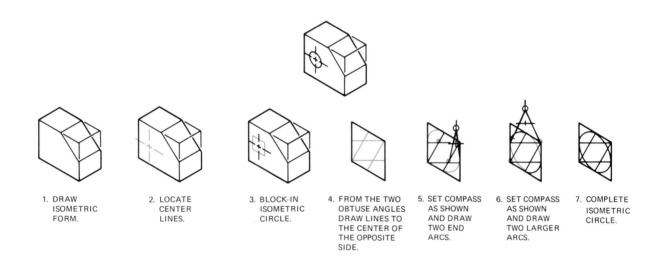

1. DRAW
   ISOMETRIC
   FORM.

2. LOCATE
   CENTER
   LINES.

3. BLOCK-IN
   ISOMETRIC
   CIRCLE.

4. FROM THE TWO
   OBTUSE ANGLES
   DRAW LINES TO
   THE CENTER OF
   THE OPPOSITE
   SIDE.

5. SET COMPASS
   AS SHOWN
   AND DRAW
   TWO END
   ARCS.

6. SET COMPASS
   AS SHOWN
   AND DRAW
   TWO LARGER
   ARCS.

7. COMPLETE
   ISOMETRIC
   CIRCLE.

FIG. 11-15   Construction of an isometric circle

CENTERING AN ISOMETRIC DRAWING

Centering an isometric drawing will save layout time. By following the steps in Figure 11-10 you can center an isometric drawing within a specific drawing area.

NONISOMETRIC LINES

Nonisometric lines are lines that are not parallel to the isometric axes (Figure 11-11). All lines that are parallel to the isometric axes are isometric lines and are drawn true size. Therefore a nonisometric line must be an angular line. An angular line on an isometric drawing cannot be measured with a protractor (unless an isometric protractor like the one shown in Figure 11-12 is used). The fastest way to draw an angular line is to measure the location of its ends on the isometric lines (Fig. 11-13). An oblique surface can be drawn in the same manner (Fig. 11-14).

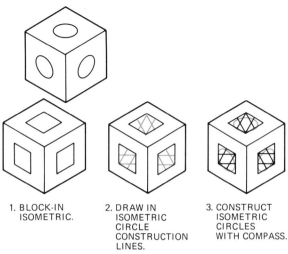

1. BLOCK-IN
ISOMETRIC.

2. DRAW IN
ISOMETRIC
CIRCLE
CONSTRUCTION
LINES.

3. CONSTRUCT
ISOMETRIC
CIRCLES
WITH COMPASS.

FIG. 11-16   Isometric circles on other surfaces

ISOMETRIC CIRCLES

Isometric circles are required on many drawings. A circle in an isometric drawing will appear as a 35° 16' ellipse. The procedure for constructing an isometric circle is shown in Figure 11-15. This procedure can be used to construct isometric circles on any of the three isometric faces of a drawing (Fig. 11-16). Drawing a circle on a nonisometric surface will require plotting several points and using a French curve to darken in the circle (Fig. 11-17).

Constructing an isometric circle is time consuming. Since industry is concerned with cutting costs and reducing drawing time, many drafting

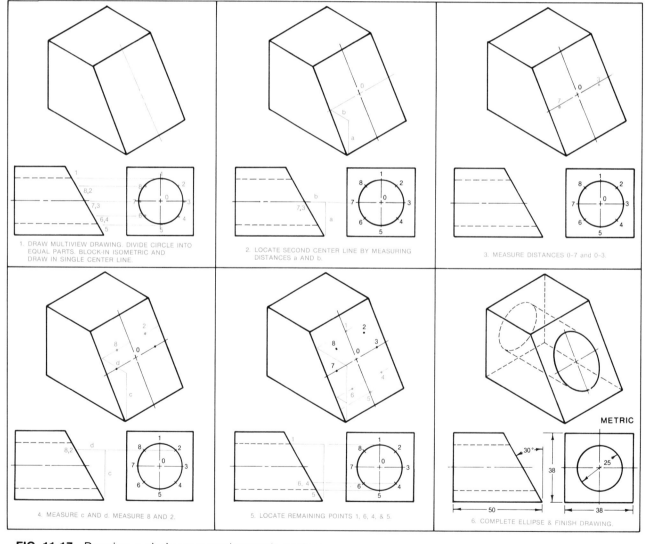

1. DRAW MULTIVIEW DRAWING. DIVIDE CIRCLE INTO EQUAL PARTS. BLOCK-IN ISOMETRIC AND DRAW IN SINGLE CENTER LINE.

2. LOCATE SECOND CENTER LINE BY MEASURING DISTANCES a AND b.

3. MEASURE DISTANCES 0-7 and 0-3.

4. MEASURE c AND d. MEASURE 8 AND 2.

5. LOCATE REMAINING POINTS 1, 6, 4, & 5.

6. COMPLETE ELLIPSE & FINISH DRAWING.

FIG. 11-17   Drawing a circle on a nonisometric surface

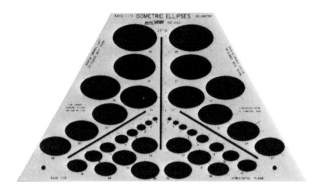

FIG. 11-18 An isometric circle template
(RapiDesign, Inc.)

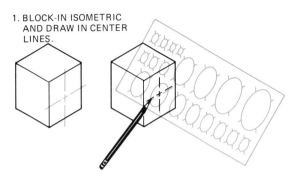

1. BLOCK-IN ISOMETRIC AND DRAW IN CENTER LINES.

2. LINE UP CENTER LINE MARKS ON ISOMETRIC TEMPLATE WITH CENTER LINES ON THE DRAWING. DRAW IN THE CIRCLE.

Note:
The major axis of the ellipse is always perpendicular to the axis of the hole.

FIG. 11-19 Use of the isometric template for circles

technicians use an isometric circle template (Fig. 11-18). To draw an isometric circle with a template simply line up the center lines on the drawing with the center lines on the template (Fig. 11-19). Semi-circles and quarter rounds are prepared and drawn in the same manner, using the same isometric template (Fig. 11-20). Draw only the part of the isometric circle that is required.

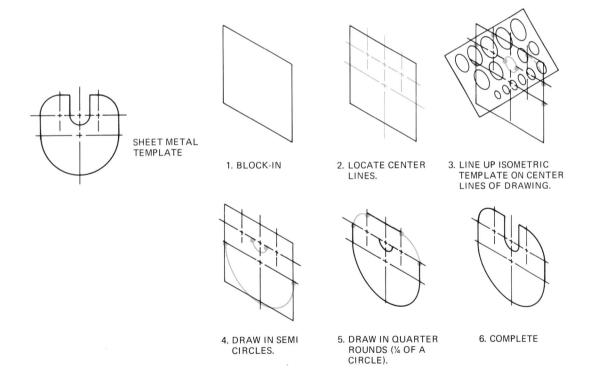

SHEET METAL TEMPLATE

1. BLOCK-IN

2. LOCATE CENTER LINES.

3. LINE UP ISOMETRIC TEMPLATE ON CENTER LINES OF DRAWING.

4. DRAW IN SEMI CIRCLES.

5. DRAW IN QUARTER ROUNDS (¼ OF A CIRCLE).

6. COMPLETE

FIG. 11-20 Use of the isometric template for arcs

WALL SHELF (ISOMETRIC DRAWING)

1. LAY OUT CONVENIENT SIZE METRIC GRID.

2. MARK INTERSECTIONS OF GRID AND IRREGULAR LINE.

3. LAY OUT HORIZONTAL AND VERTICAL DISTANCES OF ALL POINTS ON ISOMETRIC SURFACE.

4. DARKEN IN WITH A FRENCH CURVE.

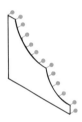

5. LAY OUT A SERIES OF 30° LINES THE THICKNESS OF MATERIAL.

6. DARKEN IN.

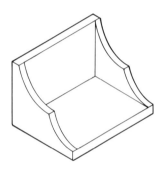

7. COMPLETE.

FIG. 11-21 Drawing isometric irregular curves

ISOMETRIC IRREGULAR CURVES

To draw an irregular curve on an isometric drawing, plot the points of the curve on the orthographic view, and then transfer them to the isometric surface. The irregular curve is plotted from points measured on the isometric lines (Fig. 11-21). The more points plotted, the truer the curve. The points are connected by sketching them lightly and darkening them with a French curve.

ISOMETRIC DRAWING POSITIONS

An isometric drawing may be placed so that the object can be viewed in different positions (Fig. 11-22). The axes will never change their angles, only their position, and will always remain at 120 degree included angles.

ISOMETRIC DIMENSIONS

There are two methods of dimensioning an isometric drawing, the aligned isometric dimensioning system (Fig. 11-23) and the unidirectional isometric dimensioning system (Fig. 11-24). Note that the dimension lines are always parallel to the edge they are dimensioning.

Care must be taken to draw arrowheads within an isometric frame (Fig. 11-25). Isometric arrowheads can be drawn freehand with practice. The back of isometric arrowheads should be parallel to the edge of the isometric object.

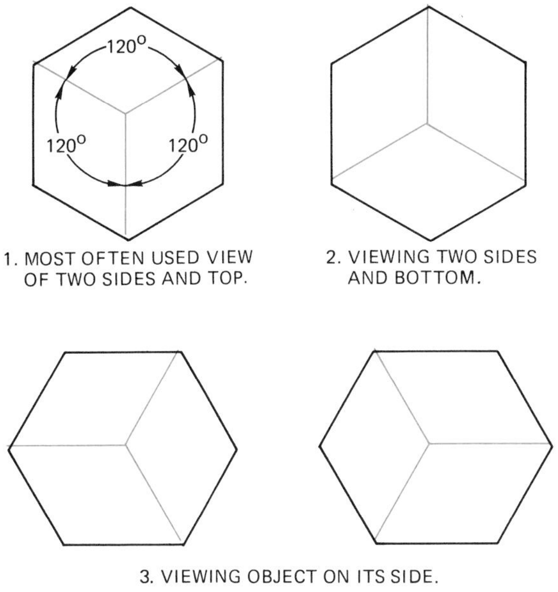

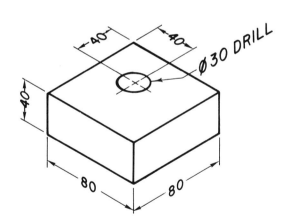

120°
120°
120°
1. MOST OFTEN USED VIEW OF TWO SIDES AND TOP.
2. VIEWING TWO SIDES AND BOTTOM.
3. VIEWING OBJECT ON ITS SIDE.

FIG. 11-22 Different views of an isometric object

40
40
40
Ø30 DRILL
80
80

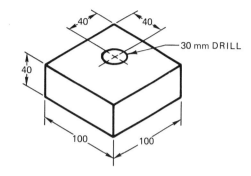

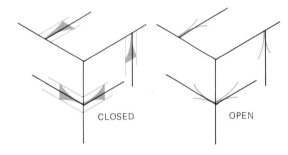

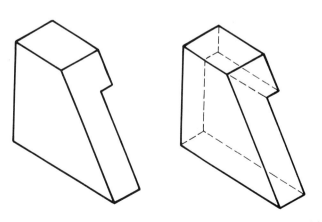

FIG. 11-23 Aligned isometric dimensions

ISOMETRIC HIDDEN LINES

Hidden lines are not used in pictorial drawings unless they are necessary to identify part of the object that cannot be seen (Fig. 11-26).

REVIEW OF ISOMETRIC DRAWING PROCEDURE

The normal procedure for making an isometric drawing is shown in Figure 11-27. Following these steps will ensure fast and accurate isometric drawings. (Nonisometric arrowheads are acceptable on isometric dimensions.)

40
40
30 mm DRILL
40
100
100

FIG. 11-24 Unidirectional isometric dimensions

CLOSED
OPEN

FIG. 11-25 Isometric arrowheads

FIG. 11-26 Isometric hidden lines

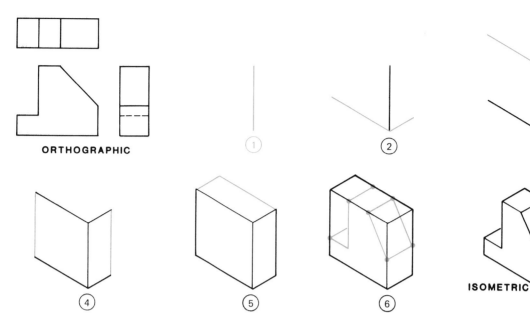

FIG. 11-27   Basic drawing steps for an isometric drawing

ORTHOGRAPHIC

ISOMETRIC

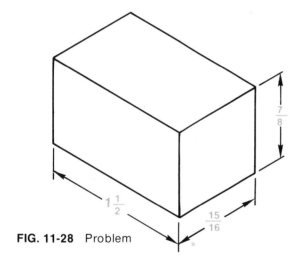

FIG. 11-28   Problem

IN.	mm
0.9	23
1.6	41
1.7	43
1.8	46
3.0	76

FIG. 11-29   Draw the isometric, cabinet, and cavalier with millimetre units of measure

PROBLEMS

Follow the instructions given for each problem. Check with your instructor as to dimensioning, additions, or deletions to the instructions. For additional problems, see the problems in Chapter 12.

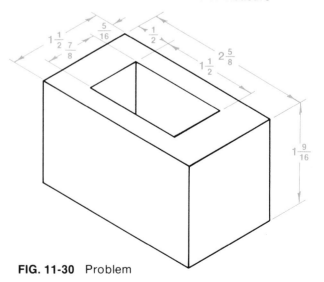

FIG. 11-30   Problem

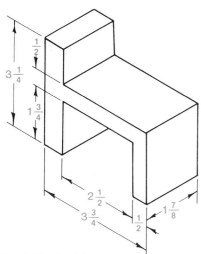

FIG. 11-31   Problem

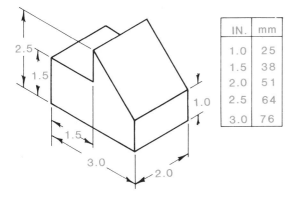

IN.	mm
1.0	25
1.5	38
2.0	51
2.5	64
3.0	76

FIG. 11-32   Draw the isometric with the inch/decimal system
Draw the cabinet and cavalier with metric units

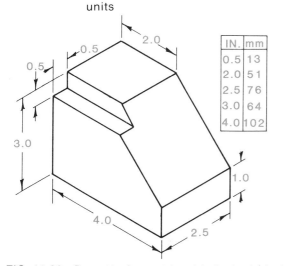

IN.	mm
0.5	13
2.0	51
2.5	76
3.0	64
4.0	102

FIG. 11-33   Draw the isometric with the inch/decimal system
Draw the cabinet and cavalier with metric units

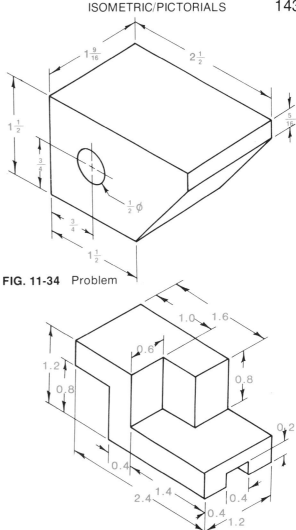

FIG. 11-34   Problem

FIG. 11-35   Draw the isometric, cabinet, and cavalier

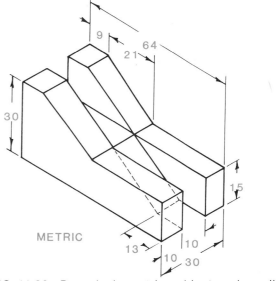

METRIC

FIG. 11-36   Draw the isometric, cabinet, and cavalier

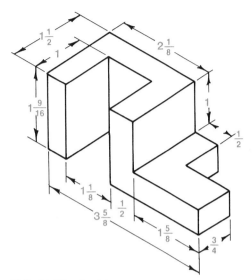

FIG. 11-37 Problem

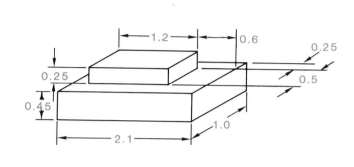

FIG. 11-39 Problem

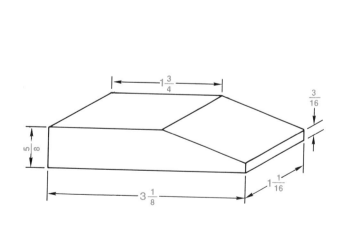

FIG. 11-38 Problem

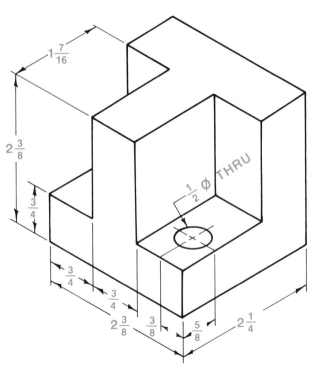

FIG. 11-40 Problem

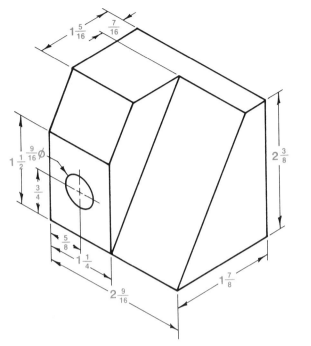

FIG. 11-41   Problem

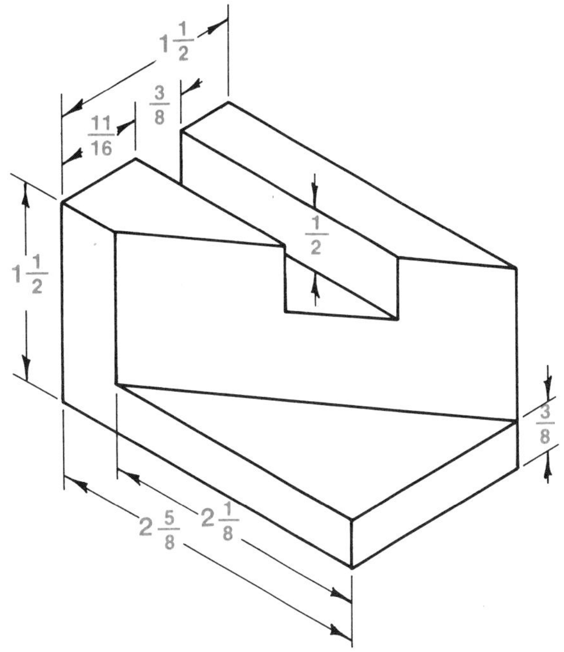

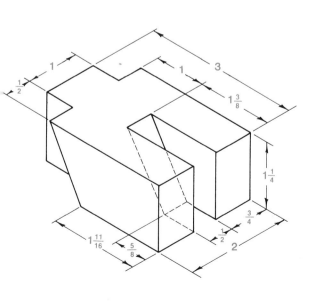

FIG. 11-42   Problem

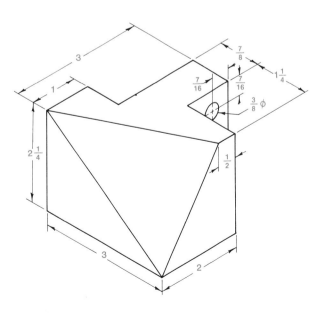

FIG. 11-43   Problem

FIG. 11-44   Problem

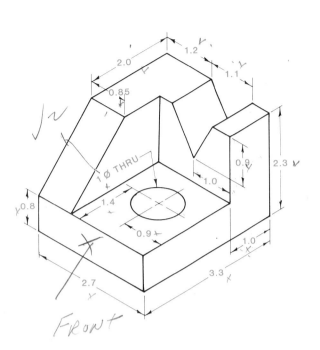

FIG. 11-45 Problem

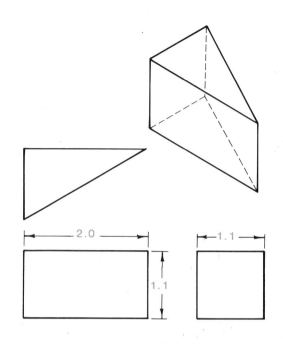

FIG. 11-47 Problem

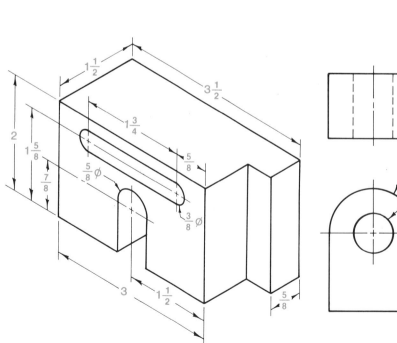

FIG. 11-46 Problem

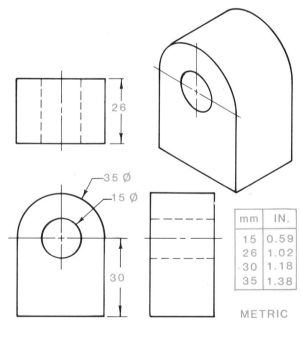

FIG. 11-48 Problem

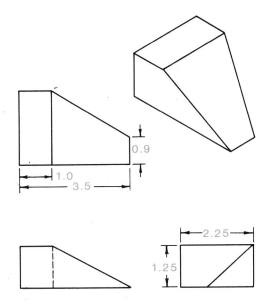

FIG. 11-49 Problem

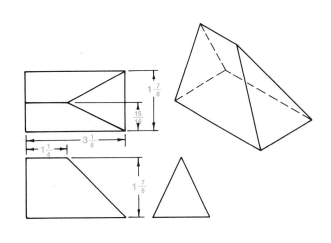

FIG. 11-51 Problem

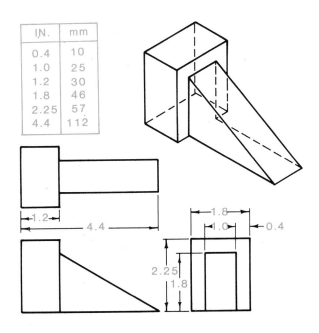

FIG. 11-50 Problem

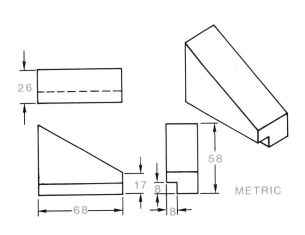

FIG. 11-52 Problem

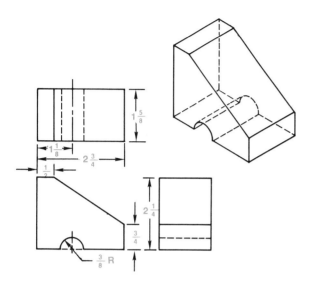

FIG. 11-53   Problem

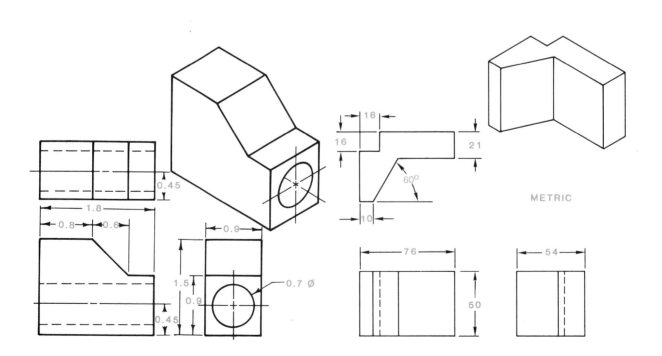

FIG. 11-54   Problem

FIG. 11-55   Problem

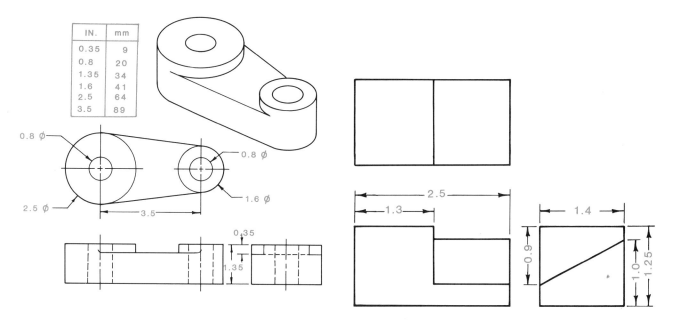

IN.	mm
0.35	9
0.8	20
1.35	34
1.6	41
2.5	64
3.5	89

FIG. 11-56   Problem

FIG. 11-58   Problem

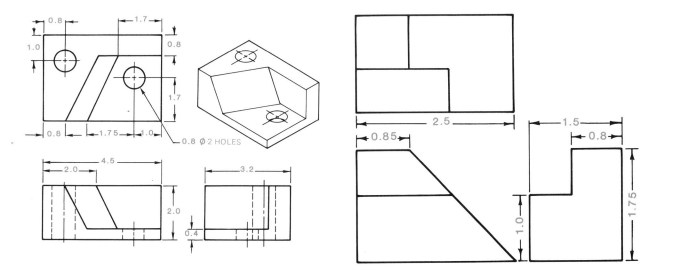

FIG. 11-57   Problem

FIG. 11-59   Problem

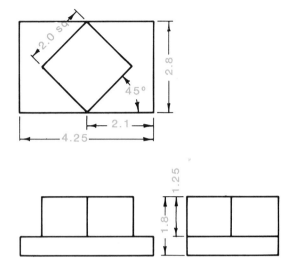

FIG. 11-60   Problem

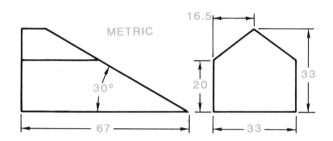

FIG. 11-61   Problem

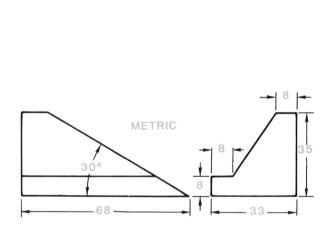

FIG. 11-62   Problem

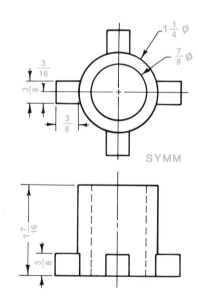

FIG. 11-63   Problem

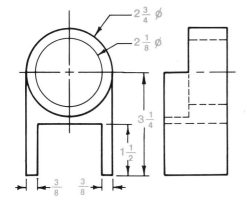

FIG. 11-64   Problem

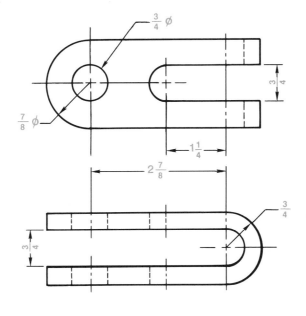

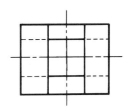

FIG. 11-65   Problem

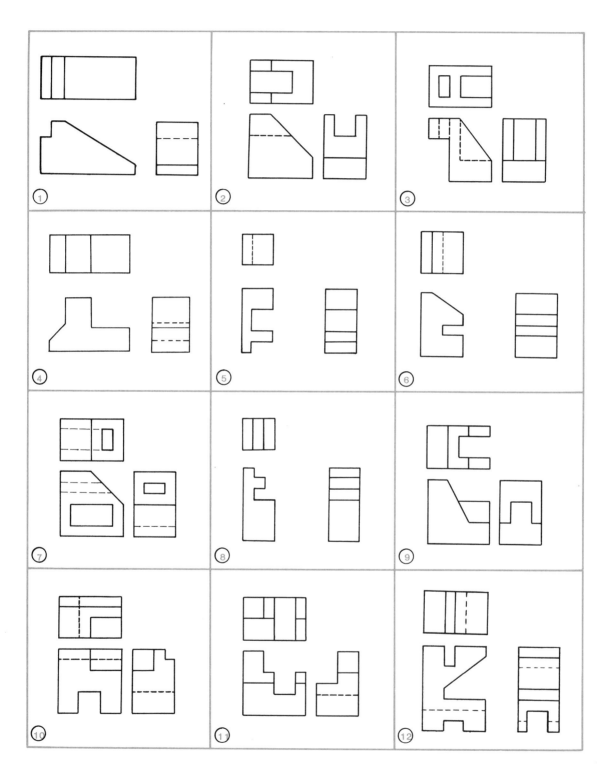

FIG. 11-66 Problem

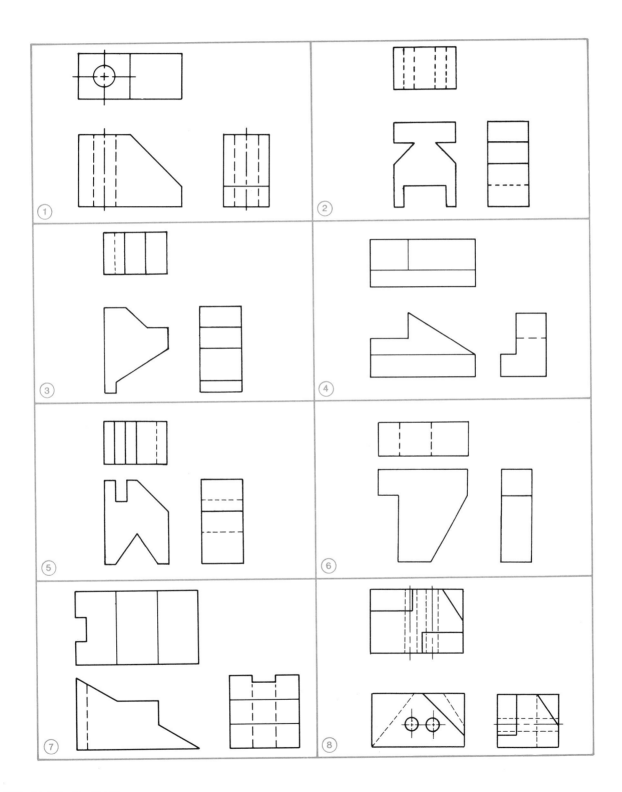

FIG. 11-67   Problem

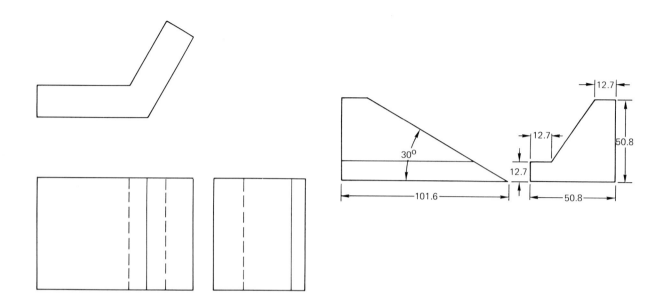

FIG. 11-73   Problem

FIG. 11-75   Problem

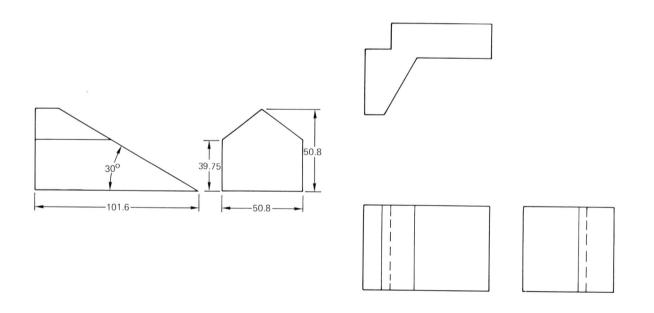

FIG. 11-74   Problem

FIG. 11-76   Problem

12

Chapter

Orthographic Projection

built. A simple flat object can be represented by one view (Fig. 12-2). A complicated object will require more views. This type of drawing is called a multiview or orthographic drawing. The views are at right angles to each other and directly in line with each other (Fig. 12-3).

As you study this chapter, you should learn the following concepts:

1. Planes of projection.
2. Visualizing views of an object to determine the best method of presenting them graphically.
3. Selection of views for a two-view drawing.
4. Selection of views for a three-view drawing.
5. Making a multiview drawing using the third angle of projection.
6. Making a multiview drawing using the first angle of projection.
7. Methods of projecting multiview drawings.
8. Proper use of line conventions on a multiview drawing.
9. Dimensioning a multiview drawing (Chapter 13).

INTRODUCTION

A *working drawing* is a communication between the designer and the fabricator of the object to be

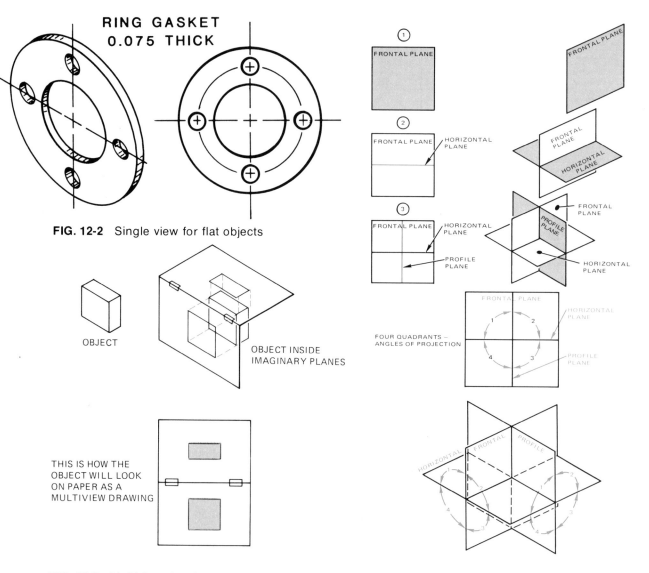

FIG. 12-2   Single view for flat objects

RING GASKET
0.075 THICK

OBJECT

OBJECT INSIDE
IMAGINARY PLANES

THIS IS HOW THE
OBJECT WILL LOOK
ON PAPER AS A
MULTIVIEW DRAWING

FIG. 12-3   Multiview drawing

FRONTAL PLANE

FRONTAL PLANE / HORIZONTAL PLANE

FRONTAL PLANE / HORIZONTAL PLANE / PROFILE PLANE

FRONTAL PLANE / HORIZONTAL PLANE / PROFILE PLANE

FOUR QUADRANTS —
ANGLES OF PROJECTION

FRONTAL PLANE / HORIZONTAL PLANE / PROFILE PLANE

HORIZONTAL / FRONTAL / PROFILE

FIG. 12-4   Three planes of orthographic projection

ORTHOGRAPHIC PROJECTION

The projection of orthographic views was developed by a Frenchman named Gaspard Monge. He developed a system called Monge's planes of projection around 1795. The planes are frontal, horizontal, and profile (Fig. 12-4). Monge's system of multiview drawings is an arrangement of orthographic views on a single plane, represented by the drawing paper. The principle is that any two adjacent views lie on perpendicular planes of projection. All adjacent planes are at right angles to each other. The two arrangements of views used in mechanical drawing are *first-angle projection* (Fig. 12-5) and *third-angle projection* (Fig. 12-6). In first-angle projection the object is viewed in *front* of the planes and projected back to the planes (Fig. 12-7). In third-angle projection the object is viewed from *behind* and projected forward to the planes (Fig. 12-8).

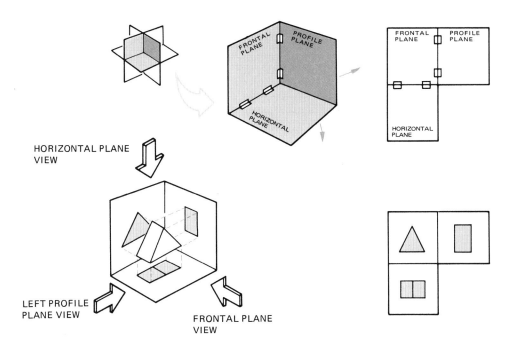

FIG. 12-5 First-angle projection—3 views

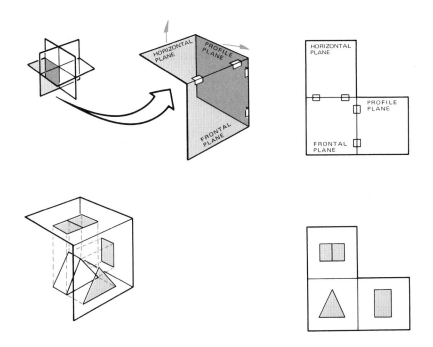

FIG. 12-6 Third-angle projection—3 views

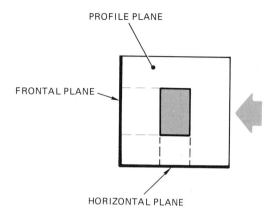

FIG. 12-7 First-angle projection views the object from in front of the planes

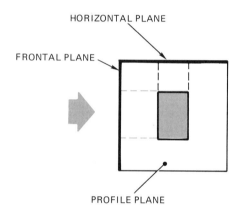

FIG. 12-8 Third-angle projection views the object from behind the planes

Orthographic projection is a method of representing an object by a line drawing on a projection plane that is perpendicular to parallel projectors (Fig. 12-9). The adjacent planes are at 90 degrees. The parallel projectors are at 90 degrees to the planes.

MULTIVIEW DRAWINGS

Multiview drawings in first or third-angle of projection can have two to six views (Fig. 12-10). The views most often used in first-angle projection are

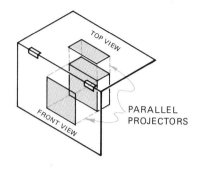

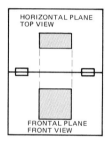

FIG. 12-9 Planes are perpendicular to projectors

front, top, and left profile views (Fig. 12-11). The views most often used in third-angle projection are the front, top, and right profile views (Fig. 12-12). To define which type of projection is used for a multiview drawing, a small explanatory sketch is placed on the drawing paper (Fig. 12-13).

The majority of engineers and draftspersons in the United States will use the third-angle projection method because of its clarity. The objection to first-angle projection is that when the planes are revolved away from the object, the top view appears below the front view, the right side view appears to the left of the front view, and the left side view appears to the right of the front view. This makes visualization difficult. Most multiview drawings in this text will be third-angle projection.

The theory of projecting from an object to a projection plane, then swinging the plane to a flat surface is important to understand. Most draftspersons find it faster and easier to draw each view as an actual picture as seen with a line of sight perpendicular to each particular plane of projection. To see the front view, using third-angle projection, the draftsperson should imagine he is in front of the

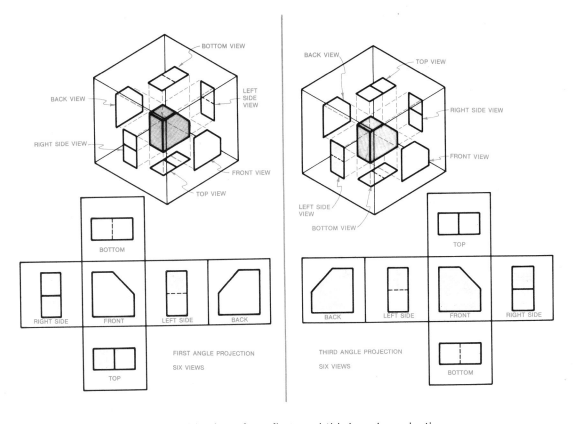

FIG. 12-10 Six possible orthographic views from first- and third-angle projections

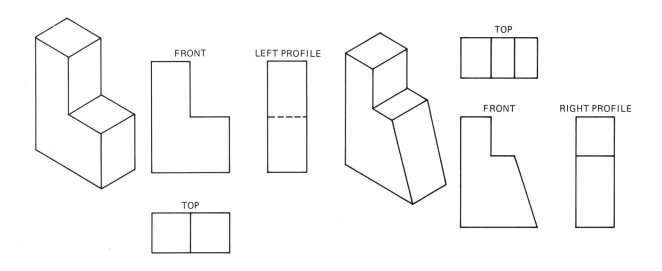

FIG. 12-11 Common views used in first-angle projection

FIG. 12-12 Common views used in third-angle projection

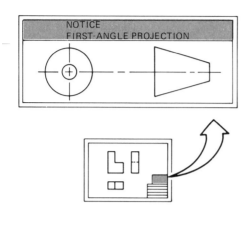

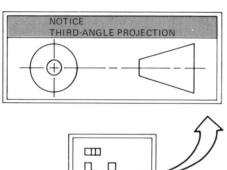

FIG. 12-13   Note block for projection angle

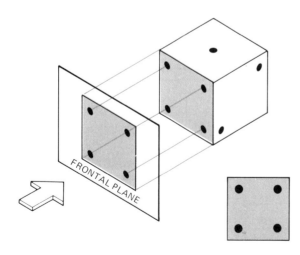

FIG. 12-14   Front view of a third-angle projection

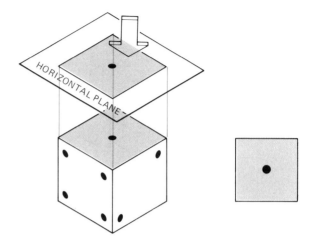

FIG. 12-15   Top view of a third-angle projection

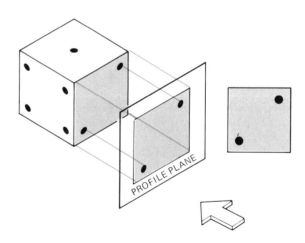

FIG. 12-16   Right side view of a third-angle
projection

object, so that the line of sight is perpendicular to
the frontal projection plane (Fig. 12-14). To visualize the top view, the observation should be a
bird's-eye view perpendicular to the horizontal
plane (Fig. 12-15). To visualize the right profile
view, the observation should be from the right side
(Fig. 12-16). All three views in third-angle projection are shown in Figure 12-17. The same process
using first-angle projection is shown in Figure 12-18.

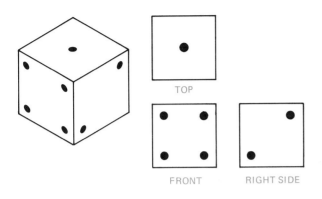

FIG. 12-17 Three-view orthographic with third-angle projection

INCREASING DRAWING SPEED

To reduce drawing time, the third view of an orthographic drawing can be projected as shown in Figure 12-19. The end view is developed by projecting lines from the top and front views. The process can be reversed by projecting from the front and profile views to develop the top view (Fig. 12-20).

LINE CONVENTION

Line conventions must be used correctly in working drawings so the fabricator will have no difficulty

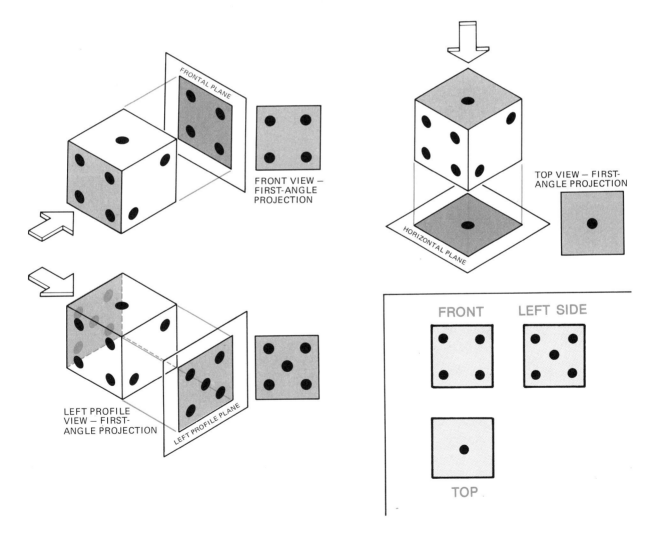

FIG. 12-18 Three-view orthographic with first-angle projection

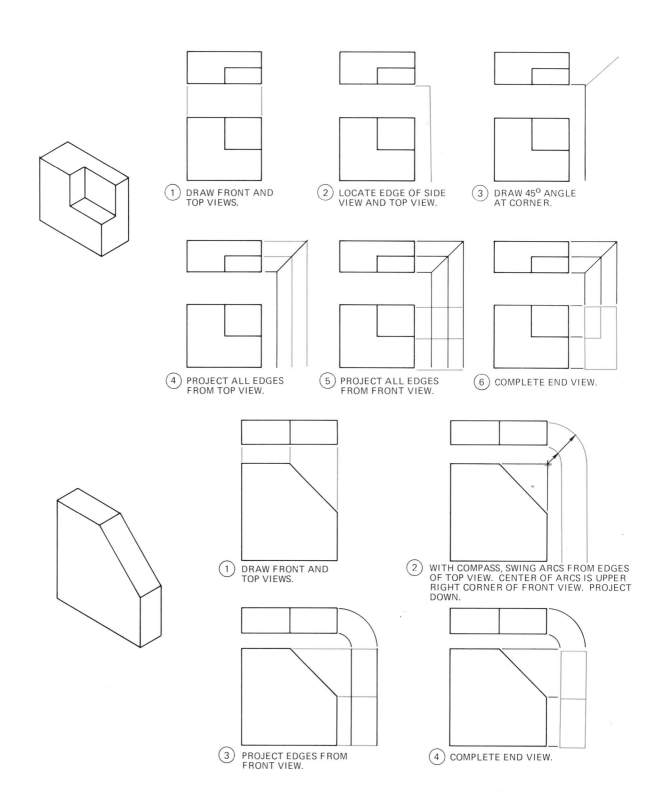

1. DRAW FRONT AND TOP VIEWS.

2. LOCATE EDGE OF SIDE VIEW AND TOP VIEW.

3. DRAW 45° ANGLE AT CORNER.

4. PROJECT ALL EDGES FROM TOP VIEW.

5. PROJECT ALL EDGES FROM FRONT VIEW.

6. COMPLETE END VIEW.

1. DRAW FRONT AND TOP VIEWS.

2. WITH COMPASS, SWING ARCS FROM EDGES OF TOP VIEW. CENTER OF ARCS IS UPPER RIGHT CORNER OF FRONT VIEW. PROJECT DOWN.

3. PROJECT EDGES FROM FRONT VIEW.

4. COMPLETE END VIEW.

FIG. 12-19 Projecting lines for the end view

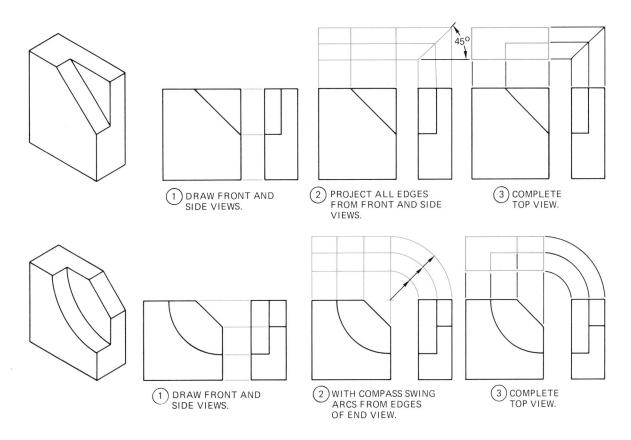

① DRAW FRONT AND SIDE VIEWS.

② PROJECT ALL EDGES FROM FRONT AND SIDE VIEWS.

③ COMPLETE TOP VIEW.

① DRAW FRONT AND SIDE VIEWS.

② WITH COMPASS SWING ARCS FROM EDGES OF END VIEW.

③ COMPLETE TOP VIEW.

FIG. 12-20 Projecting lines for the top view

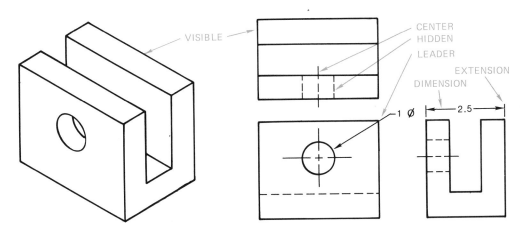

FIG. 12-21 Line conventions

reading the drawing (see Chapter 6). Figure 12-21 is an example of the language of lines used in an orthographic drawing. Carefully note the weight of the lines. It is important that the lines be of uniform thickness and dark enough to reproduce well.

Hidden Lines

Hidden lines are used to show parts of an object that are not visible. It is important for an orthographic drawing to have concise and neat hidden

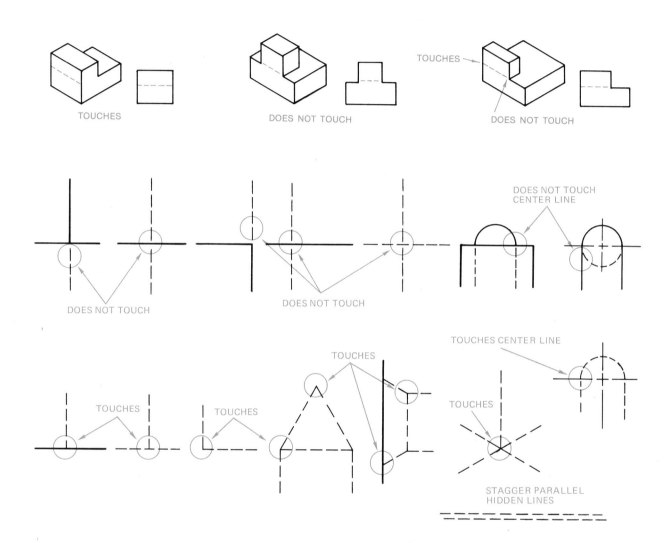

FIG. 12-22 Hidden line standards

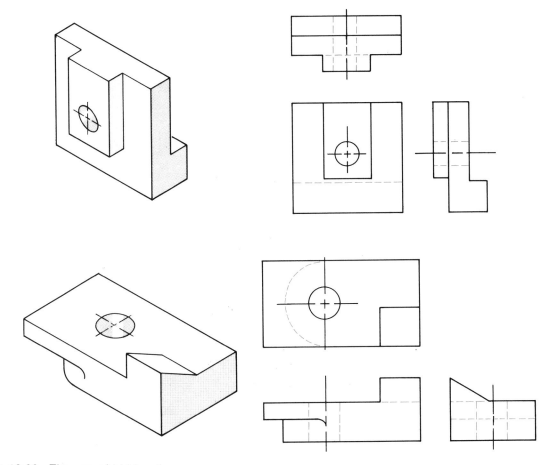

FIG. 12-23   The use of hidden lines

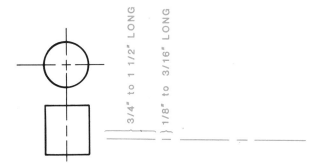

FIG. 12-24   Center line dimensions

lines that conform to drafting standards (Fig. 12-22).
Hidden lines should be a series of dashes drawn
⅛ inch long and spaced ¹/₃₂ inch apart, or a near
visual estimate of that. Hidden lines should be
heavy, but thinner than object lines. The proper use
of hidden lines is shown in Figure 12-23.

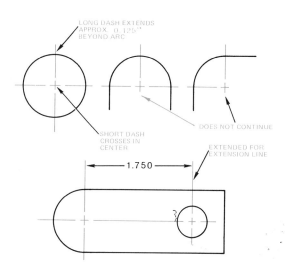

FIG. 12-25   The proper use of center lines

CYLINDRICAL FORMS

CONICAL FORMS

SYMMETRICAL SURFACES

$\frac{9}{16}$

CURVED SURFACES

FIG. 12-26 Applications of center lines

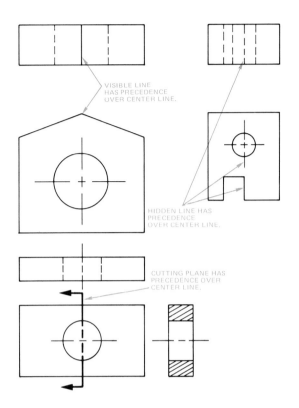

FIG. 12-27 Overlapping lines

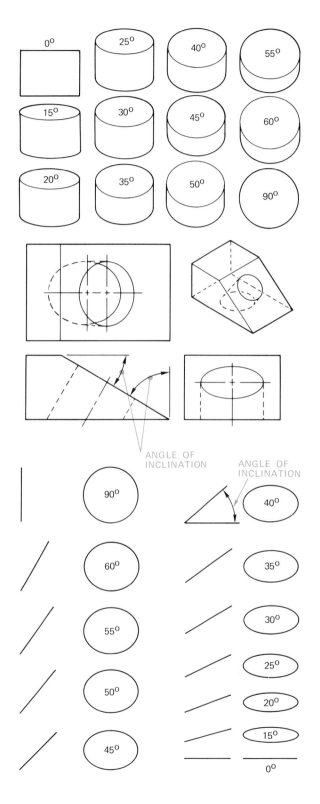

FIG. 12-28 Ellipses at various angles of inclination

Center Lines

Center lines locate the centers of symmetrical round objects. They are a series of thin long and short dashes (Fig. 12-24). The proper use of center lines is shown in Figure 12-25. Center lines may be used to show the centers of many forms (Fig. 12-26).

Overlapping Lines

The precedence of overlapping lines in any view follows the rule of showing the darker of the two (Fig. 12-27).

Ellipses

A circle that is not projected at a right angle will appear as an ellipse (Fig. 12-28). Methods of constructing an ellipse are shown in Figure 12-29. To save time, an ellipse template may be used as shown in Figure 12-30. Ellipse templates come in many sizes and angles.

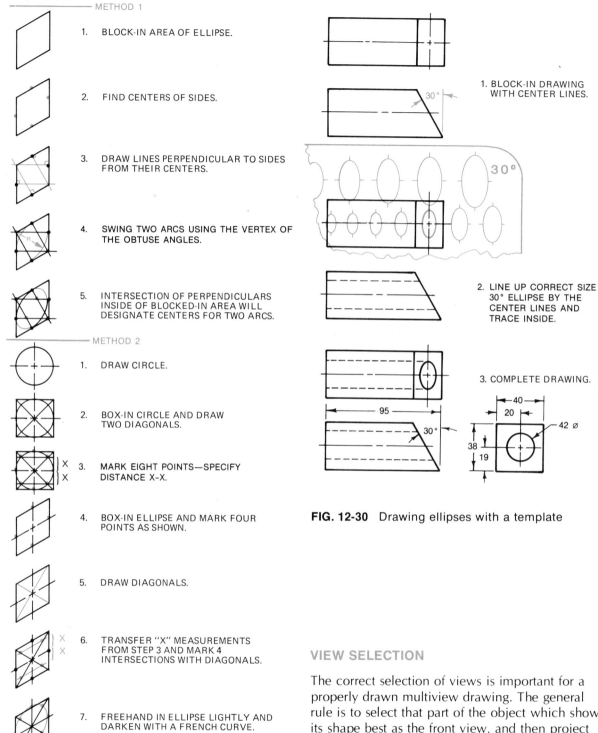

METHOD 1

1. BLOCK-IN AREA OF ELLIPSE.

2. FIND CENTERS OF SIDES.

3. DRAW LINES PERPENDICULAR TO SIDES FROM THEIR CENTERS.

4. SWING TWO ARCS USING THE VERTEX OF THE OBTUSE ANGLES.

5. INTERSECTION OF PERPENDICULARS INSIDE OF BLOCKED-IN AREA WILL DESIGNATE CENTERS FOR TWO ARCS.

METHOD 2

1. DRAW CIRCLE.

2. BOX-IN CIRCLE AND DRAW TWO DIAGONALS.

3. MARK EIGHT POINTS—SPECIFY DISTANCE X-X.

4. BOX-IN ELLIPSE AND MARK FOUR POINTS AS SHOWN.

5. DRAW DIAGONALS.

6. TRANSFER "X" MEASUREMENTS FROM STEP 3 AND MARK 4 INTERSECTIONS WITH DIAGONALS.

7. FREEHAND IN ELLIPSE LIGHTLY AND DARKEN WITH A FRENCH CURVE.

FIG. 12-29   Methods of drawing ellipses

1. BLOCK-IN DRAWING WITH CENTER LINES.

2. LINE UP CORRECT SIZE 30° ELLIPSE BY THE CENTER LINES AND TRACE INSIDE.

3. COMPLETE DRAWING.

FIG. 12-30   Drawing ellipses with a template

VIEW SELECTION

The correct selection of views is important for a properly drawn multiview drawing. The general rule is to select that part of the object which shows its shape best as the front view, and then project other views of the multiview drawing (Fig. 12-31).

Because of the value of saving time in industry, it is important to not draw unnecessary views. Draw only those views necessary to communicate

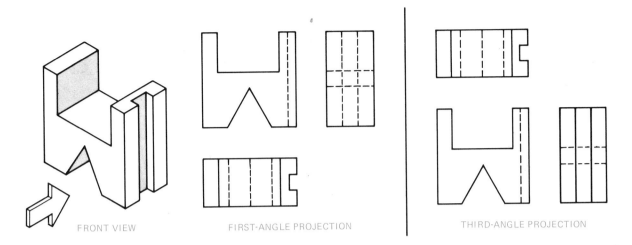

FIG. 12-31   Selecting the best views for orthographic drawings

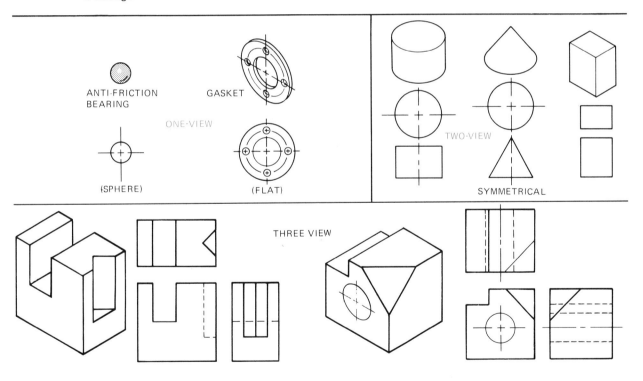

FIG. 12-32   Most objects require three views

to the fabricator how the object is to be constructed (Fig. 12-32). A multiview working drawing rarely needs more than three views to communicate all of the necessary information to the fabricator. However, if the object is extremely complex, more than three views may be used to eliminate errors in production (Fig. 12-33).

A typical industrial multiview drawing is shown in Figure 12-34. See Chapter 30 for more detailed multiview drawings.

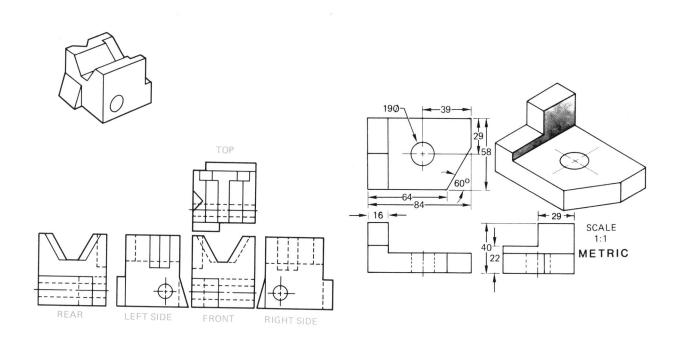

FIG. 12-33 Complex objects may require more views

FIG. 12-34 Typical industrial uses of multiview drawings

USE THE DIRECTION OF THE ARROW FOR SIGHTING THE FRONT VIEW. SKETCH ONLY THE FRONT VIEWS. WHAT DO YOUR SKETCHES REVEAL?

FIG. 12-35 Problem

PROBLEMS

Follow the instructions given by your instructor for each problem. Check with your instructor about dimensioning, additions, or deletions to the instructions. For additional problems, refer to the problems in Chapter 11.

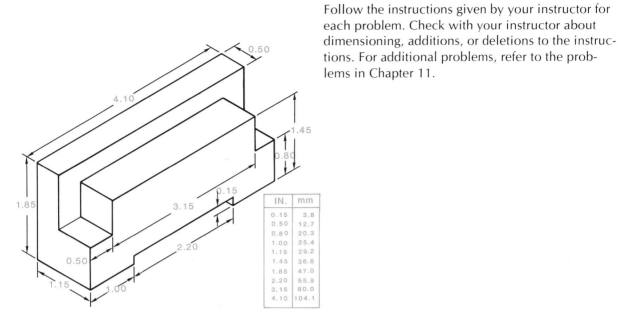

FIG. 12-36   Problem

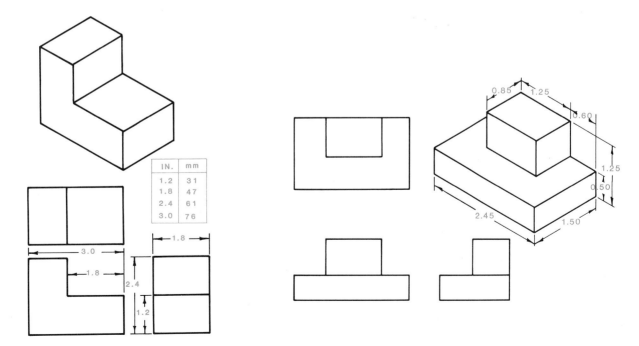

FIG. 12-37   Problem

FIG. 12-38   Problem

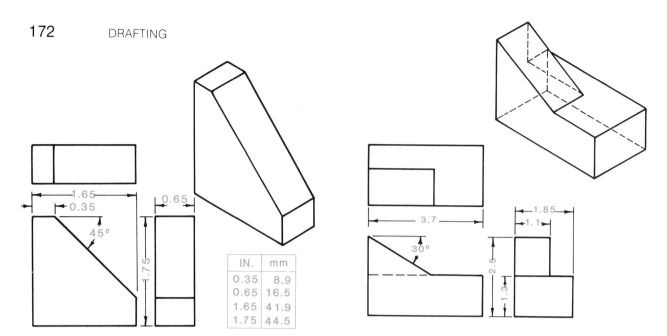

IN.	mm
0.35	8.9
0.65	16.5
1.65	41.9
1.75	44.5

FIG. 12-39 Problem

FIG. 12-40 Problem

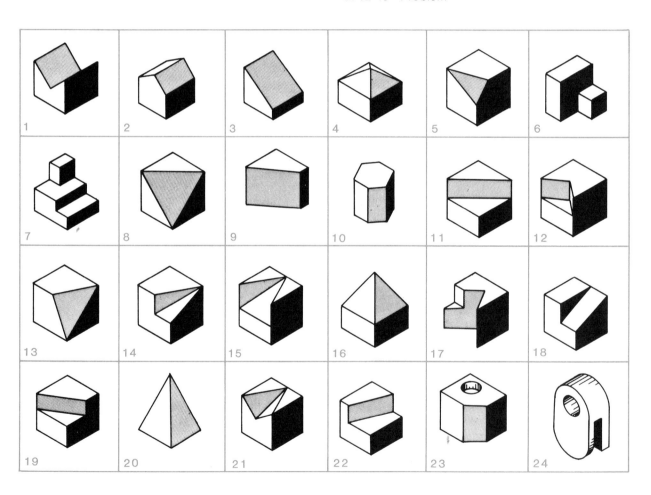

FIG. 12-41 Problem

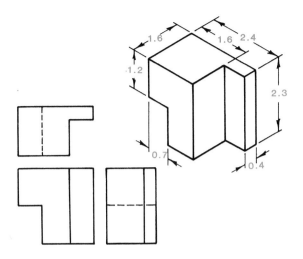

FIG. 12-42 Problem

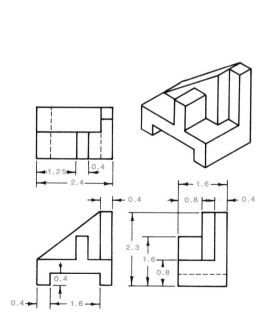

FIG. 12-43 Problem

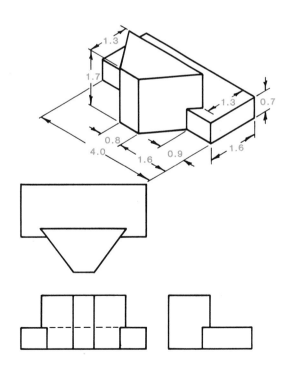

FIG. 12-44 Problem

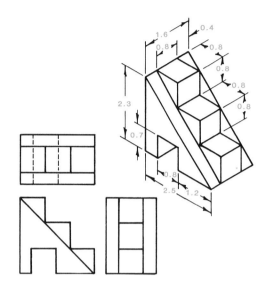

FIG. 12-45   Problem

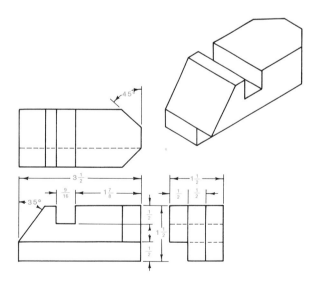

FIG. 12-46   Problem

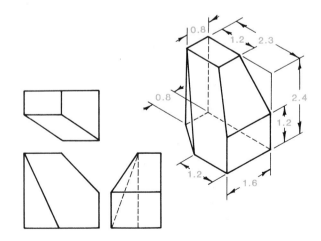

FIG. 12-47   Problem

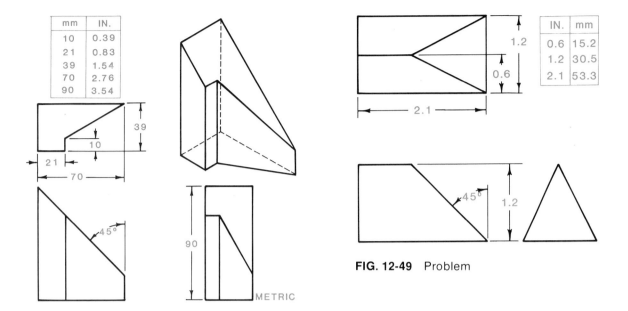

mm	IN.
10	0.39
21	0.83
39	1.54
70	2.76
90	3.54

METRIC

FIG. 12-48   Problem

IN.	mm
0.6	15.2
1.2	30.5
2.1	53.3

FIG. 12-49   Problem

IN.	mm
1/4	6.4
5/16	7.9
9/16	14.3
5/8	15.9
7/8	22.2
1 7/16	36.5
1 5/8	41.3
1 3/4	44.4
3 5/8	92.1

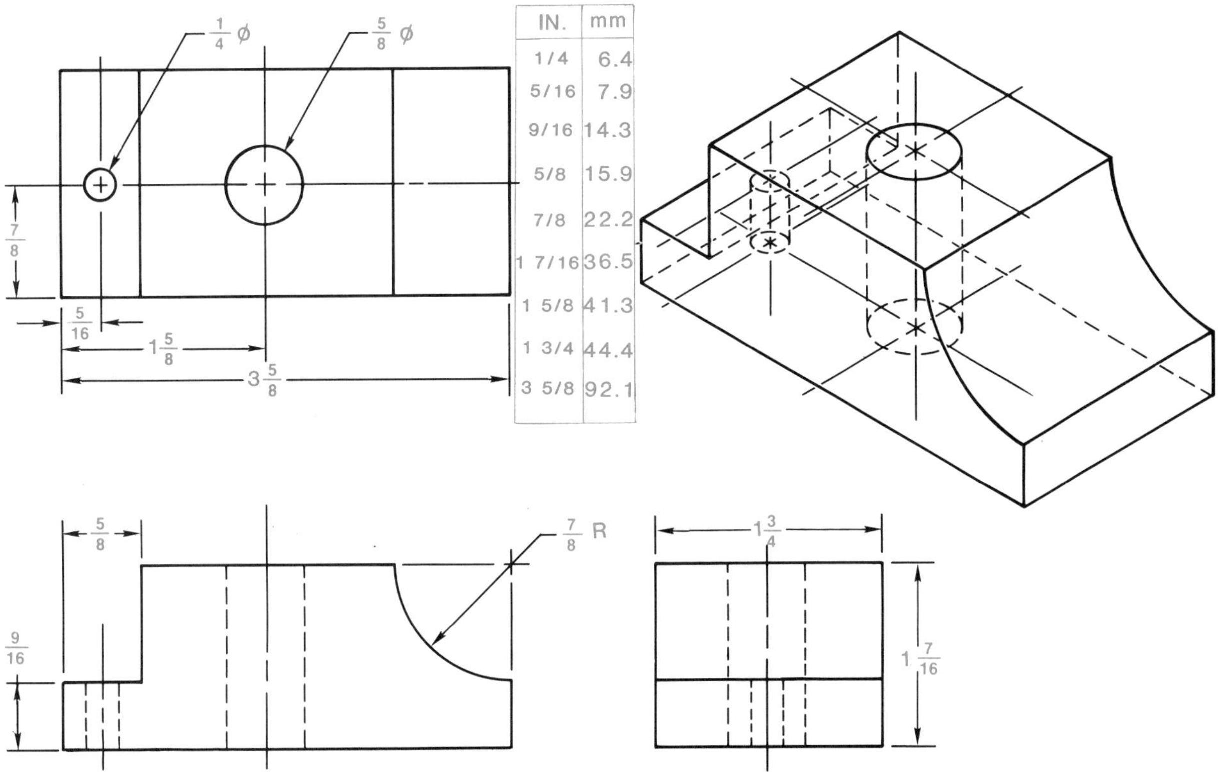

FIG. 12-50   Problem

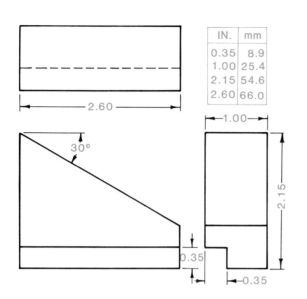

IN.	mm
0.35	8.9
1.00	25.4
2.15	54.6
2.60	66.0

FIG. 12-51   Problem

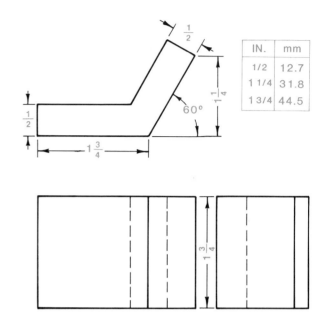

IN.	mm
1/2	12.7
1 1/4	31.8
1 3/4	44.5

FIG. 12-52   Problem

13
Chapter

Dimensioning

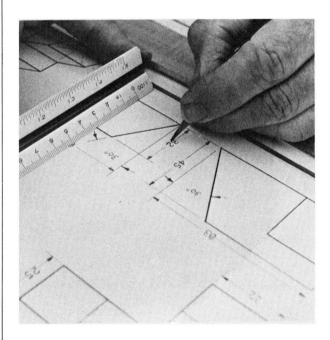

INTRODUCTION

Before a product can be manufactured, a working drawing describing its shape must be drawn (Fig. 13-2), and dimensions must be placed on it specifying sizes, locations, and general information for all the drawn parts. (Fig. 13-3).

This chapter defines the rules, principles, and methods of dimensioning mechanical drawings. Most of the dimensioning standards were developed by ANSI (American National Standards Institute) and ISO (International Organization for Standardization). As you study this chapter, you should learn the dimensioning symbols, become skillful in dimensioning techniques, understand the dimensioning of geometric forms, and understand how to dimension and note machine operations.

Dimensioning is a system of lines, symbols, numerical values, and notations. Because the number of shapes and forms of industrial products are unlimited, basic rules for dimensioning are needed. Each dimension should be logically placed, using the principles of dimensioning as guidelines.

A finished drawing is a complete set of instructions to manufacture an item. Additional instructions should not be necessary. Most dimensioning can be placed into four categories. These categories are:

1. **Dimensioning Symbols**
   Dimension lines
   Arrowheads

Dots
Extension lines
Center lines
Crossing extension lines
Leaders
Leader to a circle
Numbers
Diameter
Square
Radius

2. **Dimensioning Techniques**
Types of dimensions
Size dimensions
Location dimensions
Expression of dimensions
Typical scales
Dimension placement
Dimension placement for holes
Dimension spacing
Staggered dimensions
Point location
Oblique dimensions
Unilateral dimensions
Aligned dimensions
Repetitive dimensions
Reference dimensions
Baseline dimensions
Datum
Datum—irregular curve
Small areas
Tabular dimensions
Direction
Rectangular coordinate dimensions
Polar coordinate dimensions
Grid dimensions

3. **Dimensioning Geometric Forms**
Prisms
Cylinders
Cones
Pyramids
Inclined surfaces
Fillets
Rounds
Diameters
Spheres
Angles
Arcs
Dimensioning arcs, chords, and angles
Radii

CAN THIS PART BE MANUFACTURED WITHOUT DIMENSIONS?

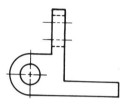 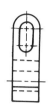

FIG. 13-2   A nondimensioned part

Round ends
Partial round ends
Unlocated centers of radii
Foreshortened radii
True radius
Tangent curves
Symmetrical parts
Broken symmetrical parts

4. **Dimensioning Machine Operations**
Holes
Chamfers
Internal chamfers
Countersink
Counterbore
Counterdrill
Spot face
Drill and ream
Bolt circle
Necks
Undercuts
Knurling
Keyway, keyseat, and key
Machining centers
Taper

PROBLEMS

Double or triple the size of each drawing problem with a divider on page 198. Draw and dimension the problems. Leave enough space between views for dimension placement.

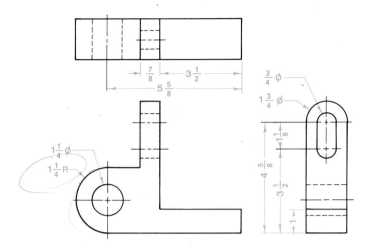

FIG. 13-3 A fully dimensioned part insures accurate manufacturing

FIG. 13-4 Dimension Lines: Dimension lines are thin, sharp lines with a break in the middle for dimensions. The line is the exact length of the dimension.

FIG. 13-6 Dots: A dot is placed on the end of a leader to specify a large area. It is approximately $1/16$ inch in diameter.

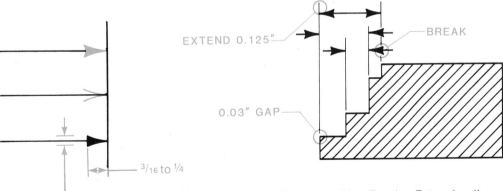

FIG. 13-5 Arrowheads: Arrowheads indicate the ends of dimension lines. Arrowheads are also used on the end of a leader to point to a specific item. The arrowhead is drawn $3/16$ inch to $1/4$ inch long by $3/32$ inch to $1/8$ inch wide, depending on the drawing size. Arrowheads can be open or closed, but must be narrow, pointed, and sharp.

FIG. 13-7 Extension Line Breaks: Extension lines are thin, sharp lines that extend from the edges of the drawing. They mark the point from which a dimension line begins. They show the exact distance of the dimension. They are almost always drawn at right angles to dimension lines. They are drawn with a small gap between them and the outline of the drawing. Extension lines may be gapped so they do not cross arrowheads.

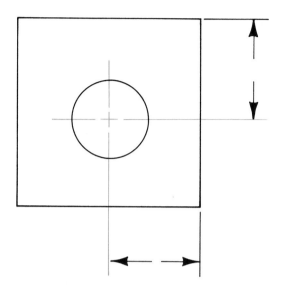

FIG. 13-8 Center Lines: An extended center line can be used as an extension line.

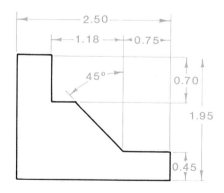

FIG. 13-9 Crossing Extension Lines: Do not cross extension or dimension lines unless it is unavoidable. To keep crossing to a minimum, the shortest dimension line should be next to the object line.

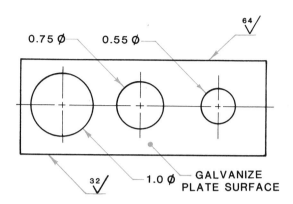

FIG. 13-10 Leaders: A leader is used to show the exact location of a dimension, symbol, or note for a specific feature on the part. The leader will start with a short horizontal line extending from the information, change to a 45 degree angle, and terminate with an arrowhead or dot. Leaders should be parallel to each other if possible.

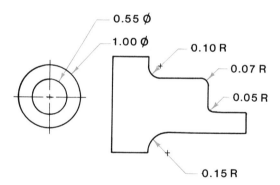

FIG. 13-11 Leader to a Circle: A leader directed to an arc of a circle must point to the center. The arrow should touch the arc. The leader should be at approximately 45 degrees.

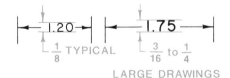

FIG. 13-12 Numbers: Numbers are usually ⅛ inch tall for average drawing sizes. For headings or very large drawing formats the numbers should be 3/16 inch to ¼ inch tall.

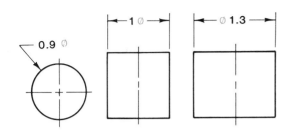

FIG. 13-13   Diameter: The symbol for diameter is ⌀. It may be placed before or after the numerical value. It must always be used on the edge view of a circle, but its use on the face of a circle is optional. Its height is approximately 1/8 inch.

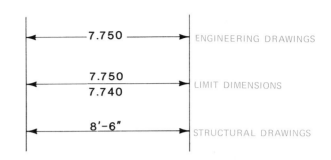

FIG. 13-16   Types of Dimensions: Dimensions can be divided into three general groups: basic dimensions for engineering drawings; basic limit dimensions for engineering drawings; dimensions for structural (architectural) drawings.

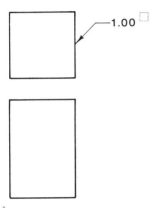

FIG. 13-14   Square: The symbol for square is □.

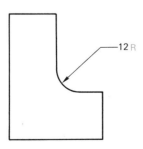

FIG. 13-15   Radius: The symbol for radius is R. It is placed after the numerical value of the radius.

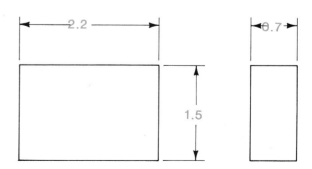

FIG. 13-17   Size Dimensions: Dimensions that give the sizes of a form or feature are called size dimensions.

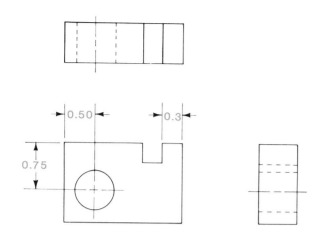

25 679.8I mm

LOWER CASE LETTERS

SPACE

DOT FOR DECIMAL POINT
NO COMMA OR SPACE
(SPACE IS OPTIONAL OVER
5 FIGURES)

FIG. 13-19 Expression of Metric Dimensions: Use
only millimetres for all dimensions, pref-
erably even millimetres. When possible, use
whole millimetres. A value less than one
millimetre will be written with a zero to
the left of the decimal point (0.51).

FIG. 13-18 Location Dimensions: Dimensions that
locate the positions of shapes and forms
are called location dimensions.

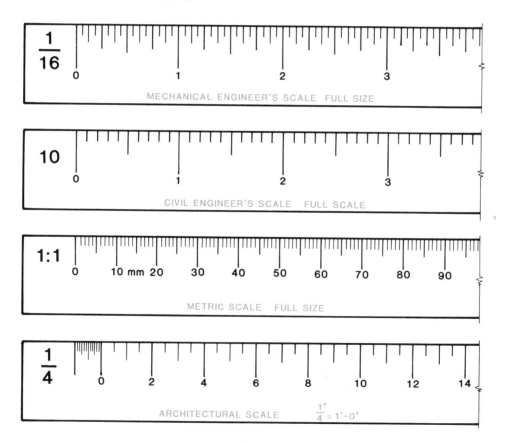

FIG. 13-20 Measuring Scales: Units used are: Inch/
fraction, inch/decimal, millimetres,
foot/inch.

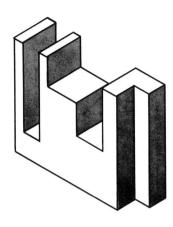

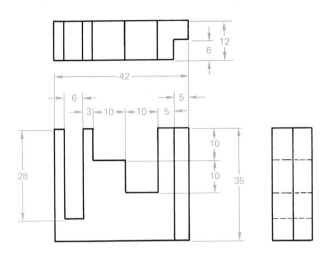

FIG. 13-21 General Rules for Dimension Placement: 1. Place dimensions between views whenever possible. 2. Dimension the most descriptive part or view of the item. 3. Extension lines may cross object lines, but do not break them across there. 4. Dimension inside a recess if it improves communication. 5. Dimension to visible lines only, not to hidden lines. 6. Extension lines may cross each other if necessary, but dimension lines should not cross. 7. Keep dimensions off the objects. 8. Dimension true length lines and surfaces only. 9. Do not use a dimension line for an extension line. 10. Place running dimensions in direct line with each other when possible.

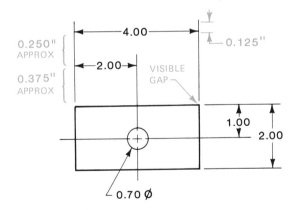

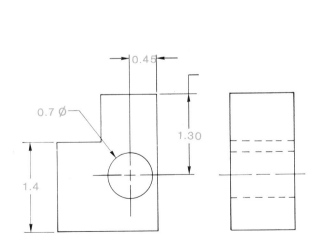

FIG. 13-22 Dimension Location for Holes: Show hole location dimensions and hole sizes on the circular view. Do not dimension to hidden lines.

FIG. 13-23 Dimension Spacing: Proper spacing of dimensions will eliminate crowding. Measure dimension line spacing for the first drawing. A close estimation from then on should be satisfactory. The first dimension is ⅜ inch from the object line, while all others have a ¼ inch spacing. Note the extension line gap at the object line, and a ⅛ inch (approximate) extension past the dimension line.

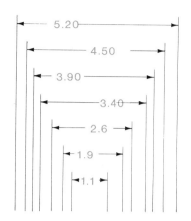

FIG. 13-24   Staggered Dimensions: Dimension values are usually placed in the center of the dimension line. If there are additional values above, they should be staggered so they do not line up. This makes the numbers easier to read.

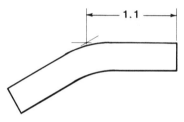

FIG. 13-25   Point Location: To dimension a point on a curve or bend, it is necessary to continue the surface edge with extension lines. The point location is their intersection. The extension lines should pass through the point.

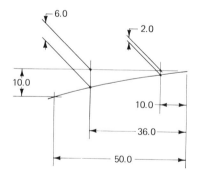

FIG. 13-26   Oblique Extension Lines: Extension lines are usually drawn perpendicular to the dimension lines. In a crowded or limited space, it is permissible to draw extension lines at an oblique angle. A small dot should be used at the point of intersection.

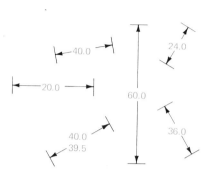

FIG. 13-27   Unidirectional Dimensions: In the unidirectional dimensioning system all dimensions are horizontal regardless of the direction of the dimension line.

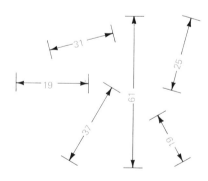

FIG. 13-28   Aligned Dimensions: In the aligned dimensioning system all dimensions are perpendicular to their dimension lines. *Use only one system of dimensioning on a drawing.*

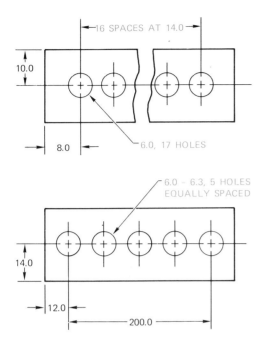

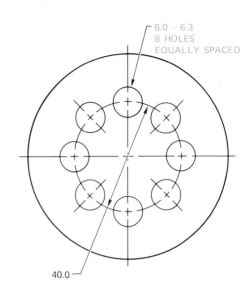

FIG. 13-29 Repetitive Dimensions: Repetitive dimensions for a series of equally spaced features can be given in a note with the regular dimensions for the part. A note may give the number of spaces and their distances, or a note "equally spaced" can be used.

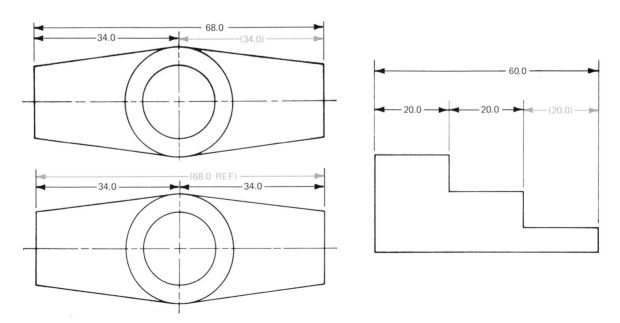

FIG. 13-30 Reference Dimensions: Reference dimensions can be specified by the abbreviation "REF" or by bracketing the dimension. They should be used only for reference or for dimensional checking. Reference dimensions are not used in manufacturing.

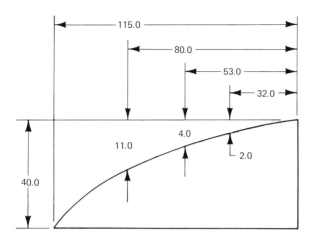

FIG. 13-31 Baseline Dimensioning: If a number of dimensions in a view originate from the same baseline, the dimension can be individually extended to the baseline.

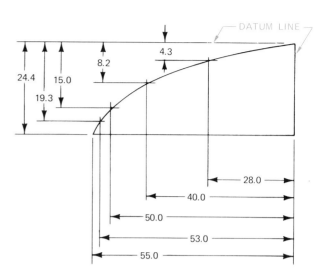

FIG. 13-33 Dimensioning Irregular Curves from Datum: One method used to dimension an irregular curve is to locate a series of points from two datum lines. The number of points will depend on how accurate the irregular curve must be.

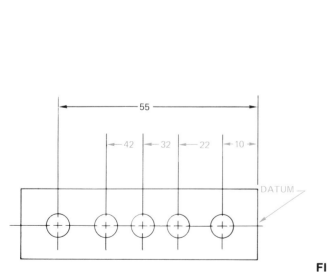

FIG. 13-32 Datum Dimensioning—in Line: Datum are exact starting places used to locate features.

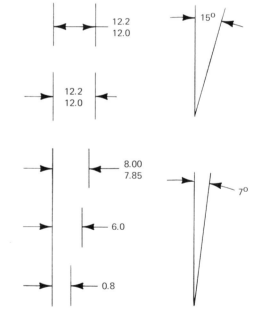

FIG. 13-34 Dimensions in Small Areas: Dimensioning in small areas will necessitate placing the dimensions or arrowheads outside of the extension lines.

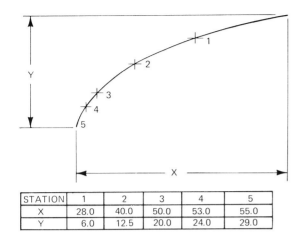

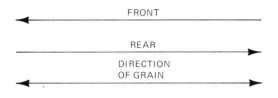

STATION	1	2	3	4	5
X	28.0	40.0	50.0	53.0	55.0
Y	6.0	12.5	20.0	24.0	29.0

FIG. 13-35 Tabulated Curves: A curve can be dimensioned by the coordinate or offset method. If all the horizontal and vertical coordinates are listed in a table, it is called a tabulated curve.

FIG. 13-36 Showing Direction: An arrowhead, leader, and note are used to show direction of motion.

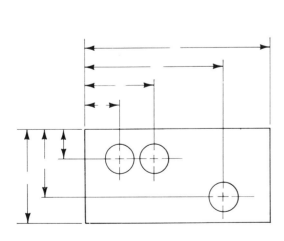

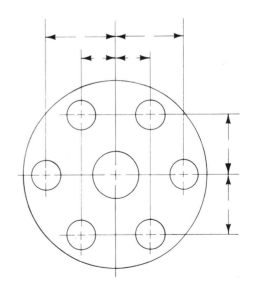

FIG. 13-37 Rectangular Coordinates Dimensioning: Rectangular coordinate dimensioning is the practice of originating all dimensions from two or more perpendicular surfaces. Extension lines, dimension lines, and surfaces from which the dimensions originate form rectangular areas.

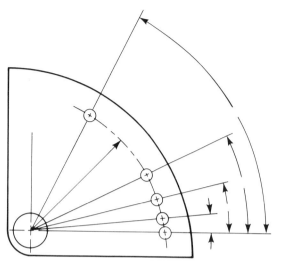

FIG. 13-38   Polar Coordinate Dimensions: Polar coordinate dimensions use a single reference point for all angular dimensions.

FIG. 13-40   Dimensioning Prisms: The prism is a basic form which always has three dimensions: length, width, and height.

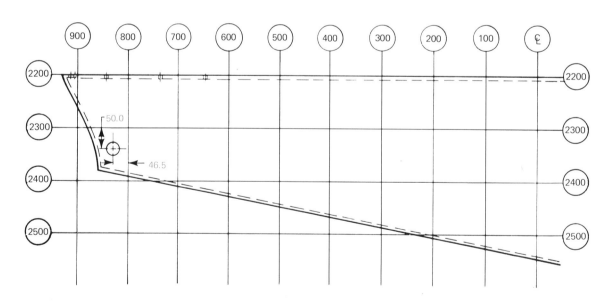

FIG. 13-39   Grid Dimensioning: Large parts can be dimensioned by grid lines. Additional dimensions can be added to locate and detail the part. All added dimensions should be placed so the dimension value can be added to the nearest grid line value.

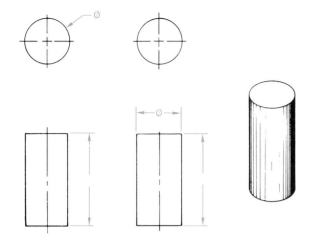

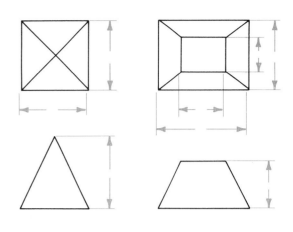

FIG. 13-41   Dimensioning Cylinders: Cylinders and cylindrical holes may be dimensioned on their circular or rectangular views. Two dimensions will accurately measure a cylinder.

FIG. 13-43   Dimensioning Pyramids: Pyramids must show the three basic dimensions of length, width, and height. The truncated pyramid must have additional dimensions for the truncated area.

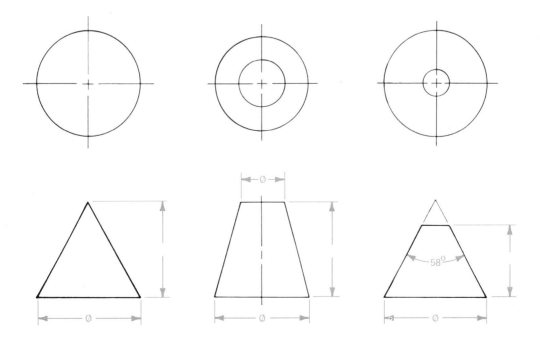

FIG. 13-42   Dimensioning Cones: A cone can be measured with two dimensions. A truncated cone will require three dimensions.

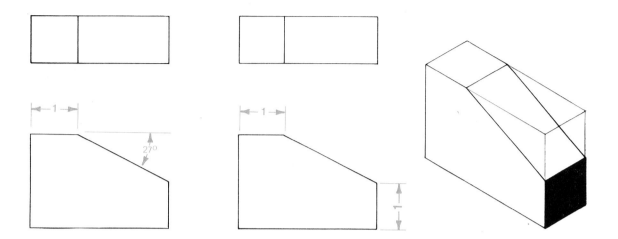

FIG. 13-44 Dimensioning an Inclined Surface: An inclined surface may be dimensioned by locating one corner of the inclined surface and its angle. Another method is to locate both corners of the inclined surface.

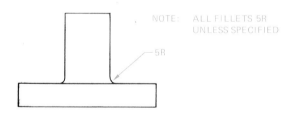

FIG. 13-45 Dimensioning a Fillet: Fillets can be dimensioned on the drawing or specified with a note.

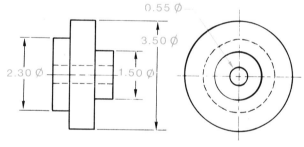

FIG. 13-47 Diameters: Dimensions indicating the diameter of a circle are shown with the symbol $\oslash$. The symbol may be before or after the value. If the dimension is on the circular view the use of the symbol is optional.

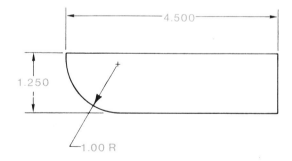

FIG. 13-46 Dimensioning Rounded Corners: When corners are rounded, dimensions should locate the edges so the rounded corner can be drawn tangent.

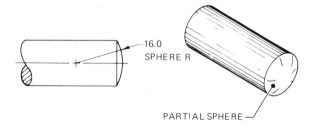

FIG. 13-48 Spherical Radius: Dimensioning a sphere or partial sphere can be done by giving the radius, followed by "SPHERE R"

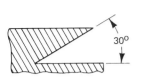
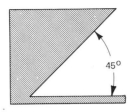

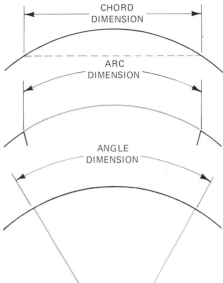

FIG. 13-49 Location of Angular Dimensions: To dimension a small angle, extend the extension lines until there is enough space for the dimension. A large angle can be dimensioned inside the angle. Use the vertex of the angle for the compass point and swing the dimension line, leaving a space for the number of degrees or a note. Do not let the dimension line touch a corner of the angle.

FIG. 13-51 Dimensioning Arcs, Chords, and Angles: A segment of a circle can be dimensioned as an arc or a chord, or with an angular dimension.

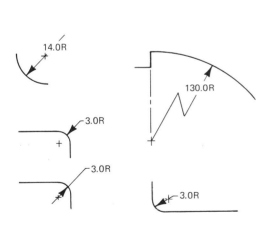
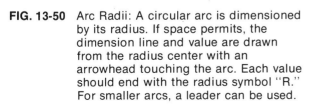

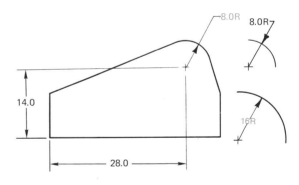

FIG. 13-50 Arc Radii: A circular arc is dimensioned by its radius. If space permits, the dimension line and value are drawn from the radius center with an arrowhead touching the arc. Each value should end with the radius symbol "R." For smaller arcs, a leader can be used.

FIG. 13-52 Dimensioning Radii from Centers: When space permits, the dimension for a radius should have a dimension line starting from the marked center, gapped in the middle for the value, and end with an arrowhead touching the arc. If the space is too small, the leader should be continued and the radius value placed in a convenient location. With this method the arrowhead may be placed on the other side of the arc.

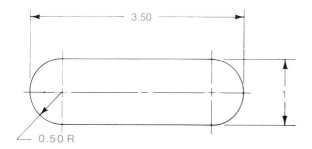

FIG. 13-53 Dimensioning Round Ends: Parts that have full rounded ends should be dimensioned with an overall dimension, a radius dimension, and the overall height dimension.

FIG. 13-54 Dimensioning Partially Rounded Ends: Dimensioning an object with partially rounded ends requires the overall dimensions and the radius (R).

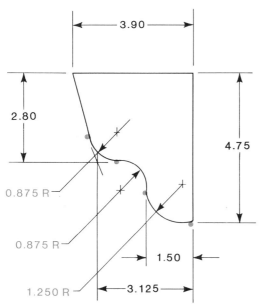

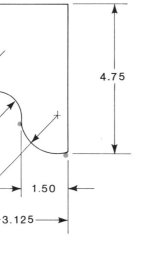

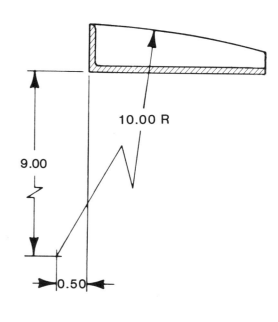

FIG. 13-55 Dimensioning Radii From Unlocated Centers: A radius may be located by the intersecting points of the arcs' tangents rather than their centers. This method is used for radii with centers outside or inside the drawing area.

FIG. 13-56 Foreshortened Radii: When the center of a radius is off the drawing paper, or it interferes with another view, the radius dimension line can be broken and shortened to an open area. The arrowhead part of the dimension line should be pointing toward the true center.

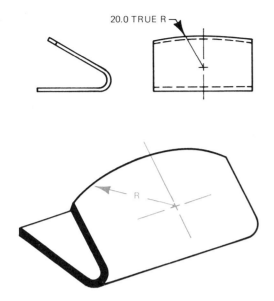

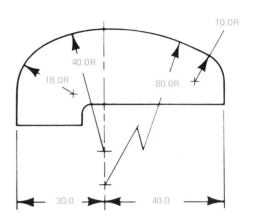

FIG. 13-57 True Radius: To avoid drawing an auxiliary view for a radius that does not show the true shape of the radius, the elliptical view of the radius may be dimensioned by "TRUE R."

FIG. 13-58 Dimensioning Tangent Curves: A curve consisting of a series of circular tangent arcs should be dimensioned with the radius and center location for each arc.

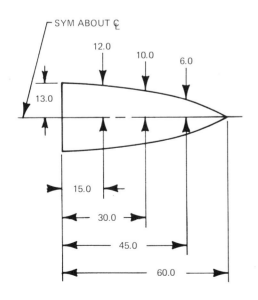

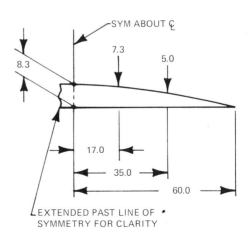

FIG. 13-59 Dimensioning Symmetrical Outlines: Symmetrical outlines are usually dimensioned on only one side of the axis of symmetry.

FIG. 13-60 Dimensioning Broken Symmetrical Outlines: A symmetrical part can be dimensioned on only one side of its symmetry axis. The call-out for the axis is shown by the note "SYM ABOUT ℄." (℄ is the symbol for center line.)

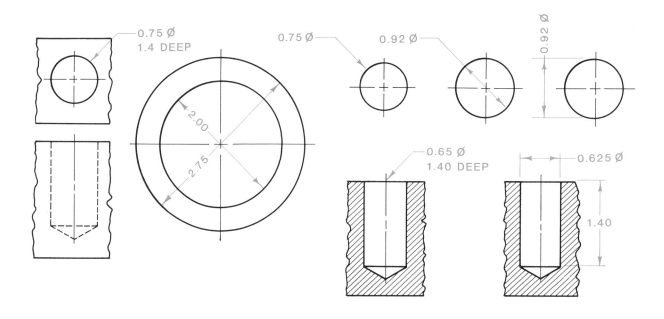

FIG. 13-61 Dimensioning Round Holes: Regular holes can be dimensioned in the various ways shown.

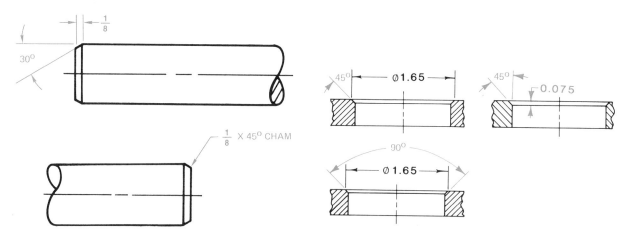

FIG. 13-62 Dimensioning Chamfers: A chamfer requires either an angle and depth dimension, or a note with the same information.

FIG. 13-63 Dimensioning Internal Chamfers. An inside chamfer is dimensioned by using the outside diameter of the chamfer and its angle. Another method is to dimension the depth and angle of the chamfer the same as outside chamfers.

FIG. 13-64 Dimensioning Countersunk Holes: A countersunk hole is dimensioned by a note giving the diameter and angle of the countersink. To dimension through holes, an abbreviation "THRU" should follow the dimension. Note that the depth of a blind hole is dimensioned to the shoulder, not to the point of the drilled hole.

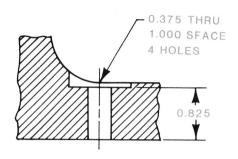

FIG. 13-67 Dimensioning Spot-Faced Holes: Dimensions required for a spot-faced hole are the diameter and its depth. The remaining material can be dimensioned instead of the depth.

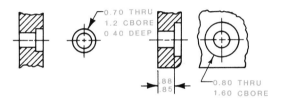

FIG. 13-65 Dimensioning Counterbored Holes: A counterbored hole is dimensioned by a note giving the diameter and depth. If the remaining thickness of material is more important than the depth of the hole, it should be dimensioned rather than the depth of the counterbore.

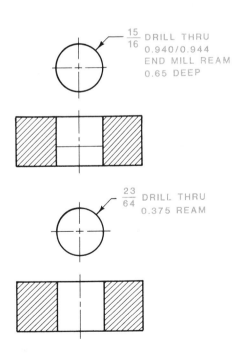

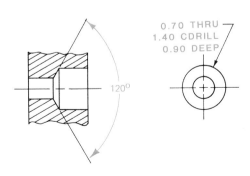

FIG. 13-66 Dimensioning Counterdrilled Holes: A counterdrilled hole is dimensioned by indicating the diameter, depth, and angle.

FIG. 13-68 Dimensioning a Drilled and Reamed Hole: Drilling and reaming operations are dimensioned with a note and leader.

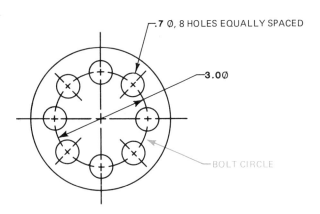

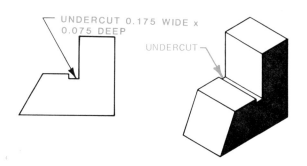

FIG. 13-69 Dimensioning a Bolt Circle: A bolt circle is a circular center line on which holes are located. The diameter of the bolt circle is given by a dimension. The holes are specified by a note.

FIG. 13-71 Dimensioning an Undercut: An undercut is dimensioned with a leader and a note giving the width and depth. An undercut is a recess at the intersection of two perpendicular planes that ensures a flush fit against the shoulder of a mating part.

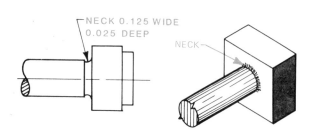

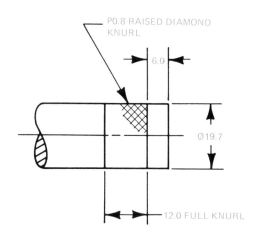

FIG. 13-70 Dimensioning a Neck: A neck is dimensioned with a leader and a note giving the width and depth. A neck is a recess in a cylindrical part that ensures a flush fit against the shoulder of the adjacent part. A neck can be round or square.

FIG. 13-72 Dimensioning Knurls: Knurls are dimensioned by giving the dimetral pitch (DP), type, grade (coarse, medium, or fine), length, and diameter of the shaft. In the drawing the knurled surface can be fully or partially drawn or completely omitted.

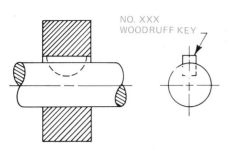
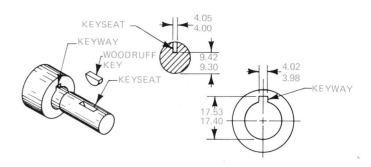

FIG. 13-73 Dimensioning a Keyseat, Keyway, and Key: The only dimension required for a key is the number of the particular key taken from engineering tables that have all the size dimensions necessary for machining. Keyseats and keyways are dimensioned by width, depth, and location. The depth can be dimensioned from the bottom or the opposite side of the shaft or hole.

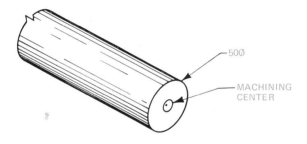

FIG. 13-74 Dimensioning Machining Centers: Machining centers may be needed for turning materials in a lathe. When the machining centers are to remain on the finished part, they should be so indicated or dimensioned on the drawing.

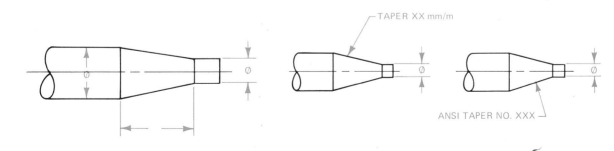

FIG. 13-75 Dimensioning a Taper: Tapers are dimensioned as a cone or a ratio, or noted from a prepared engineering table of tapers.

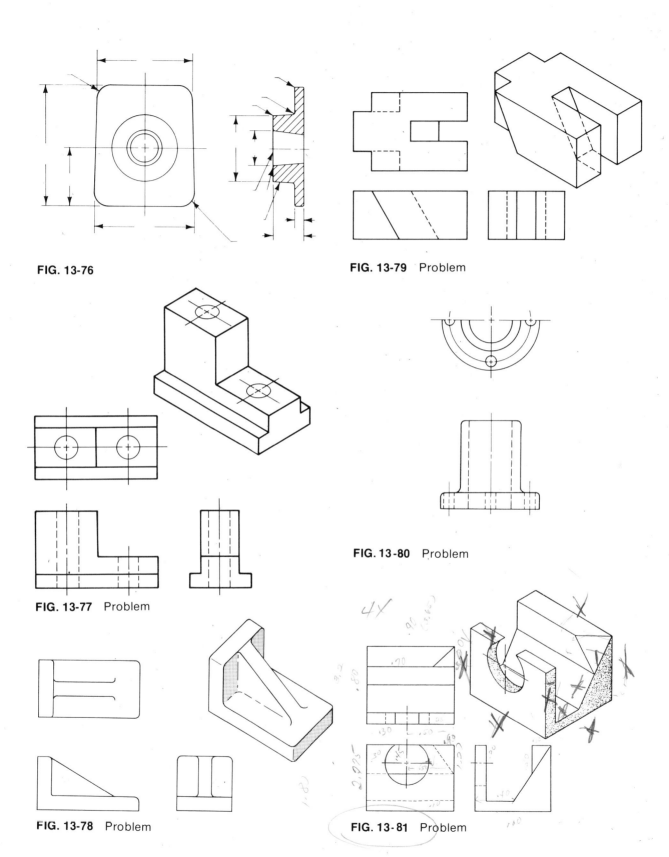

FIG. 13-76

FIG. 13-79   Problem

FIG. 13-77   Problem

FIG. 13-80   Problem

FIG. 13-78   Problem

FIG. 13-81   Problem

Tolerancing

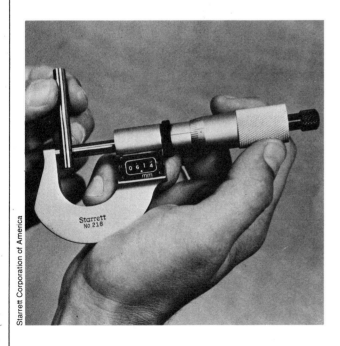

Starrett Corporation of America

INTRODUCTION

If an item is to be manufactured only once, a single dimension can be given to any two parts that must fit together. An added note would be required to tell how close the parts needed to fit (Fig. 14-2). The fabricator could simply machine each separate part until the required size and fit are reached.

With mass production, however, all parts must be interchangeable. Interchangeability of parts is the key to the manufacture of inexpensive industrial products. Interchangeability is accomplished by allowing limited variation of dimensions during manufacturing. It is a practical way of achieving the precision needed for mass production. Effective size control reduces costs while controlling the size and position of the parts, and ensuring that the parts will fit together properly.

The critical question in assembling parts is how closely they should fit together. The parts of a bicycle do not have to fit as closely as the parts of a turbine engine, and variations in the degree of accuracy will affect production costs. Designers must determine the permissible size variations while keeping in mind the function of the mating parts and the production costs.

As you study this chapter, you should understand the basic concepts and terminology of tolerancing, be able to read a blueprint with tolerance dimensions, and be able to make a drawing using tolerance dimensions.

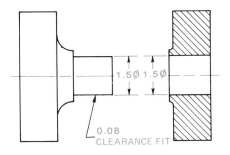

FIG. 14-2 The fit of matching parts

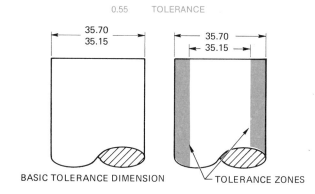

FIG. 14-3 Dimensions with tolerances

TOLERANCING

A tolerance is a precise statement of the variations that can be permitted within a dimension (Fig. 14-3).

BASIC, DATUM, and REFERENCE dimensions do not have tolerances. However, every other dimension on a technical drawing should have a tolerance unless it is not related to the actual size and shape of the part. Tolerances can be expressed on a drawing (see Figure 14-4), or they can be called out by a note such as: NOTE: ALL TOLERANCES ± 0.003 INCH UNLESS SPECIFIED. For the majority of parts, tolerances will run between 0.001 inch and 0.100 inch. Nonmating surfaces can usually be left rough, or can be machined to large tolerances because these surfaces have little or no effect on the finished working part. Tolerances on a drawing of an external part will have the larger tolerance as the top figure. Tolerances for an internal part will be drawn with the small tolerance on top. Therefore, the rule is that the top figure in the tolerance expression is the first size to be reached when the part is machined. The tolerance expression should be in the same form and with the same number of decimal places as the dimension.

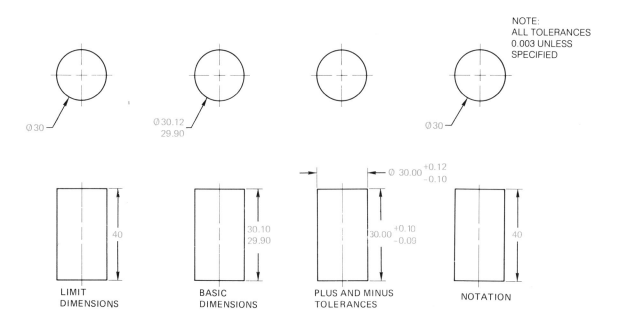

FIG. 14-4 Different methods of tolerancing

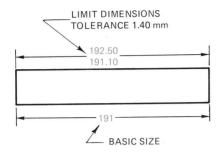

FIG. 14-5 Limit dimensions

There are several quick ways to indicate tolerancing. One of them is to add a note similar to the following to the drawing.

Tolerances for Finish Dimensions Not Otherwise Specified

From 0″ to 3″   ± 0.005
Over 3″ to 10″ ± 0.010
Over 10″ to 20″ ± 0.020
Over 20″   ± 0.040

The sizes and tolerances will vary from one company to another and by the type of industrial fabrication used.

Limit Dimensions

A limit dimension indicates the two limits allowed for one dimension. It is the minimum and maximum allowed for one part (Fig. 14-5).

The nominal size is an approximate size used to identify the general size of a part. It is given to the closest millimetre.

The basic size is the exact theoretical size from which the limit dimensions are derived. The basic size is used when working with tolerancing tables to obtain required fits.

The actual size is the measured finished size of the object.

Allowance is the difference between sizes of the mating parts (Fig. 14-6). It is the tightest fit of two mating parts, or the smallest hole size and the largest shaft size.

FIT

Fit is the term used to specify the range of tightness that will occur with the application of tolerances and allowances in mating parts. There are a large number of recommended fits. The selection of fits is made by the designer from tables such as the one

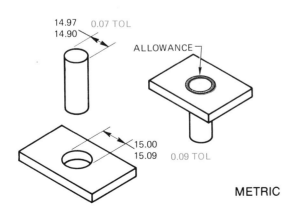

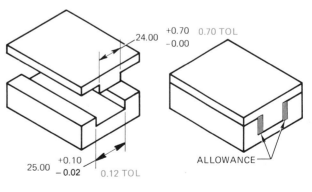

METRIC

FIG. 14-6 Allowances between parts

STANDARD TOLERANCES

Nominal Size Range in Inches	Slip Gauges, Production and Wear Tolerances of Gauges and Measuring Instruments		Parts Subject to Very Tight Tolerances, Precision Bearings, Precision Assemblies		Precision Engineered Designs (General)		Engineering Work in General, Giving Scope for Wider Tolerances		Rough Work, Steel Structures, Castings, Agricultural Machinery	
Over - To	Grade 4	Grade 5	Grade 6	Grade 7	Grade 8	Grade 9	Grade 10	Grade 11	Grade 12	Grade 13
0.04 - 0.12	0.00015	0.00020	0.00025	0.0004	0.0006	0.0010	0.0016	0.0025	0.004	0.006
0.12 - 0.24	0.00015	0.00020	0.0003	0.0005	0.0007	0.0012	0.0018	0.0030	0.005	0.007
0.24 - 0.40	0.00015	0.00025	0.0004	0.0006	0.0009	0.0014	0.0022	0.0035	0.006	0.009
0.40 - 0.71	0.0002	0.0003	0.0004	0.0007	0.0010	0.0016	0.0028	0.0040	0.007	0.010
0.71 - 1.19	0.00025	0.0004	0.0005	0.0008	0.0012	0.0020	0.0035	0.0050	0.008	0.012
1.19 - 1.97	0.0003	0.0004	0.0006	0.0010	0.0016	0.0025	0.0040	0.006	0.010	0.016
1.97 - 3.15	0.0003	0.0005	0.0007	0.0012	0.0018	0.0030	0.0045	0.007	0.012	0.018
3.15 - 4.73	0.0004	0.0006	0.0009	0.0014	0.0022	0.0035	0.005	0.009	0.014	0.022
4.73 - 7.09	0.0005	0.0007	0.0010	0.0016	0.0025	0.0040	0.006	0.010	0.016	0.025
7.09 - 9.85	0.0006	0.0008	0.0012	0.0018	0.0028	0.0045	0.007	0.012	0.018	0.028
9.85 - 12.41	0.0006	0.0009	0.0012	0.0020	0.0030	0.0050	0.008	0.012	0.020	0.030
12.41 - 15.75	0.0007	0.0010	0.0014	0.0022	0.0035	0.006	0.009	0.014	0.022	0.035
15.75 - 19.69	0.0008	0.0010	0.0016	0.0025	0.004	0.006	0.010	0.016	0.025	0.040
19.69 - 30.09	0.0009	0.0012	0.0020	0.003	0.005	0.008	0.012	0.020	0.030	0.050
30.09 - 41.49	0.0010	0.0016	0.0025	0.004	0.006	0.010	0.016	0.025	0.040	0.060
41.49 - 56.19	0.0012	0.0020	0.003	0.005	0.008	0.012	0.020	0.030	0.050	0.080
56.19 - 76.39	0.0016	0.0025	0.004	0.006	0.010	0.016	0.025	0.040	0.060	0.100
76.39 - 100.9	0.0020	0.003	0.005	0.008	0.012	0.020	0.030	0.050	0.080	0.125
100.9 - 131.9	0.0025	0.004	0.006	0.010	0.016	0.025	0.040	0.060	0.100	0.160
131.9 - 171.9	0.003	0.005	0.008	0.012	0.020	0.030	0.050	0.080	0.125	0.200
171.9 - 200	0.004	0.006	0.010	0.016	0.025	0.040	0.060	0.100	0.160	0.250

FIG. 14-7   Recommended system of limits and fits

The standard tolerances shown in Fig. 14-7 are typical manufacturing tolerances in industry. This series of tolerance grades provides a suitable range from which appropriate tolerances for holes and shafts can be selected. These grades will permit the use of standard checking gauges.

The designer must also consider the accuracy that can be achieved with the available industrial forming equipment as shown in Figure 14-8. This chart can be used as a guide to determine the matching process that will produce the items within the prescribed tolerance grade.

Fit has three types of mating connections, which are classified as clearance, interference, and transition.

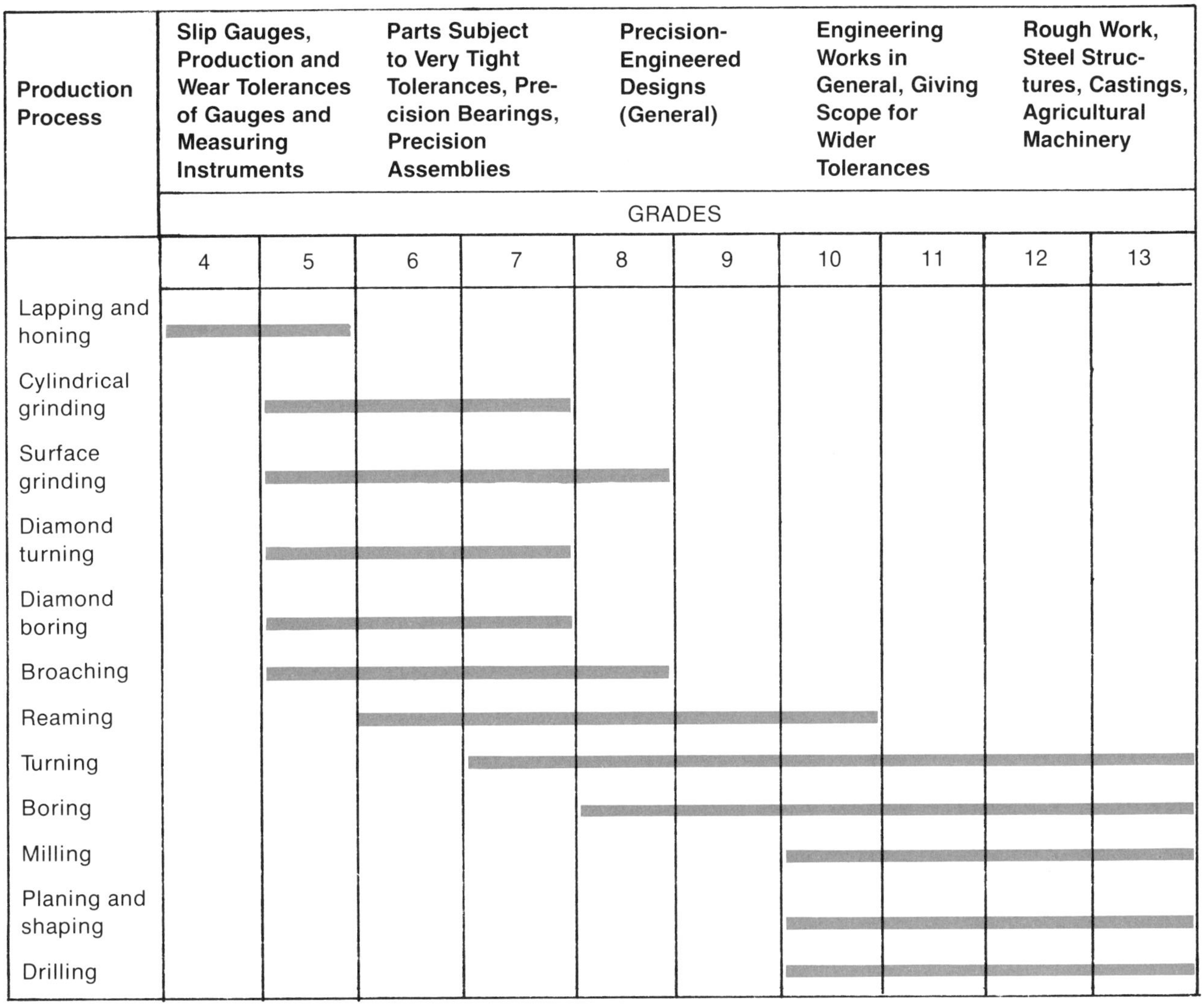

Production Process	Slip Gauges, Production and Wear Tolerances of Gauges and Measuring Instruments		Parts Subject to Very Tight Tolerances, Precision Bearings, Precision Assemblies		Precision-Engineered Designs (General)		Engineering Works in General, Giving Scope for Wider Tolerances		Rough Work, Steel Structures, Castings, Agricultural Machinery	
	GRADES									
	4	5	6	7	8	9	10	11	12	13
Lapping and honing	▬	▬								
Cylindrical grinding		▬	▬	▬						
Surface grinding		▬	▬	▬	▬					
Diamond turning		▬	▬	▬						
Diamond boring		▬	▬	▬						
Broaching		▬	▬	▬	▬					
Reaming			▬	▬	▬	▬	▬			
Turning				▬	▬	▬	▬	▬	▬	▬
Boring					▬	▬	▬	▬	▬	▬
Milling							▬	▬	▬	▬
Planing and shaping							▬	▬	▬	▬
Drilling							▬	▬	▬	▬

FIG. 14-8   Possible accuracy from industrial forming equipment

Clearance Fit

A clearance fit is one in which a positive allowance, or air space, is left between mating parts. It is defined as the maximum difference between two limit dimensions (Fig. 14-9).

Interference Fit

Interference fit creates a negative allowance between mating parts. It is defined as the amount of overlap. Two examples of interference fits are shown in Figure 14-10.

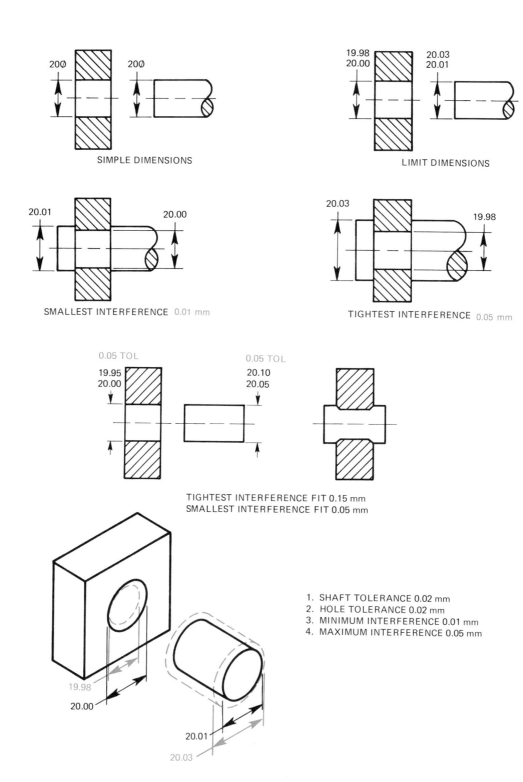

SIMPLE DIMENSIONS

LIMIT DIMENSIONS

SMALLEST INTERFERENCE 0.01 mm

TIGHTEST INTERFERENCE 0.05 mm

TIGHTEST INTERFERENCE FIT 0.15 mm
SMALLEST INTERFERENCE FIT 0.05 mm

1. SHAFT TOLERANCE 0.02 mm
2. HOLE TOLERANCE 0.02 mm
3. MINIMUM INTERFERENCE 0.01 mm
4. MAXIMUM INTERFERENCE 0.05 mm

FIG. 14-9 Clearance fit

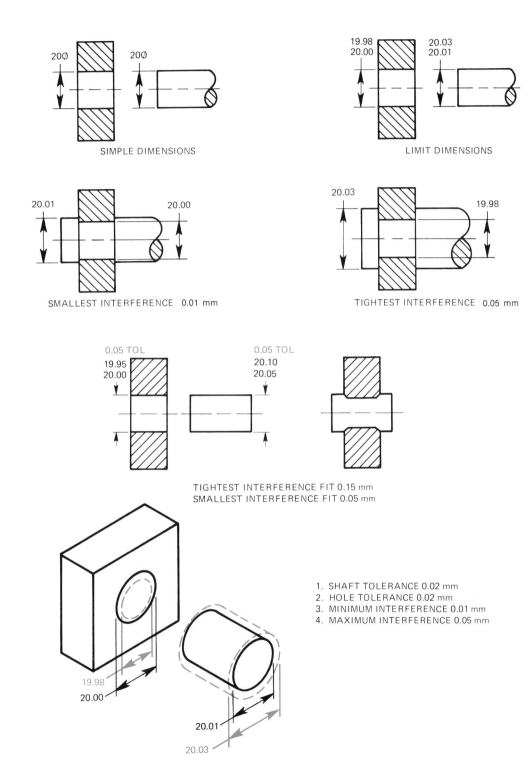

SIMPLE DIMENSIONS

LIMIT DIMENSIONS

SMALLEST INTERFERENCE 0.01 mm

TIGHTEST INTERFERENCE 0.05 mm

TIGHTEST INTERFERENCE FIT 0.15 mm
SMALLEST INTERFERENCE FIT 0.05 mm

1. SHAFT TOLERANCE 0.02 mm
2. HOLE TOLERANCE 0.02 mm
3. MINIMUM INTERFERENCE 0.01 mm
4. MAXIMUM INTERFERENCE 0.05 mm

FIG. 14-10   Interference fit

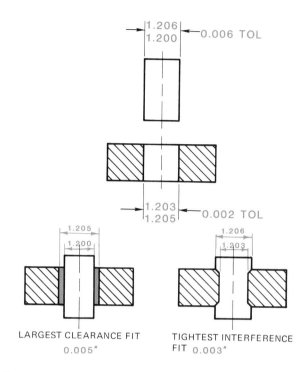

FIG. 14-11 Transition fit is a combination of clearance and interference fits

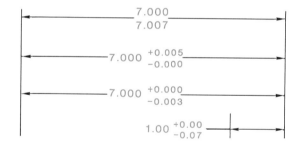

FIG. 14-12 Unilateral tolerances

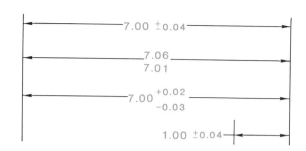

FIG. 14-13 Bilateral tolerances

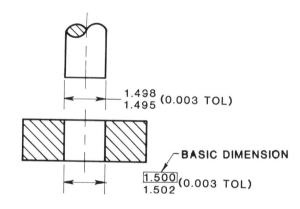

FIG. 14-14 Basic hole system for tolerancing

Transitional Fit

Transitional fit has limits that will allow either a clearance fit or an interference fit for mating parts (Fig. 14-11).

Unilateral Tolerance

Unilateral tolerances allow dimensional variations in only one direction. The tolerance will be all plus (+) or it will be all minus (−) as shown in Figure 14-12. The direction of no variation is noted with zeros.

Bilateral Tolerance

Bilateral tolerances allow dimensional variations in both directions. These tolerances can be equal or unequal in plus (+) and minus (−) directions (Fig. 14-13).

BASIC SIZE SYSTEMS

Basic Hole System

The basic hole system of designing mating cylindrical parts uses the basic dimension of the shaft hole for its first limit dimension (Fig. 14-14). (Note the placement of the crowded limit dimensions in

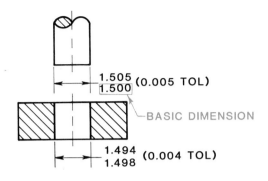

FIG. 14-15 Basic shaft system for tolerancing

gauge. These advantages make the basic hole system of design more efficient than the basic shaft system.

Basic Shaft System

The basic shaft system of design uses the basic dimension of the shaft itself as the first limit dimension (Fig. 14-15). This system is efficient when several different items such as bearings or collars with different classes of fit are to be used on the shaft.

Figure 14-4). One advantage of using the basic hole system is that hole cutters such as drills, reamers, and broaches are manufactured to standard sizes. Also, external mating parts can be machined more easily than internal openings. Another advantage is the ease of checking hole size with a standard hole

Accumulative Tolerances

Since each dimension has a tolerance, it is important not to allow these tolerances to build up and accumulate a large total tolerance (Fig. 14-16). The best method to control a string of dimensions is with a datum as shown in Figure 14-17.

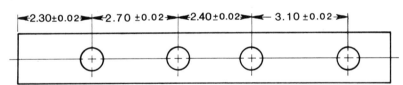

FIG. 14-16 A large tolerance accumulation is not permitted in designing

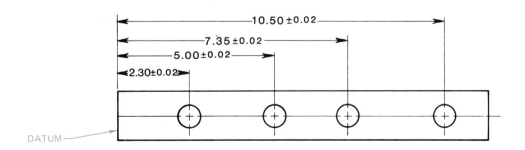

FIG. 14-17 This method of dimensioning does not have a tolerance accumulation

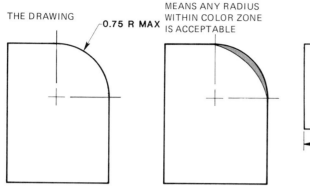

FIG. 14-18   Maximum material limit

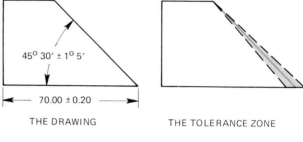

FIG. 14-20   Angle tolerances

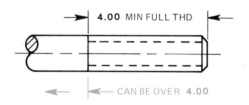

FIG. 14-19   Minimum material limit

Single Limits

It is not always necessary to state both limits. Minimum (MIN) or maximum (MAX) limits can be used alone when the opposite single limit dimension is not critical to the design. Single limit dimensions are typically used with hole depths, thread lengths, chamfers, and corner radii.

Maximum Material Limit

The maximum material limit is a statement of the maximum size a particular form can be. If a radius is specified as "0.75 inch R MAX," the radius can be no larger than 0.75 inch, but it can be any smaller size (Fig. 14-18).

Minimum Material Limit

The minimum material limit is a statement of the minimum size a part can be. If the length of a threaded rod is "4.00 inch MIN," the threads cannot be less than 4.00 inches long; however, the threads can be more than 4.00 inches long (Fig. 14-19).

Angle Tolerances

Tolerances for angles are usually noted on the drawing as follows: NOTE: ALL ANGLES ± 0°−30' UNLESS SPECIFIED. The tolerance can also be shown with the angle (see Figure 14-20).

Selective Assembly

Selective assembly is the process of selecting and measuring mating parts by inspecting parts until the required close tolerance is found. While the process of selective assembly is slow, it may be less expensive than machining all the parts to extremely close tolerances.

PROBLEMS

1. Answer the questions in Figure 14-21.
2. Answer the questions in Figure 14-22.
3. What is the tolerance for a Grade 5 limit fit in Figure 14-23?
4. What is the tolerance for a Grade 9 limit fit in Figure 14-23?

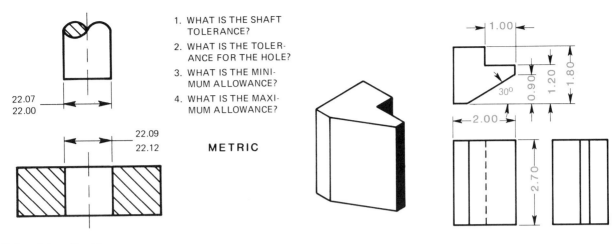

1. WHAT IS THE SHAFT TOLERANCE?
2. WHAT IS THE TOLERANCE FOR THE HOLE?
3. WHAT IS THE MINIMUM ALLOWANCE?
4. WHAT IS THE MAXIMUM ALLOWANCE?

METRIC

FIG. 14-21 Problem

FIG. 14-24 Problem

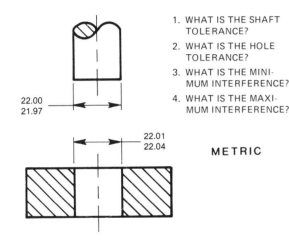

1. WHAT IS THE SHAFT TOLERANCE?
2. WHAT IS THE HOLE TOLERANCE?
3. WHAT IS THE MINIMUM INTERFERENCE?
4. WHAT IS THE MAXIMUM INTERFERENCE?

METRIC

FIG. 14-22 Problem

5. What is the tolerance for a Grade 11 limit fit in Figure 14-23?

6. Show a bilateral tolerance for a Grade 3 fit for Figure 14-23.

7. Show a unilateral tolerance for a Grade 3 fit for Figure 14-23.

8. Show a basic hole system tolerance for a Grade 4 fit in Figure 14-23.

9. Show a basic shaft system tolerance for a Grade 7 fit in Figure 14-23.

10. Using a Grade 8 limit from Figure 14-7, add the tolerances to Figure 14-24. Use limit tolerances.

11. Using a Grade 4 limit from Figure 14-7, add the tolerances to Figure 14-25. Use plus and minus tolerances.

12. Using a Grade 12 limit from Figure 14-7, add the tolerances to Figures 14-26 through 14-28.

13. Using the table in Figure 14-8, list the types of machines that can be used to machine the required tolerances in problems 10, 11, and 12.

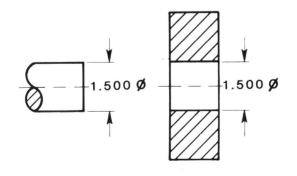

FIG. 14-23 Problem

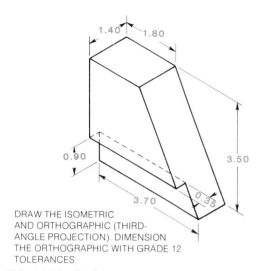

1.40   1.80

0.90

3.50

3.70   0.35

DRAW THE ISOMETRIC
AND ORTHOGRAPHIC (THIRD-
ANGLE PROJECTION). DIMENSION
THE ORTHOGRAPHIC WITH GRADE 12
TOLERANCES

FIG. 14-25   Problem

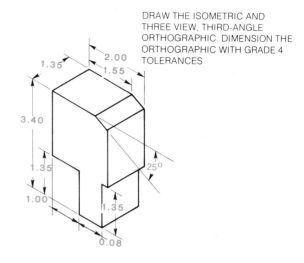

DRAW THE ISOMETRIC AND
THREE VIEW, THIRD-ANGLE
ORTHOGRAPHIC. DIMENSION THE
ORTHOGRAPHIC WITH GRADE 4
TOLERANCES

1.35   2.00   1.55

3.40

1.35   25°

1.00   1.35

0.08

FIG. 14-26   Problem

15
Chapter

Positional Tolerancing

INTRODUCTION

Positional tolerancing and geometric form tolerancing are extensions of dimensioning.

Dimensioning will control the size and shape of an object (Fig. 15-2). However, positional tolerancing and geometric form tolerancing are more accurate ways of describing how a designer or draftsperson wants a part to be produced.

Geometric form tolerancing controls the form or shape of an object (Fig. 15-3). Positional tolerancing controls the position or location of the various holes and protrusions (Fig. 15-4).

As you study this chapter, you should gain an understanding of the symbolism of positional tolerancing. You should also learn to apply elementary positional symbols to a drawing.

DATUM

A datum is a reference location for the form tolerance. It is considered to be a theoretically perfect plane, line, or point. The selection of a datum or datums for a part is very important. The datum should be an accurately machined surface. It should be easy to locate and measure on this surface.

The datum is called out on a drawing with a datum identifying symbol. Any letter of the alphabet except I, O, and Q may be used as a datum refer-

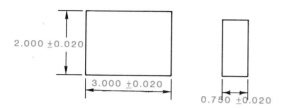

FIG. 15-2   Dimensions are used to control the size of an object

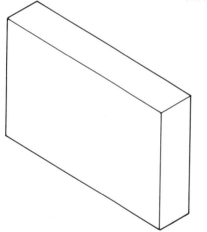

- HOW PARALLEL MUST EDGES BE?
- HOW PARALLEL MUST SURFACES BE?
- HOW SQUARE MUST CORNERS BE?
- HOW FLAT SHOULD A SURFACE BE?
- HOW STRAIGHT SHOULD A SURFACE BE?

FIG. 15-3   Factors affecting the form of a rectangular block

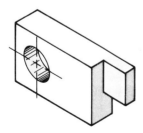

FIG. 15-4   Positional tolerancing locates holes and protrusions

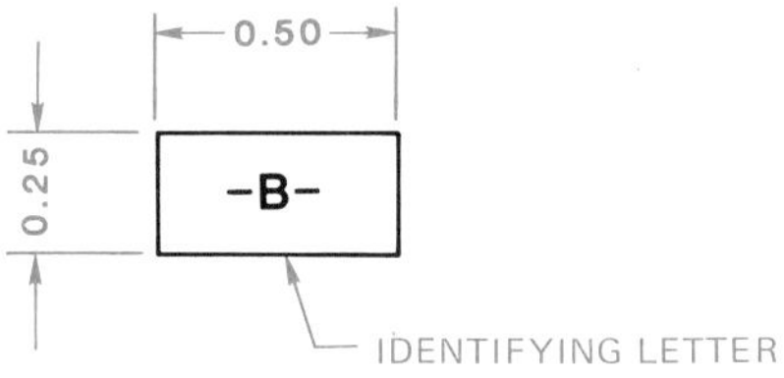

FIG. 15-5   Size of datum identifying symbol

ence letter. The size of the box enclosing the datum letter is shown in Figure 15-5. Figure 15-6 shows the typical placement of the datum on a drawing.

GEOMETRIC CHARACTERISTIC SYMBOLS

After a datum is located and noted, geometric characteristic symbols may be placed (Fig. 15-7). Figure 15-7 shows that the size of the block is to be accurate within ± 0.020. The geometric characteristic symbol indicates that the top surface is to be parallel to the bottom surface within a total of 0.007. This is an illustration of how form tolerancing is used to control the shape or form of an object.

Figure 15-8 shows the use of multiple datum callouts.

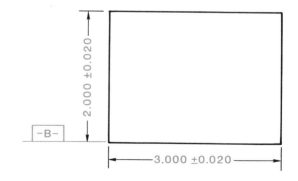

FIG. 15-6   Placement of the datum

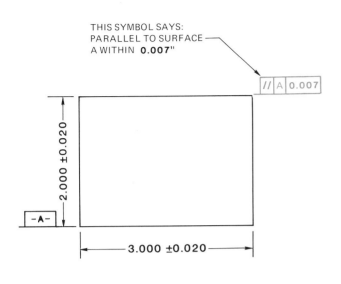

THIS SYMBOL SAYS:
PARALLEL TO SURFACE
A WITHIN **0.007**"

// | A | 0.007

2.000 ±0.020

-A-

3.000 ±0.020

FIG. 15-7 Use of geometric characteristic symbols

GEOMETRIC CHARACTERISTIC SYMBOLS		
GROUP	CHARACTERISTIC	SYMBOL
A	FLAT	▱
	STRAIGHT	—
	ROUND	○
	CYLINDRICAL	⌀
	PROFILE OF A LINE	⌒
	PROFILE OF A SURFACE	◠
B	PERPENDICULAR	⊥
	PARALLEL	//
	ANGULAR	∠
C	TRUE POSITION	⌖
	SYMMETRICAL	≡
	CONCENTRIC	◎
D	RUNOUT	↗

FIG. 15-9 Geometric characteristic symbols

Figure 15-9 shows the geometric characteristic symbols. They are classified into four groups.

Group A   These symbols need not refer to a datum. For example, a surface may be specified as being flat, but it is not flat compared to another surface.

Group B   These symbols require a datum reference. For example, to call out a form of parallelism, the surface must be parallel to some other surface (the datum).

Group C   These symbols indicate locational tolerances. They locate features. True position locates or shows the positional tolerance of a hole or slot.

Group D   Runout tolerance controls a circular tolerance zone for cylindrical shapes.

Interpretation of Symbols

The following illustrations are examples of some of the commonly used symbols:

Flatness   Every point on the surface must lie between two parallel planes a specified distance apart. Note that no specification of datum is necessary (Fig. 15-10).

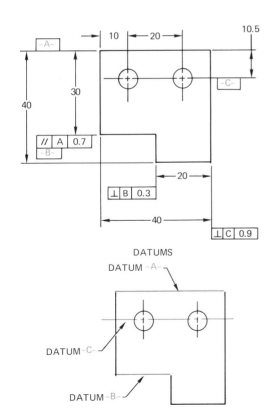

FIG. 15-8 The use of multiple datum callouts

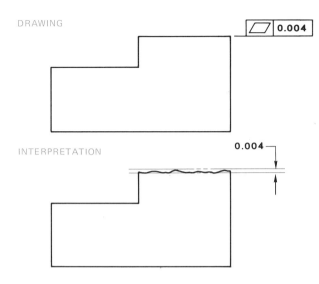

FIG. 15-10   Flatness of a surface

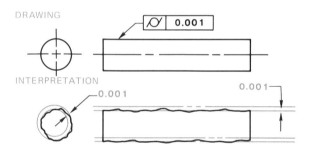

FIG. 15-11   Roundness of a cylindrical surface

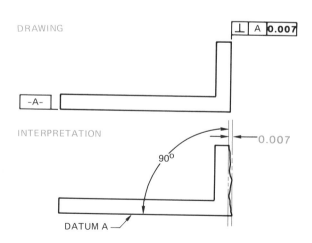

FIG. 15-12   Surface perpendicular to a datum

Roundness   The cylindrical surface must be within the specified tolerance of size and must lie between two concentric cylinders. This is illustrated by the phantom lines. Note that no datum call out is necessary as roundness is in group A (Fig. 15-11).

Perpendicular   The specified surface must lie between two parallel planes (0.007 apart in the example) that are perpendicular to the datum plane (Fig. 15-12).

Parallel   The specified surface must lie between two planes that are parallel to the datum plane (Fig. 15-13).

Angular   The surface must lie between two parallel planes that are inclined at the specified angle (30 degrees in the example) to the datum plane (Fig. 15-14).

True Position   The center line of the hole called out by the ⊕ symbol may lie anywhere in a circular tolerance zone specified (0.030 in the example). This is a circular locational tolerance zone (Fig. 15-15).

INDUSTRIAL APPLICATION

Figure 15-16 is an example of geometric form and true positional dimensioning. This example is followed by a detailed explanation of the drawing (Fig. 15-17 through 15-21).

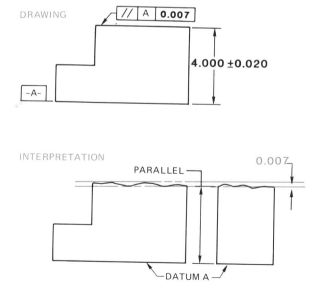

FIG. 15-13   Surface parallel to a datum

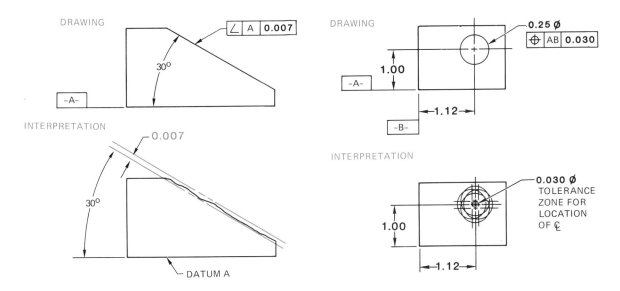

FIG. 15-14 Surface at an angle to a datum

FIG. 15-15 True position of a circle to two datums

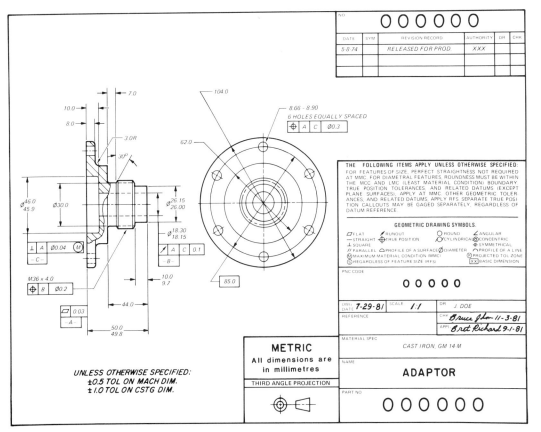

FIG. 15-16 An industrial application
(General Motors Corp.)

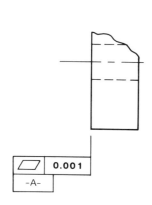

DATUM –A– IS THE FLAT SURFACE ON THE BACK OF THE
ADAPTOR. THIS IS SPECIFIED TO BE FLAT WITHIN **0.001** .

FIG. 15-17   Detail of flatness

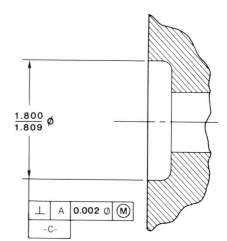

DATUM –C– IS THE **1.800–1.809** HOLE THAT IS COUNTERBORED
IN THE BACK OF THE ADAPTOR. THE SIDES OF THIS COUNTER-
BORE ARE PERPENDICULAR TO DATUM –A– WITHIN **0.002″** ⌀
WHEN THE HOLE IS AT **1.800″** (MAXIMUM MATERIAL
CONDITION).

FIG. 15-19   Detail of perpendicularity

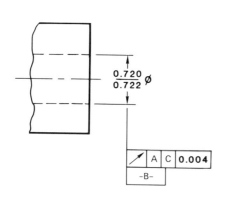

DATUM –B– IS THE **.720 –.722** ⌀ HOLE THROUGH THE CENTER
OF THE ADAPTOR. THIS FEATURE HAS A RUNOUT TOLERANCE
OF **.004″** WITH RESPECT TO DATUMS A AND C.

FIG. 15-18   Detail of runout

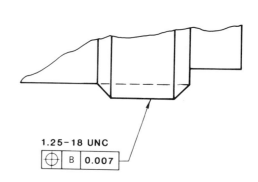

THE THREADED PORTION OF THE ADAPTOR **1.25–18UNC** IS
LOCATED WITH RESPECT TO DATUM –B– WITHIN **0.007″** ⌀

FIG. 15-20   Detail of true position

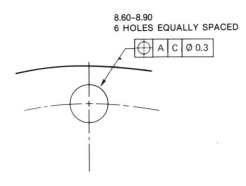

8.60–8.90
6 HOLES EQUALLY SPACED

⊕ | A | C | ∅ 0.3

THE SIX HOLES AROUND THE FLANGE OF THE ADAPTOR ARE
LOCATED WITH RESPECT TO DATUMS –A– AND –C– WITHIN
0.010 DIAMETER.

FIG. 15-21   Detail of true position

PROBLEMS

1. Interpretation of the index plate (Fig. 15-22)
   A. Sketch the index plate and show the lo-
      cations of the datums with bold lines.
   B. Explain in words the meaning of:
   a. ⊥ | A | 0.3
   b. ⊕ | C | B | 0.2
   c. ▱ 0.6
   d. // | B | 0.8
   C. Draw a sketch showing the meaning of
      ⊥ | A | 0.3 .   See the interpretation of
      perpendicularity in Fig. 15-12.

2. Draw a multiview drawing of Figure 15-23.
   Add the following geometric characteristic
   symbols:
   A. The top surface is flat, within 0.004 inch.
   B. The bottom surface is perpendicular to
      datum A within 0.015 inch.
   C. The bottom is parallel to datum B within
      0.010 inch.

3. Draw a multiview drawing of Figure 15-24.
   Add the following geometric characteristic
   symbols:
   A. The inclined surface's angle is within
      0.020 inch to datum A.

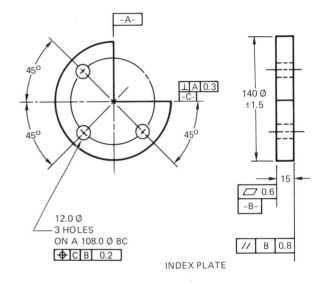

12.0 ∅
3 HOLES
ON A 108.0 ∅ BC

⊕ | C | B | 0.2

INDEX PLATE

140 ∅
±1.5

15

▱ 0.6
–B–

// | B | 0.8

FIG. 15-22   Problem

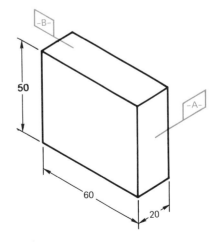

–B–

50

–A–

60

20

FIG. 15-23   Problem

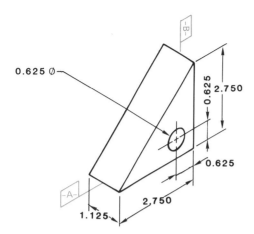

0.625 Ø

0.625

2.750

0.625

2,750

1.125

-B-

-A-

FIG. 15-24   Problem

0.75 Ø

3.25

-A-

FIG. 15-25   Problem

B.   The true position of the circle's center lines is within a tolerance zone of 0.030 inch to datums A and B.

4.   Draw a multiview drawing of Figure 15-25. Add the following geometric characteristic symbols:

A.   The roundness is within 0.030 inch.

B.   One end is parallel to datum A within 0.004 inch.

5.   Draw a multiview drawing and detail drawings of the automobile cam follower in Figure 15-26. Dimension and change the written instructions to geometric characteristic symbols.

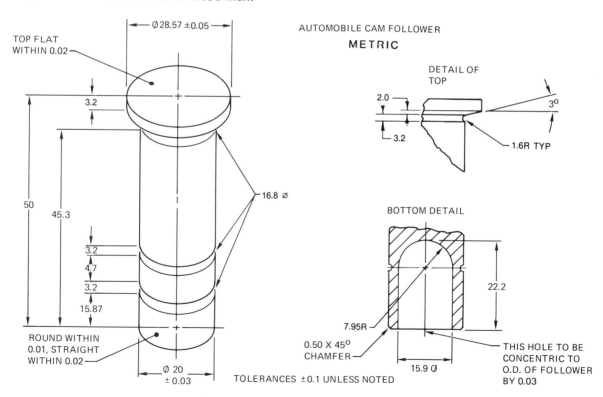

TOP FLAT WITHIN 0.02

Ø 28.57 ±0.05

3.2

50

45.3

3.2

4.7

3.2

15.87

16.8 Ø

ROUND WITHIN 0.01, STRAIGHT WITHIN 0.02

Ø 20 ± 0.03

TOLERANCES ±0.1 UNLESS NOTED

AUTOMOBILE CAM FOLLOWER

METRIC

DETAIL OF TOP

2.0

3.2

3°

1.6R TYP

BOTTOM DETAIL

7.95R

0.50 X 45° CHAMFER

22.2

15.9 Ø

THIS HOLE TO BE CONCENTRIC TO O.D. OF FOLLOWER BY 0.03

FIG. 15-26   Problem

Surface Finishes

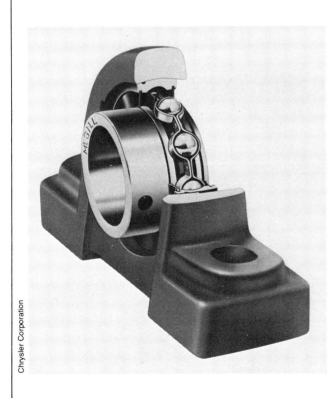

Chrysler Corporation

drawing refer to the required finish that must be machined on any particular surface. As you study this chapter, you should learn to note and read finished surfaces on a drawing.

CONTROLLED SURFACES

Surface finish is important for the proper function of mating surfaces (Fig. 16-2). Parts such as pistons, gears, bearings, and seals require a close finish control. One reason for the indication of finished surfaces on a drawing is to show where extra material must be provided for the finishing processes. The degree of surface finish is a factor in cost control during manufacturing.

NONCONTROLLED SURFACES

Many surfaces do not need further machining after they are initially formed by casting, forging, or machining. The texture will not affect the function of the finished product if these surfaces do not make contact with other surfaces (Fig. 16-3).

INTRODUCTION

Surface texture or surface finish is the amount of geometric irregularity produced on the surface of an object during fabrication. Surface finish notes on a

SURFACE FINISH CHARACTERISTICS

An absolutely smooth surface is not possible. All surfaces have irregularities. However, the amount of

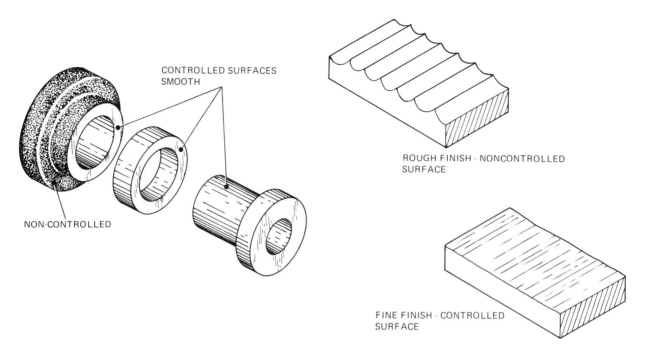

FIG. 16-2 Surface finish on mating parts

FIG. 16-3 Extremes of surface finish

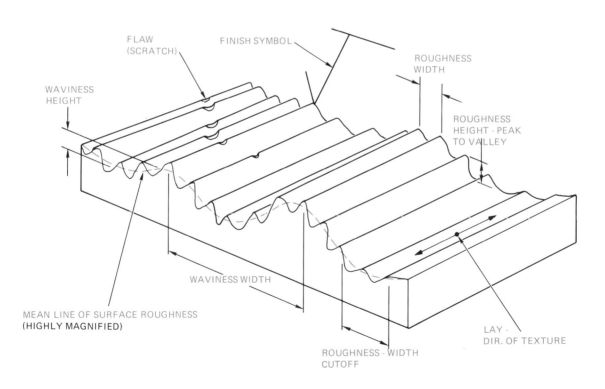

FIG. 16-4 Surface finish terminology

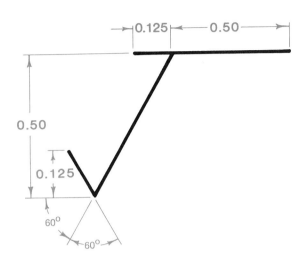

FIG. 16-5 Approximate dimensions for the surface texture symbol

DIMENSIONAL CALLOUTS

Dimensions on the finish symbol are given in micro-inches and inches. Units read in microinches are roughness-height. Units read in inches are waviness-height, waviness-width, roughness-width, and roughness-width cutoff.

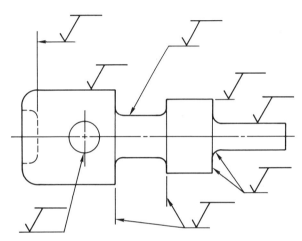

FIG. 16-6 Placement of finish symbols

this surface texture can be controlled during manu-facturing. Surface textures are classified as roughness, waviness, lay, and flaws (Fig. 16-4).

SURFACE QUALITY DESIGNATIONS

The surface quality designation is specified on a drawing with a finish symbol (Fig. 16-5), which in-dicates the limits of surface texture. The point of the finish symbol must touch the line representing the surface it controls (Fig. 16-6), or the finish may be indicated in a note.

Control information for the finished surface is placed around the finish symbol (Fig. 16-7). Each surface to be finished will be indicated only once. The symbol controls the entire surface it is touch-ing. The transitional area to the next surface (fillets, chamfers, or curves) will conform to the roughest adjacent finish (Fig. 16-8).

To specify the type of machine finish, a note is placed alongside the roughness-width part of the symbol (Fig. 16-9). Typical notes are hone, polish, rough grind, fine grind, file, etc.

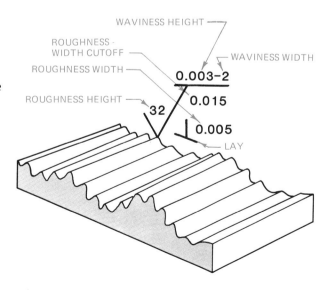

FIG. 16-7 Placement of information on the symbol

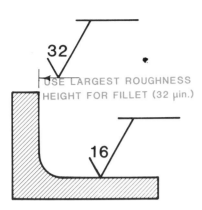

FIG. 16-8   Curves and fillets with adjacent surfaces will be finished to the roughest finish

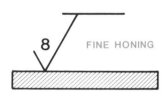

FIG. 16-9   Note for specified machine finish

ROUGHNESS

All smooth surfaces have some small peaks and valleys caused by machine cutting operations. These finely spaced surface irregularities are called roughness (Fig. 16-10).

Roughness-Height

The roughness-height is an arithmetical average of the roughness deviation and is measured in microinches. The measured distance is the variation from ridge to valley in cut marks (Fig. 16-11). The rating value placed on the finish symbol indicates the maximum value, and any lesser value is acceptable. The specification of maximum and minimum values indicates the permissible range of roughness-height permitted (Fig. 16-12). Recommended roughness values are shown in Figure 16-13.

Roughness-Width

The roughness-widths depend on the machine, cutting tool, and feed. The roughness-width distance is parallel to the nominal surface and is measured in inches (Fig. 16-14). Any lesser value is acceptable.

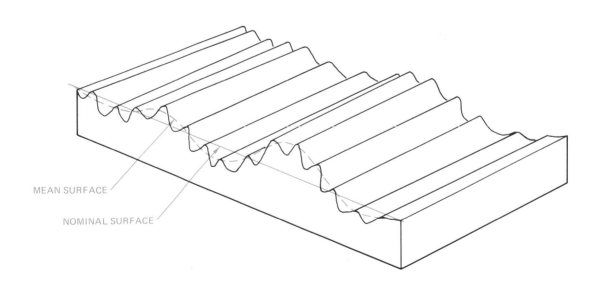

FIG. 16-10   Surface roughness

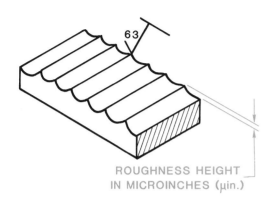

ROUGHNESS HEIGHT
IN MICROINCHES (μin.)

FIG. 16-11 Roughness height measurement

**Recommended Roughness
Average Rating Values
Microinches, μin   [Micrometres, μm] ***

μin	μm	μin	μm	μin	μm	μin	μm
1	[0.025]	10	[0.25]	50	[1.25]	250	[6.3]
2	**[0.050]**	13	[0.32]	**63**	**[1.6]**	320	[8.0]
3	[0.075]	**16**	**[0.40]**	80	[2.0]	400	[10.0]
4	**[0.100]**	20	[0.50]	100	[2.5]	**500**	**[12.5]**
5	[0.125]	25	[0.63]	**125**	**[3.2]**	600	[15.0]
6	[0.15]	**32**	**[0.80]**	160	[4.0]	800	[20.0]
8	**[0.20]**	40	[1.00]	200	[5.0]	1000	[25.0]

* Boldface values preferred

FIG. 16-12 Recommended roughness values

Roughness Height Microinches (μ inch)	Surface Description	Machining Processes
1000	Very rough	Saw or torch cutting, forging or sand casting
500	Rough machining	Coarse feeds and heavy cuts in machining
250	Coarse	Coarse surface grind, medium feeds and average cuts in machining
125	Medium	Sharp tools, light cuts, fine feeds, high speeds with machining
63	Good finish	Sharp tools, light cuts, extra fine feeds, light cuts with machining
32	High grade finish	Very sharp tools, very fine feeds and cuts
16	Higher grade finish	Surface grinding, coarse honing, coarse lapping
8	Very fine machine finish	Fine honing and fine lapping
3	Extremely smooth machine finish	Extra fine honing and lapping

FIG. 16-13 Description and process of roughness values

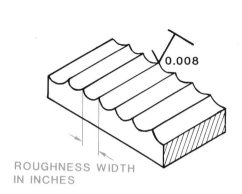

FIG. 16-14   Roughness width

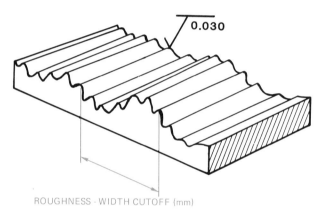

FIG. 16-15   Roughness-width cutoff

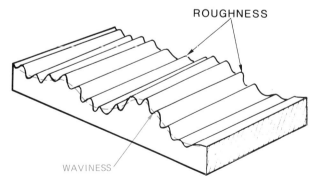

FIG. 16-16   Waviness

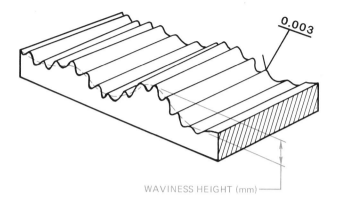

FIG. 16-17   Waviness height

Recommended Waviness Height Values Inches, in. Millimetres, mm *					
in.	mm	in.	mm	in.	mm
.00002	**[0.0005]**	.0003	[0.008]	**.005**	**[0.12]**
.00003	[0.0008]	**.0005**	**[0.012]**	.008	[0.20]
.00005	**[0.0012]**	.0008	[0.020]	**.010**	**[0.25]**
.00008	[0.0020]	**.0010**	**[0.025]**	.015	[0.38]
.00010	**[0.0025]**	**.002**	**[0.05]**	**.020**	**[0.50]**
.0002	**[0.005]**	.003	[0.08]	.030	[0.80]

*Boldface values preferred

FIG. 16-18   Recommended waviness height values

Roughness-Width Cutoff

The distance used to figure the arithmetical average deviation is called the roughness-width cutoff (Fig. 16-15). It is measured in inches. When no value is shown, a distance of approximately 0.030 inch is assumed.

WAVINESS

Waviness irregularities are longer roughness variations that may result from imperfections, vibration in the machining equipment, work deflections, warping, heat treatment, or strained material. Roughness is the texture on the wavy surface (Fig. 16-16).

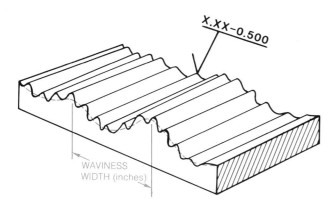

X.XX—0.500

WAVINESS
WIDTH (inches)

FIG. 16-19 Waviness width

Waviness-Height

Waviness-height is the distance from the peak of the wave to its valley, and is measured in inches (Fig. 16-17). Recommended waviness-height values are shown in Figure 16-18. Any lesser value is acceptable.

Waviness-Width

Waviness-width is the spacing between successive waves, and is measured in inches (Fig. 16-19). Any lesser value is acceptable.

SYMBOL	DESCRIPTION OF LAY	EXAMPLE OF APPLICATION
‖	THE LAY IS PARALLEL TO THE LINE REPRESENTING THE SURFACE TO WHICH THE SYMBOL IS APPLIED.	DIRECTION OF TOOL MARKS
⊥	THE LAY IS PERPENDICULAR TO THE LINE REPRESENTING THE SURFACE TO WHICH THE SYMBOL IS APPLIED.	DIRECTION OF TOOL MARKS
X	THE LAY IS ANGULAR OR OBLIQUE IN TWO DIRECTIONS TO THE LINE REPRESENTING THE SURFACE TO WHICH THE SYMBOL IS APPLIED.	DIRECTION OF TOOL MARKS
M	THE LAY IS MULTIDIRECTIONAL OR RANDOM.	
C	THE LAY IS APPROXIMATELY CIRCULAR RELATIVE TO THE CENTER OF THE SURFACE REPRESENTED BY THE LINE TO WHICH THE SYMBOL IS APPLIED.	
R	THE LAY IS APPROXIMATELY RADIAL RELATIVE TO THE CENTER OF THE SURFACE REPRESENTED BY THE LINE TO WHICH THE SYMBOL IS APPLIED.	
P	THE LAY IS NONDIRECTIONAL, PITTED, OR PROTUBERANT TO THE LINE REPRESENTING THE SURFACE TO WHICH THE SYMBOL IS APPLIED.	

FIG. 16-20 Lay symbols

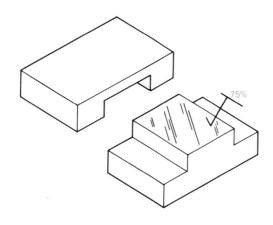

75%

FIG. 16-21 Specifiying the contact area

LAY

Lay is the primary direction of the surface pattern made by machine tool marks. Seven symbols are given for the lay patterns produced by different manufacturing processes (Fig. 16-20).

CONTACT AREA

For design quality, the designer may specify an amount of contact area with the other mating surface. The function and durability of mating parts are dependent on their surface smoothness. The amount of contact area is specified with percent values. Typical values are 50%, 75%, and 90% (Fig. 16-21).

FLAWS

Flaws are infrequent irregularities that occur at random places on a machined part. The most common flaws are cracks, scratches, gouges, and checks. There are no finish symbols for flaws. A supplementary note must be used on the drawing when a surface must be free of flaws, or the number and size of flaws must be limited.

ISO FINISH SYMBOLS

Another system used by ISO for designating surface finish, Standard R - 1302, is shown in Figure 16-22. The roughness is coded by grade numbers.

ISO Surface Finish Symbols	
Symbol	Meaning
√	Basic symbol. It may only be used alone when its meaning is explained by a note.
N7/ ∨	Removal of material is optional; N7 roughness
N7/ ∨	Removal of material is obligatory; N7 roughness
N7/ ○	No removal of material is permitted. Surface is to be left in the state from a preceding process; N7 roughness

ISO Roughness Values	
Roughness Grade Number	Max. Surface Roughness (Arithmetic Average) Micrometres (μm)
N12	50
N11	25
N10	12.5
N9	6.3
N8	3.2
N7	1.6
N6	.8
N5	.4
N4	.2
N3	.1
N2	.05
N1	.025

FIG. 16-22 ISO surface finish symbols and values

SURFACE TEXTURE ROUGHNESS

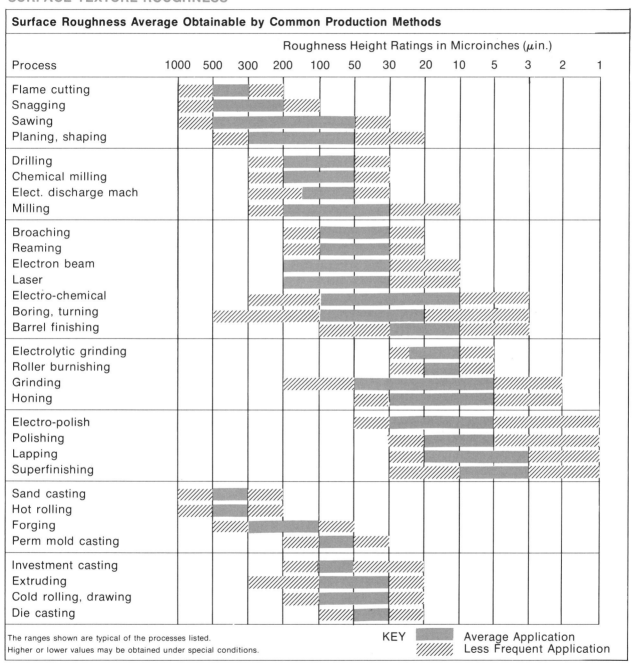

Surface Roughness Average Obtainable by Common Production Methods

Roughness Height Ratings in Microinches (μin.)

Process	1000	500	300	200	100	50	30	20	10	5	3	2	1

Flame cutting
Snagging
Sawing
Planing, shaping

Drilling
Chemical milling
Elect. discharge mach
Milling

Broaching
Reaming
Electron beam
Laser
Electro-chemical
Boring, turning
Barrel finishing

Electrolytic grinding
Roller burnishing
Grinding
Honing

Electro-polish
Polishing
Lapping
Superfinishing

Sand casting
Hot rolling
Forging
Perm mold casting

Investment casting
Extruding
Cold rolling, drawing
Die casting

The ranges shown are typical of the processes listed.
Higher or lower values may be obtained under special conditions.

KEY   Average Application   Less Frequent Application

FIG. 16-23   Approximate surface finish obtainable by common production methods

PRODUCTION PROCEDURES

To specify the proper surface finish of a machined part, the designer must know the equipment available and the type of surface qualities each machine produces (Fig. 16-23). The surface finish left by manufacturing equipment is controlled by:

1. Type of machine
2. Cutting speed
3. Feed speed
4. Condition of cutting tool
5. Type of material being cut
6. Type of coolant
7. Type of abrasive surface

Close mating parts need a smoother finish than nonmating surfaces. The smoother the finish the costlier the production. If the same finish must be used for the whole part, then a note is placed on the drawing: FINISH ALL OVER or FAO. If a finish mark is placed on a surface with no size specification, any finish will be satisfactory. If no finish symbol is specified, it will be assumed that the surface produced by the machining operation will be satisfactory. If parts are to be painted or plated, the parts must show the roughness value before and/or after the coating.

MEASUREMENT OF SURFACE ROUGHNESS

Surface roughness can be measured with a profilometer. A profilometer is an electronic surface analyzer that has a tracer with a stylus, which is moved over the surface by a motor. As the stylus rises and falls with the surface roughness, an electric signal is generated in a magnetic coil which is enclosed in the tracer. This signal is amplified and read on a scale in microinches.

When close accuracy is not required, a method of sight and touch comparison can be used. A set of standard roughness blocks with finish textures from smooth to coarse can be used to make an approximate comparison of the part and its finish. Although this method is not accurate it is fast and inexpensive.

PROBLEMS

Follow the directions in Figures 16-24 and 16-25.

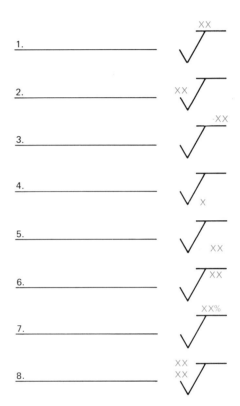

1. _____
2. _____
3. _____
4. _____
5. _____
6. _____
7. _____
8. _____

FIG. 16-24   Name each surface designation

	μm	Roughness Grade Number
1	0.8	
2	0.05	
3	25	
4		N3
5		N10

FIG. 16-25   With the ISO roughness values, provide the missing values

17
Chapter

Sections

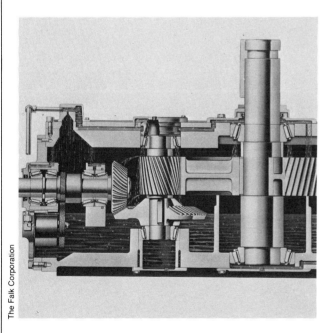

INTRODUCTION

When the internal shape of an object is so complicated that hidden lines will not show its form clearly, the interpretation of the drawing will be difficult (Fig. 17-2). The best solution is to draw a section of it (also called a sectional view or a cross-sectional view). A sectional view reveals the internal construction of the object within the outline of its external shape (Fig. 17-3).

An imaginary plane divides or cuts the object (Fig. 17-4). The front part of the object is removed to show the internal construction with visible lines. The idea of a cross-sectional view is simple. It is easy to draw and read. A sectional view can be an orthographic, isometric, or oblique drawing (Fig. 17-5). There are many different types of sections that reveal the interior details of an object (Fig. 17-6). As you study this chapter, you should gain the knowledge and skills required to make sectional drawings.

CUTTING PLANES

The function of the cutting plane is to indicate the location of an imaginary cut through the object (Fig. 17-7). The front portion is then removed, permitting a direct view of the interior details. The edge of the cutting plane shows the exact location of the sectional view.

Symbols for cutting planes are shown in Figure 17-8. The ends of the cutting planes have arrow-

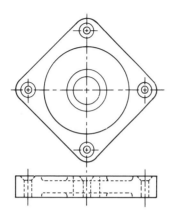

FIG. 17-2   A drawing with complicated hidden lines

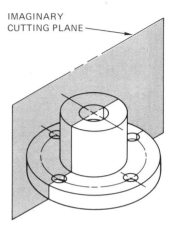

THE CUTTING PLANE DEFINES THE
LOCATION OF THE SECTION

FIG. 17-4   The cutting plane

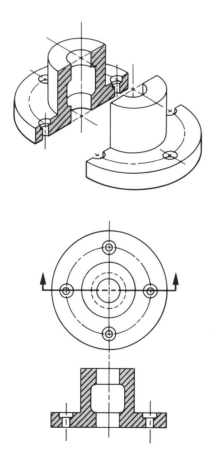

FIG. 17-3   Revealing the internal construction

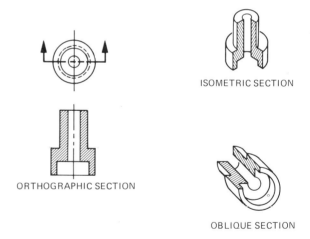

ISOMETRIC SECTION

ORTHOGRAPHIC SECTION

OBLIQUE SECTION

FIG. 17-5   Sections in three drawing types

heads to identify the direction of sight shown in the
sectional view. Letters are used to identify the cutting plane if the sectional view is not projected
directly from the cutting plane (Fig. 17-9) or if there
is more than one section (Fig. 17-10). The letters I,
O, Q, and Z are not recommended because these
letters are easily mistaken for numbers.

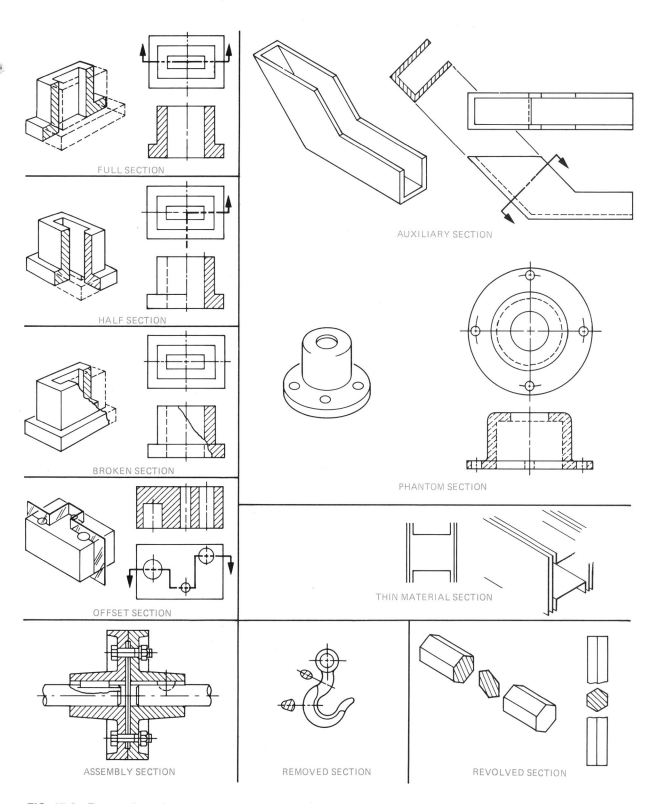

FULL SECTION

HALF SECTION

BROKEN SECTION

OFFSET SECTION

ASSEMBLY SECTION

AUXILIARY SECTION

PHANTOM SECTION

THIN MATERIAL SECTION

REMOVED SECTION

REVOLVED SECTION

FIG. 17-6   Types of sections

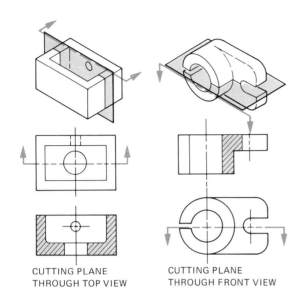

CUTTING PLANE
THROUGH TOP VIEW

CUTTING PLANE
THROUGH FRONT VIEW

FIG. 17-7   The cutting plane in two views

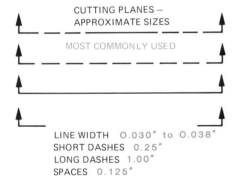

CUTTING PLANES —
APPROXIMATE SIZES

MOST COMMONLY USED

LINE WIDTH   0.030" to 0.038"
SHORT DASHES   0.25"
LONG DASHES   1.00"
SPACES   0.125"

FIG. 17-8   Cutting plane symbols

The line width of the cutting plane is 0.035 inch, and it is drawn very darkly (Fig. 17-8). It is one of the heaviest lines in drafting line conventions. It will always take precedence over lighter lines such as center lines and hidden lines. The cutting plane can be placed on an object anywhere it is necessary to show an internal section. The edges that the cutting plane intersects are projected for the sectional view. The rules of orthographic projection also apply here. If it is clearly understood that the section comes from a center line, then the cutting plane may be omitted from the drawing (Fig. 17-11).

SECTION LINES

Section lines (also called cross-hatching) show the solid material that is cut by the cutting plane. They give a shading effect to the solid surfaces of the sectional view (Fig. 17-12). Section lines are drawn with a medium-hard pencil that will produce crisp, sharp, smudge-free lines. The width of section lines is about 0.020 inch. Section line spacing varies between 0.10 inch and 0.20 inch depending on the size of the section (Fig. 17-13). The spacing and line consistency should be done by eye (Fig. 17-14).

Section lines are usually drawn at a 45 degree angle. They should not be parallel or perpendicular to any outline of the object (Fig. 17-15). Any convenient angle can be used.

All section lines of each part should be drawn in the same direction (Fig. 17-16). Section lines for

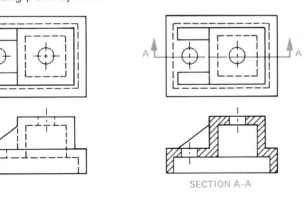

SECTION A-A

ORTHOGRAPHIC

FULL SECTION

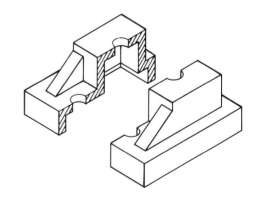

FULL ISOMETRIC SECTION

FIG. 17-9   Identification of the cutting plane

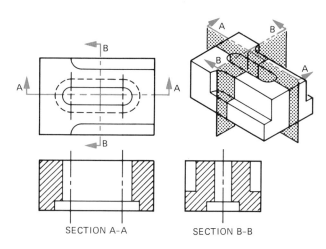

SECTION A-A   SECTION B-B

FIG. 17-10   Identification callouts are needed for more than one section

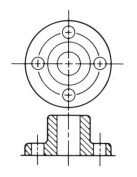

FIG. 17-11   A cutting plane may be omitted from a simple section

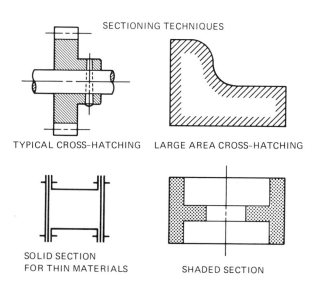

SECTIONING TECHNIQUES

TYPICAL CROSS-HATCHING   LARGE AREA CROSS-HATCHING

SOLID SECTION
FOR THIN MATERIALS   SHADED SECTION

FIG. 17-12   Drawing sectioning lines

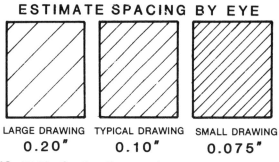

ESTIMATE SPACING BY EYE

LARGE DRAWING   TYPICAL DRAWING   SMALL DRAWING
0.20"   **0.10"**   **0.075"**

FIG. 17-13   Section line spacing

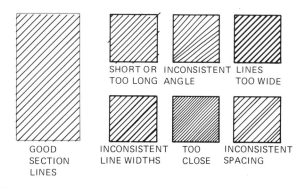

GOOD
SECTION
LINES

SHORT OR   INCONSISTENT   LINES
TOO LONG   ANGLE   TOO WIDE

INCONSISTENT   TOO   INCONSISTENT
LINE WIDTHS   CLOSE   SPACING

FIG. 17-14   Examples of poor section lines

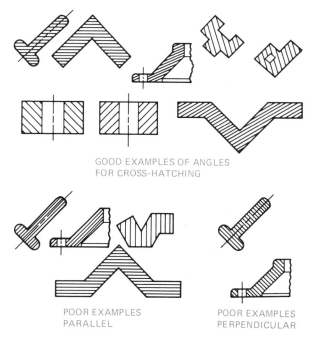

GOOD EXAMPLES OF ANGLES
FOR CROSS-HATCHING

POOR EXAMPLES
PARALLEL

POOR EXAMPLES
PERPENDICULAR

FIG. 17-15   Selection of angles for section lines

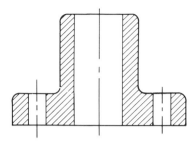

FIG. 17-16 Section lining for each part is consistent in direction

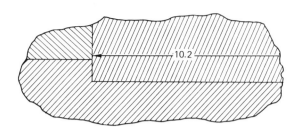

FIG. 17-18 Dimensioning technique on a sectioned surface

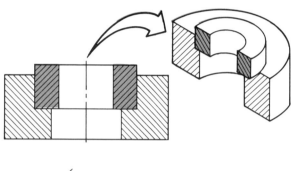

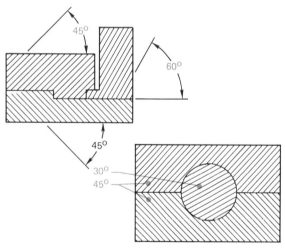

FIG. 17-17 Section lines in adjacent components

adjacent parts should be drawn at a different angle (Fig. 17-17) to make the sections easier to read.

Keep dimensions off the crosshatching if possible. When necessary, dimensions can be added on the section as shown in Figure 17-18.

Section Lining Symbols

Section lining can graphically indicate the different materials with line symbols (Fig. 17-19). Many industries use the cast iron section lining symbol for all parts and specify the type of material with a note (Fig. 17-20). The sectioning symbol for thin materials such as sheet metal, gaskets, and electric insulation is shown with a solid section (Fig. 17-21).

Full Section

A full section is a sectional view in which the cutting plane passes completely through the object. Figure 17-22 shows a full section through a round object. Figure 17-23 shows a rectangular object. In the sectional view all visible details behind the cutting plane are shown with visible lines. Figure 17-24 shows typical errors caused by leaving out visible lines in a sectional view. Sectional views can be taken from the front, top, and side views (Fig. 17-25). More than one sectional view can be taken from a single view (Fig. 17-26).

Hidden Lines

Hidden lines behind the cutting plane are usually omitted for clarity. Do not add hidden lines unless they are essential for clarification or for dimensioning purposes (Fig. 17-27).

Offset Section

The offset section is similar to the full section in that the cutting plane passes completely through the

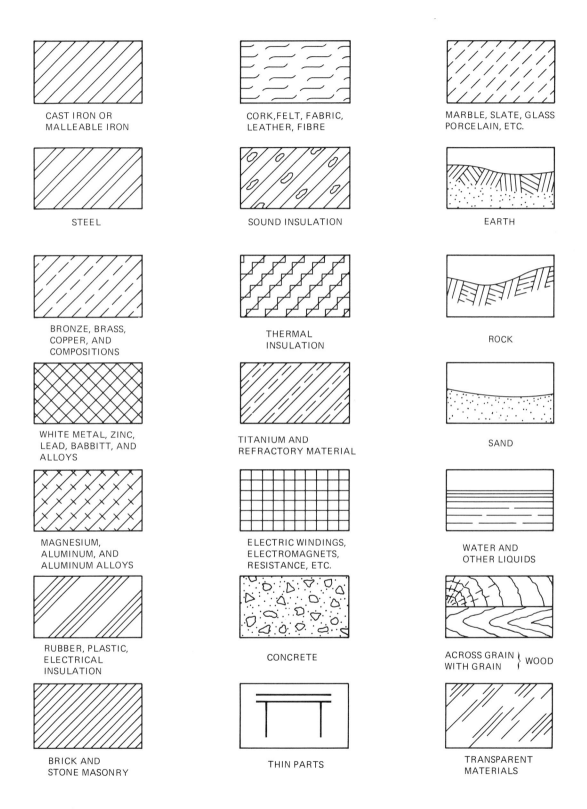

FIG. 17-19 Symbols for section lining

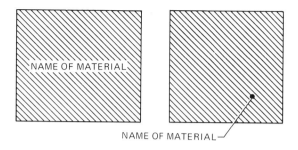

FIG. 17-20 General procedure for identifying materials

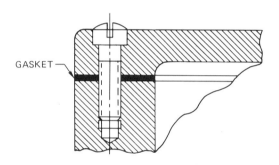

FIG. 17-21 Thin materials in section

object. However, the cutting plane for the offset section can be offset (Fig. 17-28) or bent (Fig. 17-29) to include features that are not in a straight line. This allows several planes to be drawn within one sectional view.

The offsets of the cutting plane will not show in the sectional view (Fig. 17-30). If it is necessary to show the offset cutting plane for clarity, the offset corners can be identified with letter callouts (Fig. 17-31).

Half Section

Half sections are best suited for symmetrical objects. The cutting plane passes through one quarter

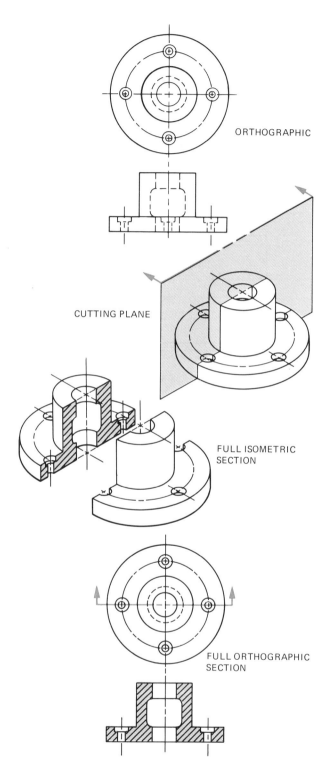

FIG. 17-22 The full section through a round object

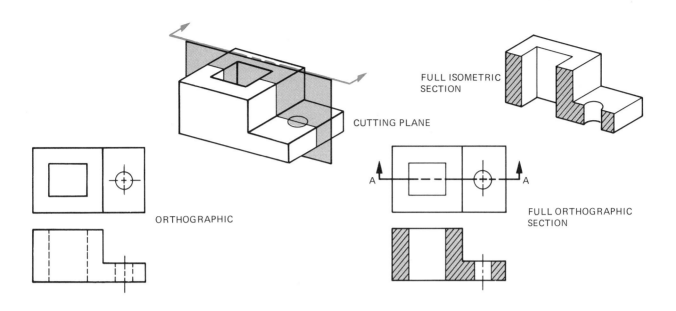

FULL ISOMETRIC
SECTION

CUTTING PLANE

ORTHOGRAPHIC

A ◀   ▶ A

FULL ORTHOGRAPHIC
SECTION

FIG. 17-23   The full section through a rectangular object

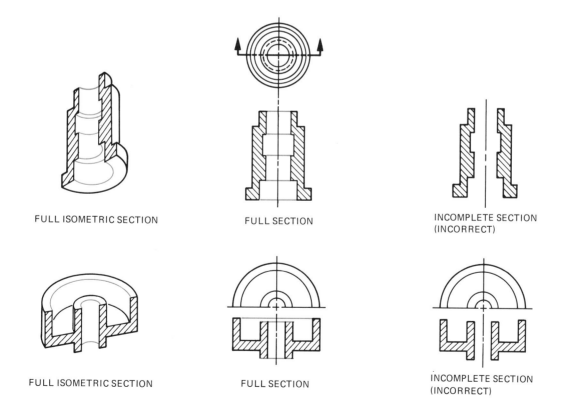

FULL ISOMETRIC SECTION

FULL SECTION

INCOMPLETE SECTION
(INCORRECT)

FULL ISOMETRIC SECTION

FULL SECTION

INCOMPLETE SECTION
(INCORRECT)

FIG. 17-24   Errors in leaving out lines

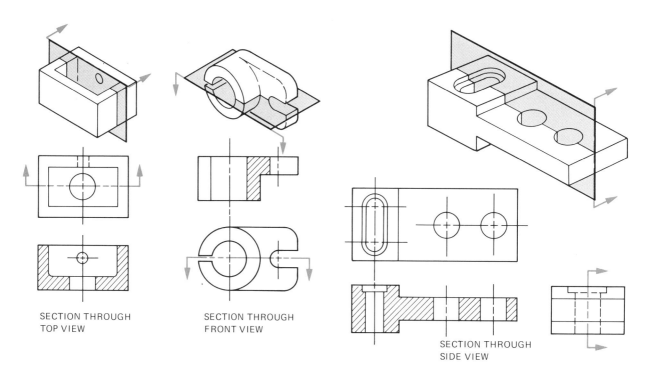

FIG. 17-25   Sections through all three views

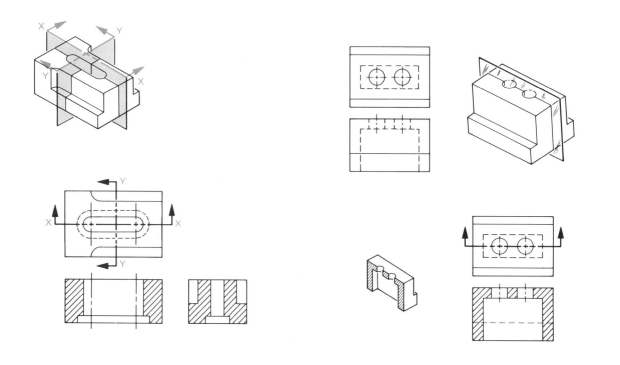

FIG. 17-26   Two cutting planes on one view

FIG. 17-27   Hidden lines where required for clarity

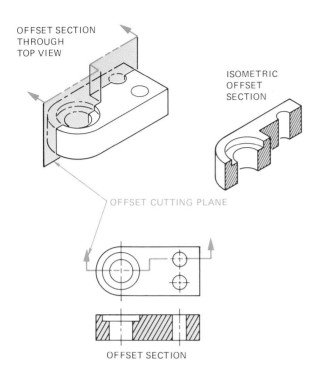

FIG. 17-28 The offset section

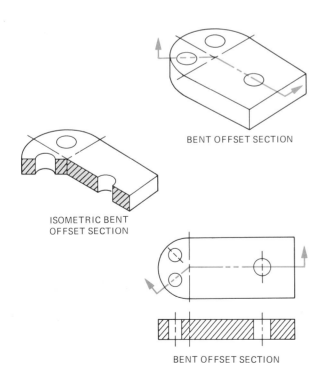

FIG. 17-29 The bent offset section

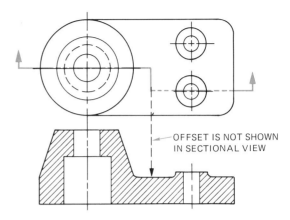

FIG. 17-30 Sometimes an offset is not shown in the sectional view

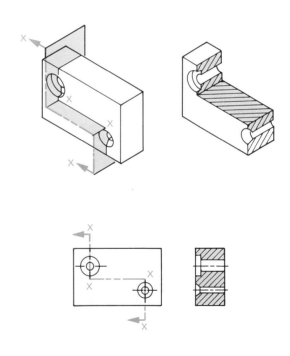

FIG. 17-31 Identifying an offset cutting plane with letters at corners

of the object (Fig. 17-32). When this quarter is removed, one half of the inside of the object can be viewed with a sectional view. Note that a center line is used to separate the sectional view from the external view (Fig. 17-33). If it is necessary to view

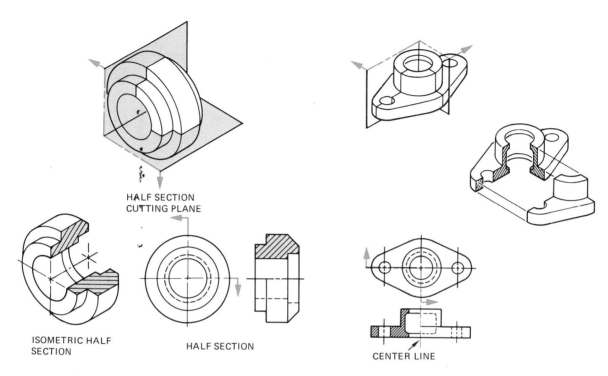

FIG. 17-32   The half section

FIG. 17-33   The center line as a section line

HALF SECTION
CUTTING PLANE

ISOMETRIC HALF
SECTION

HALF SECTION

CENTER LINE

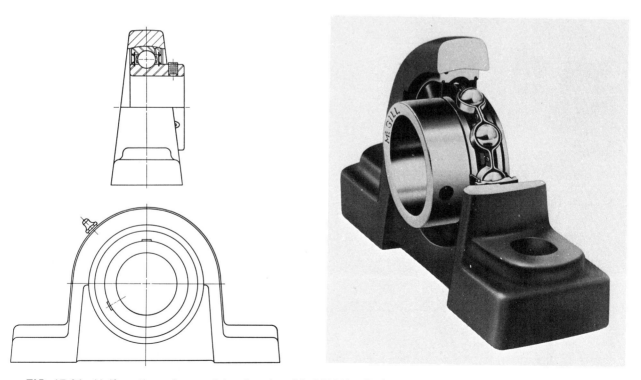

FIG. 17-34   Half section of a precision bearing (McGill Mfg. Co.)

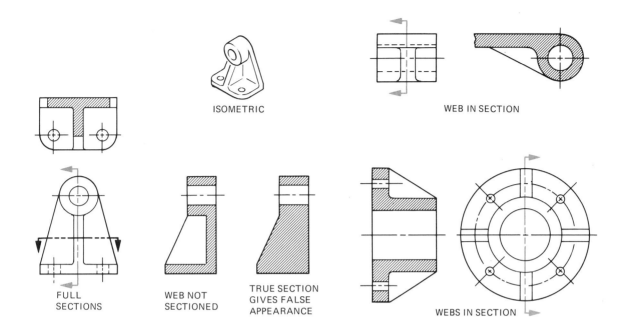

ISOMETRIC

WEB IN SECTION

FULL
SECTIONS

WEB NOT
SECTIONED

TRUE SECTION
GIVES FALSE
APPEARANCE

WEBS IN SECTION

FIG. 17-35 Nonsectioned webs

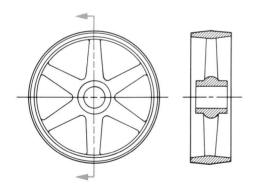

FIG. 17-36 Spokes in section

the outside of the object, a half section will provide both an internal and an external view in a single

drawing. The half section of a precision bearing is shown in Figure 17-34.

NONSECTIONED ITEMS

In certain situations the rules of projection and sectioning can be changed. The general rule is to break the rules only when doing so will add clarity to the drawing. These changes are called conventional practices. A conventional practice is to not use true sectioning when it is misleading. The webs and ribs in Figure 17-35 are examples. A true section gives a false impression of the part. Thin, noncontinuous parts such as spokes should not be sectioned (Fig. 17-36). Any continuous part such as the web in Figure 17-37 is sectioned.

Objects that have no interior details should not be sectioned when a cutting plane passes longitudinally through them (Fig. 17-38). Examples of these parts are shafts, springs, bolts, nuts, washers, keys, pins, rivets, bearings, gears, etc. (Fig. 17-39). When

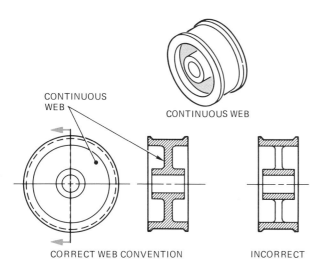

CONTINUOUS WEB

CONTINUOUS WEB

CORRECT WEB CONVENTION   INCORRECT

FIG. 17-37 Drawing convention for continuous webs

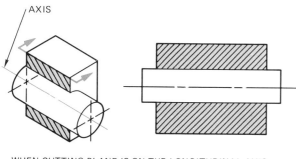

AXIS

WHEN CUTTING PLANE IS ON THE LONGITUDINAL AXIS
OF A SHAFT, THE SHAFT IS NOT SECTIONED

FIG. 17-38 Do not section the profile of a shaft

the cutting plane is perpendicular to the long axis of an object, the sectional view should be sectioned (Fig. 17-40).

Alternate Section Lining

If a nonsectioned item such as a rib must be shown for clarity, a method called alternate section lining can be used. The rib is indicated by leaving out every other section line on the rib (Fig. 17-41).

ROTATED SECTIONS

Another conventional practice for parts such as webs (Fig. 17-42), spokes (Fig. 17-43), lugs (Fig. 17-44), and holes (Fig. 17-45) that are not in line with the cutting plane is to rotate them until their dimensions are true size. These rotated sections are called aligned sections. A true projection will not give true distances. A bent cutting plane can be rotated until the dimensions in the sectional drawing are true size (Fig. 17-44). Any part in a drawing can be rotated if it will help simplify the drawing.

Revolved Sections

Revolved sections are similar to rotated sections. The conventional practice for revolved sections is to show the shape of a feature on the actual view of the part. A cutting plane is passed perpendicularly through the part being sectioned. The

NON-SECTIONING CONVENTION PRACTICES

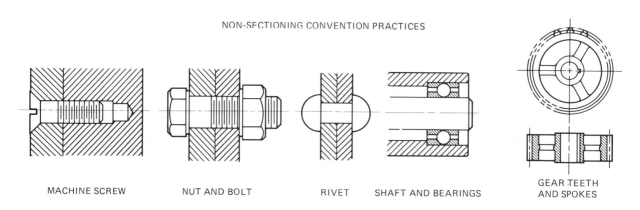

MACHINE SCREW   NUT AND BOLT   RIVET   SHAFT AND BEARINGS   GEAR TEETH AND SPOKES

FIG. 17-39 Nonsectioned components

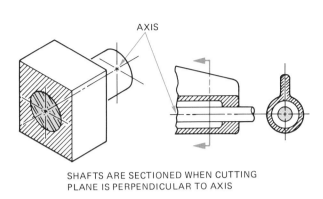

SHAFTS ARE SECTIONED WHEN CUTTING
PLANE IS PERPENDICULAR TO AXIS

FIG. 17-40 The axis of a shaft may be sectioned

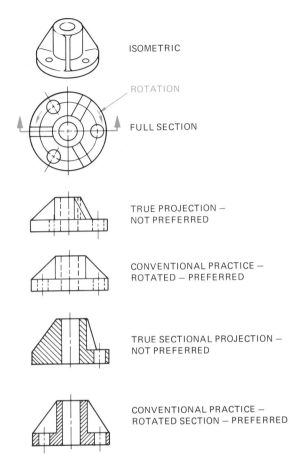

ISOMETRIC

ROTATION

FULL SECTION

TRUE PROJECTION —
NOT PREFERRED

CONVENTIONAL PRACTICE —
ROTATED — PREFERRED

TRUE SECTIONAL PROJECTION —
NOT PREFERRED

CONVENTIONAL PRACTICE —
ROTATED SECTION — PREFERRED

FIG. 17-42 Conventional practice for the rotation of
webs

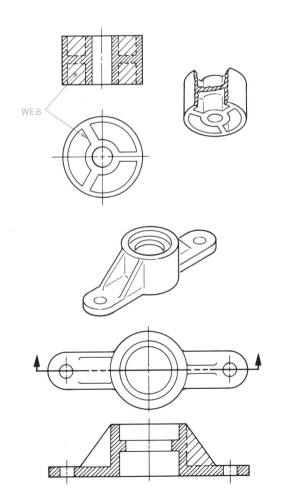

WEB

FIG. 17-41 Alternate section lines to show thin ribs

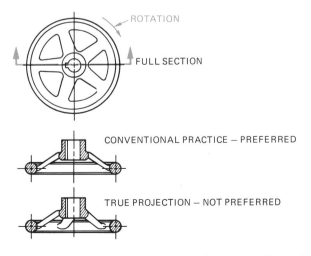

ROTATION

FULL SECTION

CONVENTIONAL PRACTICE — PREFERRED

TRUE PROJECTION — NOT PREFERRED

FIG. 17-43 Conventional practice for the rotation of
spokes

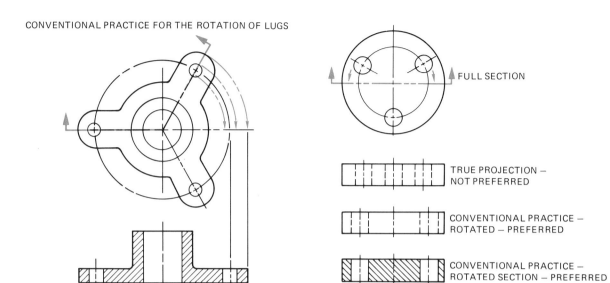

CONVENTIONAL PRACTICE FOR THE ROTATION OF LUGS

FULL SECTION

TRUE PROJECTION —
NOT PREFERRED

CONVENTIONAL PRACTICE —
ROTATED — PREFERRED

CONVENTIONAL PRACTICE —
ROTATED SECTION — PREFERRED

FIG. 17-44 Conventional practice for the rotation of lugs

FIG. 17-45 Conventional practice for the rotation of holes

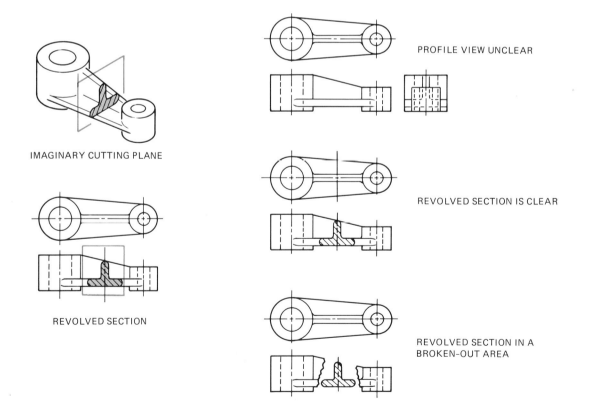

IMAGINARY CUTTING PLANE

REVOLVED SECTION

PROFILE VIEW UNCLEAR

REVOLVED SECTION IS CLEAR

REVOLVED SECTION IN A
BROKEN-OUT AREA

FIG. 17-46 Revolved sections

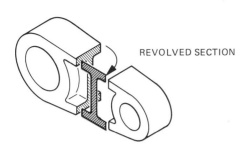

REVOLVED SECTION

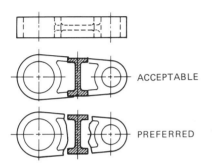

ACCEPTABLE

PREFERRED

FIG. 17-47 Breaking the outline for a revolved section

section is then rotated 90 degrees in place to show the exact shape of the object at that point (Fig. 17-46). If the drawing outline interferes with the section, the drawing outline should be broken (Fig. 17-47). When the shape of an elongated part changes in several places, revolved sections can show the varying forms (Fig. 17-48). The location of a revolved section is shown with a center line. No cutting plane is necessary.

Removed Sections

A removed section (also called an isolated or detail section) is identical to a revolved section, except for its location. A revolved section is revolved and drawn in place. A removed section is revolved and drawn next to the object (Fig. 17-49). Whenever possible the removed section should be projected perpendicular to the imaginary cutting plane (Fig. 17-50). Because the removed section is placed next

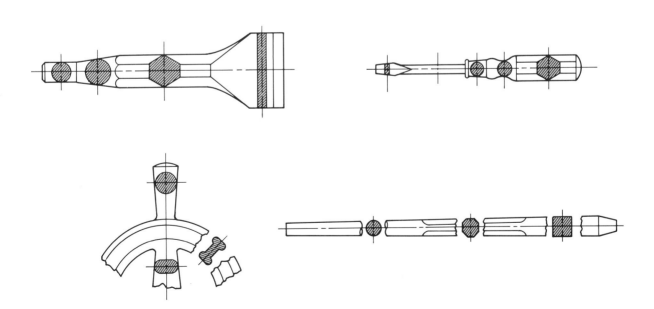

FIG. 17-48 Multiple revolved sections

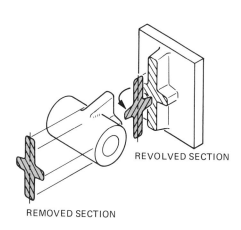

REVOLVED SECTION

REMOVED SECTION

to the object, the cutting plane and the removed section should be identified with letters (Fig. 17-51).

Broken Sections

A broken section is a convenient way to show part of an interior. The broken section (also called a broken-out or partial section) is drawn by extending the section plane far enough to show only the needed inner feature. End the section with a free-hand break line (Fig. 17-52). Cutting planes are not shown with a broken section.

The advantage of a broken section is that it is only a small portion of sectional detail. This leaves most of the exterior intact and makes it unnecessary to draw an additional exterior view (Fig. 17-53).

FIG. 17-49 Revolved and removed sections

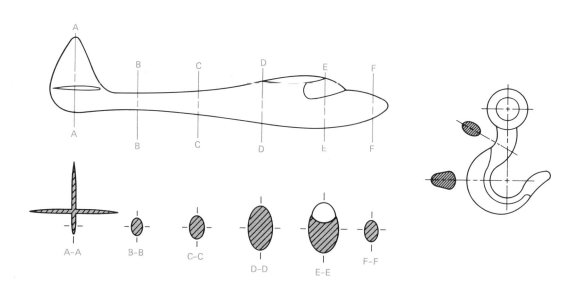

FIG. 17-50 Perpendicular removed sections

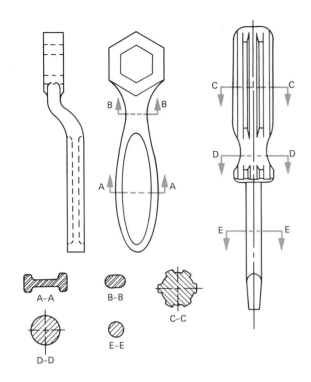

FIG. 17-51 Identifying removed sections with letters

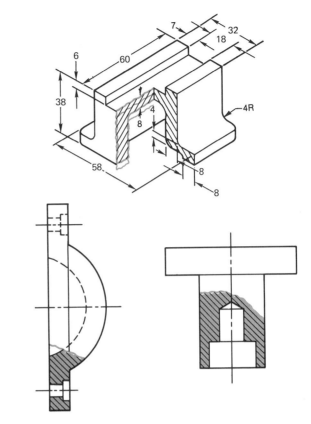

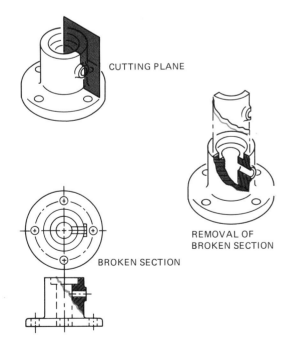

CUTTING PLANE

REMOVAL OF
BROKEN SECTION

BROKEN SECTION

FIG. 17-52 Broken sections

FIG. 17-53 Examples of broken sections (Lockheed
Missiles & Space Co.)

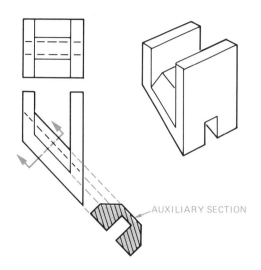

FIG. 17-54 Auxiliary section

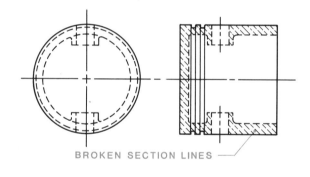

FIG. 17-55 Phantom section of a piston

Auxiliary Sections

An auxiliary section (see Chapter 18) follows all the rules of auxiliary projection. An auxiliary section is shown in Figure 17-54.

Phantom Sections

A phantom section (also called a hidden or ghost section) is used when exterior and interior features are to be shown in the same view. The section view is superimposed on the regular view without removing any part of the regular view. The section lining uses thin, sharp, evenly spaced broken lines (Fig. 17-55). The overall effect is an X ray-like view.

Pictorial Sections

A pictorial section is any sectional drawing done in a pictorial manner (Fig. 17-56 and 17-57). Standard cross-hatching is drawn at a 60 or 30 degree angle (Fig. 17-58).

Assembly Sections

An assembly section shows the relationship between the assembled components in section (Fig. 17-59). Components that are adjacent should have cross-hatching at opposite or varying angles (Fig. 17-60). Do not cross-hatch nonsectionable items such as shafts, bolts, screws, etc.

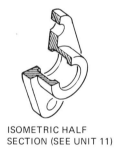

ISOMETRIC HALF
SECTION (SEE UNIT 11)

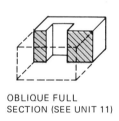

OBLIQUE FULL
SECTION (SEE UNIT 11)

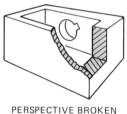

PERSPECTIVE BROKEN
SECTION (SEE UNIT 22)

FIG. 17-56 Pictorial sections

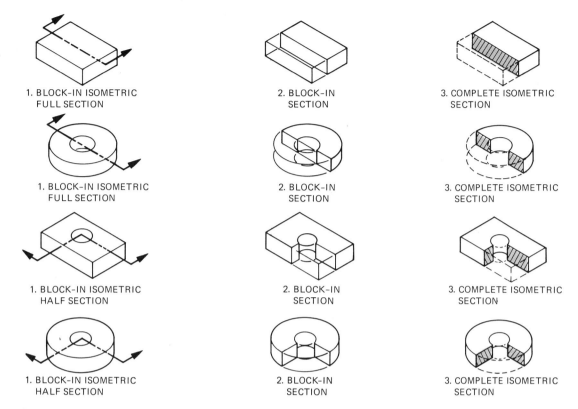

FIG. 17-57 Isometric sections

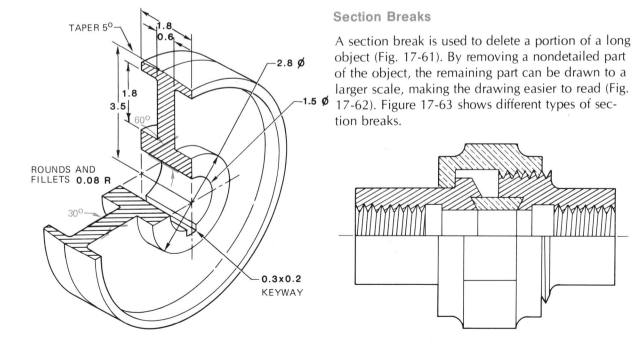

FIG. 17-58 Cross-hatching in pictorial sections

Section Breaks

A section break is used to delete a portion of a long object (Fig. 17-61). By removing a nondetailed part of the object, the remaining part can be drawn to a larger scale, making the drawing easier to read (Fig. 17-62). Figure 17-63 shows different types of section breaks.

FIG. 17-59 Half section of an assembly

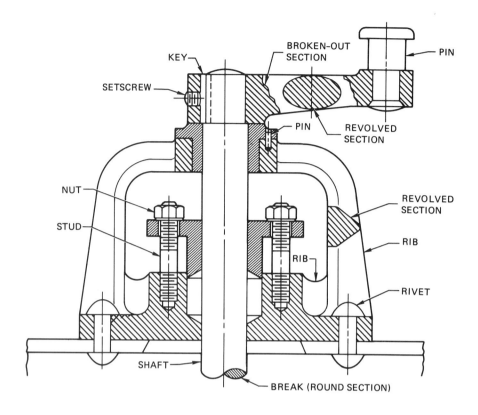

FIG. 17-60   Full section of an assembly

FIG. 17-61   A long object before a section break

PROBLEMS

Draw the sectional problems in Figures 17-64 through 17-81. Follow the instructions for each problem.

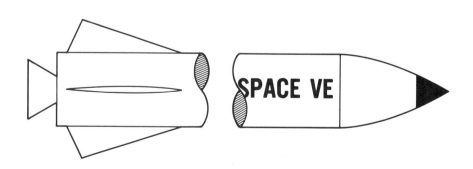

FIG. 17-62   A section break for a long object

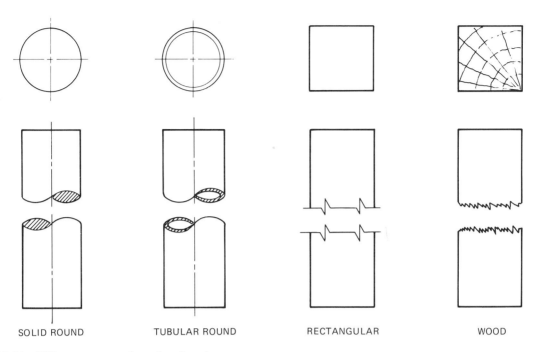

FIG. 17-63 Different types of section breaks

SOLID ROUND   TUBULAR ROUND   RECTANGULAR   WOOD

COMPLETE THE MULTIVIEW AND ISOMETRIC
FULL SECTIONS FOR EACH PROBLEM.

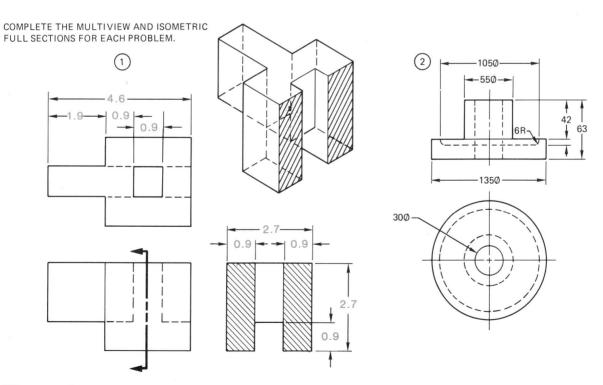

FIG. 17-64 Problem

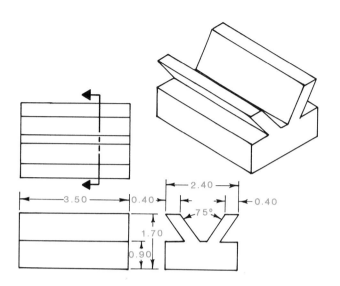

FIG. 17-65   Problem

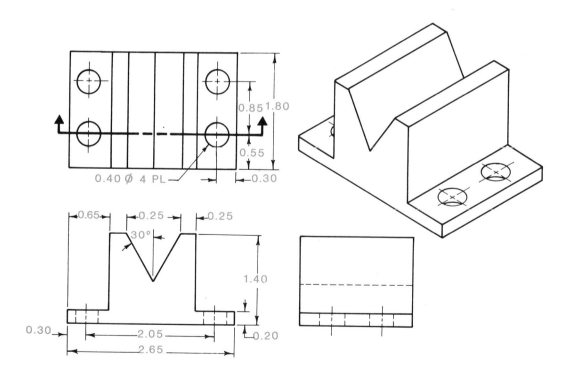

FIG. 17-66   Problem

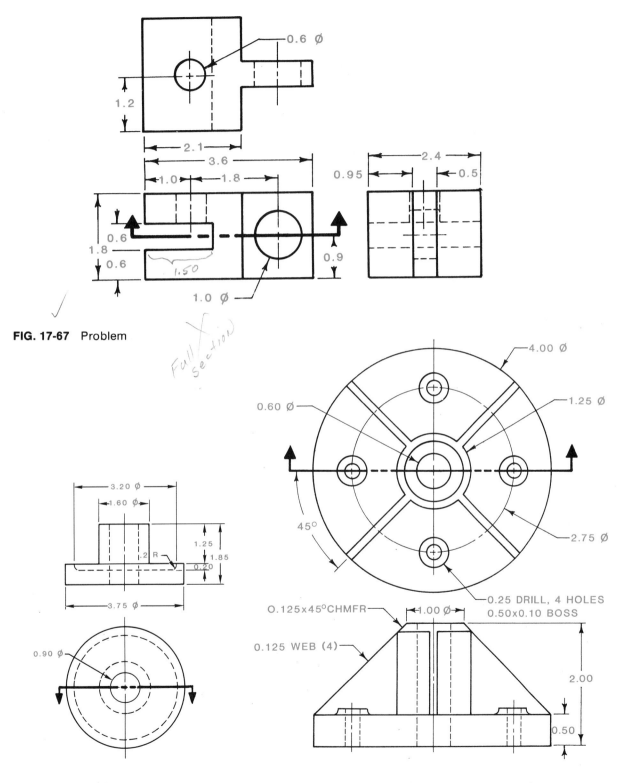

FIG. 17-67 Problem

FIG. 17-68 Problem

FIG. 17-69 Problem

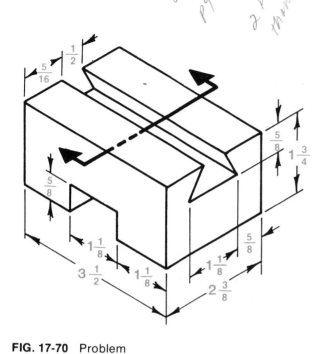

FIG. 17-70   Problem

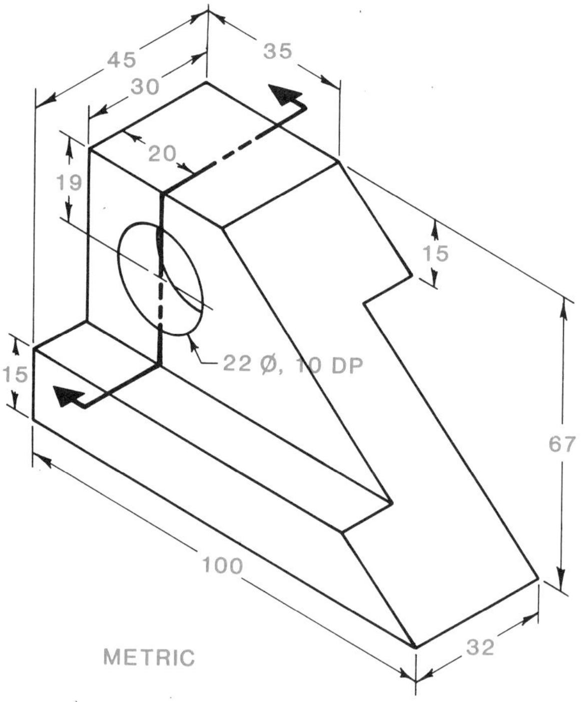

FIG. 17-72   Problem

METRIC

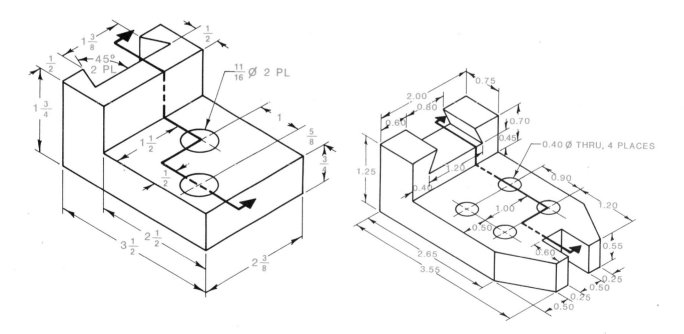

FIG. 17-71   Problem

FIG. 17-73   Problem

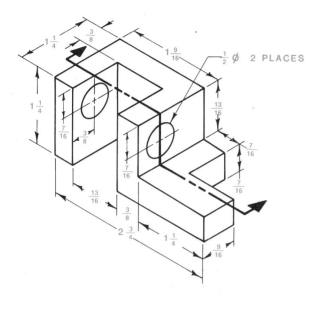

FIG. 17-74   Problem

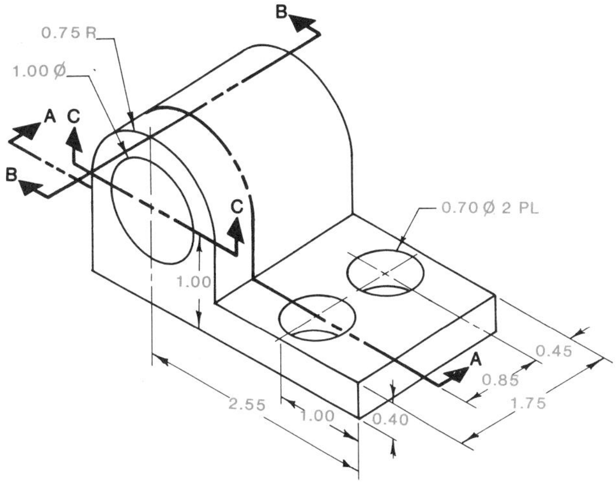

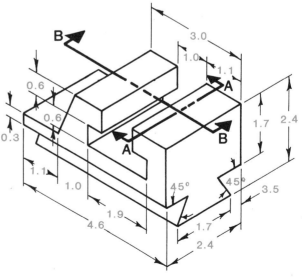

FIG. 17-76   Problem

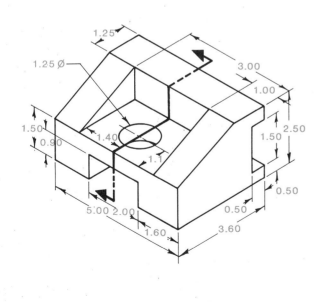

FIG. 17-75   Problem

FIG. 17-77   Problem

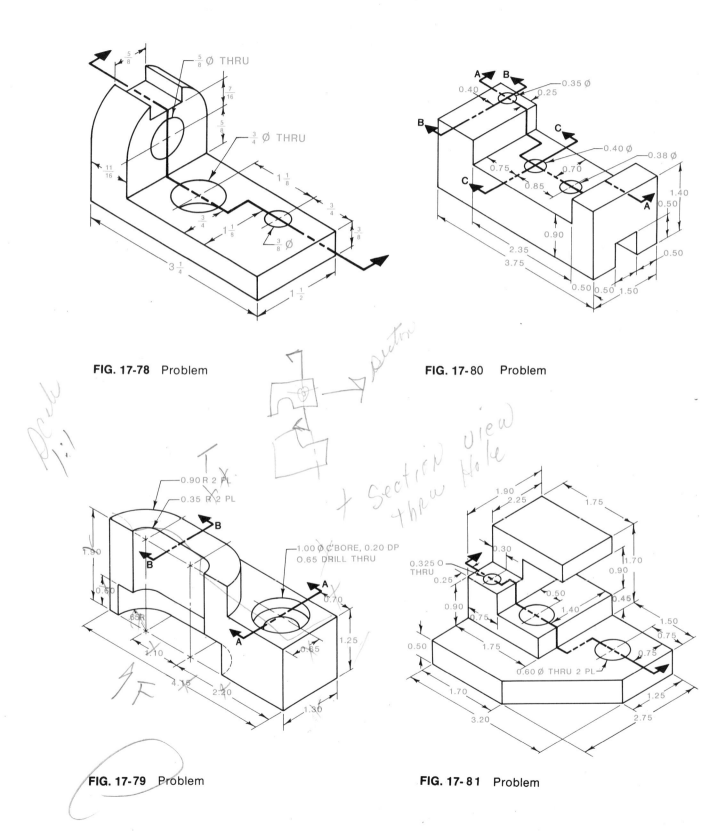

FIG. 17-78 Problem

FIG. 17-80 Problem

FIG. 17-79 Problem

FIG. 17-81 Problem

18

Chapter

Auxiliary Views

Clark E. Smith

INTRODUCTION

A drawing must clearly explain the true shape of an object. Each surface must be shown in a manner which communicates all the necessary information

to the fabricator. The orthographic multiview drawing is most often used for this communication. An auxiliary view is another view that may be added to a primary orthographic multiview drawing (Fig. 18-2). Its purpose is to make the object graphically clearer, ensuring accurate and fast fabrication of the part. As you study this chapter, you should gain the knowledge and skills required to make auxiliary drawings, and understand the difference between primary and secondary auxiliaries.

ORTHOGRAPHIC

The regular views of an orthographic drawing will describe all perpendicular surfaces in true shape. However, none of the six primary orthographic views will show an inclined surface in its true shape (Fig. 18-3). (See Chapter 12.)

AUXILIARY VIEWS

An auxiliary is a projected view in addition to the six primary orthographic views (Fig. 18-4). All the principles of orthographic projection will apply to the projection of auxiliary views. The steps in applying the principles of projection are shown in

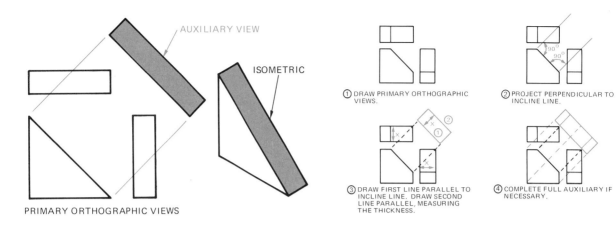

FIG. 18-2 The auxiliary view

① DRAW PRIMARY ORTHOGRAPHIC VIEWS.

② PROJECT PERPENDICULAR TO INCLINE LINE.

③ DRAW FIRST LINE PARALLEL TO INCLINE LINE. DRAW SECOND LINE PARALLEL, MEASURING THE THICKNESS.

④ COMPLETE FULL AUXILIARY IF NECESSARY.

FIG. 18-5 The steps of projection

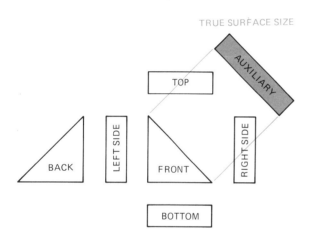

FIG. 18-3 No primary orthographic view will show the true surface

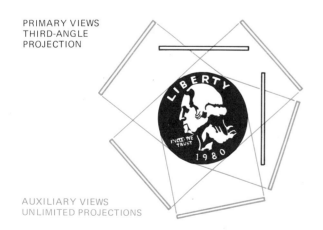

FIG. 18-4 Primary and auxiliary views

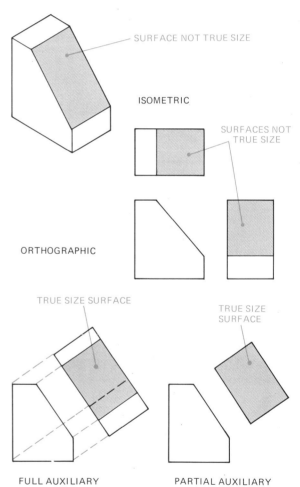

FIG. 18-6 Projection is perpendicular to an inclined surface

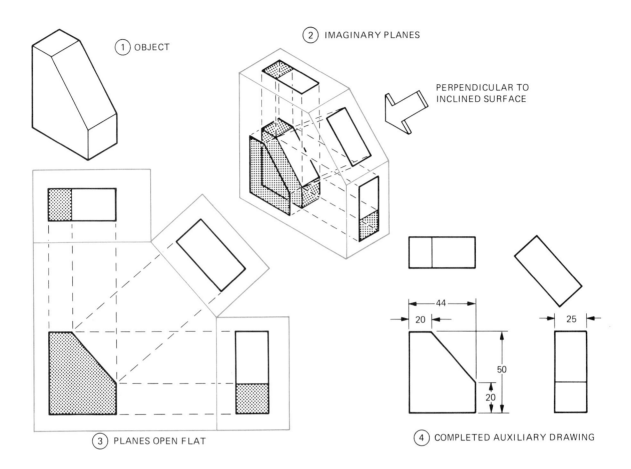

① OBJECT

② IMAGINARY PLANES

PERPENDICULAR TO
INCLINED SURFACE

③ PLANES OPEN FLAT

④ COMPLETED AUXILIARY DRAWING

FIG. 18-7   Line of sight is perpendicular to the
inclined surface

Figure 18-5. An auxiliary is simply another orthographic view projected perpendicular to the
inclined surface (Fig. 18-6). The line of sight must
be at right angles to the inclined surface, with the
edges of the auxiliary view parallel to the edge of
the inclined surface (Fig. 18-7).

TRANSFERRING AUXILIARY MEASUREMENTS

Measurements can be transferred to the auxiliary
view in two ways. One method is to transfer measurements directly from the views with a divider as
shown in Figure 18-8. The other method is to use
the reference lines from the imaginary planes of
projection as shown in Figure 18-9.

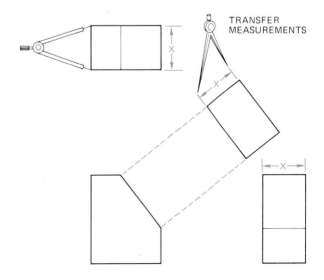

TRANSFER
MEASUREMENTS

FIG. 18-8   Transferring measurements with dividers

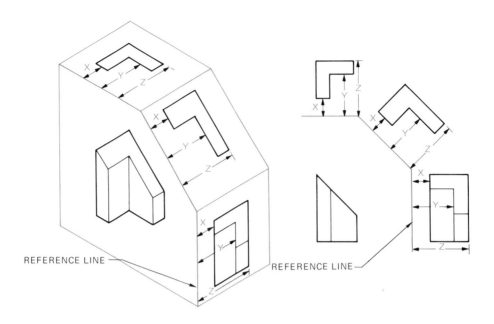

FIG. 18-9 Transferring measurements by reference lines

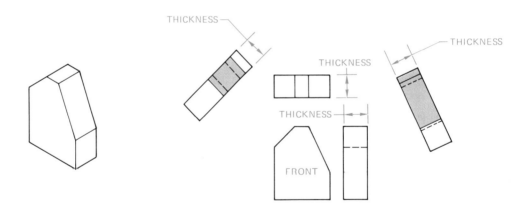

FIG. 18-10 Auxiliaries projected from the front view will show true thickness

TYPES OF AUXILIARIES

An auxiliary view can be taken from any primary orthographic view at any angle. An auxiliary taken off the front view will show the true thickness of the object (Fig. 18-10). An auxiliary taken off the top view will show the true height of the object (Fig. 18-11). An auxiliary taken off the side view will show the true length of the object (Fig. 18-12).

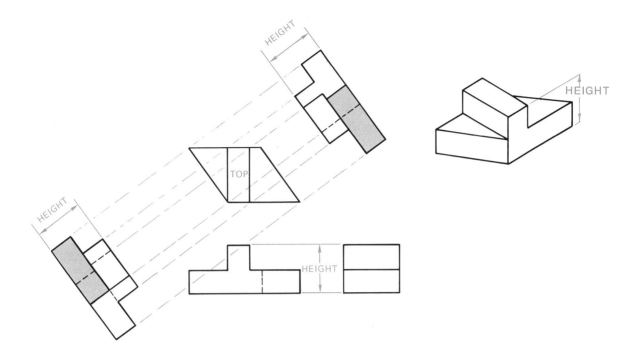

FIG. 18-11 Auxiliaries projected from the top view
will show true height

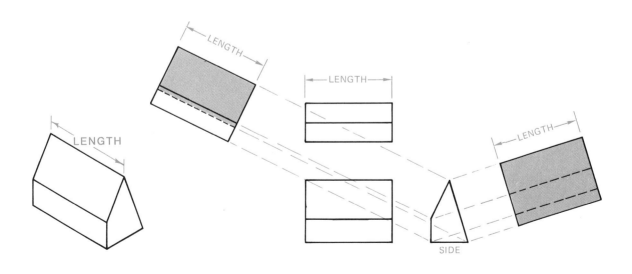

FIG. 18-12 Auxiliaries projected from the side view
will show true length

FULL AND PARTIAL AUXILIARY

Although a complete description of the part being drawn is essential, an extra drawing which does not help this communication represents wasted time, material, and money. The draftsperson must decide if a full auxiliary (Fig. 18-13), or a partial auxiliary (Fig. 18-14) is necessary. A partial auxiliary will leave off hidden lines and other details that do not make a direct contribution to the auxiliary's surface. A general rule is to draw enough information without wasting time on extra views or parts of views.

AUXILIARY SHAPE DESCRIPTION

The shapes of the objects from which auxiliary views are projected will vary. Examples of some of the different forms are:

1. Perpendicular (Fig. 18-15)
2. Round (Fig. 18-16)
3. Angular (Fig. 18-17)
4. Curved (Fig. 18-18)

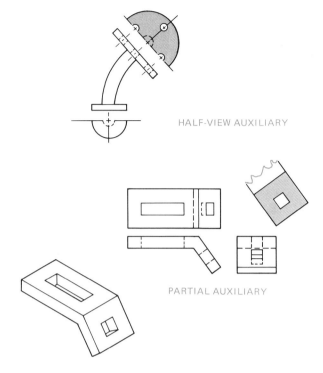

HALF-VIEW AUXILIARY

PARTIAL AUXILIARY

FIG. 18-14 Partial auxiliaries

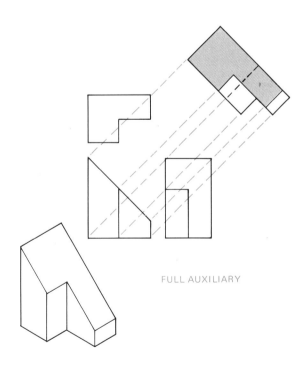

FULL AUXILIARY

FIG. 18-13 A full auxiliary

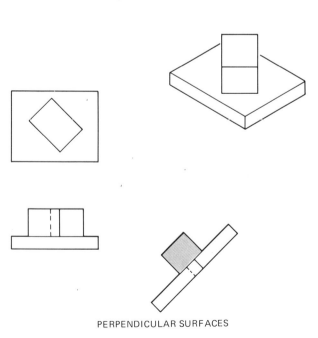

PERPENDICULAR SURFACES

FIG. 18-15 Perpendicular surface auxiliary

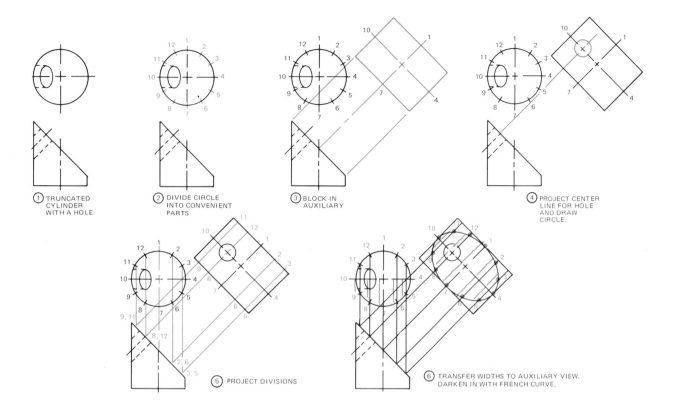

① TRUNCATED CYLINDER WITH A HOLE.

② DIVIDE CIRCLE INTO CONVENIENT PARTS

③ BLOCK-IN AUXILIARY

④ PROJECT CENTER LINE FOR HOLE AND DRAW CIRCLE.

⑤ PROJECT DIVISIONS

⑥ TRANSFER WIDTHS TO AUXILIARY VIEW. DARKEN IN WITH FRENCH CURVE.

FIG. 18-16 Round auxiliary view

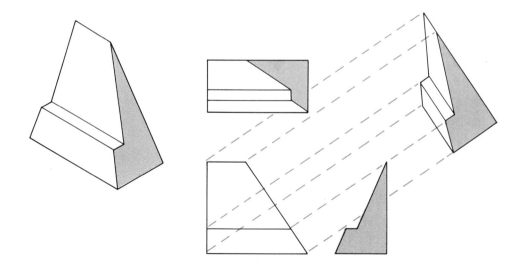

FIG. 18-17 Angular auxiliary view

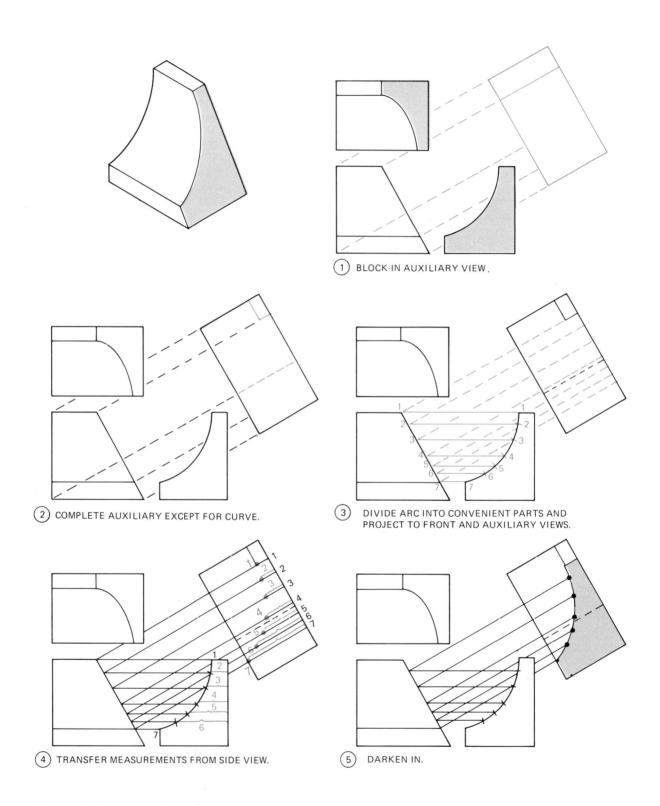

① BLOCK-IN AUXILIARY VIEW.

② COMPLETE AUXILIARY EXCEPT FOR CURVE.

③ DIVIDE ARC INTO CONVENIENT PARTS AND PROJECT TO FRONT AND AUXILIARY VIEWS.

④ TRANSFER MEASUREMENTS FROM SIDE VIEW.

⑤ DARKEN IN.

FIG. 18-18 Curved auxiliary view

NUMBER OF AUXILIARY VIEWS

The number of auxiliary views will vary with the amount of graphic information the draftsperson feels is necessary for full communication with the fabricator. The first auxiliary taken from a primary view is called the primary auxiliary. An auxiliary taken from a primary auxiliary is called a secondary auxiliary. A secondary auxiliary is sometimes necessary to obtain a true view of an oblique surface (Fig. 18-19). The reference line system of projection is used for the secondary auxiliary (see Chapter 19). If more are needed, successive auxiliaries can be drawn indefinitely (Fig. 18-20).

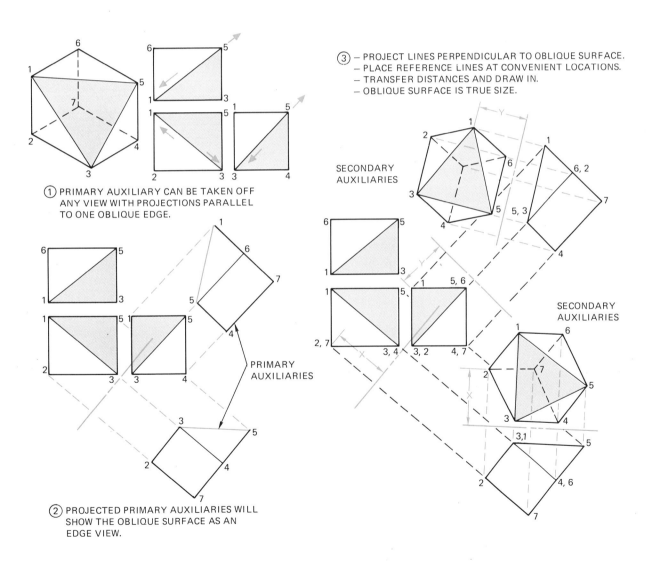

① PRIMARY AUXILIARY CAN BE TAKEN OFF ANY VIEW WITH PROJECTIONS PARALLEL TO ONE OBLIQUE EDGE.

② PROJECTED PRIMARY AUXILIARIES WILL SHOW THE OBLIQUE SURFACE AS AN EDGE VIEW.

PRIMARY AUXILIARIES

③ – PROJECT LINES PERPENDICULAR TO OBLIQUE SURFACE.
– PLACE REFERENCE LINES AT CONVENIENT LOCATIONS.
– TRANSFER DISTANCES AND DRAW IN.
– OBLIQUE SURFACE IS TRUE SIZE.

SECONDARY AUXILIARIES

SECONDARY AUXILIARIES

FIG. 18-19 Primary and secondary auxiliaries

- USE REFERENCE LINE FOR PROJECTION.
- AUXILIARIES CAN BE PROJECTED AT ANY POSITION.
- PROJECTION LINES ARE PERPENDICULAR TO REFERENCE LINE.
- SUCCESSIVE AUXILIARIES CAN BE CONTINUED INDEFINITELY.

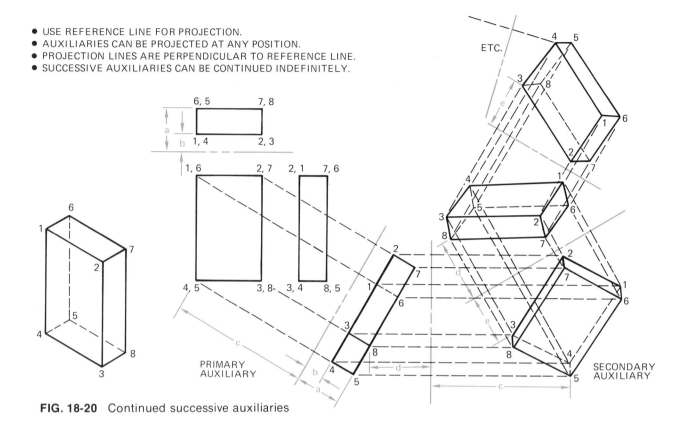

FIG. 18-20 Continued successive auxiliaries

PROBLEMS

Draw the orthographic, auxiliary, isometric views as shown for (Figures 18-21 through 18-30). Show the auxiliary projection lines lightly.

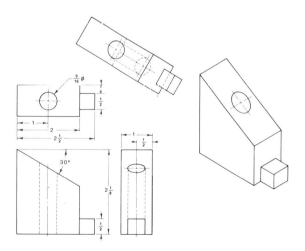

FIG. 18-21 Problem

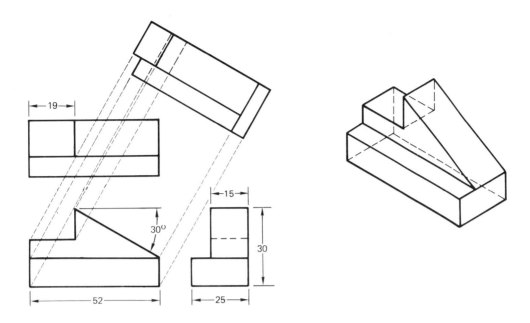

FIG. 18-22   Problem

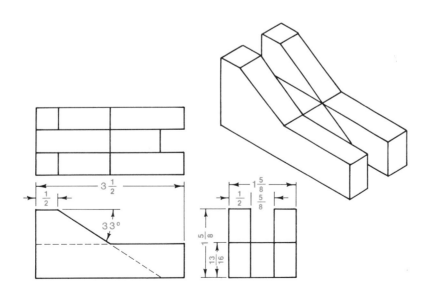

FIG. 18-23   Problem

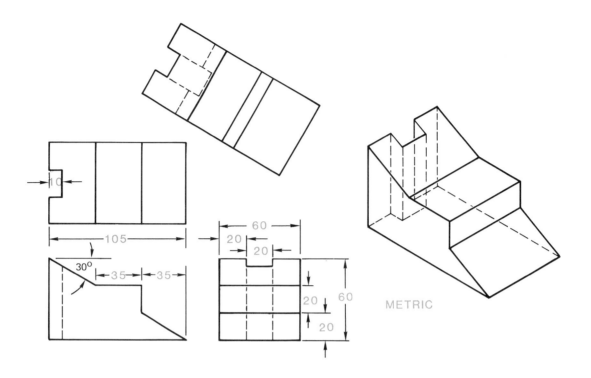

FIG. 18-24 Problem

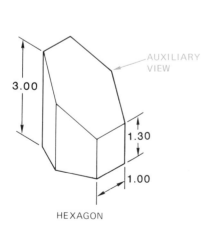

FIG. 18-25 Problem

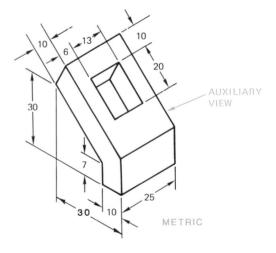

FIG. 18-26 Problem

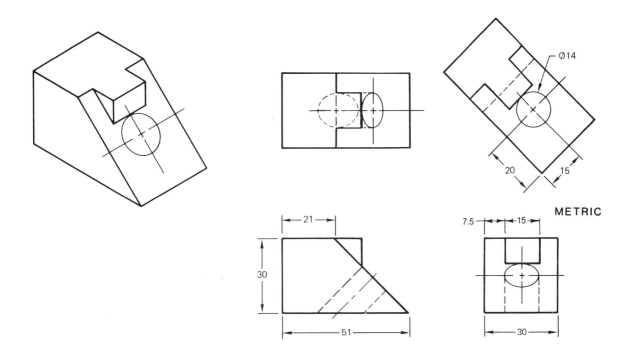

FIG. 18-27   Problem

METRIC

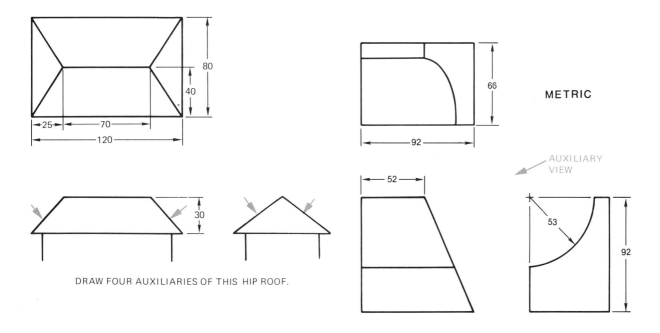

DRAW FOUR AUXILIARIES OF THIS HIP ROOF.

METRIC

AUXILIARY VIEW

FIG. 18-28   Problem

FIG. 18-29   Problem

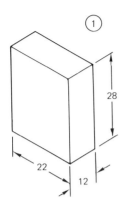

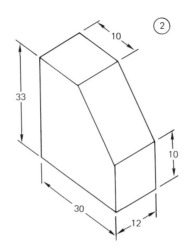

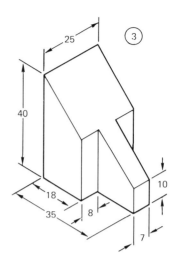

METRIC

1. DRAW ISOMETRIC AND NUMBER ALL CORNERS
2. DRAW ORTHOGRAPHIC AND NUMBER ALL CORNERS.
3. SELECT AND DRAW A PRIMARY AUXILIARY
   (USE REFERENCE PLANE).
4. SELECT AND DRAW A SECONDARY AUXILIARY
   (USE REFERENCE PLANE).
5. SELECT AND DRAW TWO SUCCESSIVE AUXILIARIES
   (USE REFERENCE PLANES).

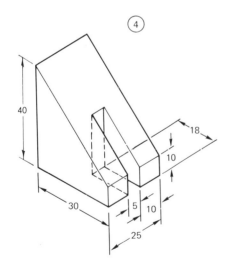

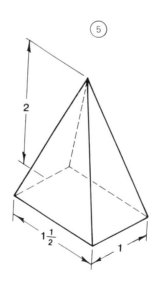

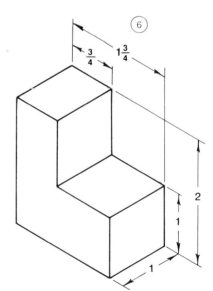

FIG. 18-30 Problem

Revolutions

Jacqueline Marsall

enough detail about the part. A revolution drawing should be made if more detail is required. A revolution drawing is similar to an auxiliary drawing. Its purpose is to show the shape of the part in clear detail. Since revolution and auxiliary drawings serve the same purpose, the drawing should be of the type that is the fastest and most convenient to draw. With an auxiliary drawing, the inclined surface will be projected to an auxiliary plane. The primary orthographic views remain the same and a new plane is added for the auxiliary (Fig. 19-3).

A revolution drawing will turn one of the primary orthographic views to a new position from which other views can be projected (Fig. 19-4). No new planes of projection are added. As you study this chapter, you should learn to draw revolutions, and to understand the concepts of axes, primary revolution, successive revolution, and true length lines.

AXIS

Any view can be revolved around an axis to any position on a 360 degree circle (Fig. 19-5). The axis can be drawn in any position (Fig. 19-6), and the views can be revolved clockwise or counterclockwise. A revolved view does not change its

INTRODUCTION

Many manufactured products have parts that revolve about a shaft or pivot point (Fig. 19-2). A primary orthographic drawing will usually show

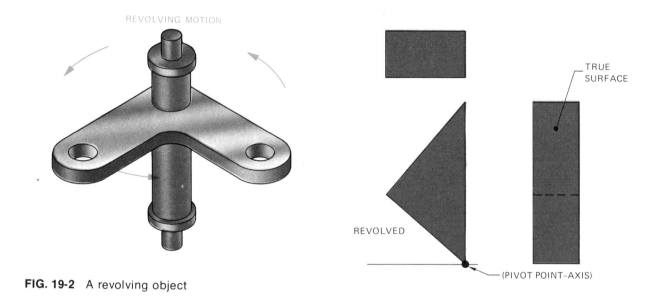

FIG. 19-2 A revolving object

FIG. 19-4 Revolution of an orthographic view needs no extra planes

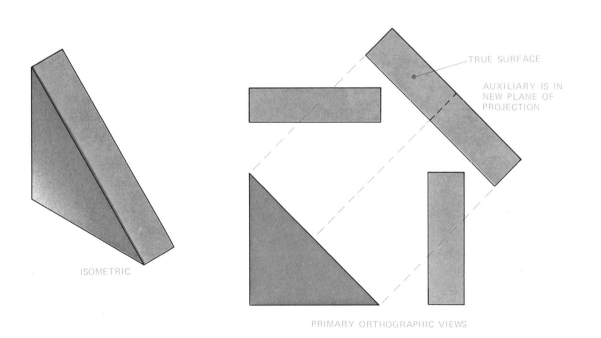

FIG. 19-3 Extra view required for auxiliary projection

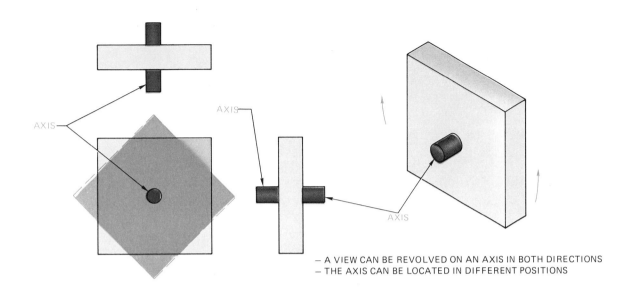

FIG. 19-5   The axis of revolution

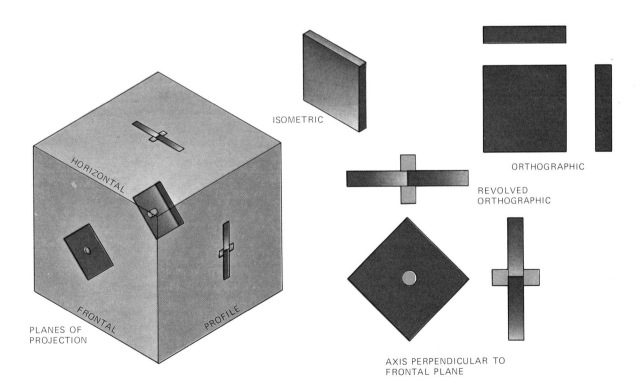

FIG. 19-6   Possible positions for an axis of revolution

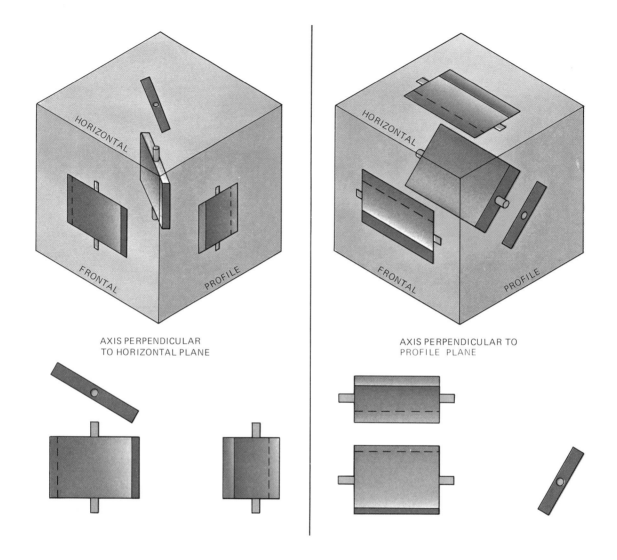

AXIS PERPENDICULAR
TO HORIZONTAL PLANE

AXIS PERPENDICULAR TO
PROFILE PLANE

form or size, only its position. The other views will change in appearance (Fig. 19-7).

If a true-size surface is desired, the view must be revolved until the inclined surface is parallel to the plane to which it is being projected. This will present the true surface (Fig. 19-8).

PRIMARY REVOLUTION

The first revolved view is called the primary revolution. A primary revolution can be taken from the frontal plane (Fig. 19-9), the horizontal plane (Fig. 19-10), or the profile plane (Fig. 19-11). The primary revolution never changes its shape, only its position. Note the identical true surfaces of the auxiliaries and revolutions in Figure 19-12.

SUCCESSIVE REVOLUTIONS

Each revolution after the primary revolution is called a successive revolution, which performs the same function as a secondary auxiliary (Fig. 19-13).

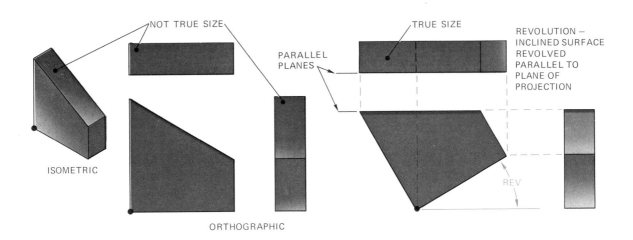

FIG. 19-7 Revolving an inclined surface

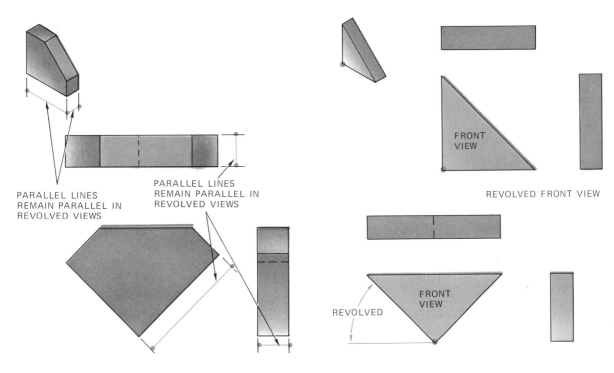

FIG. 19-8 Parallel lines remain parallel in revolved views

FIG. 19-9 Revolved front view

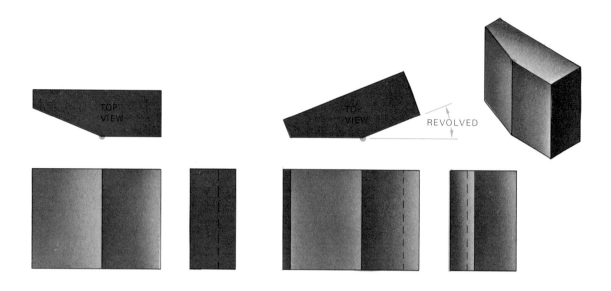

FIG. 19-10   Revolved top view

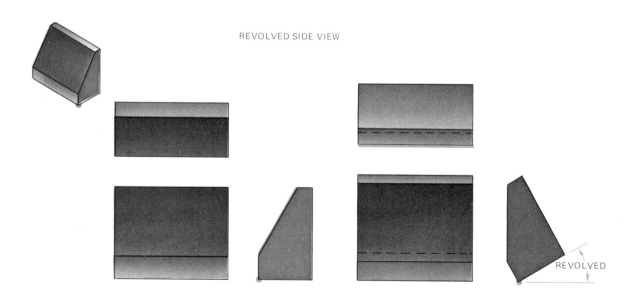

FIG. 19-11   Revolved side view

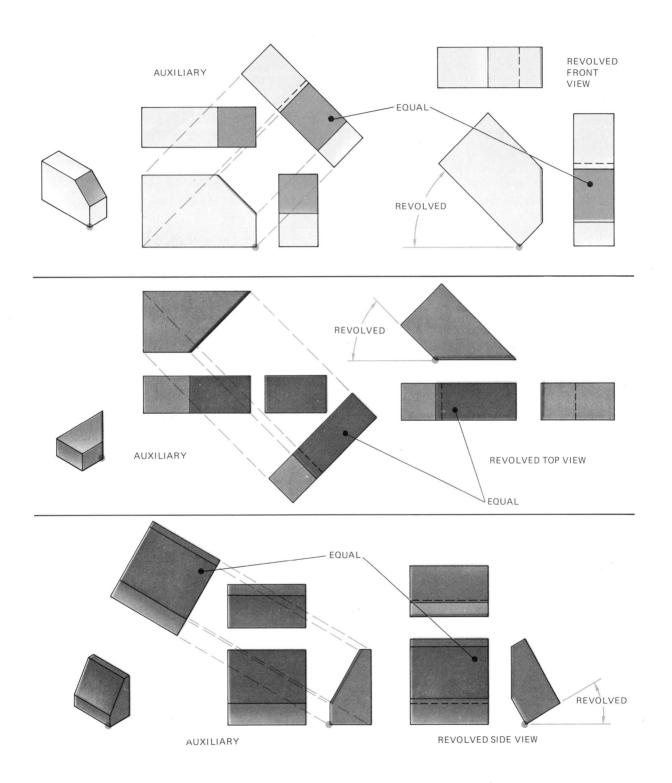

FIG. 19-12 Identical surfaces from auxiliary or revolved views

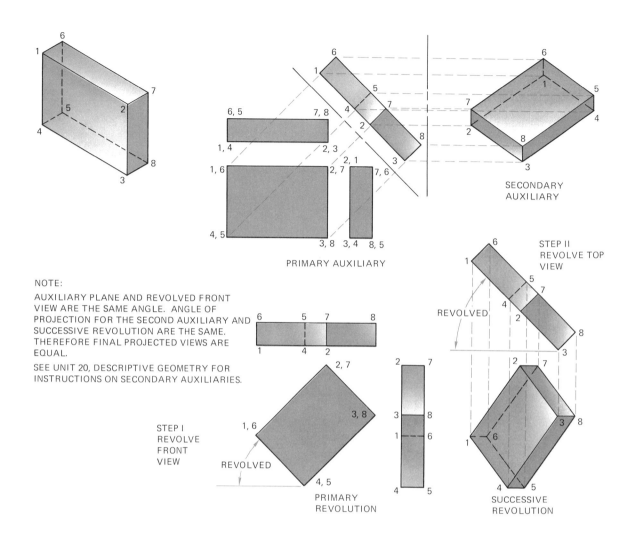

FIG. 19-13 Successive revolutions

Successive revolutions can continue indefinitely as shown in Figures 19-14 and 19-15.

To find the true size of an oblique surface the following steps are necessary (Fig. 19-16):

1. Revolve one view so one edge of the oblique surface will be parallel to one of the other primary planes. The oblique surface will appear as an edge view in one of the other orthographic views.

2. Revolve the edge view of the oblique surface so it is parallel to a primary plane. The projec-tion lines will be perpendicular, so the oblique surface will appear in true size.

Note that in Figure 19-17 the drawing of the secondary auxiliary is the same as the drawing of the successive revolution in Figure 19-16.

TRUE LENGTH LINES

At times it is only necessary to know the true length of a single line. Instead of revolving the whole part, revolve only the single line (Fig. 19-18). The line must be revolved so it is parallel to a projection

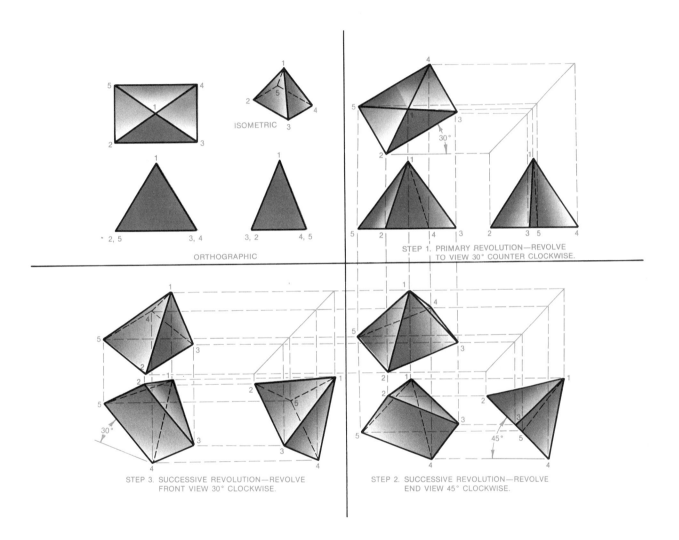

ISOMETRIC

ORTHOGRAPHIC

STEP 1. PRIMARY REVOLUTION—REVOLVE
TO VIEW 30° COUNTER CLOCKWISE.

STEP 3. SUCCESSIVE REVOLUTION—REVOLVE
FRONT VIEW 30° CLOCKWISE.

STEP 2. SUCCESSIVE REVOLUTION—REVOLVE
END VIEW 45° CLOCKWISE.

FIG. 19-14 Steps in successive revolutions

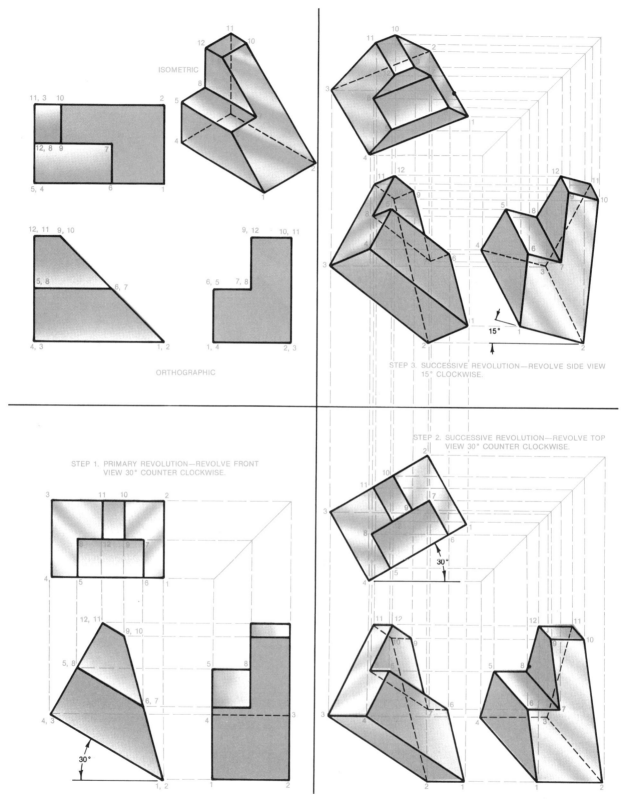

FIG. 19-15 Steps in successive revolutions

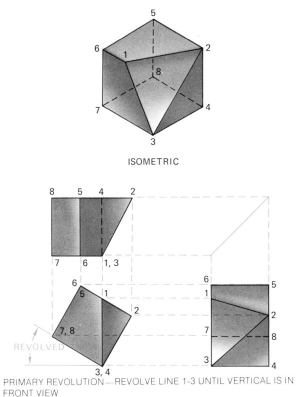

ISOMETRIC

PRIMARY REVOLUTION—REVOLVE LINE 1-3 UNTIL VERTICAL IS IN FRONT VIEW

FIG. 19-16   Steps in successive revolutions

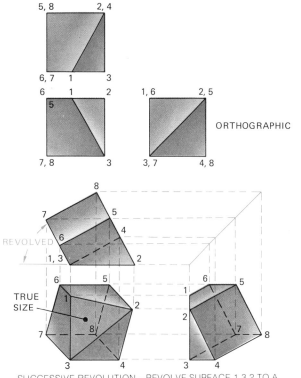

ORTHOGRAPHIC

TRUE SIZE

SUCCESSIVE REVOLUTION—REVOLVE SURFACE 1,3,2 TO A HORIZONTAL POSITION IN THE TOP VIEW. FRONT VIEW WILL SHOW TRUE SIZE FOR SURFACES 1,3,2.

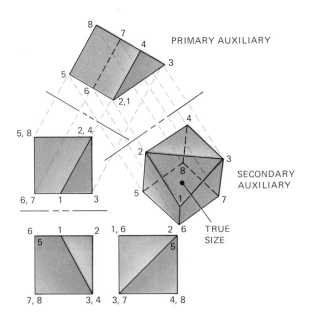

PRIMARY AUXILIARY

SECONDARY AUXILIARY

TRUE SIZE

FIG. 19-17   An auxiliary identical to the previous revolution

plane. The line will then appear as a true length line in the projected view.

CIRCLES IN REVOLUTIONS

A circle in an orthographic view will appear elliptical in a revolved view (Fig. 19-19). Once the size of the ellipse is determined it can be either constructed (which is time consuming) or drawn in with an ellipse template (recommended).

REVOLUTION CONVENTIONS

Revolving only a part of a drawing will reduce drawing time. This procedure may simplify many drawings. An example is shown in Figure 19-20.

PROBLEMS

Follow the instructions in Figures 19-21 through 19-28 for drawing objects in revolutions.

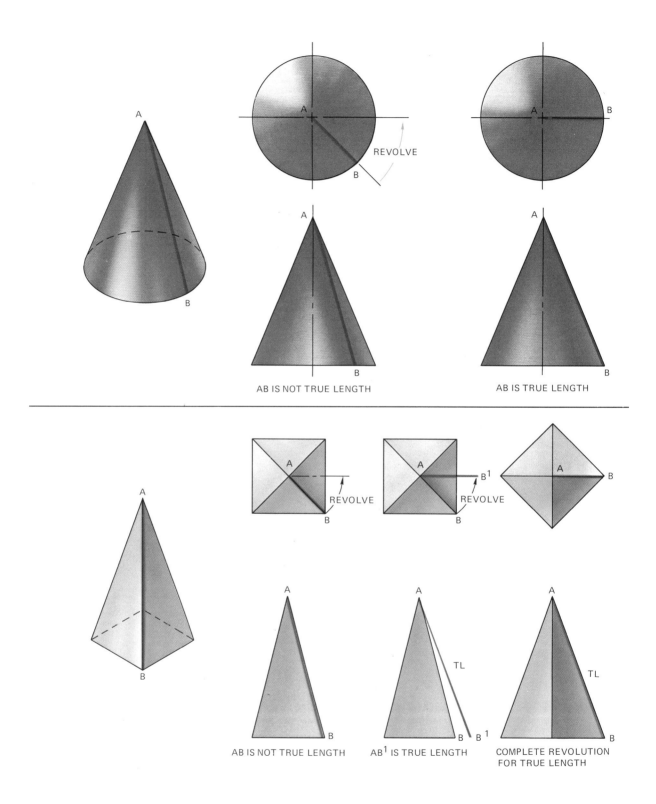

REVOLVE

AB IS NOT TRUE LENGTH

AB IS TRUE LENGTH

REVOLVE

REVOLVE

AB IS NOT TRUE LENGTH

AB^1 IS TRUE LENGTH

COMPLETE REVOLUTION
FOR TRUE LENGTH

TL

TL

FIG. 19-18   Finding the true length of single lines

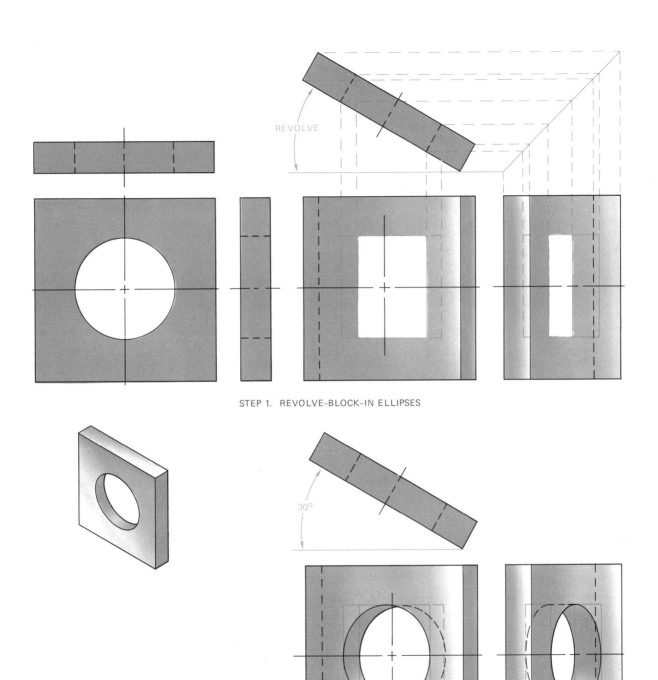

STEP 1. REVOLVE-BLOCK-IN ELLIPSES

STEP 2. DRAW ELLIPSES —
USE ELLIPSE TEMPLATES

60° ELLIPSE

30° ELLIPSE

FIG. 19-19 Circles in revolutions

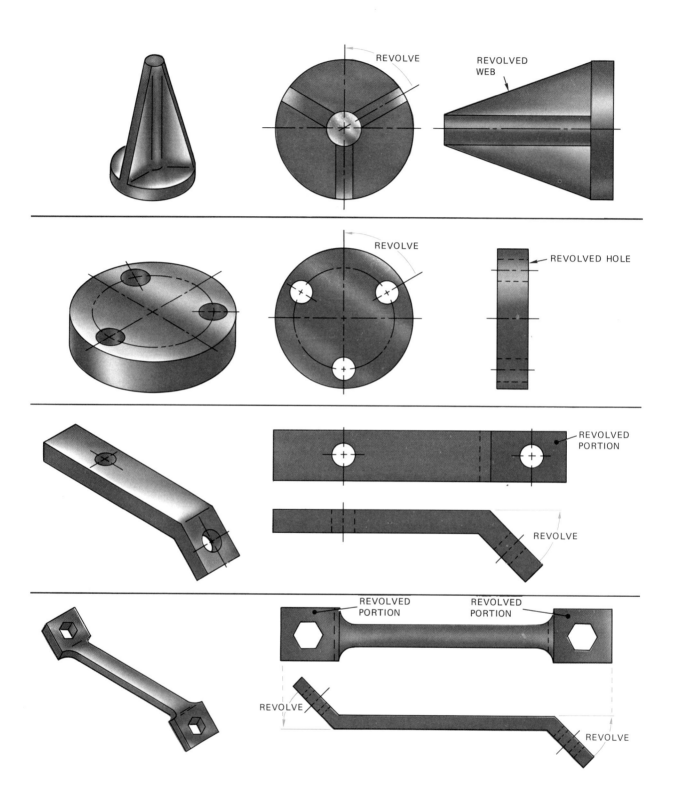

FIG. 19-20 Revolution conventions for drawings

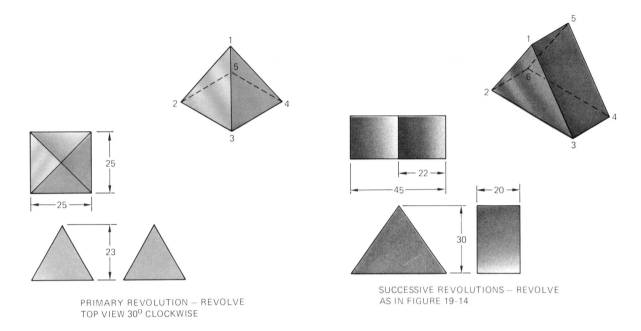

PRIMARY REVOLUTION – REVOLVE
TOP VIEW 30° CLOCKWISE

FIG. 19-21   Problem

SUCCESSIVE REVOLUTIONS – REVOLVE
AS IN FIGURE 19-14

FIG. 19-23   Problem

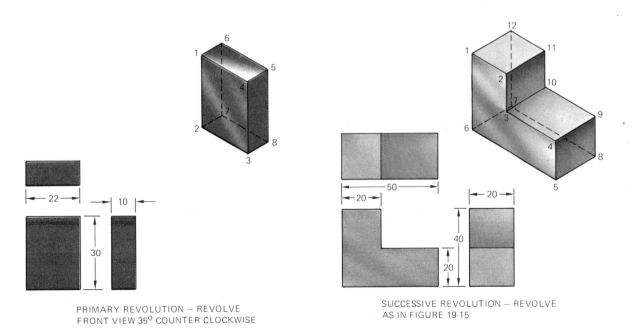

PRIMARY REVOLUTION – REVOLVE
FRONT VIEW 35° COUNTER CLOCKWISE

FIG. 19-22   Problem

SUCCESSIVE REVOLUTION – REVOLVE
AS IN FIGURE 19-15

FIG. 19-24   Problem

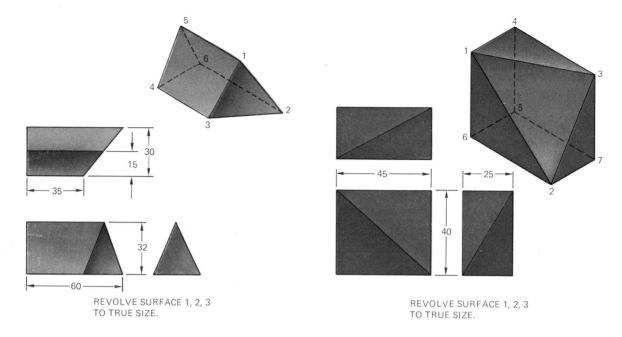

REVOLVE SURFACE 1, 2, 3
TO TRUE SIZE.

FIG. 19-25   Problem

REVOLVE SURFACE 1, 2, 3
TO TRUE SIZE.

FIG. 19-26   Problem

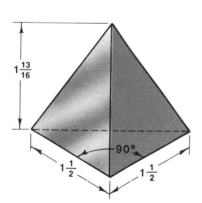

FIG. 19-27   Revolve as in Figure 19-14

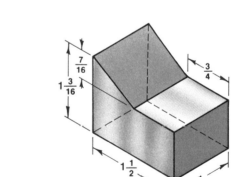

FIG. 19-28   Revolve as in Figure 19-15

Descriptive Geometry

Langley/Monkmeyer

INTRODUCTION

Descriptive geometry is the study of points, lines, and surfaces that may be located in a three-dimensional space. It is a graphic procedure for solving problems of a geometric nature. Many geo-

metric problems that are solved with mathematics can be solved graphically just as fast and accurately. Therefore, the process is an important tool for a draftsperson or designer who is working with geometric forms.

The principles of descriptive geometry are the same as those of orthographic and auxiliary projection. Descriptive geometry permits graphic solutions of problems involving points, lines, surfaces, and solids. Difficult angles and shapes can be drawn to true size and form quickly and accurately.

These principles were developed by a Frenchman named Gaspard Monge. Monge took difficult mathematical concepts and solved them with graphics, or with what we call descriptive geometry.

Because of the relationship of descriptive geometry to orthographic, auxiliary, and revolved views, it is important to have a good understanding of these drafting concepts. In this chapter, only the most basic concepts of descriptive geometry will be covered thoroughly. Students with engineering or math majors should plan on advanced descriptive geometry courses.

As you study this chapter, you should gain the knowledge and skills needed to solve simple descriptive problems.

PLANES OF PROJECTION

Planes of projection are used to visualize an object for a line drawing (Fig. 20-2). The projection planes used in descriptive geometry are orthographic projection planes (see Chapter 12) and auxiliary projection planes (see Chapter 18).

FIRST AND THIRD ANGLES OF PROJECTION

The types of orthographic projection used in descriptive geometry are first and third angles of projection (Fig. 20-3). Chapter 12 defines both methods in detail. The third angle of projection will be used for most of the work in descriptive geometry.

REFERENCE LINES

Reference lines, also called reference plane lines or folding lines, are the corners of imaginery planes of projection (Fig. 20-4). It is very important that the draftsperson have a good understanding of these reference lines, as they are the heart of descriptive geometry.

The orthographic planes (frontal, horizontal, and profile) are shown and labeled in Figure 20-5. These planes are always perpendicular to each other, and the object's projection lines are perpendicular to a plane as shown in Figure 20-2. It is not necessary to draw the outline of each plane. Only the reference line is necessary (Fig. 20-6). Each plane should be labeled for identification.

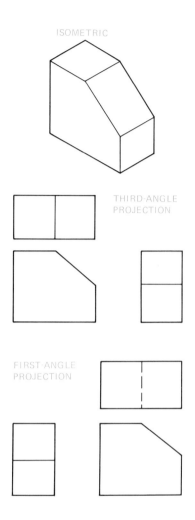

FIG. 20-3 First- and third-angle projection

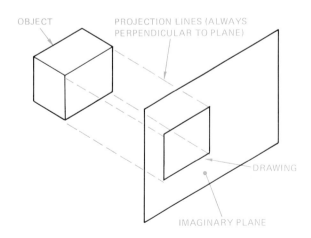

FIG. 20-2 A plane of projection

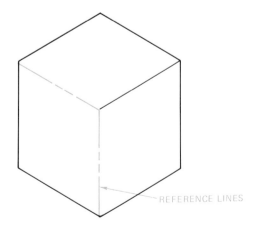

FIG. 20-4 Reference lines

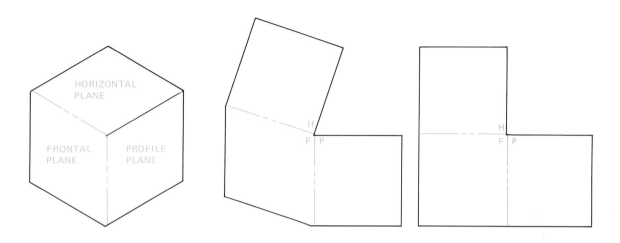

FIG. 20-5 Standard projection planes

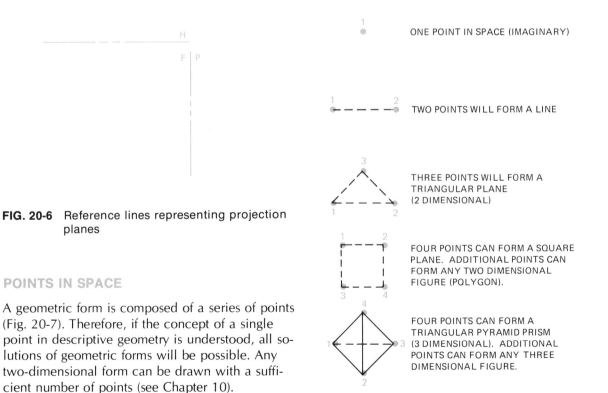

FIG. 20-6 Reference lines representing projection planes

ONE POINT IN SPACE (IMAGINARY)

TWO POINTS WILL FORM A LINE

THREE POINTS WILL FORM A TRIANGULAR PLANE (2 DIMENSIONAL)

FOUR POINTS CAN FORM A SQUARE PLANE. ADDITIONAL POINTS CAN FORM ANY TWO DIMENSIONAL FIGURE (POLYGON).

FOUR POINTS CAN FORM A TRIANGULAR PYRAMID PRISM (3 DIMENSIONAL). ADDITIONAL POINTS CAN FORM ANY THREE DIMENSIONAL FIGURE.

POINTS IN SPACE

A geometric form is composed of a series of points (Fig. 20-7). Therefore, if the concept of a single point in descriptive geometry is understood, all solutions of geometric forms will be possible. Any two-dimensional form can be drawn with a sufficient number of points (see Chapter 10).

A point in space is shown in orthographic projection planes with step-by-step development to the finished descriptive geometry drawing in Figure 20-8. To understand descriptive geometry, it is most

FIG. 20-7 Geometry as arrangements of points

important to understand steps 6 and 7 in Figure 20-8. These steps demonstrate the rule that the set-back distances "X" from a plane will be equal on two or more adjacent planes (Fig. 20-9).

To avoid errors and to make reading descriptive geometry drawings easier, each point should be identified and labeled to the plane it is on. All examples will identify points in this manner. Examples of single point descriptive geometry problems and their solutions are shown in Figure 20-10.

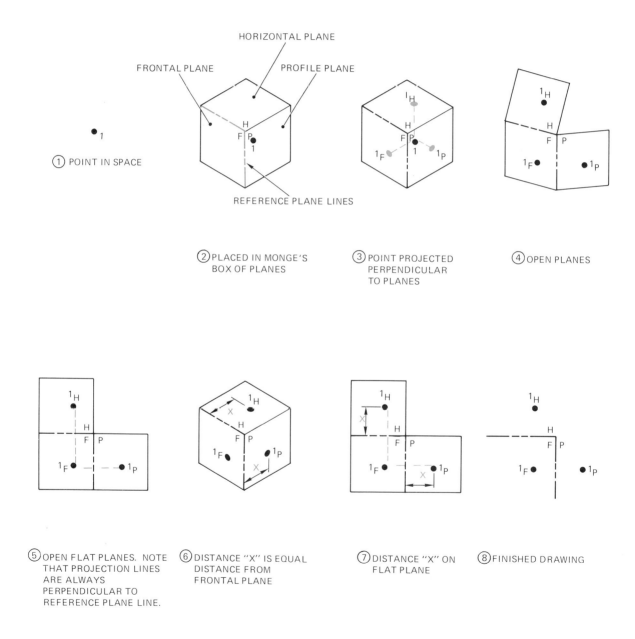

FIG. 20-8   Development of a point in space

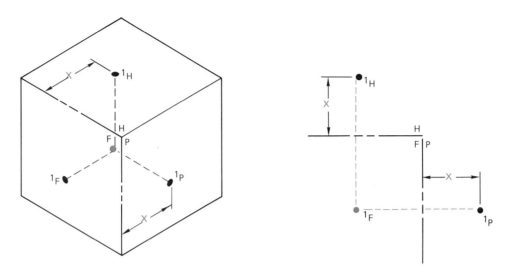

THE SET-BACK DISTANCES "X" FROM THE FRONTAL PLANE ARE IDENTICAL ON THE
HORIZONTAL AND PROFILE PLANE. PLANES P AND H ARE ADJACENT TO PLANE F.

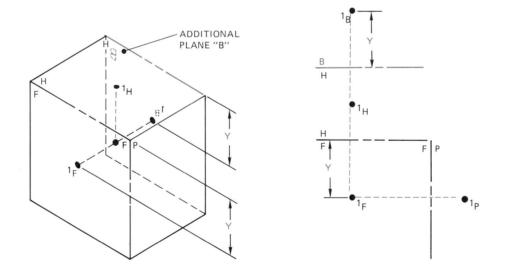

PLANES B AND F ARE ADJACENT TO PLANE H.
THEREFORE DISTANCES "Y" ARE EQUAL.

FIG. 20-9 Locating a point in space

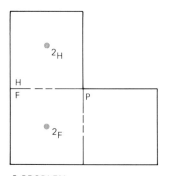

I PROBLEM:

DRAW THE POINT IN THE
PROFILE PLANE

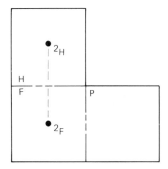

SOLUTION:

① DRAW PERPENDICULAR
CONNECTORS

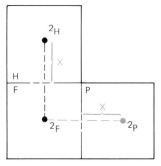

② DRAW PERPENDICULAR
TRANSFER DISTANCE "X"

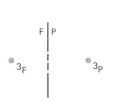

II. PROBLEM:

DRAW THE POINT IN THE
HORIZONTAL PLANE

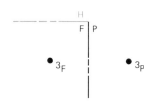

SOLUTION:

① DRAW A HORIZONTAL REFERENCE
LINE AT A CONVENIENT LOCATION

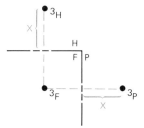

② DRAW PERPENDICULAR
CONNECTORS AND
TRANSFER "X"

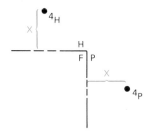

III. PROBLEM:

DRAW THE POINT IN THE
FRONTAL PLANE

NOTE: IT IS CRITICAL TO
REMEMBER THAT "X" DISTANCES
ARE ALWAYS EQUAL.

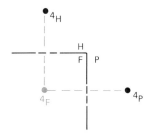

SOLUTION:

① PROJECT THE POINTS PERPENDICULAR
TO THE FRONTAL PLANE. THE POINT IS
LOCATED AT THE INTERSECTION.

FIG. 20-10 Problems of projecting a point in space

EXAMPLES OF DESCRIPTIVE GEOMETRY WITH
A SINGLE POINT

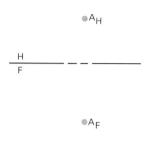

IV. PROBLEM:

    FIND POINT "A" ON
    THE PROFILE PLANE

SOLUTION:

①   FOR POINT "A" ON THE PROFILE PLANE

NOTE: REFERENCE LINE FP IS PERPENDICULAR
         TO REFERENCE LINE HF. FP MAY BE
         PLACED AT ANY CONVENIENT DISTANCE.

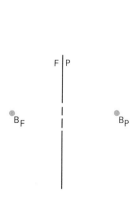

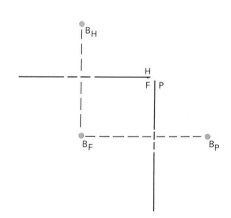

V. PROBLEM

    FIND POINT "B" ON THE
    HORIZONTAL PLANE

SOLUTION:

①   FOR POINT "B" ON THE HORIZONTAL PLANE

FIG. 20-10   Problems of projecting a point in space (Cont.)

LINE IN SPACE

A line is formed by two points. By projecting the ends of the lines as two points, problems of a line in space can be solved (Fig. 20-11). The steps necessary to complete the finished descriptive drawing are shown in Figure 20-12. All of the concepts used for a single point are used for two points (a line). Examples of descriptive geometry problems for a line in space and their solutions are shown in Figure 20-13.

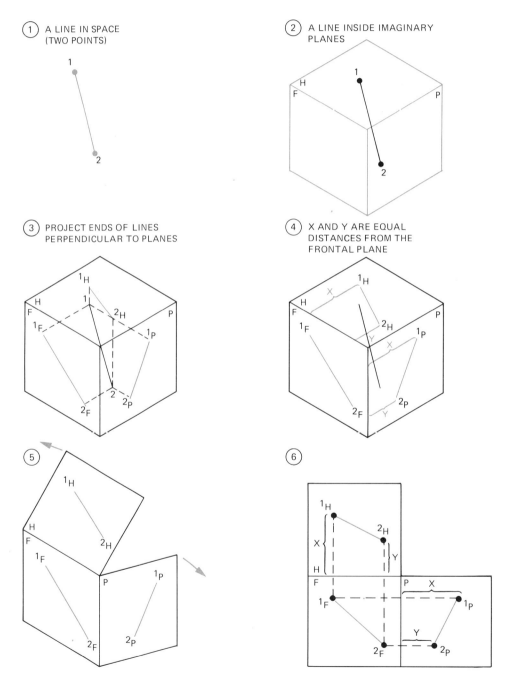

FIG. 20-11   Projecting a line in space

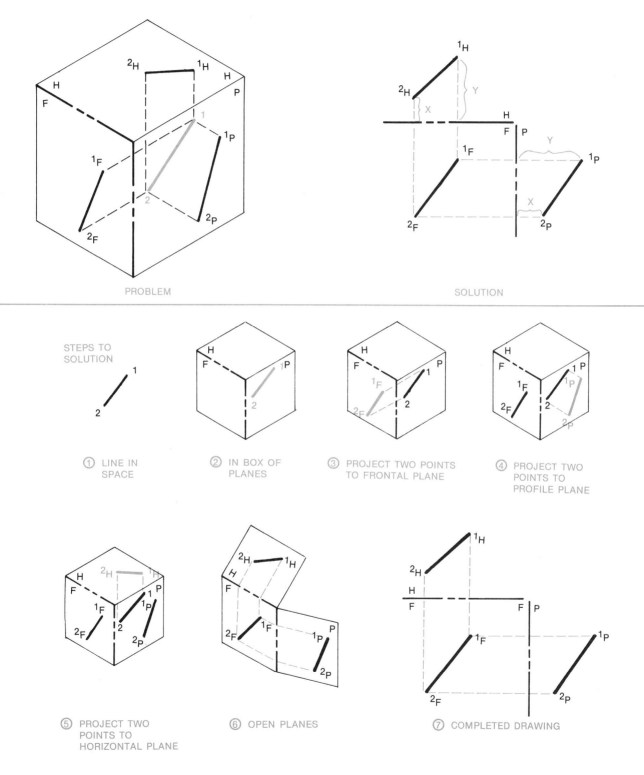

PROBLEM   SOLUTION

STEPS TO
SOLUTION

① LINE IN
   SPACE

② IN BOX OF
   PLANES

③ PROJECT TWO POINTS
   TO FRONTAL PLANE

④ PROJECT TWO
   POINTS TO
   PROFILE PLANE

⑤ PROJECT TWO
   POINTS TO
   HORIZONTAL PLANE

⑥ OPEN PLANES

⑦ COMPLETED DRAWING

FIG. 20-12   Steps in projecting a line

EXAMPLES OF DESCRIPTIVE GEOMETRY WITH ONE LINE

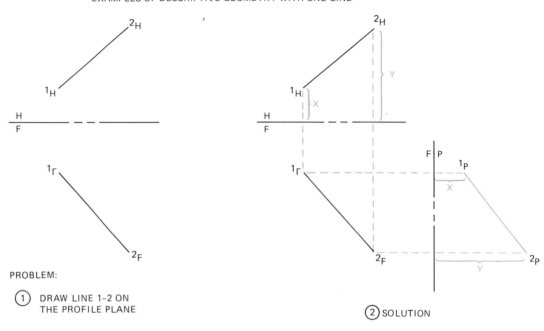

PROBLEM:

① DRAW LINE 1–2 ON
THE PROFILE PLANE

② SOLUTION

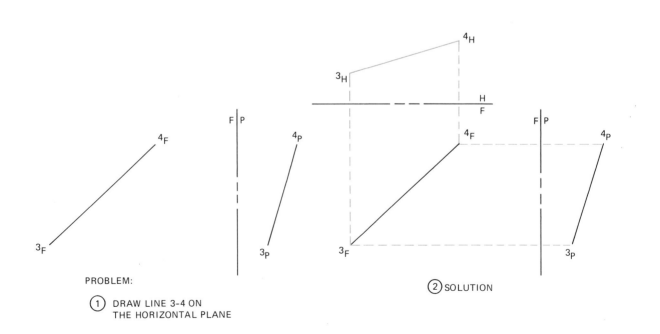

PROBLEM:

① DRAW LINE 3-4 ON
THE HORIZONTAL PLANE

② SOLUTION

FIG. 20-13 Examples of projecting a line

AUXILIARY PLANES

Up to this point, descriptive geometry has been described with the orthographic planes of projection: frontal, horizontal, and profile. For the flexibility of being able to draw a point in any position, an auxiliary plane of projection may be used. An auxiliary plane will have its reference line at an angle to the orthographic reference lines (Fig. 20-14). The auxiliary plane will always be perpendicular to the adjacent plane (Fig. 20-15).

Figure 20-16 shows an auxiliary plane next to the frontal plane. Figure 20-17 shows an auxiliary plane next to the horizontal plane. The number of auxiliary planes that can be selected from any adja-

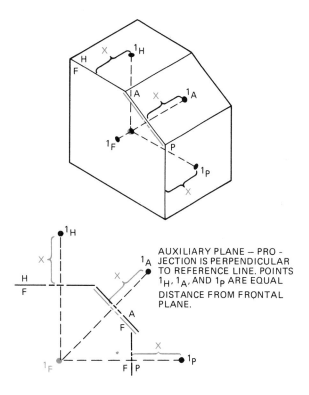

AUXILIARY PLANE – PROJECTION IS PERPENDICULAR TO REFERENCE LINE. POINTS 1_H, 1_A, AND 1_P ARE EQUAL DISTANCE FROM FRONTAL PLANE.

FIG. 20-16 Auxiliary plane off the frontal plane

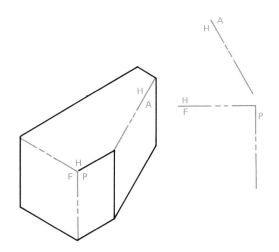

FIG. 20-14 An auxiliary plane

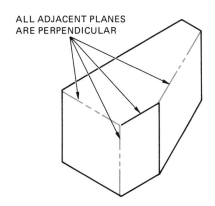

FIG. 20-15 Perpendicularity of planes

THE REFERENCE PLANE LINE CAN BE PLACED IN ANY POSITION TO GET THE REQUIRED AUXILIARY VIEW OF AN OBJECT. "X" DISTANCES ARE EQUAL.

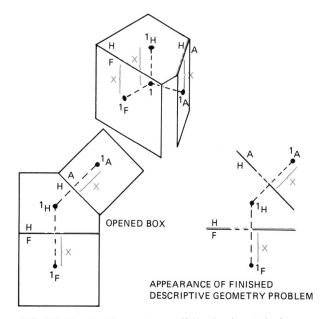

OPENED BOX

APPEARANCE OF FINISHED DESCRIPTIVE GEOMETRY PROBLEM

FIG. 20-17 Auxiliary plane off the horizontal plane

cent plane is unlimited (Fig. 20-18). This gives the draftsperson the flexibility of drawing an item in any position. Figure 20-19 shows a line (two points) with an auxiliary plane next to the horizontal plane. The concepts used for descriptive geometry with orthographic planes will apply to auxiliary planes.

Examples of auxiliary plane problems and their solutions in descriptive geometry are shown in Figure 20-20.

The placement of auxiliary or orthographic reference plane lines (frontal, horizontal, profile, and auxiliary planes) will control the location of the view in each plane. The views will always remain the same, but the draftsperson can control the location of the view on each plane (Fig. 20-21).

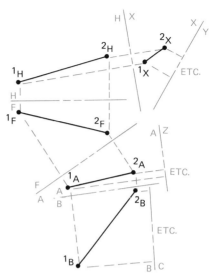

NOTE THAT IDENTICAL POINTS OFF ADJACENT VIEWS ARE ALWAYS EQUAL IN DISTANCE FROM THE REFERENCE LINES

FIG. 20-18   The number of views is unlimited

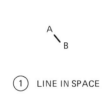

① LINE IN SPACE

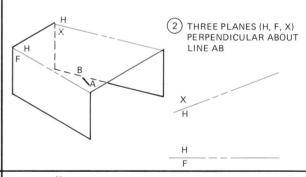

② THREE PLANES (H, F, X) PERPENDICULAR ABOUT LINE AB

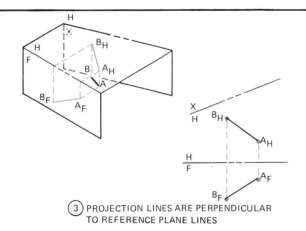

③ PROJECTION LINES ARE PERPENDICULAR TO REFERENCE PLANE LINES

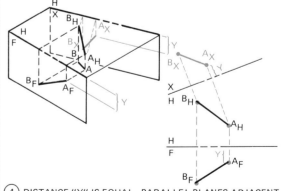

④ DISTANCE "Y" IS EQUAL. PARALLEL PLANES ADJACENT TO THE SAME PLANE WILL HAVE IDENTICAL POINTS EQUAL DISTANCE FROM THE REFERENCE LINE.

FIG. 20-19   Projecting a line off the horizontal plane

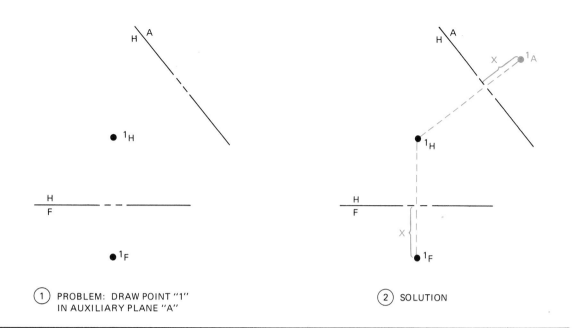

1 PROBLEM: DRAW POINT "1"
   IN AUXILIARY PLANE "A"

2 SOLUTION

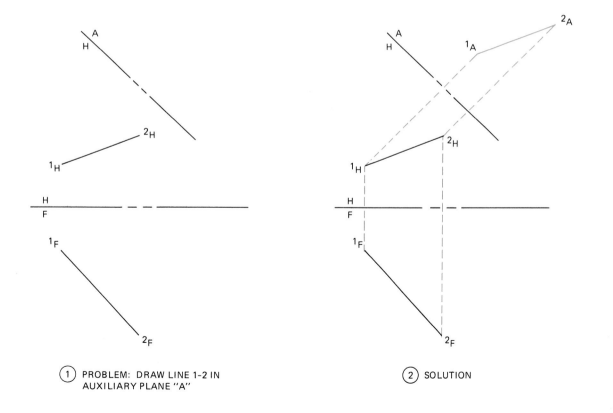

1 PROBLEM: DRAW LINE 1-2 IN
   AUXILIARY PLANE "A"

2 SOLUTION

FIG. 20-20 Examples of auxiliary plane projection

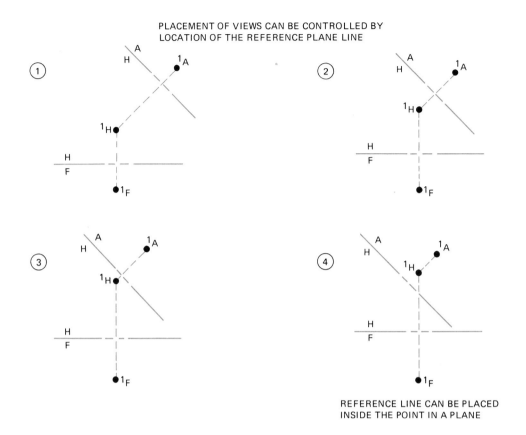

PLACEMENT OF VIEWS CAN BE CONTROLLED BY
LOCATION OF THE REFERENCE PLANE LINE

REFERENCE LINE CAN BE PLACED
INSIDE THE POINT IN A PLANE

FIG. 20-21   Placement of views

TRUE LENGTH OF A LINE

To find the true length of a line, locate the reference line parallel to the line in any plane, and project the line into the new plane. This line will be true length (Fig. 20-22). An example of a problem in finding the true length of a line is shown in Figure 20-23.

POINT VIEW OF A LINE

Once the true length of a line is drawn on a plane, the point view of the line can be determined. The point view of a line is the view looking directly at either end. This view can be drawn by placing a reference line perpendicular to the true length of a line as shown in Figure 20-24. An example of a problem in finding the point view of a line is shown in Figure 20-25.

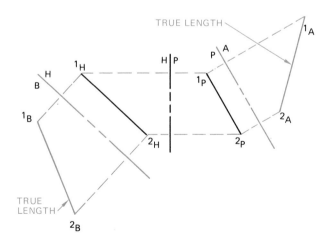

LINE 1–2 IS TRUE LENGTH AND EQUAL IN PLANES A AND B.
NOTE THAT REFERENCE LINES ARE PARALLEL TO LINE 1–2
IN THE ADJACENT PLANES.

FIG. 20-22   True length of a line

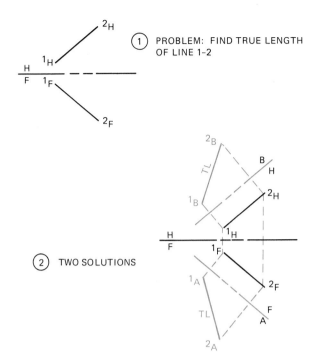

① PROBLEM: FIND TRUE LENGTH OF LINE 1–2

② TWO SOLUTIONS

FIG. 20-23 Problem in finding true length

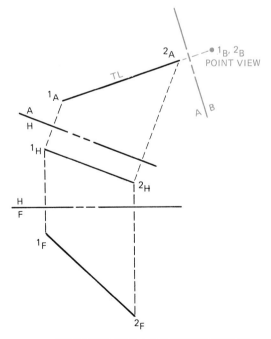

REFERENCE LINE A–B IS PERPENDICULAR TO TRUE LENGTH LINE 1$_A$–2$_A$

FIG. 20-24 Point view of a line

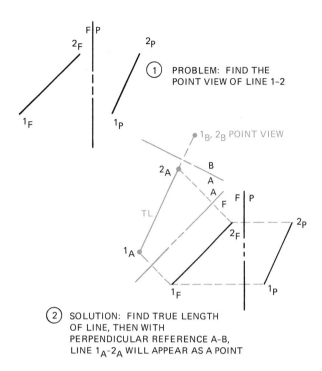

① PROBLEM: FIND THE POINT VIEW OF LINE 1–2

② SOLUTION: FIND TRUE LENGTH OF LINE, THEN WITH PERPENDICULAR REFERENCE A–B, LINE 1$_A$–2$_A$ WILL APPEAR AS A POINT

FIG. 20-25 Problem in finding the point view of a line

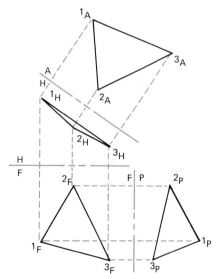

PROJECTION IN DESCRIPTIVE GEOMETRY IS THE SAME, REGARDLESS OF THE NUMBER OF POINTS

FIG. 20-26 Projection of a plane in space

PLANE IN SPACE

Any flat surface can be considered a plane. The simplest form of a plane, the triangle, is made by three points. The process of projecting a plane is identical to the process of projecting a single point and two points (a line). If each corner of a plane is considered to be a point, then any form can be drawn in any plane of descriptive geometry. Figure 20-26 shows an example of a plane in space. Each corner is treated as a point in space and projected to the adjacent plane.

Examples of triangular plane problems in descriptive geometry and their solutions are shown in Figure 20-27.

EDGE VIEW OF A PLANE

To draw the edge view of a plane in space it is necessary to place a reference line perpendicular to a true length edge of the plane (Fig. 20-28), or a true length line on the plane. A line placed on a plane parallel to a reference is called a strike line. This is a procedure for quickly obtaining a true

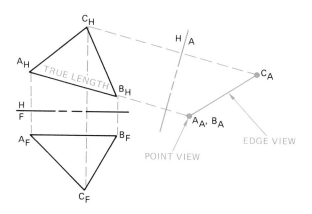

NOTE:

REFERENCE H-F IS PARALLEL TO EDGE A_F-B_F. THEREFORE LINE A_H-B_H WILL BE A TRUE LENGTH LINE.

REFERENCE LINE H-A IS PERPENDICULAR TO EDGE A_H-B_H. THEREFORE THE TRIANGULAR PLANE ABC WILL APPEAR AS AN EDGE VIEW IN PLANE "A".

EDGE A_H-B_H, TRUE LENGTH LINE WILL APPEAR AS A POINT VIEW IN PLANE "A".

FIG. 20-28 Edge view of a plane

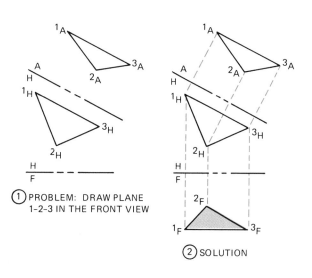

① PROBLEM: DRAW PLANE 1-2-3 IN THE FRONT VIEW

② SOLUTION

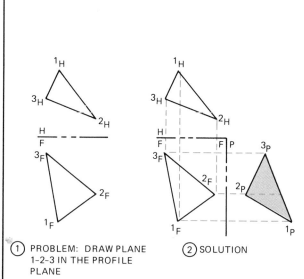

① PROBLEM: DRAW PLANE 1-2-3 IN THE PROFILE PLANE

② SOLUTION

FIG. 20-27 Problem in projecting a triangular plane

length line on a plane in space (Fig. 20-29). A reference line perpendicular to any true length line on a triangular plane will show that triangular plane as an edge view (Fig. 20-30).

Examples of problems for drawing an edge view of a plane are shown in Figure 20-31.

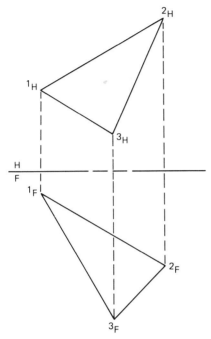

① FRONT AND HORIZONTAL VIEWS OF A PLANE

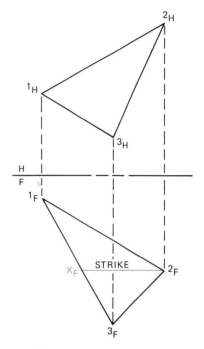

② DRAW A LINE (STRIKE LINE) PARALLEL TO THE REFERENCE LINE

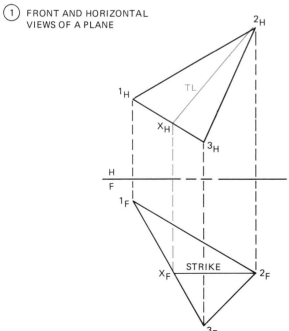

③ PROJECT STRIKE LINE TO ADJACENT PLANE FOR TRUE LENGTH LINE ON A PLANE

FIG. 20-29   Finding a true length line on a plane

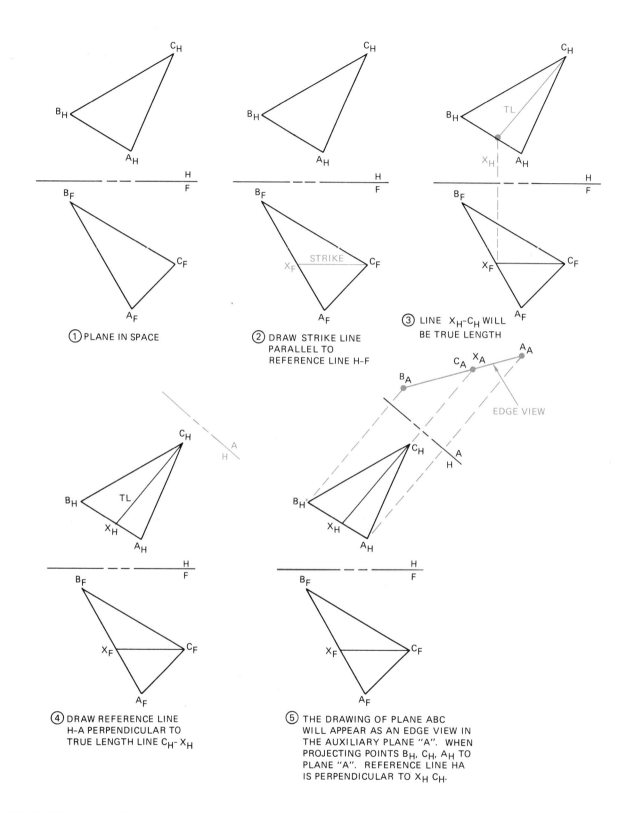

① PLANE IN SPACE

② DRAW STRIKE LINE PARALLEL TO REFERENCE LINE H–F

③ LINE X_H–C_H WILL BE TRUE LENGTH

④ DRAW REFERENCE LINE H–A PERPENDICULAR TO TRUE LENGTH LINE C_H- X_H

⑤ THE DRAWING OF PLANE ABC WILL APPEAR AS AN EDGE VIEW IN THE AUXILIARY PLANE "A". WHEN PROJECTING POINTS B_H, C_H, A_H TO PLANE "A". REFERENCE LINE HA IS PERPENDICULAR TO X_H C_H.

FIG. 20-30 Finding the edge view of a plane

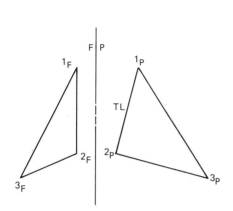

① PROBLEM: FIND EDGE VIEW OF
TRIANGLE 1-2-3. NOTE EDGE
1_F-2_F IS PARALLEL TO
REFERENCE LINE F-P.

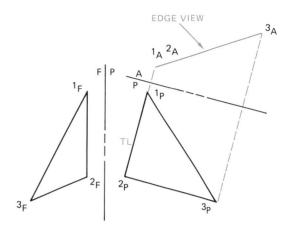

② SOLUTION: PLACE REFERENCE
LINE P-A PERPENDICULAR TO
TRUE LENGTH EDGE 1_P-2_P

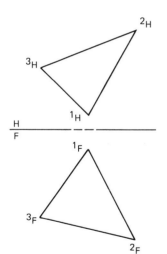

① PROBLEM: FIND EDGE VIEW
OF TRIANGLE 1-2-3

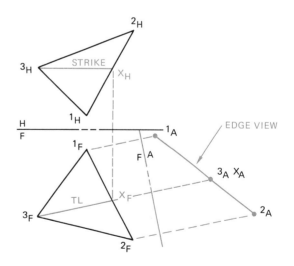

② SOLUTION: – DRAW STRIKE LINE 3_H-X_H
– DRAW TRUE LENGTH LINE 3_F-X_F
– DRAW REFERENCE LINE F-A
PERPENDICULAR TO LINE 3_F-X_F

FIG. 20-31 Problem in drawing the edge view of a
plane

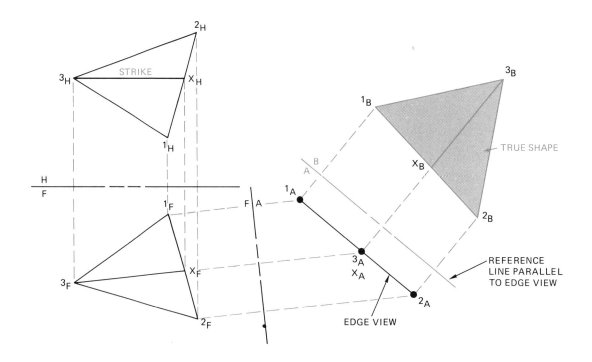

A PERPENDICULAR PROJECTION FROM AN EDGE VIEW WILL SHOW
THAT PLANE AS A TRUE SHAPE

GIVEN: PLANES H AND F

SOLUTION: 1. DRAW STRIKE LINE 3_H-X_H PARALLEL TO REFERENCE LINE H-F.
               2. PROJECT STRIKE LINE TO PLANE F (3_F-X_F).
               3. DRAW AUXILIARY REFERENCE LINE F-A PERPENDICULAR TO LINE 3_F-X_F.
               4. DRAW EDGE VIEW OF PLANE 1, 2, 3 ON PLANE A.
               5. DRAW AUXILIARY REFERENCE LINE A-B PARALLEL TO EDGE VIEW
                   1_A, 3_A, 2_A.
               6. PERPENDICULAR PROJECTION FROM THE EDGE VIEW WILL GIVE A
                   TRUE SHAPE OF PLANE 1, 2, 3.

FIG. 20-32 True shape of a plane

TRUE SHAPE OF A PLANE

The true shape of a plane can be drawn when the
line of sight is perpendicular to the edge of the
plane. This is done by placing the reference line
parallel to the edge view (Fig. 20-32). Additional
examples of true shapes are shown in Figures 20-33
and 20-34. The complete solution to find a true
shape is shown step-by-step in Figure 20-35.

Examples of problems and their solutions for
drawing the true shape of a plane are shown in Figure 20-36.

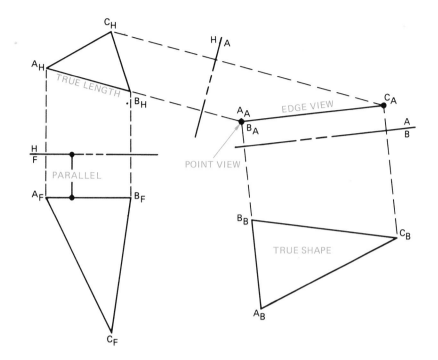

FIG. 20-33 Examples of true shapes

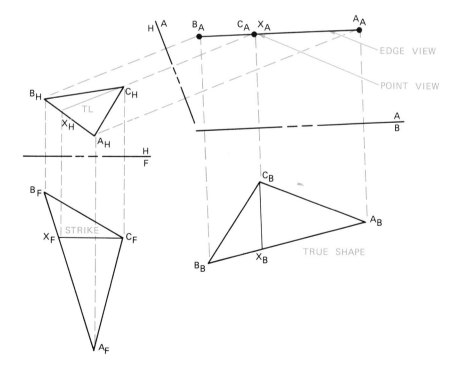

FIG. 20-34 Examples of true shapes

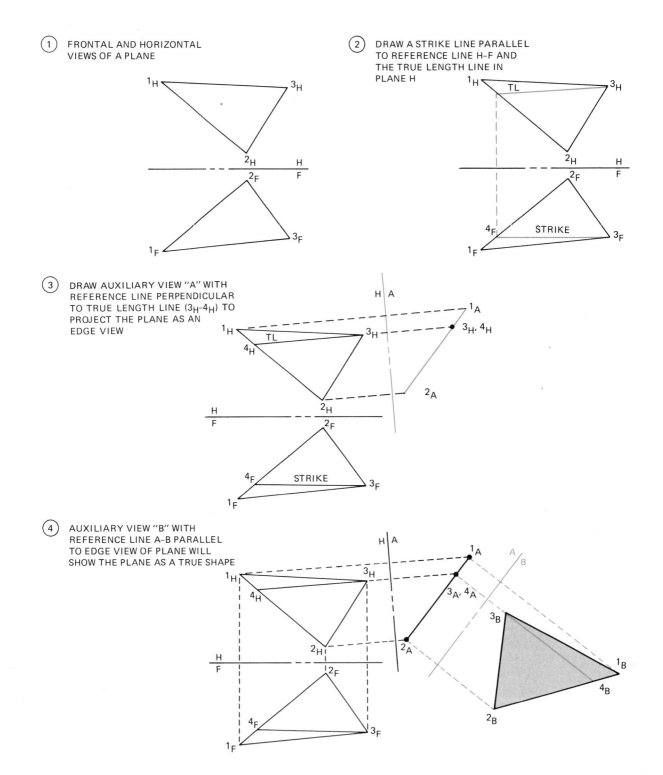

FIG. 20-35   Steps in finding the true shape of a plane

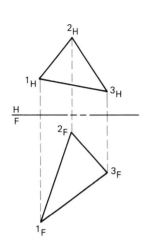

(1) PROBLEM: FIND THE
     TRUE SHAPE

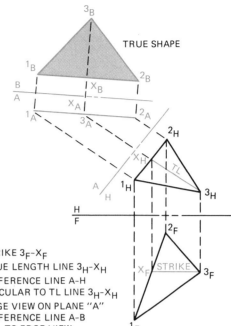

(2) SOLUTION:
1.   DRAW STRIKE 3_F-X_F
2.   DRAW TRUE LENGTH LINE 3_H-X_H
3.   PLACE REFERENCE LINE A-H
     PERPENDICULAR TO TL LINE 3_H-X_H
4.   DRAW EDGE VIEW ON PLANE "A"
5.   PLACE REFERENCE LINE A-B
     PARALLEL TO EDGE VIEW
6.   PROJECT TRIANGULAR PLANE ON
     PLANE "B". THIS WILL BE A
     TRUE SHAPE.

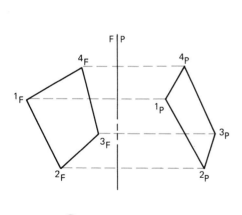

(1) PROBLEM: FIND THE
     TRUE SHAPE

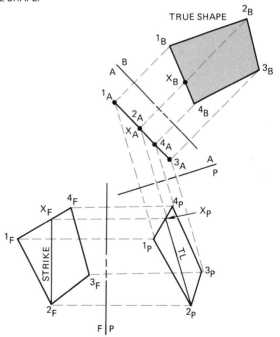

FIG. 20-36 Problems in finding a true plane

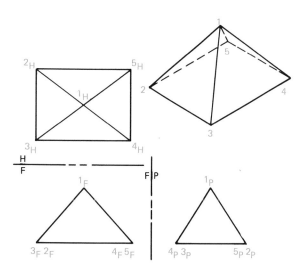

AN ORTHOGRAPHIC DESCRIPTIVE GEOMETRY DRAWING OF A
PYRAMID WITH FRONTAL, HORIZONTAL, AND PROFILE PLANES.

FIG. 20-37   A pyramid in space

SOLIDS IN SPACE

A solid is a three-dimensional form. If each corner of a solid is treated as a point, its form can be projected by using the same rules as for a single point (Fig. 20-37). The three basic orthographic planes for a prism are the frontal, horizontal, and profile planes. The number of possible auxiliary views is unlimited (Fig. 20-38).

Examples of problems and their solutions for drawing solids are shown in Figure 20-39.

VISIBILITY

With regard to lines on a prism, it is important to determine which lines are in front and will therefore be entirely visible. The lines behind the prism will be partially hidden. Using the center lines for overlapping cylinders, Figure 20-40 shows how to

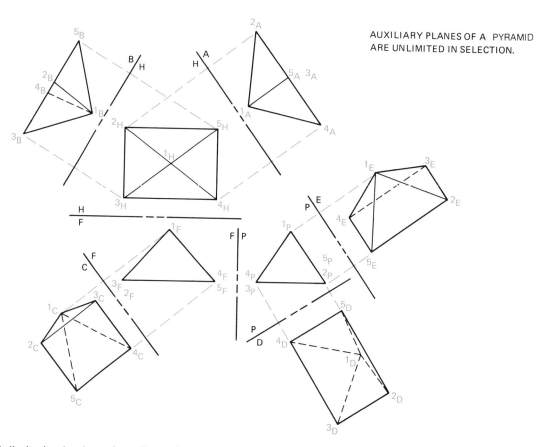

AUXILIARY PLANES OF A PYRAMID
ARE UNLIMITED IN SELECTION.

FIG. 20-38   Unlimited selection of auxiliary planes

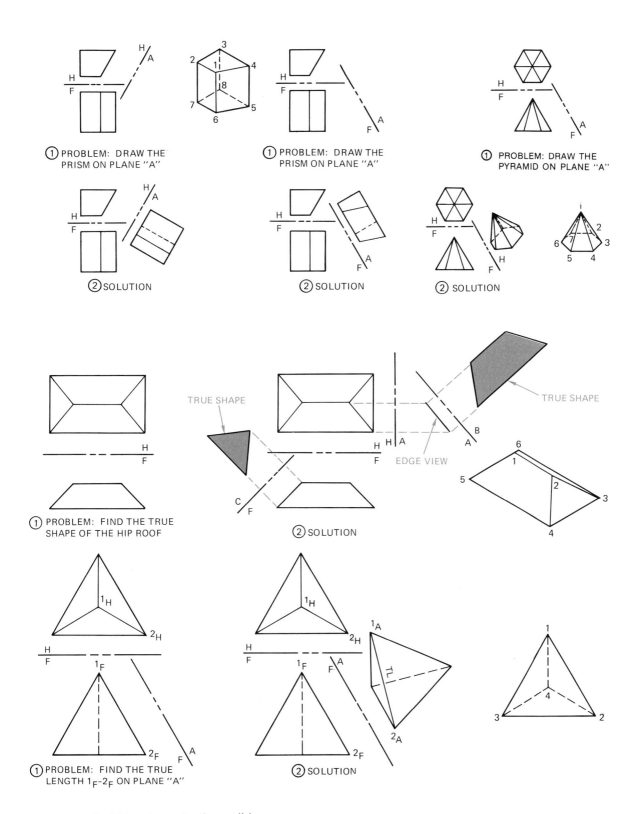

FIG. 20-39 Problems in projecting solids

determine which cylinder is in front. If the overlapping points are in direct line of projection, the lines are intersecting (Fig. 20-41).

Examples of problems and their solutions for visibility are shown in Figure 20-42.

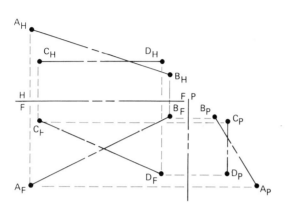

PROBLEM:

① CENTER LINES FOR OVERLAPPING CYLINDERS ARE SHOWN IN THREE PLANES. ALL DESCRIPTIVE GEOMETRY WILL APPLY.

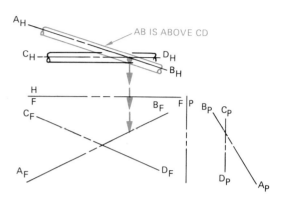

② TO FIND THE OVERLAPPING CYLINDER ON THE HORIZONTAL PLANE:

a. PROJECT "CROSS OVER POINT" ON THE HORIZONTAL PLANE TO THE FRONTAL PLANE.

b. THE CENTER LINE FIRST REACHED ON THE FRONTAL PLANE IS A_F-B_F.

c. THEREFORE CYLINDER A-B IN THE HORIZONTAL PLANE IS VISIBLE.

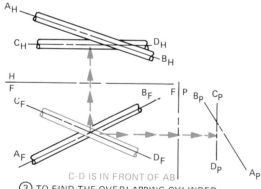

③ TO FIND THE OVERLAPPING CYLINDER ON THE FRONTAL PLANE:

a. PROJECT "CROSS OVER POINT" ON THE FRONTAL PLANE TO THE HORIZONTAL PLANE (OR TO THE PROFILE PLANE).

b. THE CENTER LINE FIRST REACHED ON THE HORIZONTAL PLANE (OR PROFILE PLANE) IS C-D.

c. THEREFORE CYLINDER C-D IN THE FRONTAL PLANE IS VISIBLE.

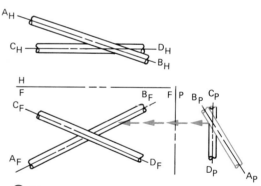

④ TO FIND THE OVERLAPPING CYLINDER ON THE PROFILE PLANE.

a. PROJECT "CROSS OVER POINT" ON THE PROFILE PLANE TO THE FRONTAL PLANE.

b. THE CENTER FIRST REACHED ON THE FRONT PLANE IS A_F-B_F.

c. THEREFORE CYLINDER A-B IN THE PROFILE PLANE IS VISIBLE.

FIG. 20-40 Finding the visibility of overlapping lines that do not intersect

LINES WILL INTERSECT
WHEN OVERLAPPING
POINT IS IN LINE.

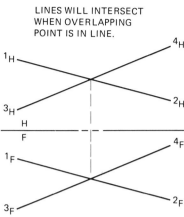

FIG. 20-41   Lines will intersect when in line

BEARING

Bearing is the direction of a line on the earth's surface. The horizontal plane is parallel to the earth's surface, therefore the bearing of a line will be on the horizontal plane and will be expressed in degrees with respect to north and south (Fig. 20-43). North will be assumed to be at the top of the drawing unless specified. Some bearings are shown in Figure 20-44.

Examples of problems and their solutions for specifying bearing are shown in Figure 20-45.

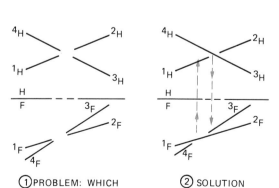

① PROBLEM: WHICH
 LINE IS VISIBLE?

② SOLUTION: BOTH,
 THEY INTERSECT.

① PROBLEM: WHICH
 LINE IS VISIBLE?

② SOLUTION

FIG. 20-42   Problems in overlapping lines

① THE TOP OF THE HORIZONTAL
 PLANE IS ALWAYS DUE NORTH

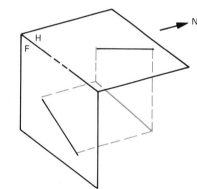

② THE BEARING IS ALWAYS FOUND
 ON THE HORIZONTAL PLANE

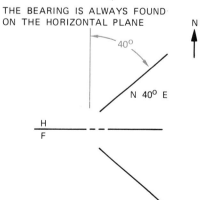

N 40° E

FIG. 20-43   Finding the bearing of a line

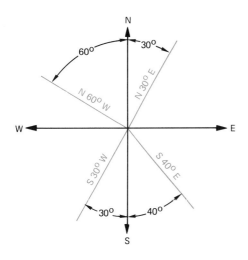

FIG. 20-44 Examples of line bearings

SLOPE

The slope of a line is the angle in degrees that the line makes with the horizontal plane. To show the slope of a line, the line must be true length and in a plane adjacent to the horizontal plane. The slope is the angle between the true length line and its horizontal reference line (Fig. 20-46). A line can be located in space by giving its slope and bearing.

To find the slope of a triangular plane, show an edge view taken off a true length line on a plane adjacent to the horizontal plane. The slope is the angle between the edge view of the triangular plane and its horizontal reference line (Fig. 20-47).

Examples of problems and their solutions for specifying slope are shown in Figure 20-48.

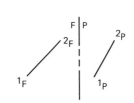

(1) PROBLEM: WHAT IS THE
BEARING OF LINE 1-2?

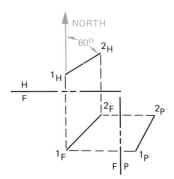

(2) SOLUTION: 60° NORTHEAST
(N 60° E)

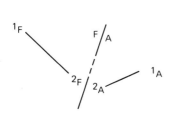

(1) PROBLEM: WHAT IS THE
BEARING OF LINE 1-2?

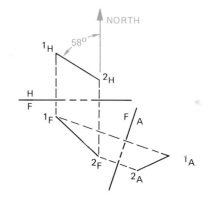

(2) SOLUTION: 58° NORTHWEST
(N 58° W)

FIG. 20-45 Problems in bearing

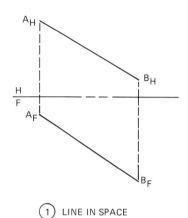

(1) LINE IN SPACE

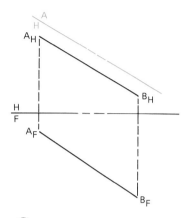

(2) DRAW REFERENCE LINE
PARALLEL TO LINE A_H-B_H

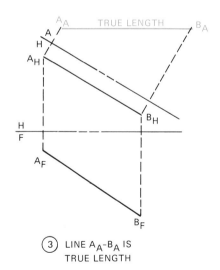

(3) LINE A_A-B_A IS
TRUE LENGTH

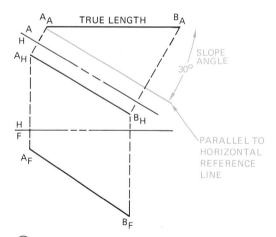

(4) THE ANGLE BETWEEN THE
TRUE LENGTH LINE AND THE
HORIZONTAL REFERENCE
LINE IS THE SLOPE. THE TRUE
SLOPE CAN ONLY BE FOUND IN
A VIEW ADJACENT TO THE
HORIZONTAL PLANE AND ONLY
WHEN THE LINE IN THE
ADJACENT VIEW IS TRUE LENGTH.

FIG. 20-46   Slope of a line

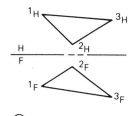

① TRIANGULAR PLANE
IN SPACE

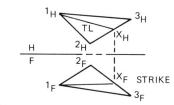

② DRAW STRIKE LINE ON FRONTAL
PLANE AND TRUE LENGTH LINE
ON HORIZONTAL PLANE

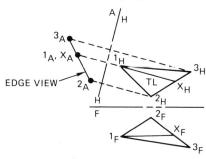

③ PLACE REFERENCE LINE A–H
PERPENDICULAR TO TRUE
LENGTH LINE 1_H–X_H. DRAW
EDGE VIEW ON PLANE "A".

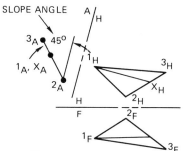

④ THE ANGLE BETWEEN THE EDGE
VIEW AND THE HORIZONTAL
REFERENCE LINE IS THE SLOPE

FIG. 20-47   Slope of a line

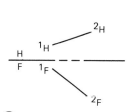

① PROBLEM: WHAT IS THE
SLOPE OF LINE 1–2?

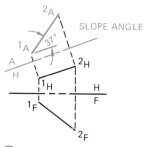

② SOLUTION: 35°

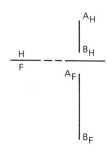

① PROBLEM: WHAT IS THE
SLOPE OF LINE A–B?

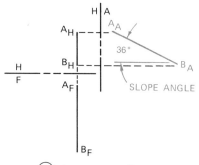

② SOLUTION: 40°

FIG. 20-48   Problem

PROBLEMS

Solve the descriptive geometry problems in Figures 20-49 through 20-60. Use dividers to transfer the problems to drawing paper, or copy the page of problems. Doubling the size of the problem will make the solution more accurate and easier to draw.

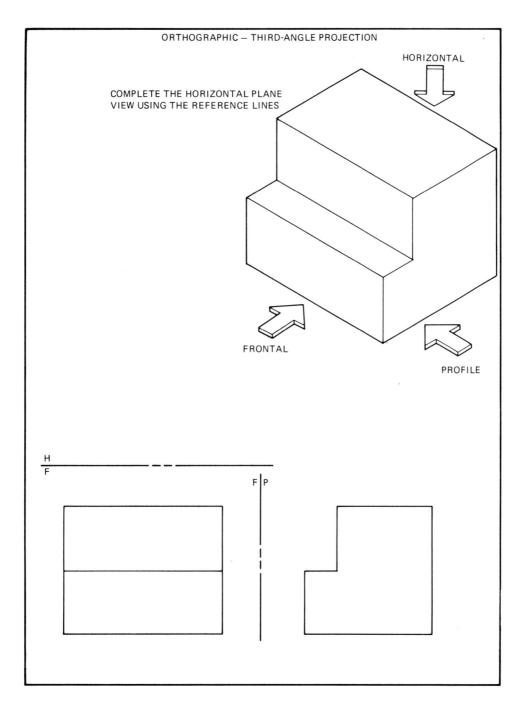

FIG. 20-49 Problem

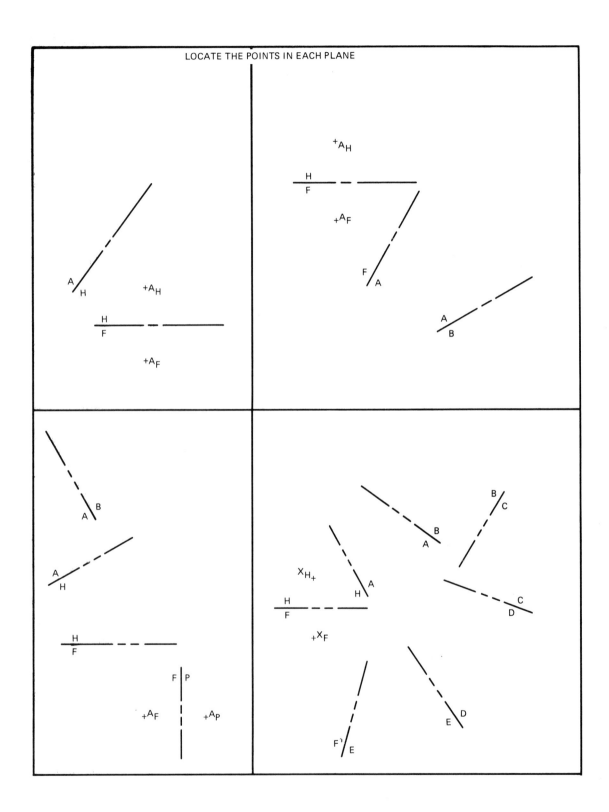

FIG. 20-50 Problem

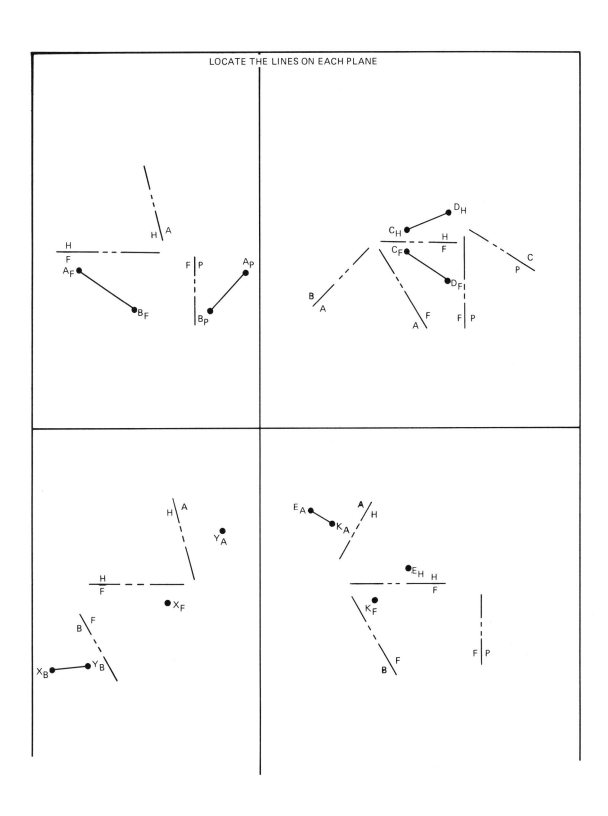

LOCATE THE LINES ON EACH PLANE

FIG. 20-51 Problem

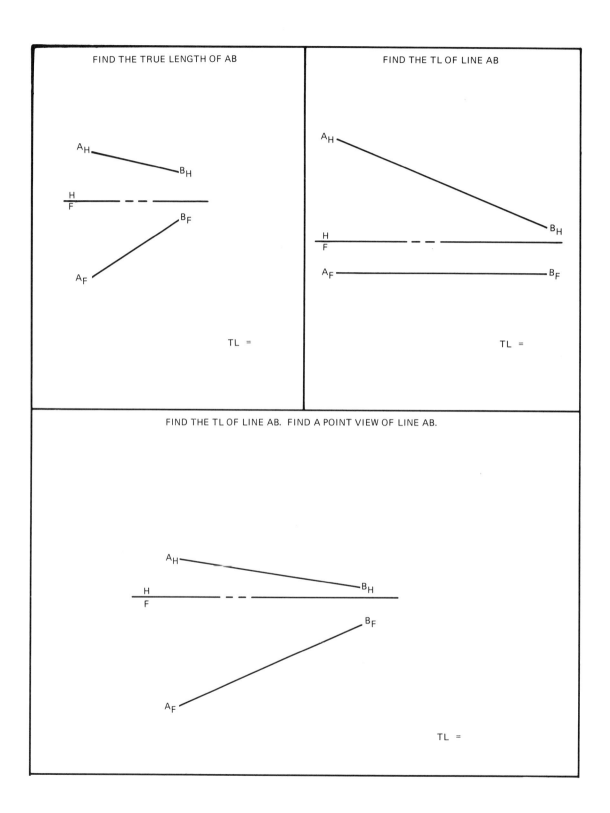

FIND THE TRUE LENGTH OF AB

TL =

FIND THE TL OF LINE AB

TL =

FIND THE TL OF LINE AB. FIND A POINT VIEW OF LINE AB.

TL =

FIG. 20-52 Problem

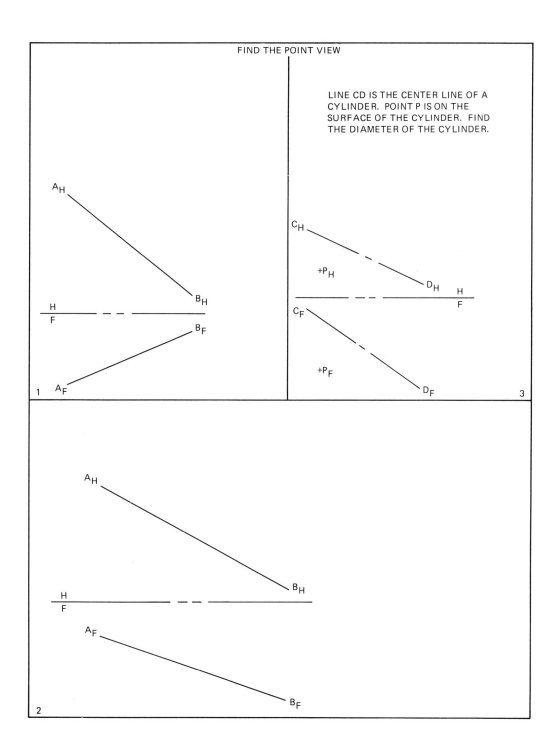

FIND THE POINT VIEW

LINE CD IS THE CENTER LINE OF A
CYLINDER. POINT P IS ON THE
SURFACE OF THE CYLINDER. FIND
THE DIAMETER OF THE CYLINDER.

FIG. 20-53 Problem

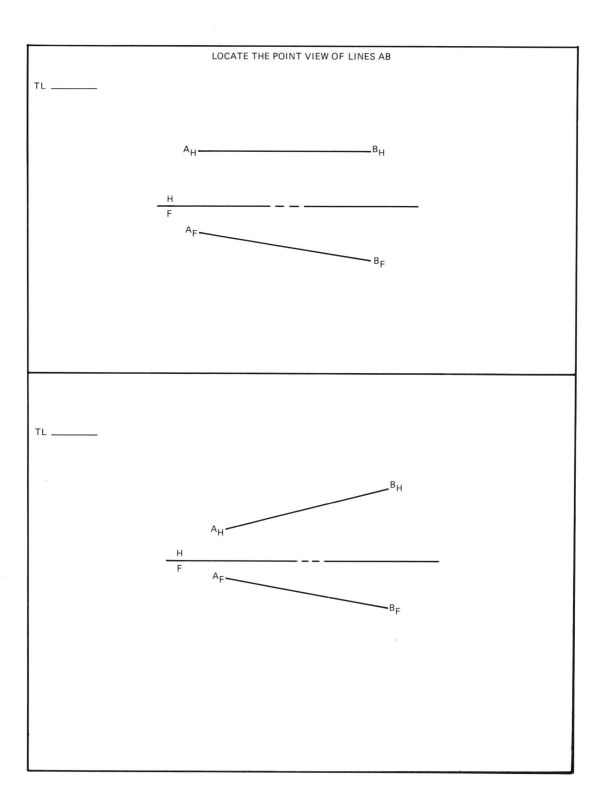

FIG. 20-54 Problem

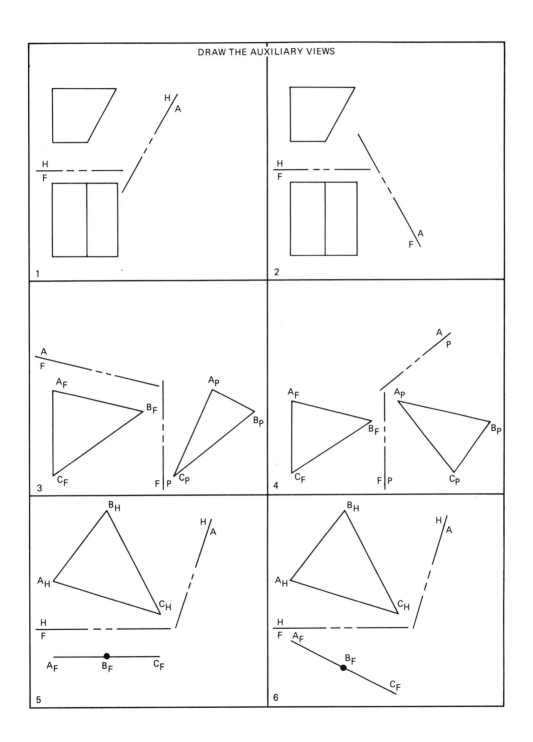

FIG. 20-55 Problem

DRAW THE EDGE VIEW (DRAW ON LARGER PAPER AND FIND THE TRUE SHAPE OF EACH PLANE)

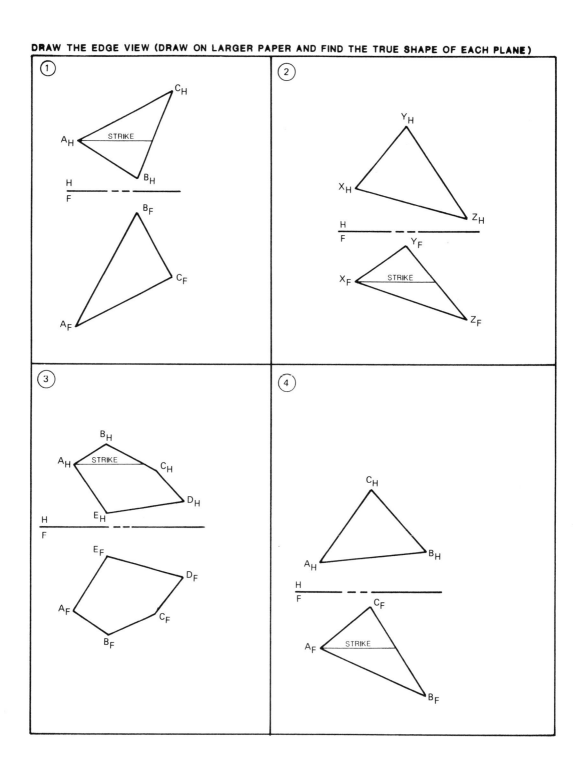

FIG. 20-56 Problem

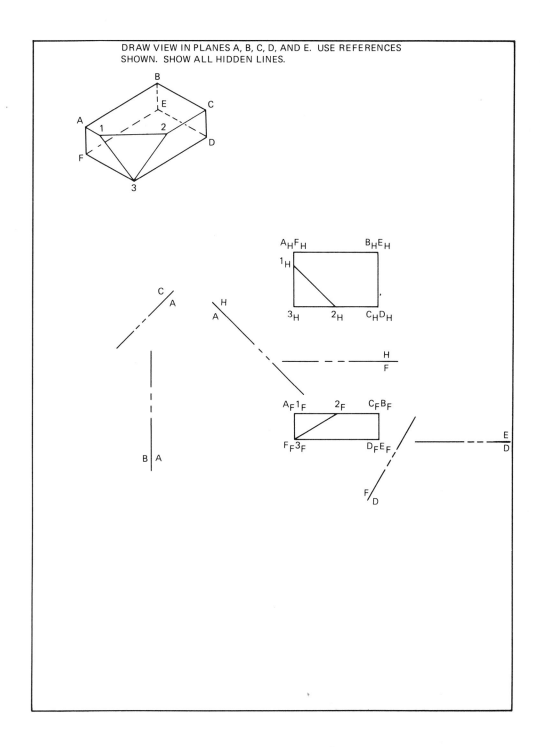

FIG. 20-57 Problem

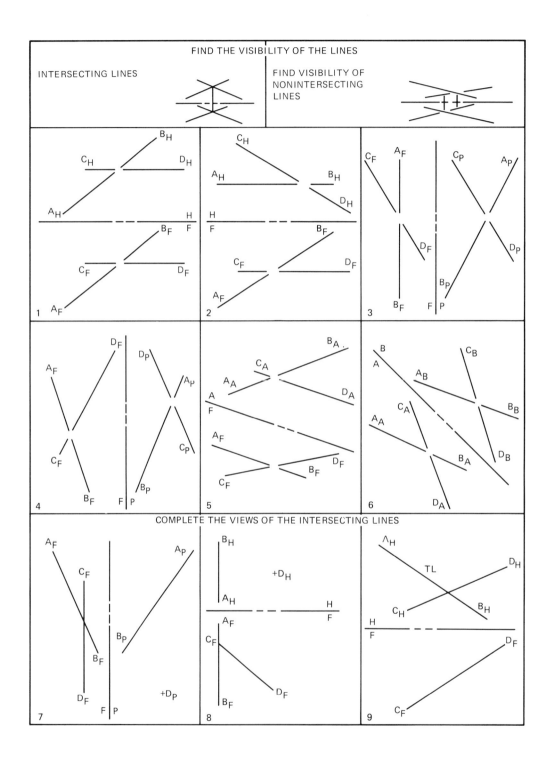

FIG. 20-58 Problem

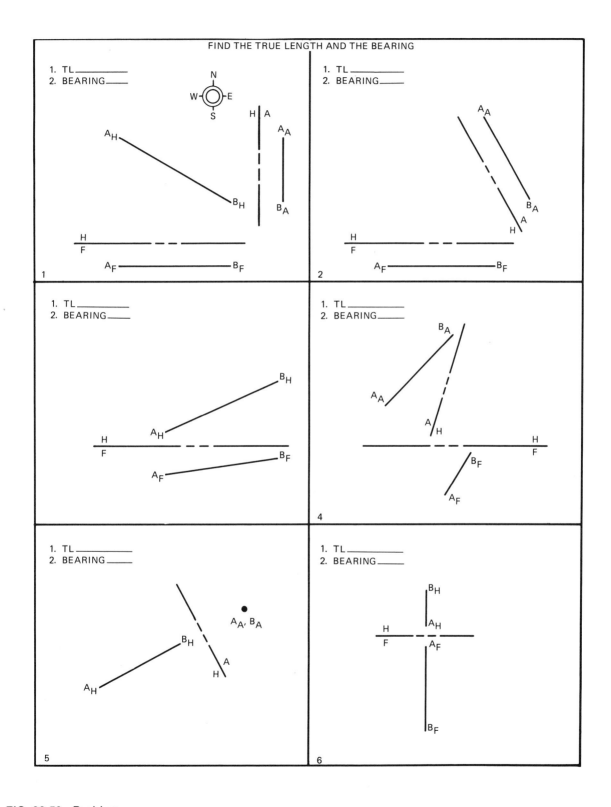

FIG. 20-59 Problem

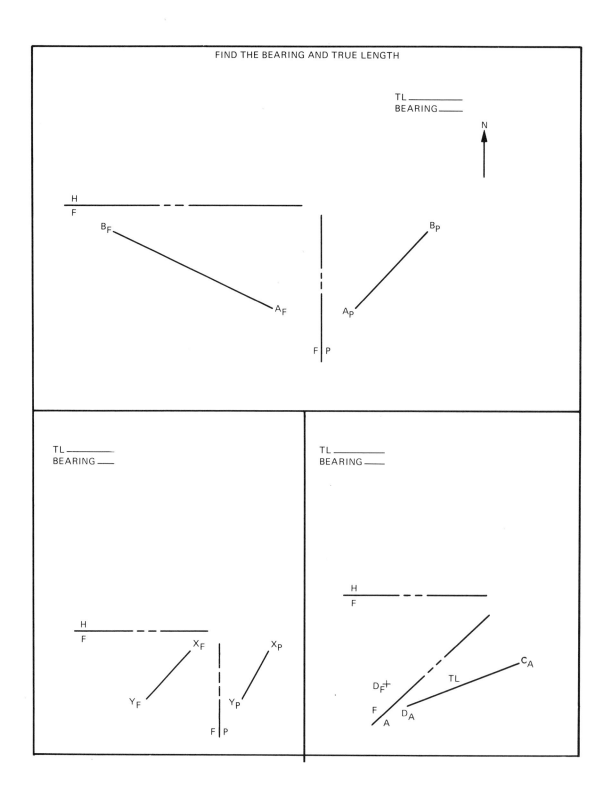

FIG. 20-60 Problem

21
Chapter

Developments

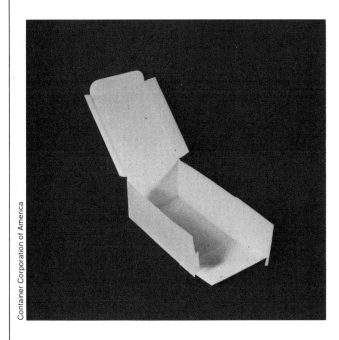

INTRODUCTION

Most industries use some form of hollow articles that are fabricated from thin materials. To fabricate these articles (Fig. 21-2), a pattern must be devel-

oped. A development is a flat-surface pattern which can be cut out and bent or rolled to form the article (Fig. 21-3).

All objects made from thin material that is folded or bent will require a pattern for the fabricators. A pattern is made from one piece that can be shaped and joined (Fig. 21-4). Drawing these patterns is called pattern drafting. The flat sheets of thin materials that are laid out to form a required shape are also called: developments; templates; stretchouts; surface developments; unrolled patterns; and unfolded patterns.

The larger industries that use patterns for their products are: sheet metal industry; packaging industry; aircraft industry; ship industry; clothing industry; and the heating and air-conditioning industry.

The steps in making a finished development are shown in Figure 21-5:

1. The object's form is drawn on paper. This is the pattern.
2. The pattern is transferred to the flat material.
3. The material is cut out.
4. The material is formed to shape.
5. The edges are joined.

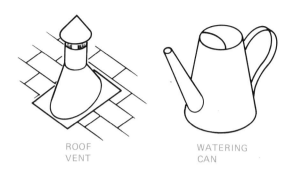

ROOF
VENT

WATERING
CAN

The process of developing patterns is important because the actual drawing may be used as a pattern on the material from which the article will be made. If the pattern is to be directly transferred, a full scale drawing will ensure accuracy. Many sheetmetal workers draw the pattern directly on the sheetmetal.

As you study this chapter, you should develop the knowledge and skills required to make development patterns and drawings of the intersections of different shapes that are joined together.

FIG. 21-2 Articles made from developed patterns

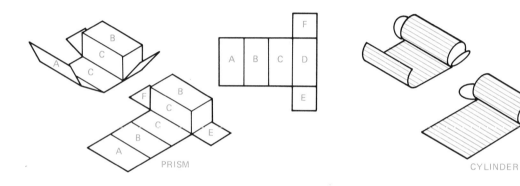

PRISM

CYLINDER

FIG. 21-3 Elementary patterns

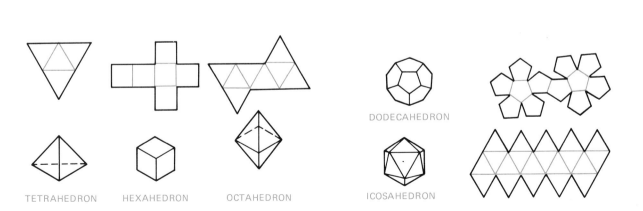

TETRAHEDRON   HEXAHEDRON   OCTAHEDRON

DODECAHEDRON

ICOSAHEDRON

FIG. 21-4 The five regular solids and their patterns

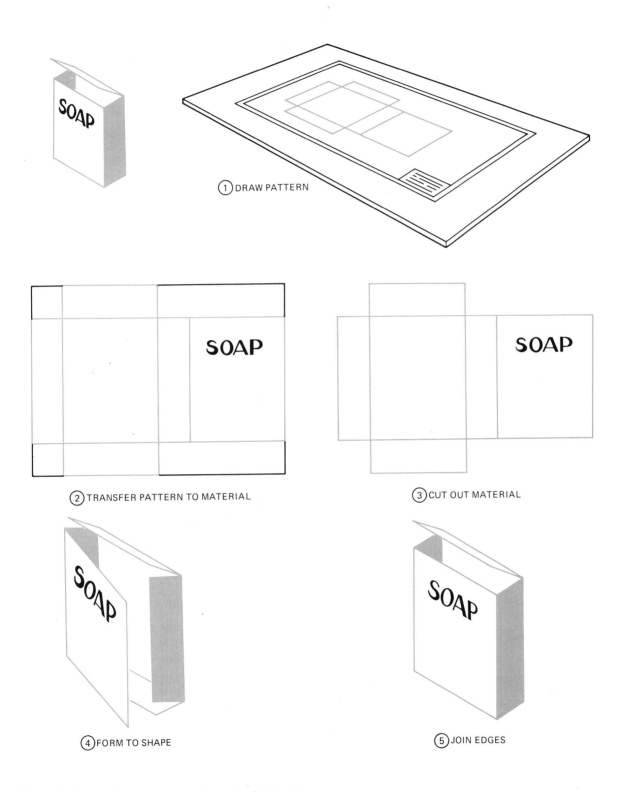

1 DRAW PATTERN

2 TRANSFER PATTERN TO MATERIAL

3 CUT OUT MATERIAL

4 FORM TO SHAPE

5 JOIN EDGES

FIG. 21-5 Production steps to pattern development

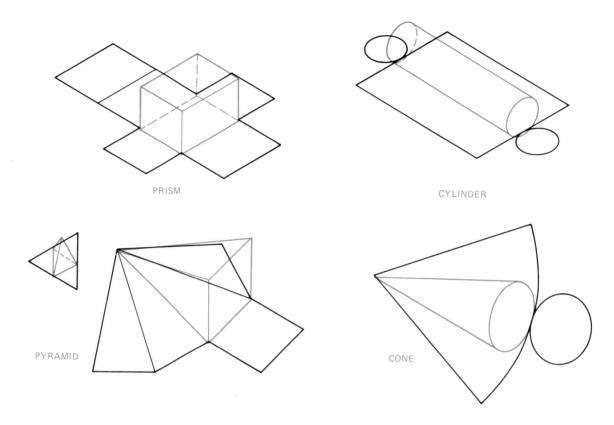

PRISM

CYLINDER

PYRAMID

CONE

FIG. 21-6 Development of patterns for four basic forms

SURFACES

As most objects are made up from combinations of four basic shapes (prisms, cylinders, cones, and pyramids), it is important that the draftsperson have an understanding of their construction (Fig. 21-6). The surface is the exterior face of any object. It is this surface that makes up the pattern.

Surfaces are classified into three general types (Fig. 21-7):

1. Plane surfaces
2. Single-curved surfaces
3. Warped surfaces

Plane and single-curved surfaces can be developed into accurate flat surface patterns. Warped surfaces such as the sphere and paraboloid cannot be developed into accurate patterns. Their development would be approximate, and they can only be fabricated by stretching or pressing the flat material with heavy equipment. This may be done by spin-ning the material on a lathe, or by pressing it between two dies in a press.

When a surface pattern is transferred to the material, it is usually marked as an inside pattern so the markings will not show. Most bending machines make inward folds.

FORMING SURFACES

When metal is folded, the material on the outside of the bend is stretched, while the material on the inside of the bend is compressed. There is a point where the material stays neutral. The material along the neutral line is not stretched or compressed. This line is about 0.44 of the thickness measured from the inside of the bend (Fig. 21-8).

Fold lines on the pattern are shown with either a thin or a dashed line. For better identification, these lines can be marked with an X or an O (Fig. 21-9). A pattern can be formed by bending, rolling, stamping, pressing, or spinning on a lathe.

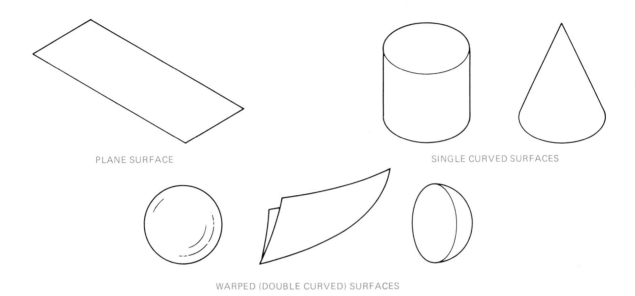

FIG. 21-7 General types of surfaces

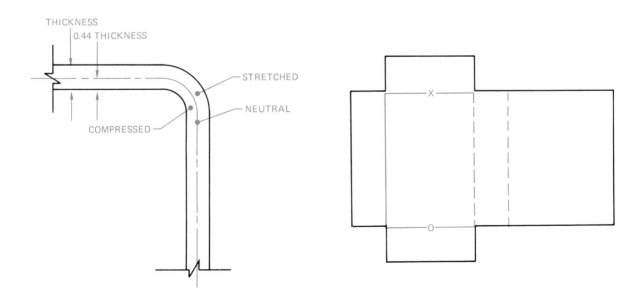

FIG. 21-8 Stretch and compression of bent materials

FIG. 21-9 Acceptable methods for representing fold lines

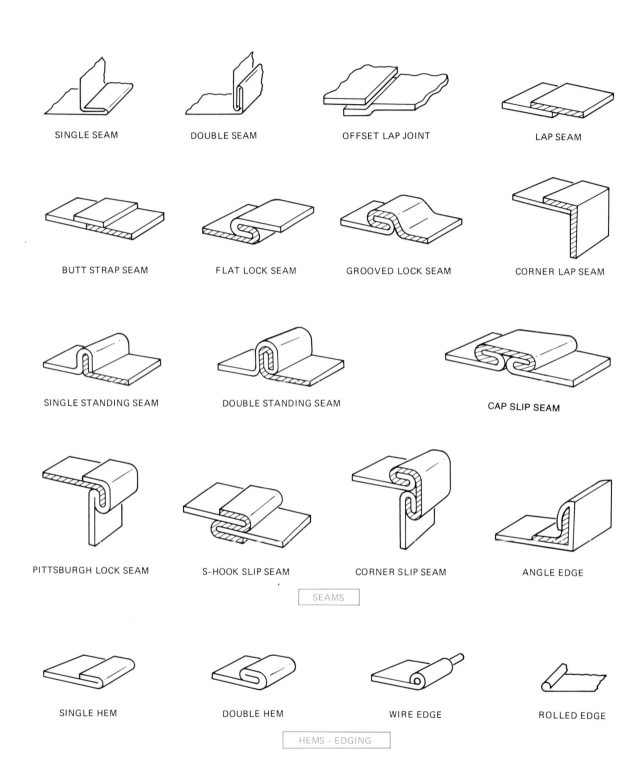

SINGLE SEAM

DOUBLE SEAM

OFFSET LAP JOINT

LAP SEAM

BUTT STRAP SEAM

FLAT LOCK SEAM

GROOVED LOCK SEAM

CORNER LAP SEAM

SINGLE STANDING SEAM

DOUBLE STANDING SEAM

CAP SLIP SEAM

PITTSBURGH LOCK SEAM

S-HOOK SLIP SEAM

CORNER SLIP SEAM

ANGLE EDGE

SEAMS

SINGLE HEM

DOUBLE HEM

WIRE EDGE

ROLLED EDGE

HEMS - EDGING

FIG. 21-10   Seams and hems

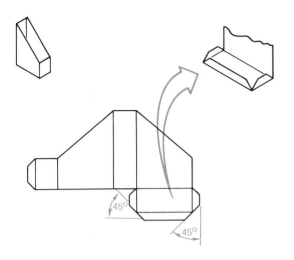

FIG. 21-11   45 degree seam corners will not overlap

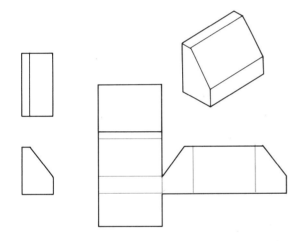

FIG. 21-12   Parallel line development

SEAMS

Before patterns are formed, allowances must be made for the amount of shortening caused by a large number of bends, especially with thicker materials. Allowances must also be made for seams that will be used to complete the fabrication. Extra material is added to the pattern for seams. The amount of seam material will depend upon:

1.   The type of seam

2.   The size of the seam

3.   The thickness of the material

There are many types of seams, including edging hems, as shown in Figure 21-10. Seams can be joined by soldering, welding, riveting, seaming, bolting, sewing, gluing, or screwing. Edging seams or exposed seams are folded or wired to strengthen them and remove sharp edges. The corners of seams are cut at a 45 degree angle, so that when they are folded they will not overlap and become too thick. Whenever possible the pattern should be developed so the seam will be on the shortest edge (Fig. 21-11). Seams for patterns are also called tabs, laps, hems, or lips. There are times when seams are not required, as when edges are butted together and sealed. To simplify the patterns in this unit, all seams will be left off.

ELEMENTS

Depending upon the basic form of the object to be developed, there are several procedures for drawing patterns. These procedures are:

1.   Parallel line development for prisms and cylinders (Fig. 21-12).

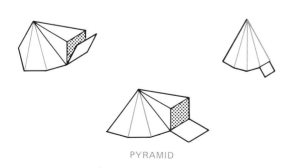

PYRAMID

FIG. 21-13   Radial line development

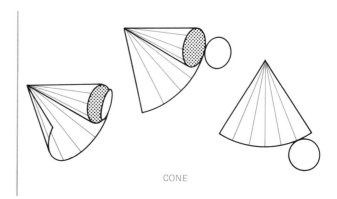

CONE

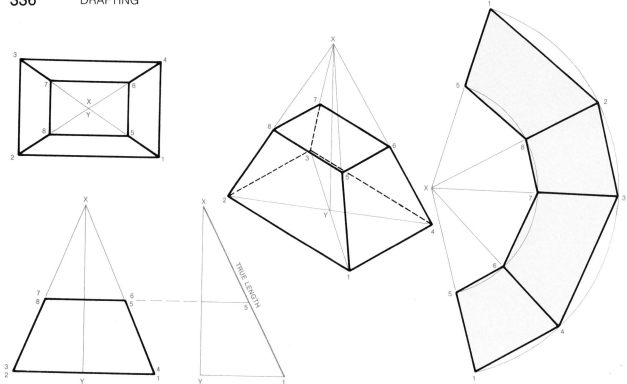

FIG. 21-14 Triangulation development

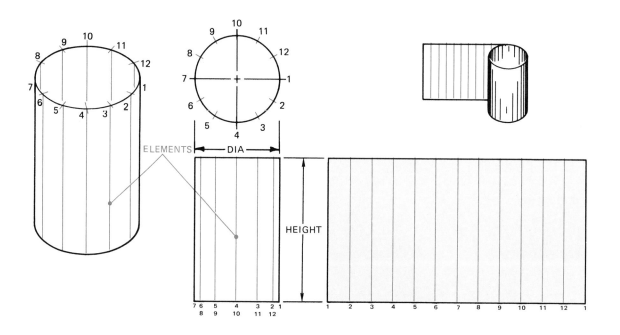

FIG. 21-15 Elements of a curved surface

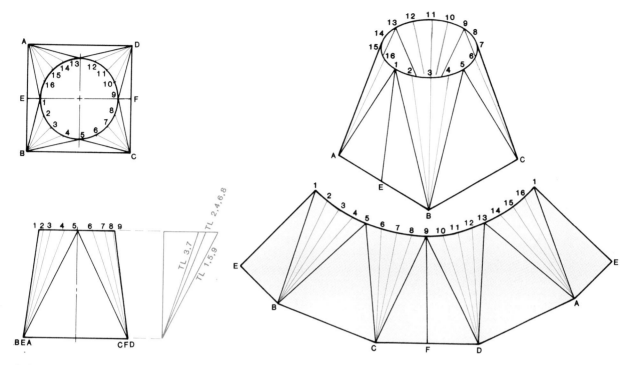

FIG. 21-16 Elements on a transitional surface

2. Radial line development for cones and pyramids (Fig. 21-13).

3. Triangulation development for oblique and complex forms (Fig. 21-14).

Some simpler forms can be developed and formed with only parallel lines as shown in Figure 21-12. Since there are no fold lines on curved surfaces, additional lines called elements are added to make the object developable (Fig. 21-15). An element is an imaginary straight line on the surface dividing the object into segments and making it easier to draw the pattern (Fig. 21-16).

TRUE LENGTH LINES

The first step in developing a pattern is to make a multiview drawing of the object. Since the pattern is made from true size lines and surfaces, it is necessary to show true length lines on the multiview drawing. True lengths can be shown by auxiliary views (Fig. 21-17), revolvements (Fig. 21-18), revolved sections (Fig. 21-19), and the right triangle method (Fig. 21-20).

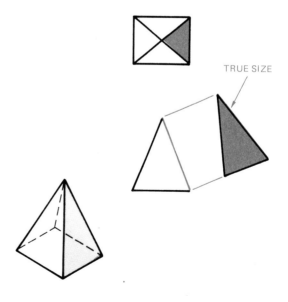

FIG. 21-17 Finding true size with an auxiliary drawing

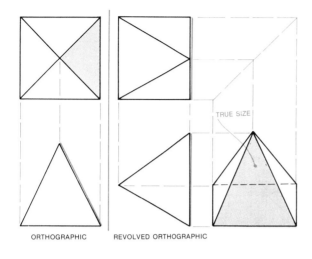

ORTHOGRAPHIC   REVOLVED ORTHOGRAPHIC

FIG. 21-18 Finding true length with a revolution

In the right triangle method the hypotenuse is the true length of the oblique line. The altitude of the right triangle is the height taken from the front view. The base of the right triangle is the length of the oblique line taken from the top view.

A draftsperson should use the fastest method of obtaining the true length of an object's edges. This is usually the revolution method in which only the line is revolved to find its true length.

PARALLEL LINE DEVELOPMENT

Parallel line development is the procedure used to develop patterns for prisms and cylinders. Right prisms and cylinders will have parallel faces or elements. All lines and surfaces will be true size (Fig. 21-21). All adjacent surfaces are perpendicular to each other. With this type of object, the pattern can

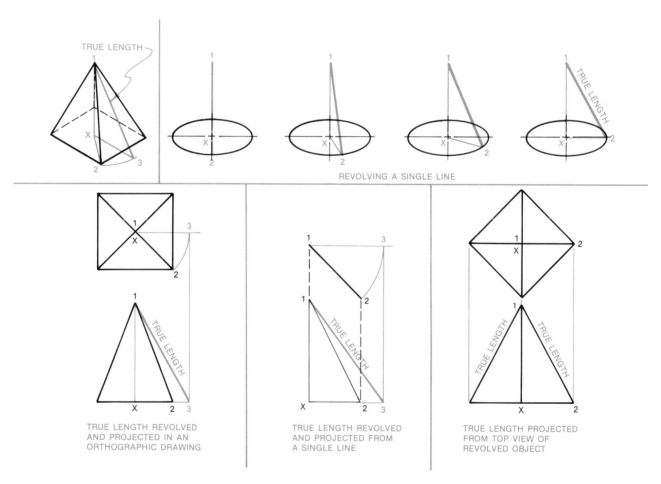

REVOLVING A SINGLE LINE

TRUE LENGTH REVOLVED AND PROJECTED IN AN ORTHOGRAPHIC DRAWING

TRUE LENGTH REVOLVED AND PROJECTED FROM A SINGLE LINE

TRUE LENGTH PROJECTED FROM TOP VIEW OF REVOLVED OBJECT

FIG. 21-19 Finding true length with a revolution

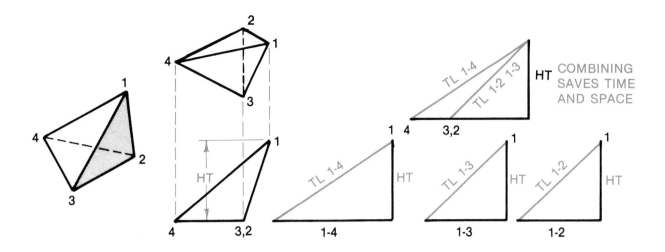

FIG. 21-20 Finding true length with the right triangle method

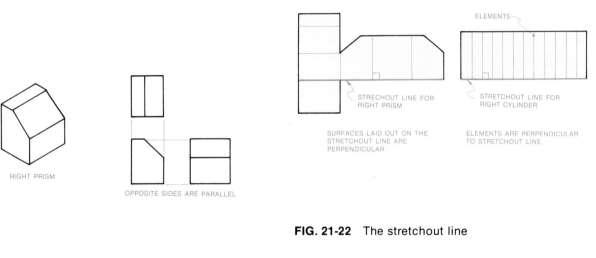

FIG. 21-21 Parallel line development

FIG. 21-22 The stretchout line

be developed by laying it out on a straight line called the stretchout line (Fig. 21-22). The edges of the faces and elements will be perpendicular to the stretchout line. Only true length lines can be measured on the stretchout line.

A cylinder can be laid out in two ways:

1. Step off the distances between the elements with dividers on the stretchout line.

2. A more accurate method is to obtain the stretchout length of the circumference mathematically (circumference equals 3.14 × diameter). Using a geometric construction

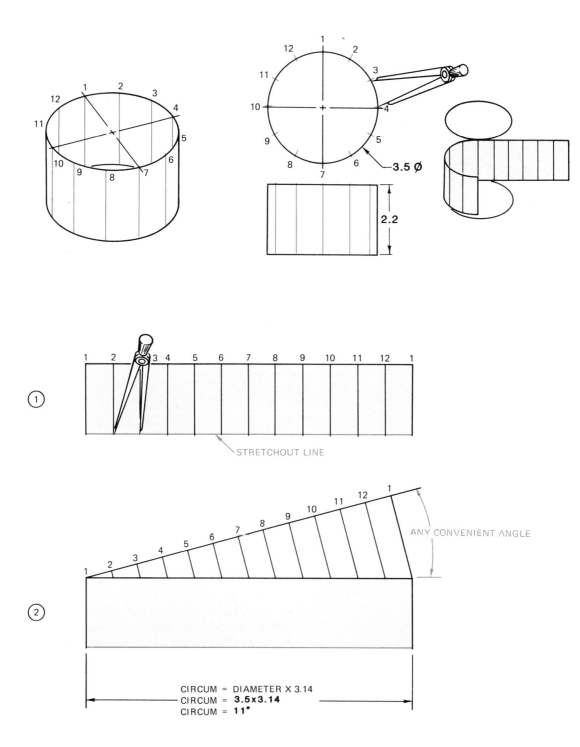

3.5 Ø

2.2

① STRETCHOUT LINE

② ANY CONVENIENT ANGLE

CIRCUM = DIAMETER X 3.14
CIRCUM = **3.5 x 3.14**
CIRCUM = **11"**

FIG. 21-23 Dividing the stretchout line with elements

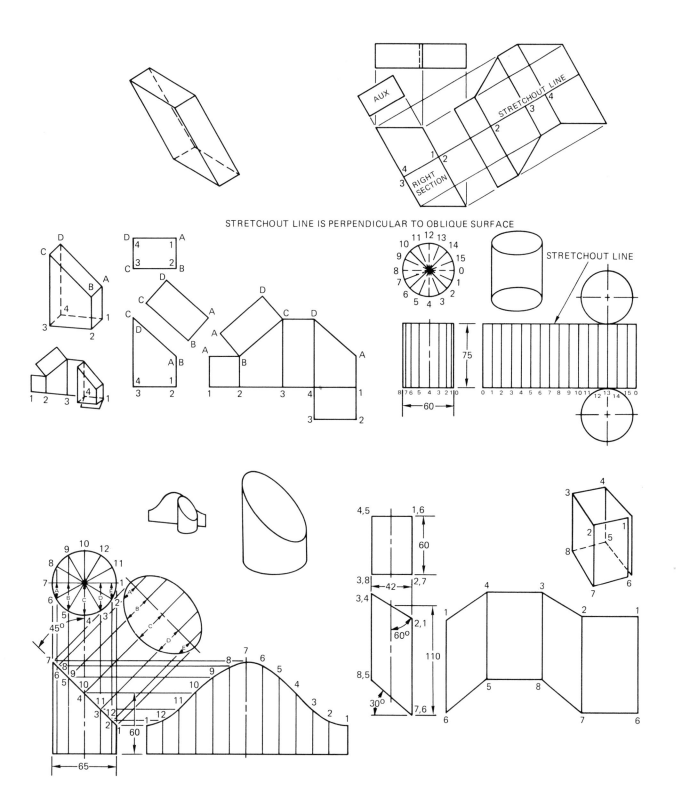

STRETCHOUT LINE IS PERPENDICULAR TO OBLIQUE SURFACE

FIG. 21-24   Examples of parallel line development

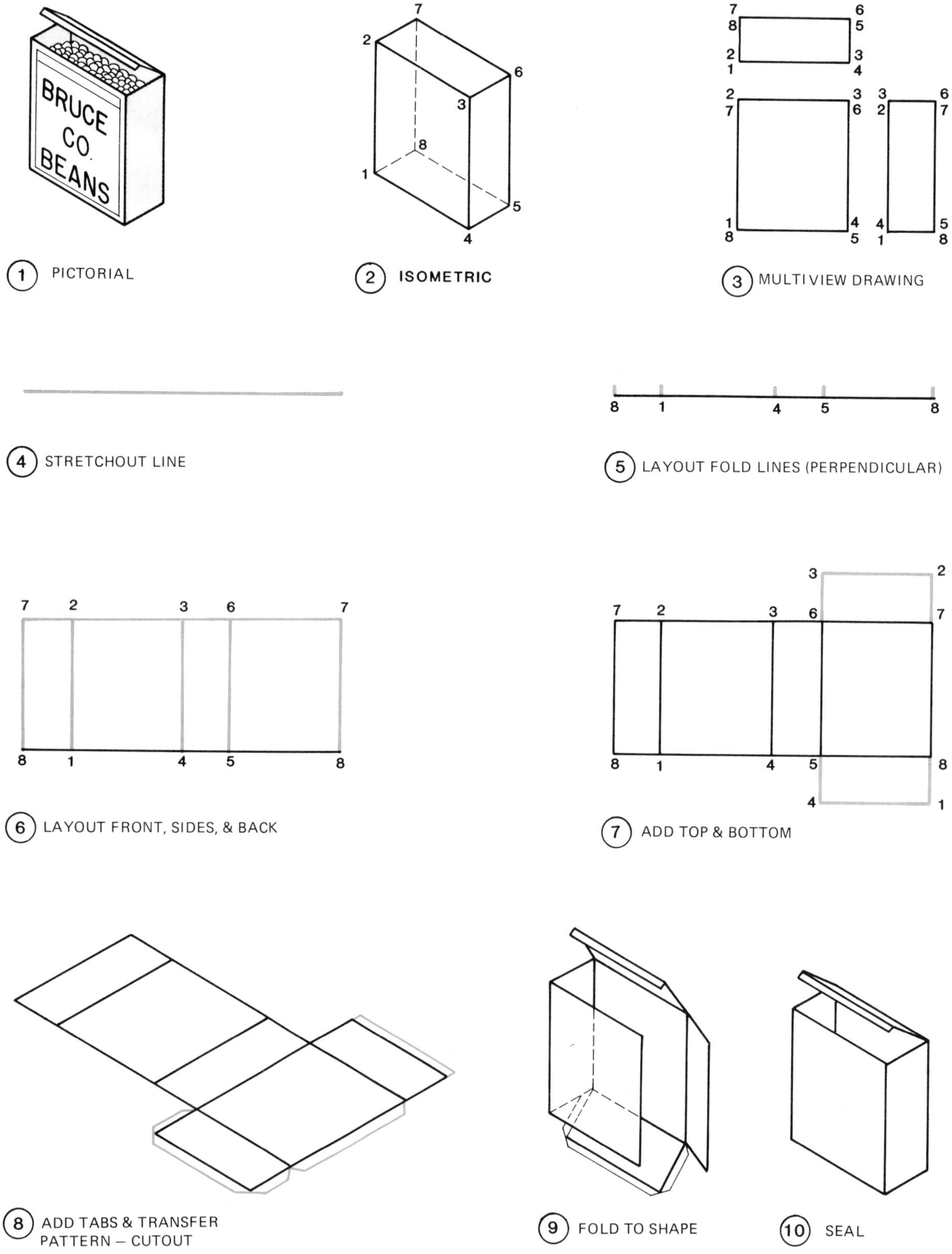
BRUCE
CO.
BEANS
1 PICTORIAL
2 ISOMETRIC
3 MULTIVIEW DRAWING
4 STRETCHOUT LINE
5 LAYOUT FOLD LINES (PERPENDICULAR)
6 LAYOUT FRONT, SIDES, & BACK
7 ADD TOP & BOTTOM
8 ADD TABS & TRANSFER PATTERN – CUTOUT
9 FOLD TO SHAPE
10 SEAL

FIG. 21-24 Examples of parallel line development (Cont.)

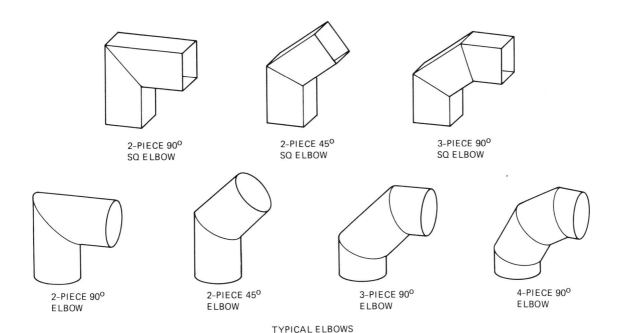

2-PIECE 90°
SQ ELBOW

2-PIECE 45°
SQ ELBOW

3-PIECE 90°
SQ ELBOW

2-PIECE 90°
ELBOW

2-PIECE 45°
ELBOW

3-PIECE 90°
ELBOW

4-PIECE 90°
ELBOW

TYPICAL ELBOWS

FIG. 21-25 Typical developed elbows used in industry

procedure, divide the stretchout line into the required number of elements (Fig. 21-23).

Examples of parallel line developments are shown in Figure 21-24. Large quantities of various types of elbows (Fig. 21-25) are fabricated in the sheet metal industry using parallel line development.

RADIAL LINE DEVELOPMENT

Radial line development is the procedure used to develop patterns for tapering objects such as pyramids and cones. The flat surfaces of a pyramid and the elements of a cone will meet at a vertex. The axis of a right pyramid or cone will be perpendicular to and centered on the base (Fig. 21-26). Before developing the pattern it is important to find all true sizes on the orthographic views.

When laying out the patterns of a cone (Fig. 21-27) and pyramid (Fig. 21-28) by radial line development, an arc is swung from the vertex. This arc is called the stretchout arc, and it is equal in length to the circumference of a cone. The chordal distances can be stepped off with dividers, or fig-

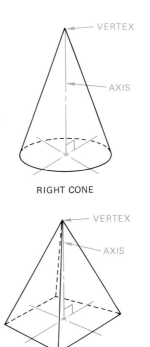

RIGHT CONE

RIGHT PYRAMID

FIG. 21-26 The right cone and right pyramid

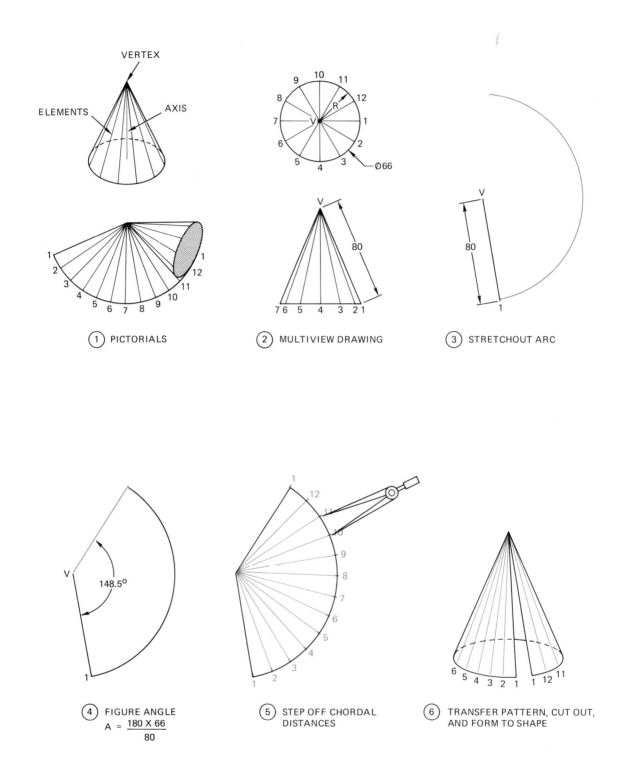

FIG. 21-27 Laying out the pattern of a cone

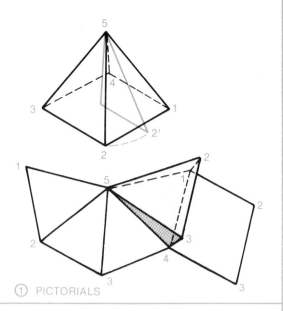

① PICTORIALS

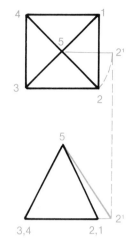

② FIND TRUE LENGTH ON ORTHOGRAPHIC DRAWING

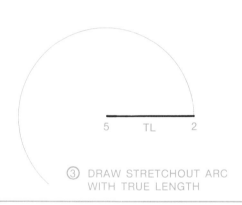

③ DRAW STRETCHOUT ARC WITH TRUE LENGTH

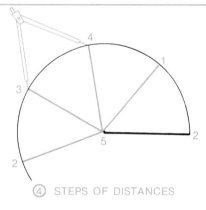

④ STEPS OF DISTANCES

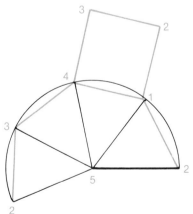

⑤ COMPLETE PATTERN, TRANS- FER PATTERN AND CUT OUT

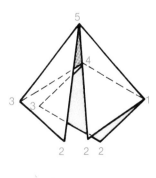

⑥ FORM TO SHAPE AND ASSEMBLE

FIG. 21-28 Laying out the pattern of a pyramid

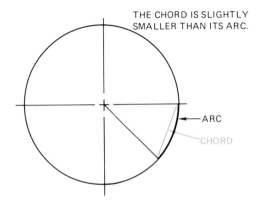

THE CHORD IS SLIGHTLY SMALLER THAN ITS ARC.

ARC

CHORD

ured more accurately mathematically (angle equals 180 times diameter divided by the length of the side of the cone). Divide the angle along the stretchout arc into the required parts with dividers. Note in Figure 21-29 that the chordal distance is shorter than its arc. The difference is small with short arcs, so that stepping off chordal distances will give reasonable accuracy.

Examples of radial developments are shown in Figure 21-30.

FIG. 21-29   The chord is slightly smaller than its arc

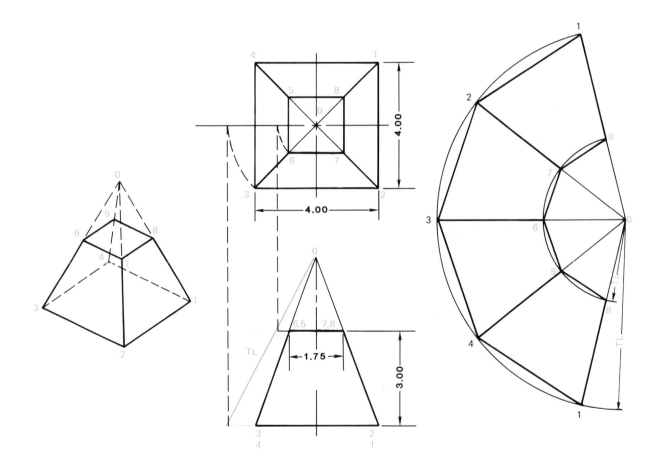

FIG. 21-30A   Example of radial line development (continued)

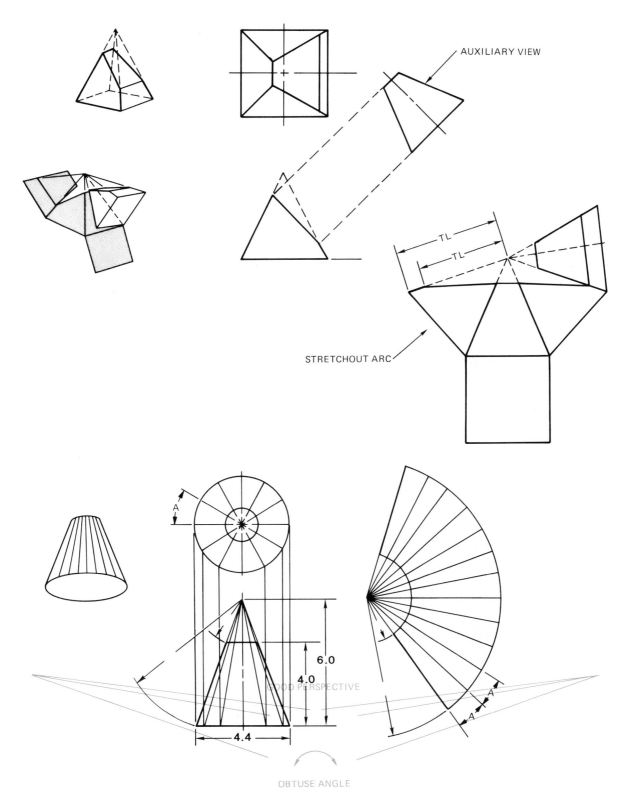

AUXILIARY VIEW

TL

TL

STRETCHOUT ARC

A

6.0

4.0

GOOD PERSPECTIVE

4.4

A

A

OBTUSE ANGLE

FIG. 21-30B Example of radial line development (continued)

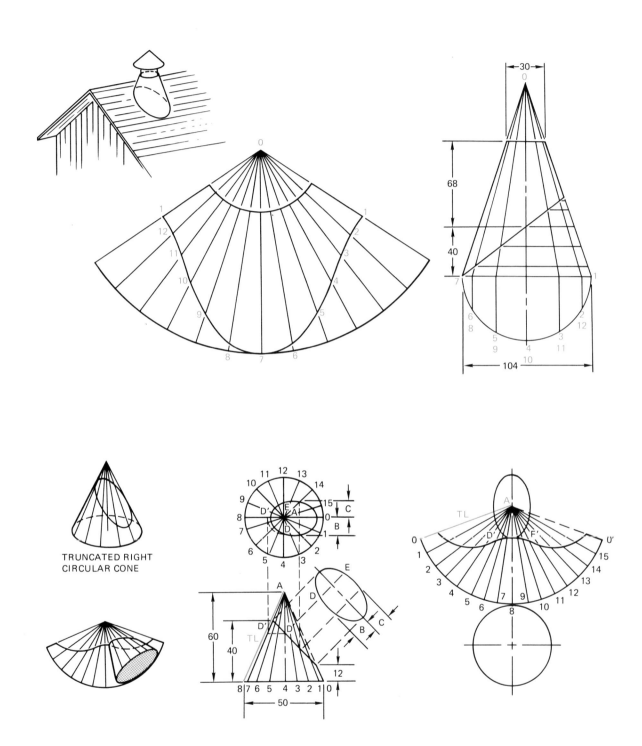

FIG. 21-30C Example of radial line development

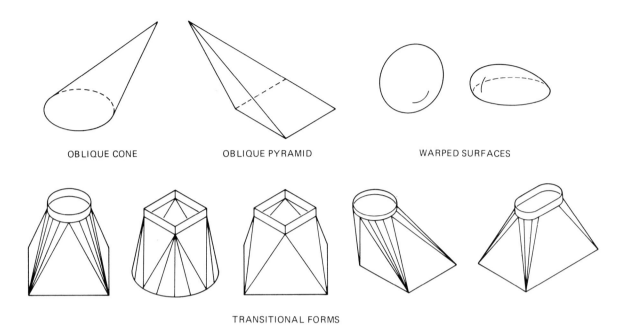

OBLIQUE CONE   OBLIQUE PYRAMID   WARPED SURFACES

TRANSITIONAL FORMS

FIG. 21-31   Triangulation development

TRIANGULATION DEVELOPMENT

Triangulation development is the procedure used to develop oblique pyramids, oblique cones, transition pieces, and other complex items that cannot be developed by the other processes (Fig. 21-31). Warped forms can be developed, but their patterns are only approximate. Triangulation development is reasonably accurate for most shapes.

The process of triangulation can be divided into three steps:

1. Divide the surfaces in the multiview drawing into a series of triangles.
2. Find the true size of each triangle by finding the true length of each of its sides.
3. Duplicate the series of adjacent true size triangles to form the pattern. The process of transferring a triangle is reviewed in Figure 21-32.

The triangulation development of an oblique pyramid is shown in Figure 21-33, and the development of an oblique cone is shown in Figure 21-34.

Examples of other triangulation developments are shown in Figure 21-35.

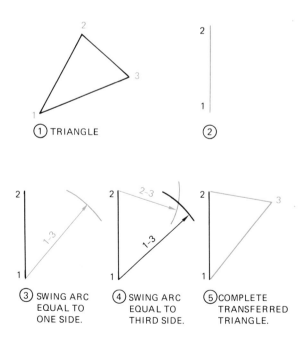

① TRIANGLE   ②

③ SWING ARC EQUAL TO ONE SIDE.   ④ SWING ARC EQUAL TO THIRD SIDE.   ⑤ COMPLETE TRANSFERRED TRIANGLE.

FIG. 21-32   Transferring a triangle

1. PICTORIAL

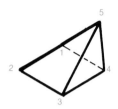

2. DRAW MULTIVIEW DRAWING. REVOLVE CORNERS AND PROJECT TO FIND THE TRUE LENGTHS.

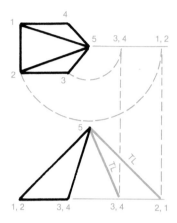

3. START FROM ANY CORNER. SWING TRUE LENGTH ARC. DRAW LINE 3-5

4. SWING ARC TL 5-2

5. SWING ARC TL 2-3 (BASE). COMPLETE TRIANGLE 2-3-5

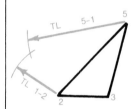

6. SWING TL ARCS 5-1 AND 1-2

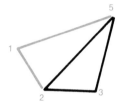

7. COMPLETE TRIANGLE 1-2-5

8. SWING TL ARCS 5-4 AND 1-4. COMPLETE TRIANGLE 1-4-5.

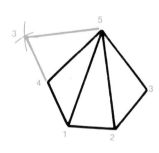

9. SWING TL ARCS 5-3 AND 4-3. COMPLETE TRIANGLE 3-4-5

10. COMPLETE BASE

FIG. 21-33 Triangulation development of an oblique pyramid

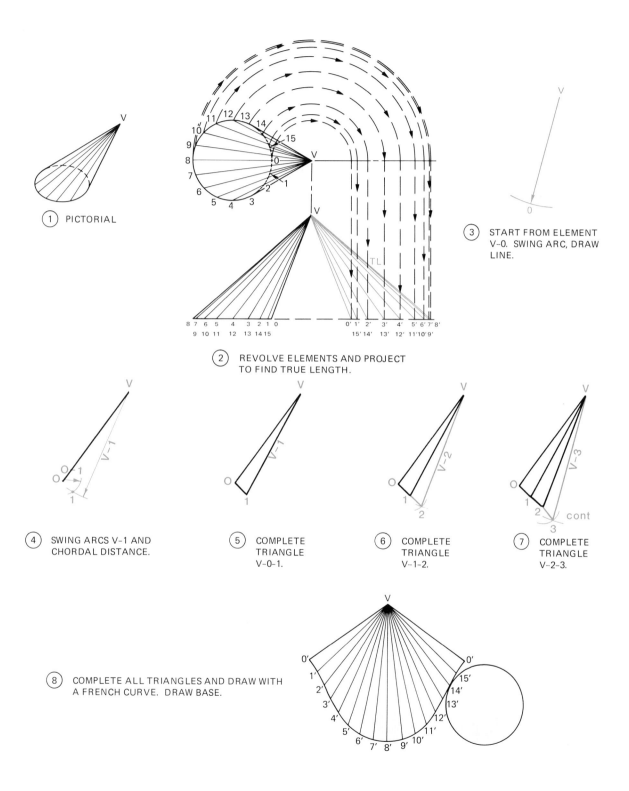

① PICTORIAL

② REVOLVE ELEMENTS AND PROJECT TO FIND TRUE LENGTH.

③ START FROM ELEMENT V–0. SWING ARC, DRAW LINE.

④ SWING ARCS V–1 AND CHORDAL DISTANCE.

⑤ COMPLETE TRIANGLE V–0–1.

⑥ COMPLETE TRIANGLE V–1–2.

⑦ COMPLETE TRIANGLE V–2–3.

⑧ COMPLETE ALL TRIANGLES AND DRAW WITH A FRENCH CURVE. DRAW BASE.

FIG. 21-34 Development of an oblique cone

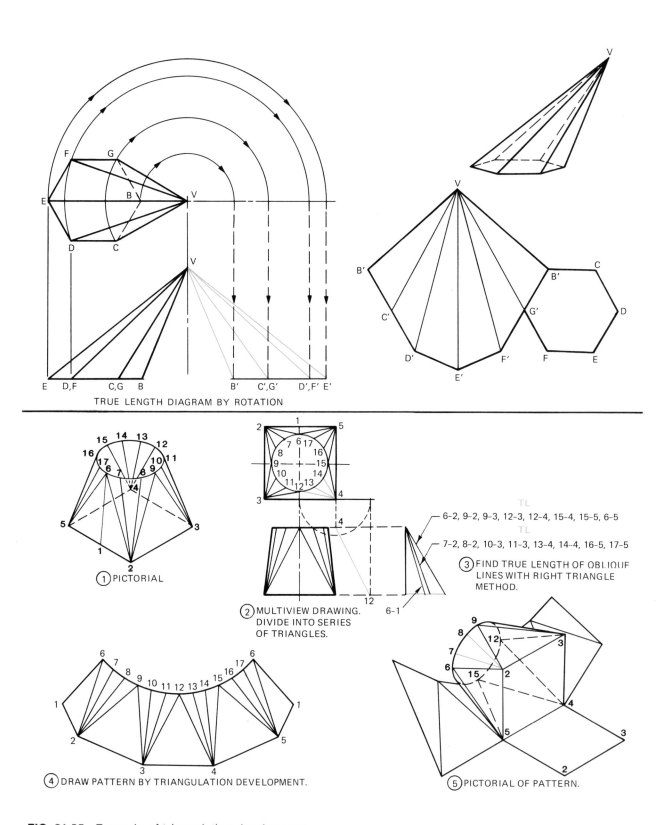

TRUE LENGTH DIAGRAM BY ROTATION

① PICTORIAL

② MULTIVIEW DRAWING. DIVIDE INTO SERIES OF TRIANGLES.

6-2, 9-2, 9-3, 12-3, 12-4, 15-4, 15-5, 6-5

7-2, 8-2, 10-3, 11-3, 13-4, 14-4, 16-5, 17-5

③ FIND TRUE LENGTH OF OBLIQUE LINES WITH RIGHT TRIANGLE METHOD.

④ DRAW PATTERN BY TRIANGULATION DEVELOPMENT.

⑤ PICTORIAL OF PATTERN.

FIG. 21-35 Example of triangulation development

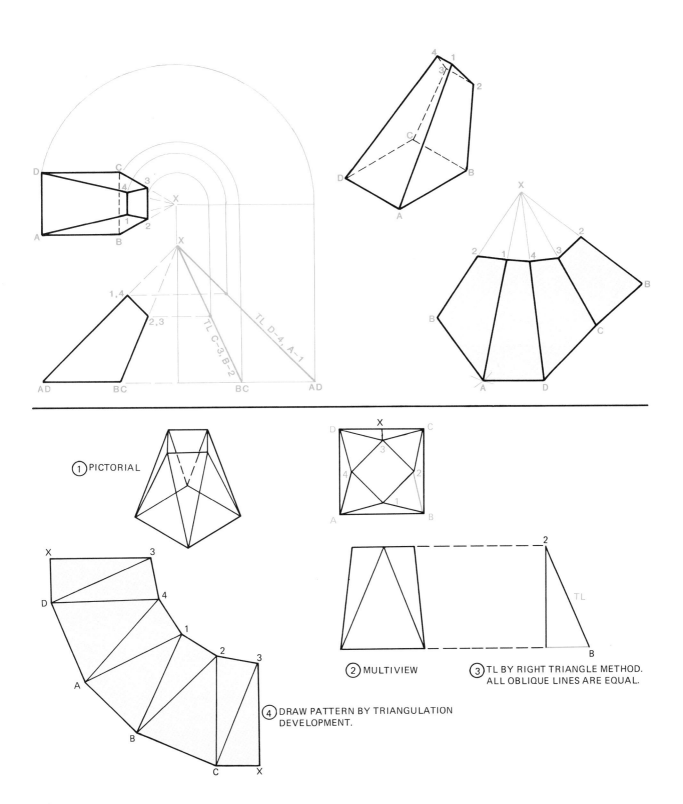

FIG. 21-35 Example of triangulation development (Cont.)

INTERSECTIONS

Up to this point, only single objects have been developed. When two objects are connected, their points of connection are called points of intersection. When a line passes through a plane, the point is called the point of intersection (Fig. 21-36). These lines may be straight or curved, depending on the geometric forms of the intersecting surfaces (Fig. 21-37). The lines of intersection must be determined on the orthographic drawing before the objects can be developed into patterns.

To plot a line of intersection, the points of intersection must be plotted on all orthographic views by projection. Each point must show in all views.

Connecting these points will give the lines of intersection. It is important to have all of the true lengths showing on the multiview drawing so that the pattern can be developed. Figure 21-38 shows straight intersecting lines, and Figure 21-39 shows curved intersecting lines.

CUTTING PLANES

A division of the circumference of a circle will act as a cutting plane through the whole object (Fig. 21-40). These planes can be placed anywhere to produce intersecting points. These points develop the intersecting surfaces.

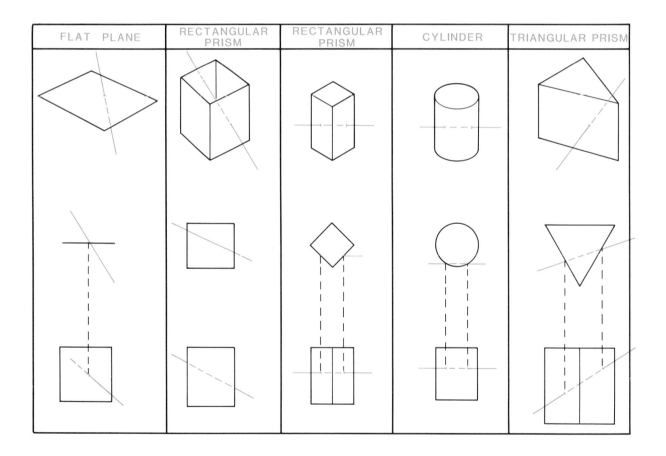

FLAT PLANE	RECTANGULAR PRISM	RECTANGULAR PRISM	CYLINDER	TRIANGULAR PRISM

FIG. 21-36 Points of intersection

FIG. 21-37 Surface intersection

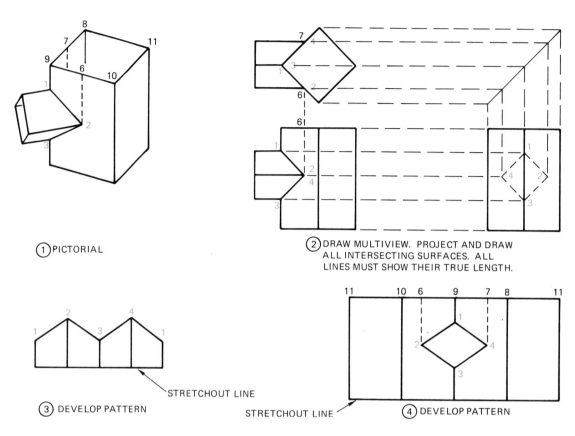

FIG. 21-38   Intersection of straight lines

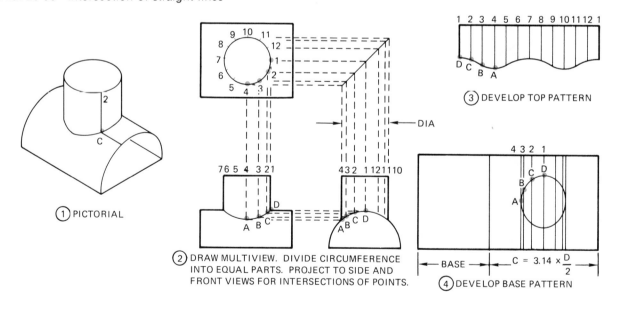

FIG. 21-39   Intersection of curved lines

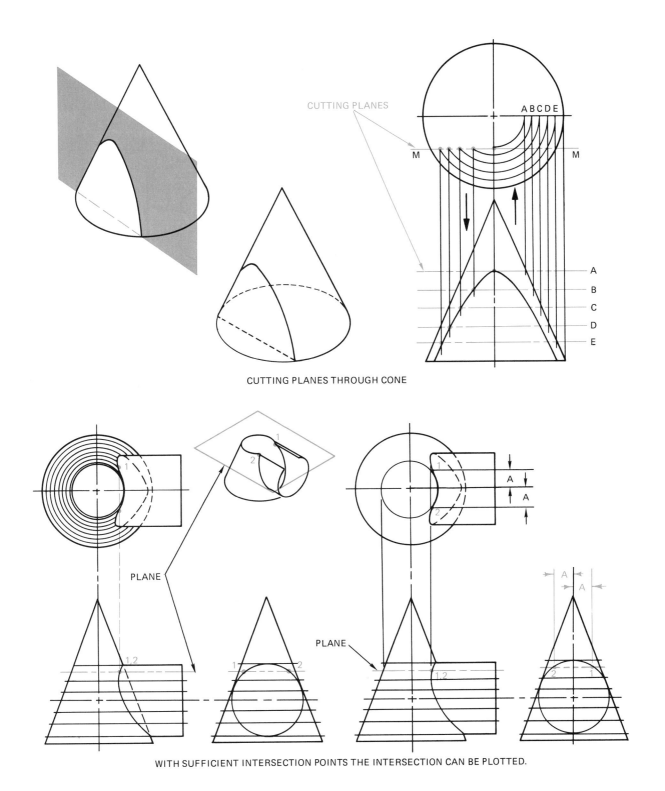

CUTTING PLANES

CUTTING PLANES THROUGH CONE

PLANE

PLANE

WITH SUFFICIENT INTERSECTION POINTS THE INTERSECTION CAN BE PLOTTED.

FIG. 21-40 Examples of how cutting planes produce points of intersection

PROBLEMS

Develop the following problems (Figures 21-41 through 21-44).

1. Make an isometric drawing of the problem you choose

2. Draw the necessary orthographic views—two or three views, first or third angle projection
3. Develop the pattern
4. Transfer the pattern to heavy paper; cut, form and tape to shape; add seams if necessary

DEVELOP THE FOLLOWING PATTERNS BY PARALLEL LINE DEVELOPMENT.

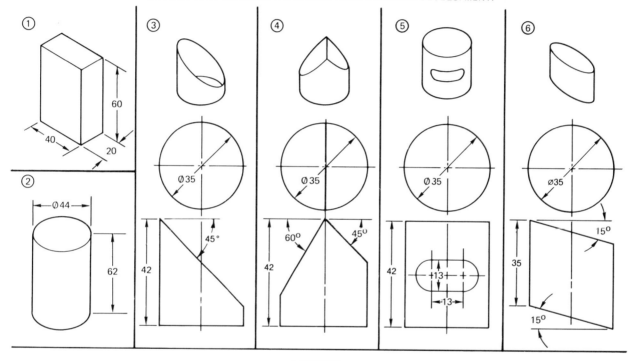

METRIC

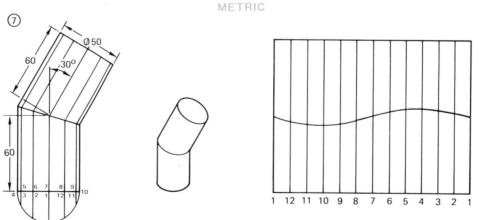

FIG. 21-41 Problem

DEVELOP THE FOLLOWING PATTERNS BY RADIAL LINE DEVELOPMENT.

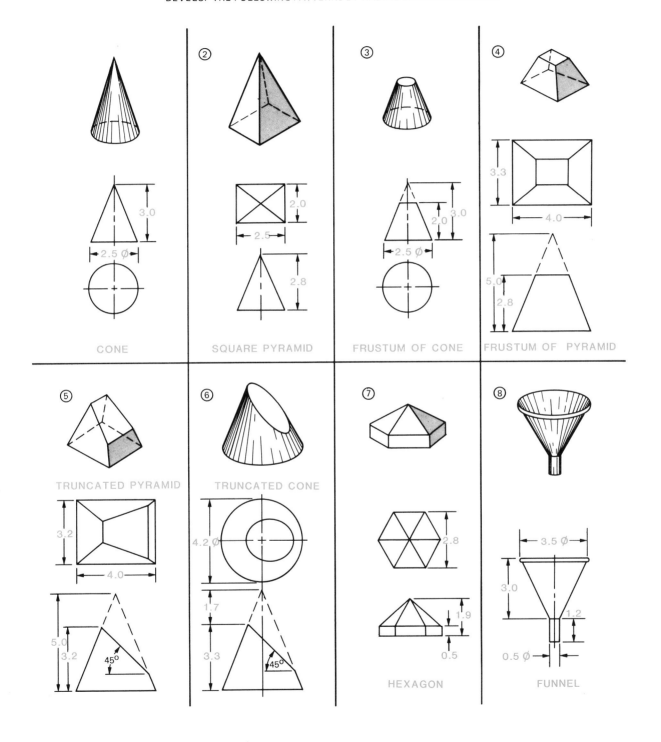

① CONE

② SQUARE PYRAMID

③ FRUSTUM OF CONE

④ FRUSTUM OF PYRAMID

⑤ TRUNCATED PYRAMID

⑥ TRUNCATED CONE

⑦ HEXAGON

⑧ FUNNEL

FIG. 21-42 Problem

DEVELOP THE FOLLOWING PATTERNS BY TRIANGULATION.

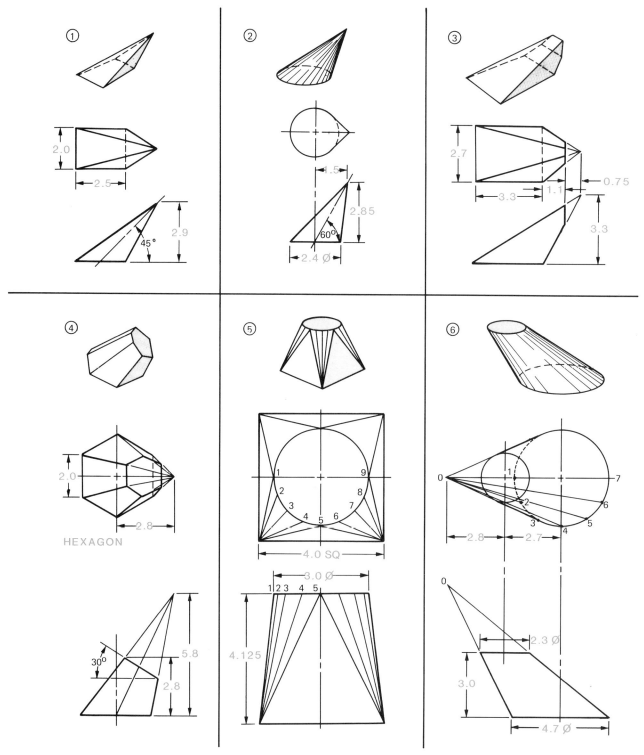

FIG. 21-43   Problem

DEVELOP PATTERNS FOR THE INTERSECTING SURFACES.
COMPLETE INTERSECTIONS IN THE MULTIVIEW DRAWINGS.

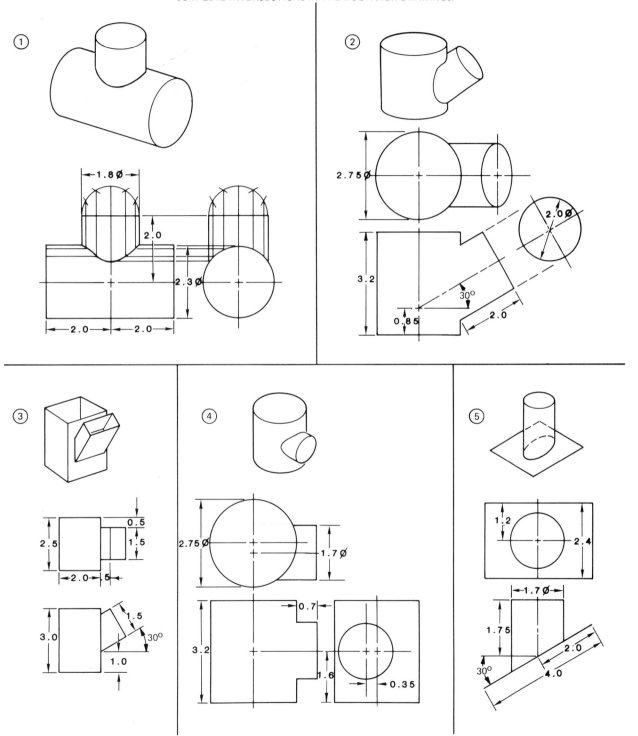

FIG. 21-44 Problems

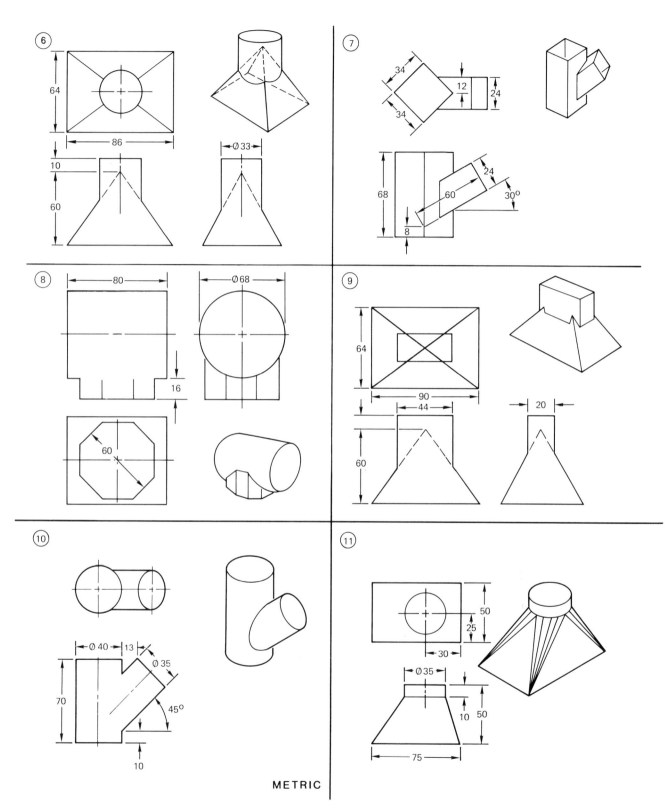

FIG. 21-44 Problems (Cont.)

Perspective

Coca-Cola Company

INTRODUCTION

Engineering drawing is primarily a method of communication between the designers and the

manufacturers of a product. However, the people who are involved with financing, development, and sales of a product do not read engineering drawings. It is advantageous for this group to have a drawing as lifelike as possible.

Perspective drawings (Fig. 22-2) help lay persons visualize the finished product. The perspective drawing is a two-dimensional representation of a three-dimensional object as it appears to the eye or as it would appear in a photograph (Fig. 22-3). As you study this chapter, you should learn to draw simple forms with one-point, two-point, and three-point types of perspective drawings.

THEORY OF PERSPECTIVE

The theory of perspective is illustrated in Figure 22-4. The projectors go from the object through the projection (picture) plane and converge at the eye. The perspective drawing is the view as seen on the projection plane. The size and position of the perspective drawing will vary with different positions of the eye, projection plane, and object. The size of the object will diminish as it recedes from the eye, which gives the perspective drawing its lifelike quality (Fig. 22-5).

MULTIVIEW
WORKING DRAWING

PERSPECTIVE

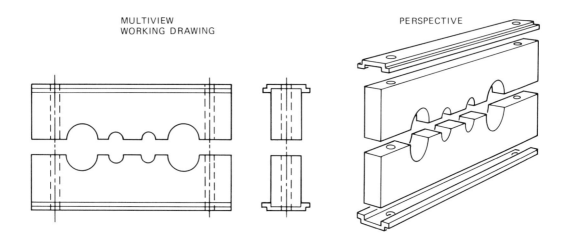

FIG. 22-2   Perspective drawing for visualization

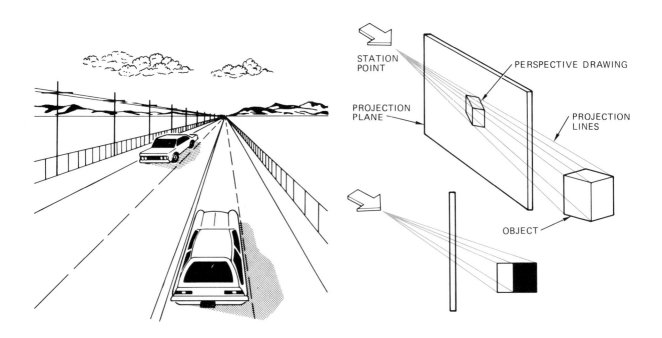

FIG. 22-3   Depth effect in perspective

FIG. 22-4   Theory of perspective

Perspective projection can be as exacting as descriptive geometry (Fig. 22-6). Exact sizes are not critical because the purpose of perspective drawing is to present lay people with a realistic image of an object, and the drawing will not be used in manufacturing. Using the rules of perspective drawing, an approximation of sizes will suffice.

PICTORIALS

Perspective drawing is pictorial drawing (see Chapter 11). The word "pictorial" comes from the word "picture." Pictorial drawings are "picture-like" drawings which show the object in one view. Pictorial drawings are placed into three groups (Fig. 22-7):

1. perspective (one-point, two-point, three-point)
2. axonometric (isometric, dimetric, trimetric)
3. oblique (cavalier, cabinet)

As the pictorials in Figure 22-7 show, the perspective drawing is more lifelike than axonometric

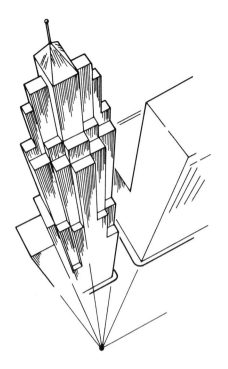

FIG. 22-5 Lifelike perspective

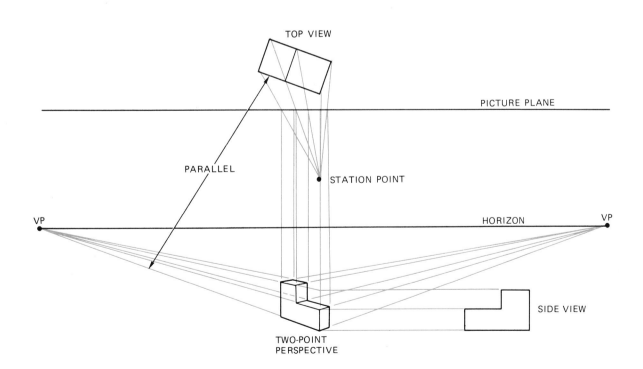

FIG. 22-6 Accuracy in perspective drawing using two-point perspective

and oblique drawings because the projectors are not parallel or perpendicular to the picture plane.

PERSPECTIVE TERMINOLOGY

A perspective drawing consists of the following parts (Fig. 22-8):

Horizon The line that separates the land (or sea) from the sky. The horizon is at the viewer's eye level.

Vanishing points The points on the horizon where lines of an object will converge.

Station point The position of the viewer's eye when viewing the object. The eye level is always level with the horizon (Fig. 22-9).

Picture plane The plane on which the

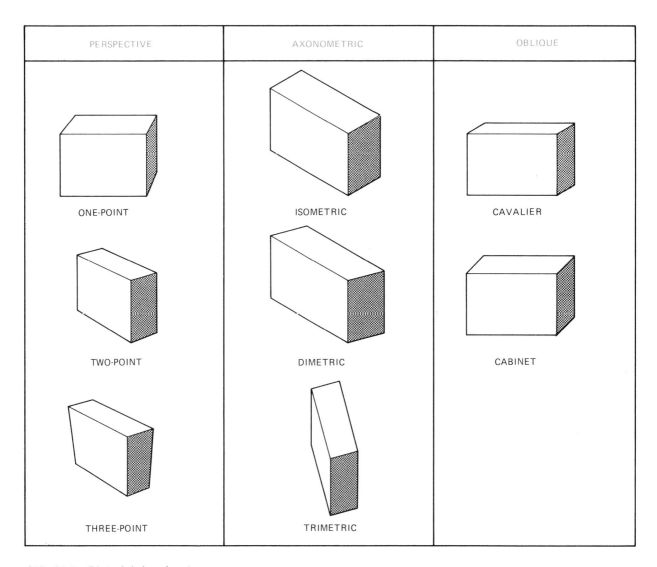

FIG. 22-7 Pictorial drawing types

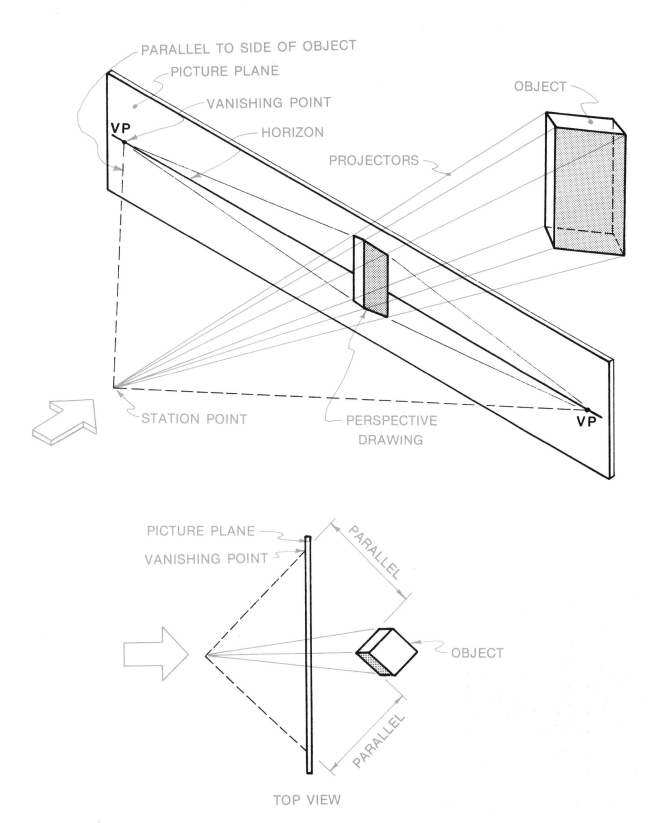

PARALLEL TO SIDE OF OBJECT
PICTURE PLANE
VANISHING POINT
OBJECT
VP
HORIZON
PROJECTORS
STATION POINT
PERSPECTIVE DRAWING
VP

PICTURE PLANE
VANISHING POINT
PARALLEL
PARALLEL
OBJECT
TOP VIEW

FIG. 22-8 Terminology of perspective drawings

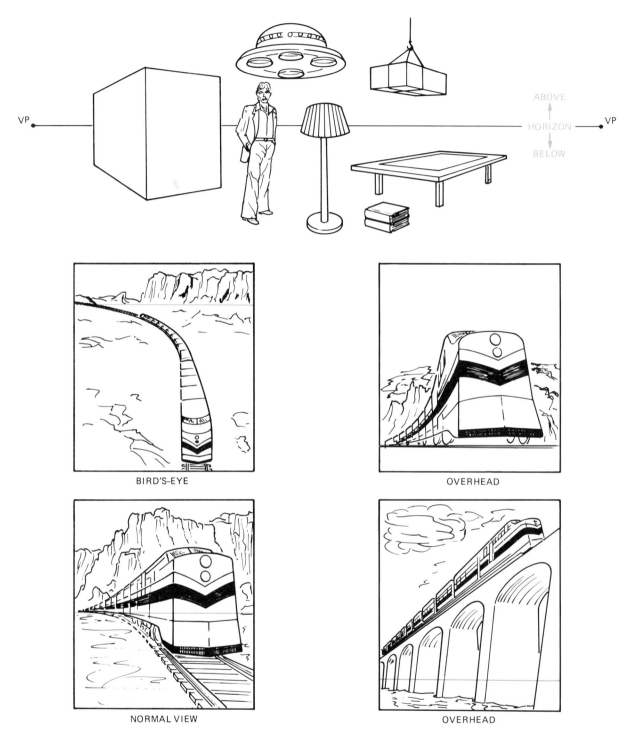

VP

ABOVE
HORIZON
BELOW

VP

BIRD'S-EYE

OVERHEAD

NORMAL VIEW

OVERHEAD

THE EYE IS ALWAYS LEVEL WITH THE HORIZON, REGARDLESS
OF WHETHER THE VIEW IS LOOKING UP OR DOWN AT THE OBJECT.

FIG. 22-9 Eye level and the horizon

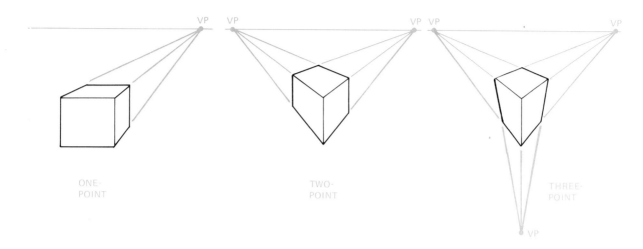

FIG. 22-10 Types of perspective

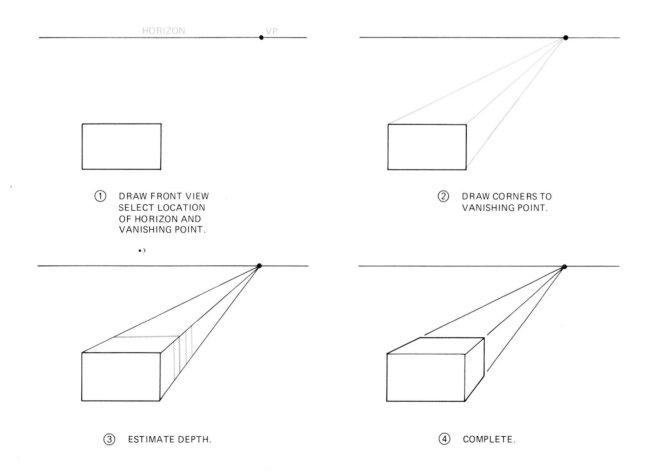

① DRAW FRONT VIEW SELECT LOCATION OF HORIZON AND VANISHING POINT.

② DRAW CORNERS TO VANISHING POINT.

③ ESTIMATE DEPTH.

④ COMPLETE.

FIG. 22-11 One-point perspective

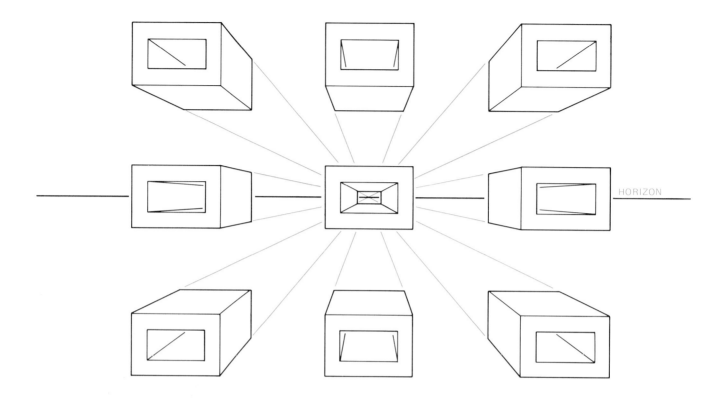

FIG. 22-12 Views in relation to the horizon with one–
point perspective

perspective image is visualized and drawn.

Projectors The lines or visual rays that converge from the object to the viewer's eye.

TYPES OF PERSPECTIVES

The draftsperson can select one-, two-, or three-point perspective (Fig. 22-10) to represent the object. The majority of perspective drawings are two-point.

ONE-POINT PERSPECTIVE

One-point perspective is also called parallel perspective because the front face of the drawing is parallel to the picture plane. As an orthographic view, this face is drawn true size. The sides of a

perspective drawing will recede to a single vanishing point (Fig. 22-11). Note in Figure 22-12 how the sides of the object are viewed as it is placed in different positions. An object drawn *above* the horizon will show a combination of its front, sides, and bottom. An object drawn *on* the horizon will show its front and sides. An object drawn *below* the horizon will show its front, sides, and top.

TWO-POINT PERSPECTIVE

Two-point perspective is also called angular perspective because both sides of the object recede at an angle from the front corner to two vanishing points. Figure 22-13 shows the steps that must be followed to draw a two-point perspective. Placing

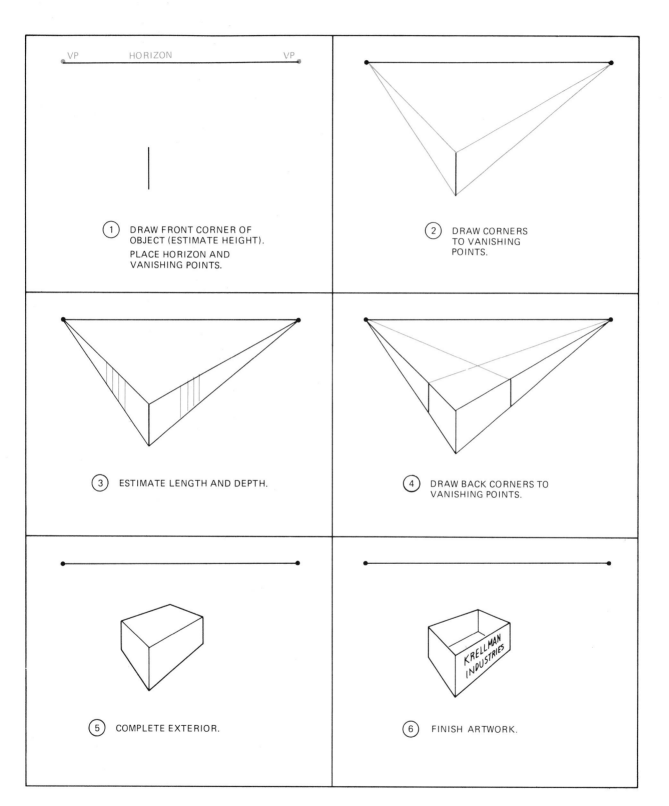

FIG. 22-13   Two-point perspective

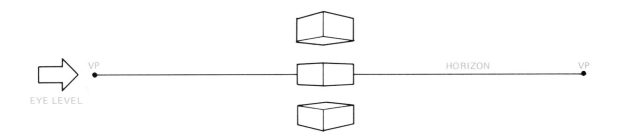

FIG. 22-14   Views in relation to the horizon with
two-point perspective

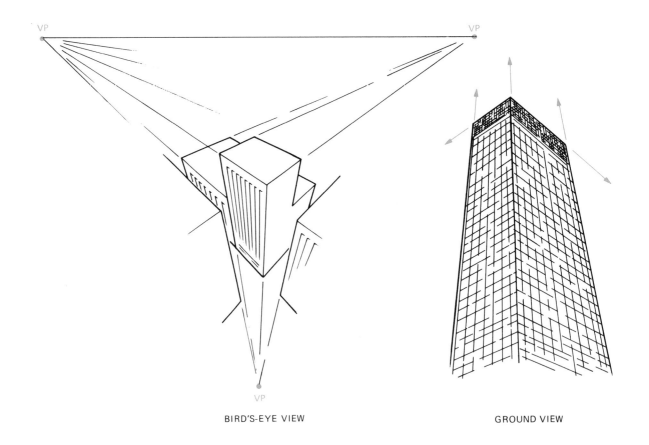

BIRD'S-EYE VIEW   GROUND VIEW

FIG. 22-15   Three-point perspective

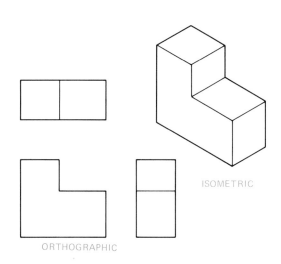

ISOMETRIC

ORTHOGRAPHIC

FIG. 22-16 True-size drawings

the object above the horizon will show its sides and bottom. The object on the horizon will show only its sides. The object below the horizon will show its sides and top (Fig. 22-14).

THREE-POINT PERSPECTIVE

Three-point perspective is used to give a more life-like quality to drawings of tall objects such as buildings (Fig. 22-15). The third vanishing point is placed in line with the station point (viewer's position). The further it is placed above or below the drawing, the less distortion there will be.

PERSPECTIVE PROPORTIONS

All objects have three dimensions: length, width, and height. These three dimensions will form a rectangular prism. Orthographic and isometric drawings are true size (Fig. 22-16). A two-point perspective drawn with true basic dimensions will appear out of proportion. The receding sides must be shortened to obtain lifelike proportions (Fig. 22-17).

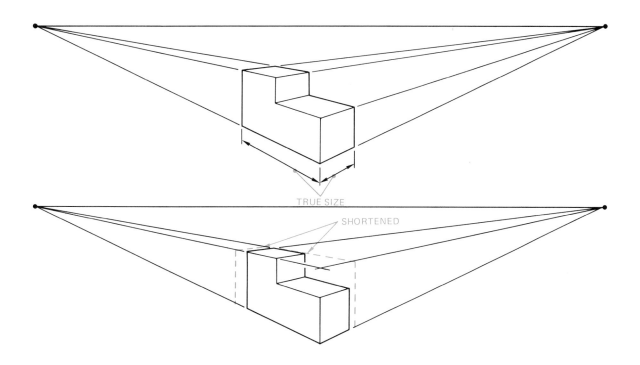

TRUE SIZE

SHORTENED

FIG. 22-17 Receding sides are shortened

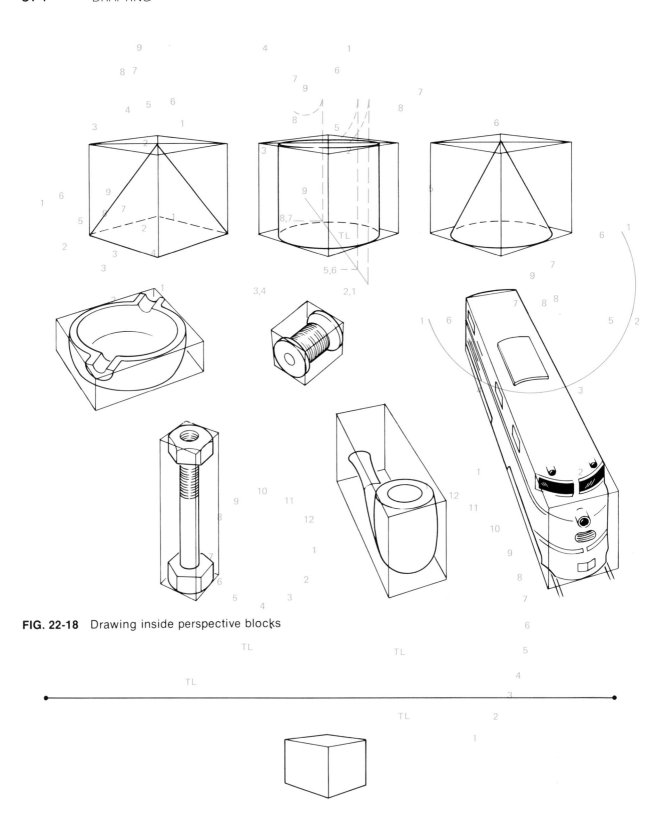

FIG. 22-18   Drawing inside perspective blocks

FIG. 22-19   Good perspective

The first step in making a perspective drawing is to draw the boundaries of the object with a perspective block and make all the necessary adjustments to length, width, and height. When the approximate proportions are achieved, the form of the object is drawn inside the block (Fig. 22-18).

DISTORTIONS

The base angle of a normal, lifelike perspective drawing must be more than ninety degrees (Fig. 22-19). An obtuse angle more than 90° will be assured when the vanishing points are spread out as far as possible and/or the object is placed close to the horizon. When the vanishing points are too close together or the object is too far away from the horizon, the base angle will be acute (less than 90°), causing distortion of the object's form (Fig. 22-20).

PERSPECTIVE PROJECTION

The science of perspective drawing is very complex. Books have been written on perspective layouts by projection of one-point (Fig. 22-21) and two-point perspective (Fig. 22-22). The concepts of these projection methods should be understood, es-

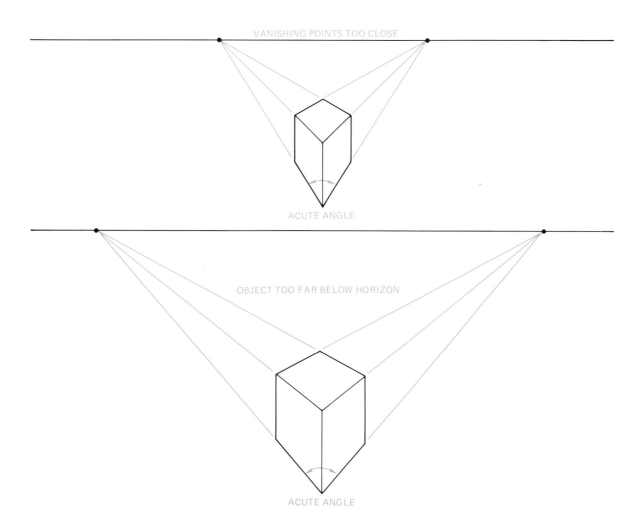

FIG. 22-20 Distortion in perspective drawings

pecially the technique of locating vanishing points (Fig. 22-23). But because perspective drawings are not used for production processes, the accurate, time consuming method of perspective drawing is not always necessary. A draftsperson with an eye for estimation can make a good perspective drawing in a fraction of the time required by the one- or two-point projection procedure.

A circle in a perspective drawing will appear as an ellipse. To draw a perspective circle, start by blocking in a perspective square the size of the circle. By sketching a circle tangent to the center of each side, an accurate appearing perspective circle can be drawn (Fig. 22-24). This sketched circle can be darkened with an irregular curve or an ellipse template.

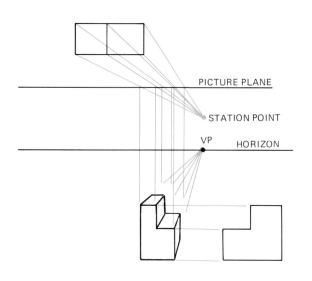

FIG. 22-21   One-point perspective projection layout

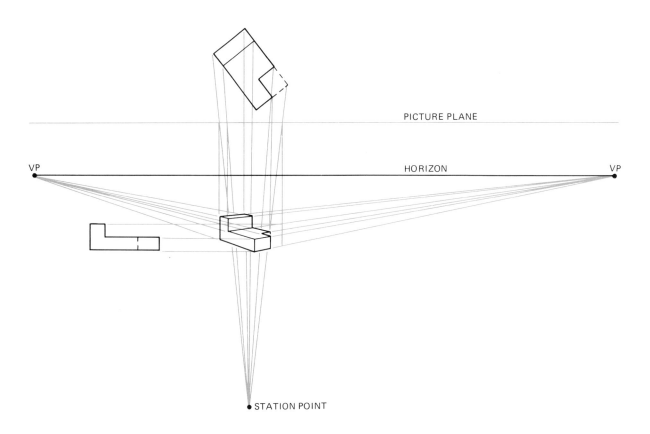

FIG. 22-22   Perspective projection

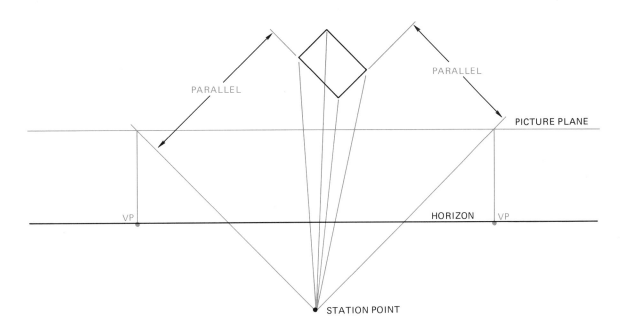

FIG. 22-23 Location of vanishing points

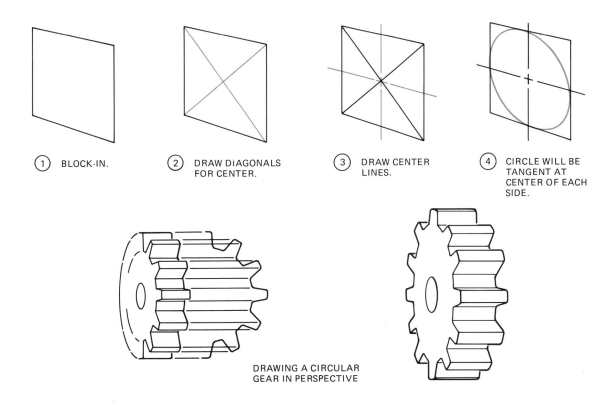

① BLOCK-IN.

② DRAW DIAGONALS FOR CENTER.

③ DRAW CENTER LINES.

④ CIRCLE WILL BE TANGENT AT CENTER OF EACH SIDE.

DRAWING A CIRCULAR GEAR IN PERSPECTIVE

FIG. 22-24 Freehand perspective circle layout

ANGULAR LINES

The location and angle of angular lines must be estimated. Lay out the ends of the angles on the perspective prism as shown in Figure 22-25.

EQUAL SPACING

As receding lines approach the vanishing point they come closer together. Therefore, equally spaced parts must be placed closer together so the spacing will look normal.

PIN IN VANISHING POINT

A perspective drawing that requires a large number of lines to be drawn to a vanishing point can be drawn much faster by placing a pin in the vanishing point (Fig. 22-26). When the straight-edge is held against the pin it will consistently be lined up with the vanishing point.

PERSPECTIVE GRID PAPER

Perspective grid paper will help the draftsperson

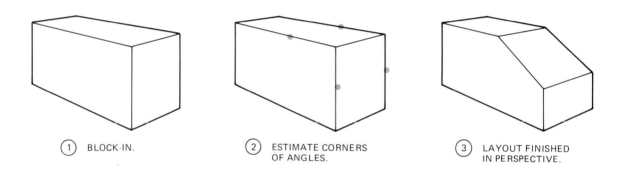

① BLOCK-IN.   ② ESTIMATE CORNERS OF ANGLES.   ③ LAYOUT FINISHED IN PERSPECTIVE.

FIG. 22-25 Angles in perspective

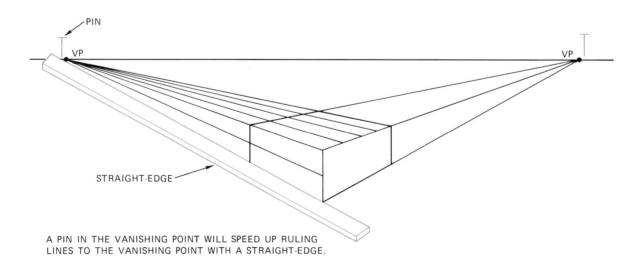

A PIN IN THE VANISHING POINT WILL SPEED UP RULING LINES TO THE VANISHING POINT WITH A STRAIGHT-EDGE.

FIG. 22-26 Use of a pin in the vanishing point

make fast, freehand perspective sketches. The main disadvantages of perspective grid paper are the extra cost and the limitation of the perspective angles of grid paper.

PERSPECTIVE BOARDS

There are several types of perspective drawing boards. They have a rotating straight edge which

will radiate from a vanishing point. These boards are expensive, but they reduce perspective drawing time.

PROBLEMS

Make perspective drawings of the objects in Figures 22-27 through 22-29.

DRAW A ONE-POINT PERSPECTIVE AND AN ISOMETRIC DRAWING FOR EACH PROBLEM.

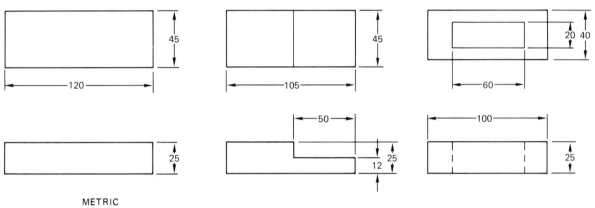

METRIC

FIG. 22-27 Problem

DRAW A TWO-POINT PERSPECTIVE AND AN ISOMETRIC DRAWING FOR EACH PROBLEM.

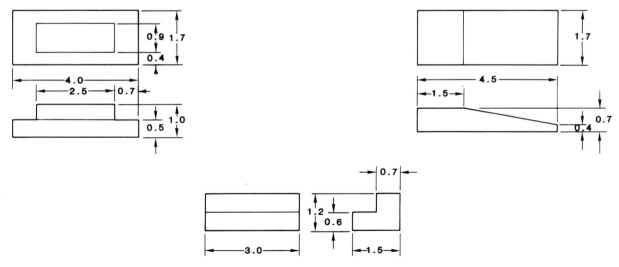

FIG. 22-28 Problem

DRAW A ONE- OR TWO-POINT PERSPECTIVE AND A TWO OR THREE VIEW
ORTHOGRAPHIC DRAWING FOR EACH PROBLEM.

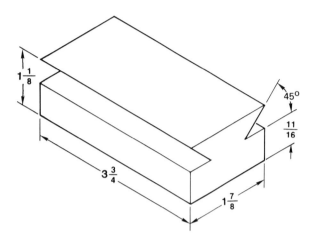

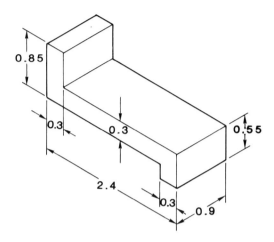

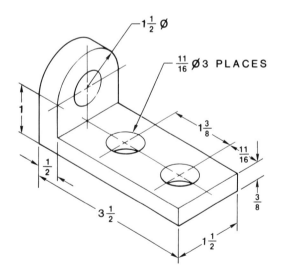

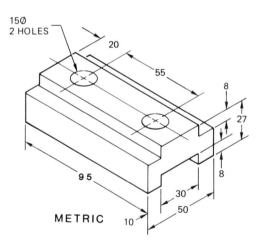

FIG. 22-29 Problems

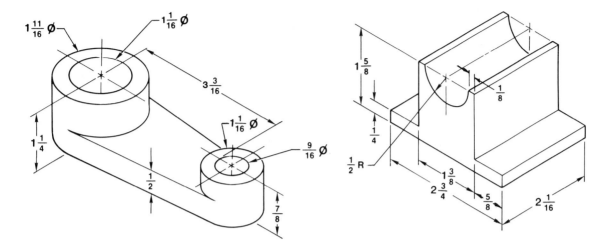

FIG. 22-29 Problems (Cont.)

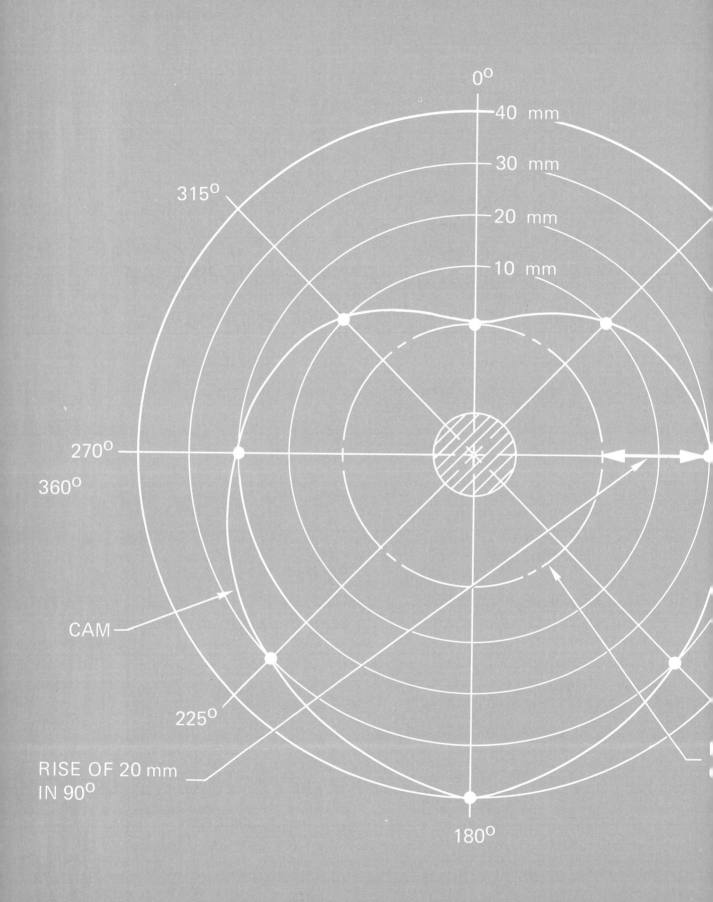

0°
40 mm
30 mm
20 mm
10 mm
315°
270°
360°
CAM
225°
RISE OF 20 mm
IN 90°
180°

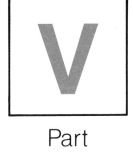

Part

Industrial Drafting

Before any item can be manufactured, engineers, designers and drafts-persons must design and draw each part of the item. The final stages of a mechanical drawing require critical thinking, creativity, initiative, and knowledge of the ever changing design concepts and drawing practices of industry. Our technological society is continually demanding better and faster tools and methods of visual communication.

Part V of this text introduces many of the specialized areas of drafting which are of value and interest to the beginner.

Chapter

Technical Illustration

INTRODUCTION

Technical illustration is the production of drawings that aid in the understanding of a concept, idea, or point of view. Technical illustrations aid the viewer in assembling and maintaining mechanical devices or structures. Three-dimensional technical illustrations as well as working drawings are used in instruction and repair manuals. Figure 23-1 is an example of a drawing used in assembly instruction manuals. Anyone who has ever put together a child's toy (Fig. 23-2) has used this type of technical illustration. Maintenance handbooks (Fig. 23-3) contain three-dimensional drawings which illustrate such general maintenance procedures as check-out, adjustment, and lubrication. Repair handbooks contain three-dimensional drawings that are arranged in step-by-step sequence. These illustrations (Fig. 23-4) may show enlarged details of a particular area of the unit being repaired.

Not all technical illustrations are three dimensional. Two-dimensional electrical schematics (Fig. 23-5) are used by technicians to assemble, maintain, and repair electrical equipment. Non-electrical schematics (Fig. 23-6) such as hydraulic diagrams are also used for this purpose.

Technical illustrations are also used as sales tools to promote a concept, idea, or point of view. A technical illustration may be a training aid (Fig. 23-7) used to train workers in a new method or procedure, may be a very artistic rendering (Fig. 23-

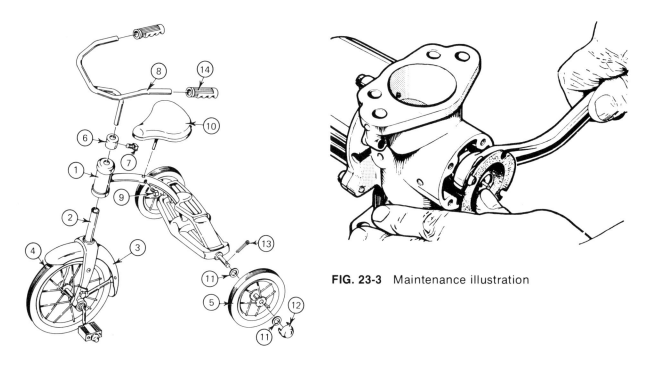

FIG. 23-2 An exploded assembly drawing

FIG. 23-3 Maintenance illustration

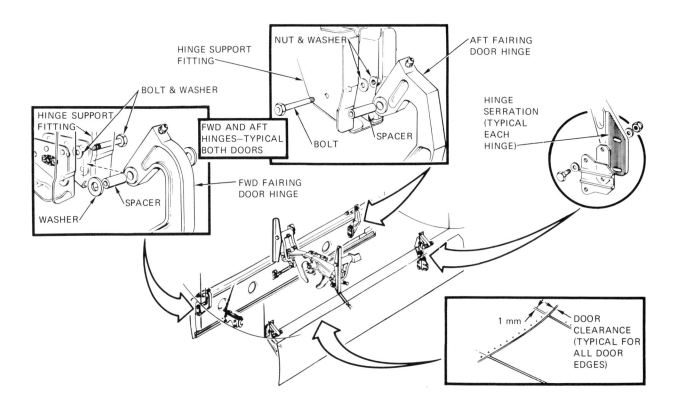

FIG. 23-4 Sequence illustrations (Boeing Commercial Airplane Co.)

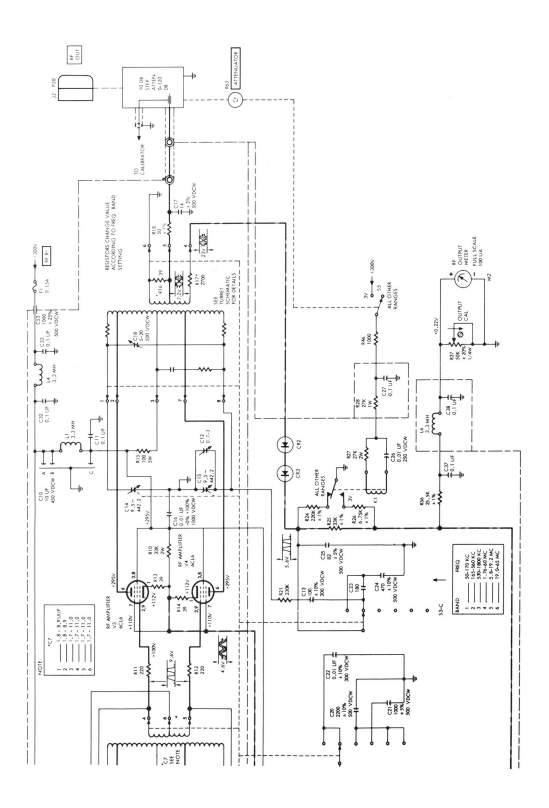

FIG. 23-5 An electrical schematic (Boeing Commercial Airplane Co.)

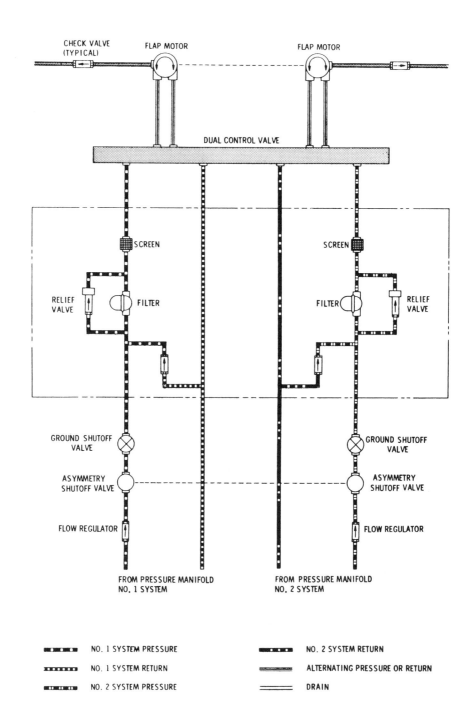

FIG. 23-6 A hydraulic schematic (Boeing Commercial Airplane Co.)

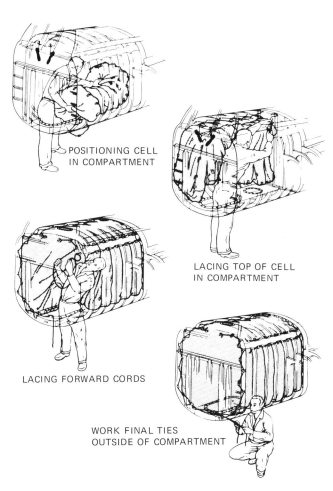

FIG. 23-7 A training aid illustration (Boeing Commercial Airplane Co.)

POSITIONING CELL
IN COMPARTMENT

LACING TOP OF CELL
IN COMPARTMENT

LACING FORWARD CORDS

WORK FINAL TIES
OUTSIDE OF COMPARTMENT

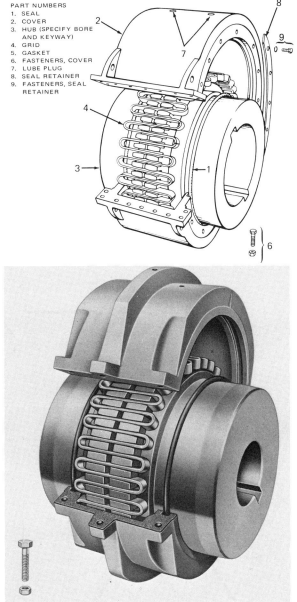

PART NUMBERS
1. SEAL
2. COVER
3. HUB (SPECIFY BORE AND KEYWAY)
4. GRID
5. GASKET
6. FASTENERS, COVER
7. LUBE PLUG
8. SEAL RETAINER
9. FASTENERS, SEAL RETAINER

FIG. 23-8 Artistic renderings (The Falk Corp.)

8) used to promote a concept or to sell a product, or may be a very simple drawing such as a chart (Fig. 23-9) or a graph (Fig. 23-10).

As you study this chapter, you should develop the skills and knowledge needed to produce drawings classified as technical illustrations. You should also develop skills with technical illustrator's instruments such as technical fountain pens, special templates, and the airbrush.

THE TECHNICAL ILLUSTRATOR

A technical illustrator is a person who produces two- and three-dimensional drawings from blueprints, technical sheets, engineering sketches, photographs, or actual parts. Because much of the illustrator's work will eventually appear in published form, it is most often done in ink for better reproduction. The technical illustrator must be able to visualize engineering data and convert it into the necessary two- or three-dimensional drawings by

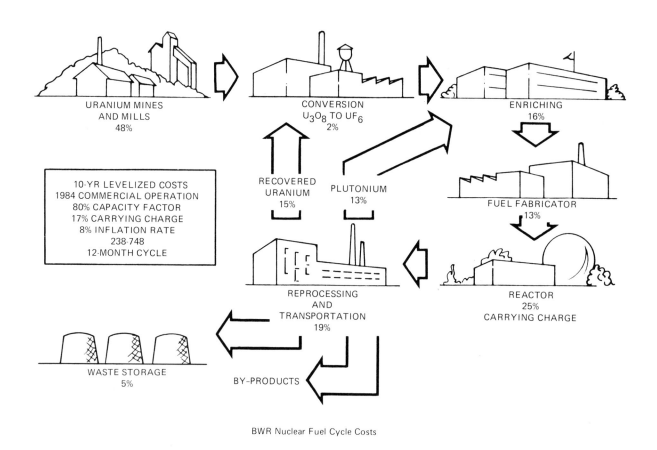

10-YR LEVELIZED COSTS
1984 COMMERCIAL OPERATION
80% CAPACITY FACTOR
17% CARRYING CHARGE
8% INFLATION RATE
238-748
12-MONTH CYCLE

URANIUM MINES
AND MILLS
48%

CONVERSION
U_3O_8 TO UF_6
2%

ENRICHING
16%

RECOVERED
URANIUM
15%

PLUTONIUM
13%

FUEL FABRICATOR
13%

REPROCESSING
AND
TRANSPORTATION
19%

REACTOR
25%
CARRYING CHARGE

WASTE STORAGE
5%

BY-PRODUCTS

BWR Nuclear Fuel Cycle Costs

FIG. 23-9   A simple flowchart

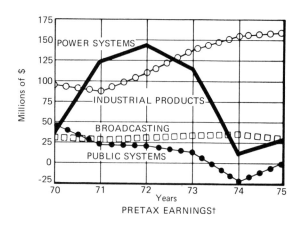

POWER SYSTEMS

INDUSTRIAL PRODUCTS

BROADCASTING

PUBLIC SYSTEMS

Millions of $

Years

PRETAX EARNINGS†

FIG. 23-10   Illustration of a graph

judging what the best method of presentation might be. The process includes planning the layout, making an accurate pencil drawing, and finishing the drawing (usually in ink) according to specifications. The normal progression of an illustration from beginning through publication is illustrated in Figure 23-11.

Technical illustrators must have knowledge and skills from both the technical and artistic fields. However, most successful technical illustrators are stronger in one area than the other (Fig. 23-12). Those who are more technically oriented usually specialize in maintenance and repair manuals, assembly and production drawings, or schematic drawings and brochures. Those who are more artistically inclined tend to produce advertising and sales drawings and the more complex technical illustrations as shown in Figure 23-13 and 23-14.

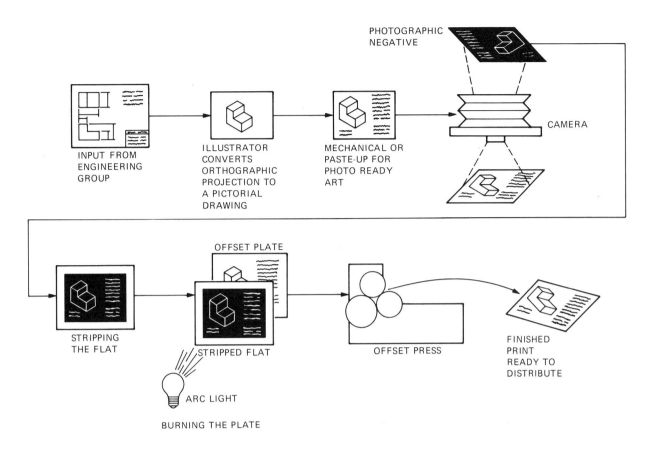

FIG. 23-11 The process of technical illustration

The basic skills and tools used by an illustrator are similar to those used by a draftsperson. Where the draftsperson is more technically oriented, the illustrator is more artistically oriented. Where the draftsperson's media is most often pencil, the illustrator's is most often ink. Where the draftsperson most often produces two-dimensional orthographic projections, the illustrator most often produces three-dimensional isometric or perspective pictorial drawings. Where the draftsperson is concerned with correct dimension callouts and material specifications, the illustrator is concerned with shading and textural illusions (Fig. 23-15).

TOOLS FOR TECHNICAL ILLUSTRATION

The basic tools used by the illustrator are similar to those used by the draftsperson. Additional equip-

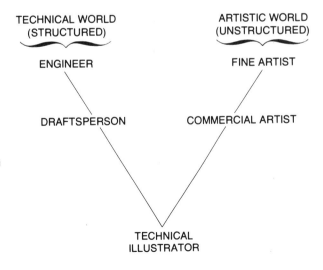

FIG. 23-12 Skill spectrum of the technical illustrator

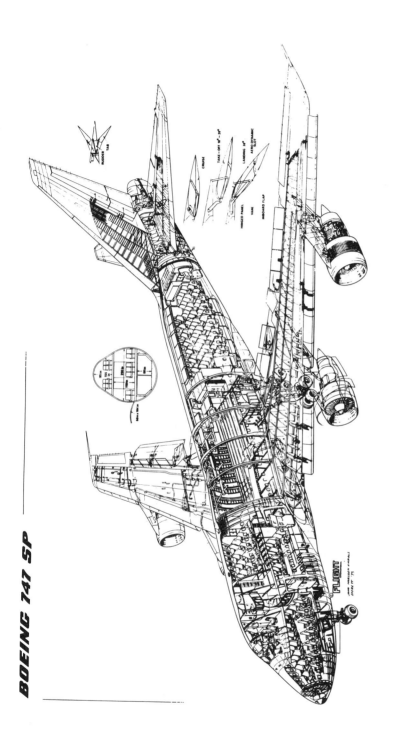

FIG. 23-13   A complex technical illustration (Boeing Commercial Airplane Co.)

FIG. 23-14 Complex renderings (Boeing Commercial Airplane Co.)

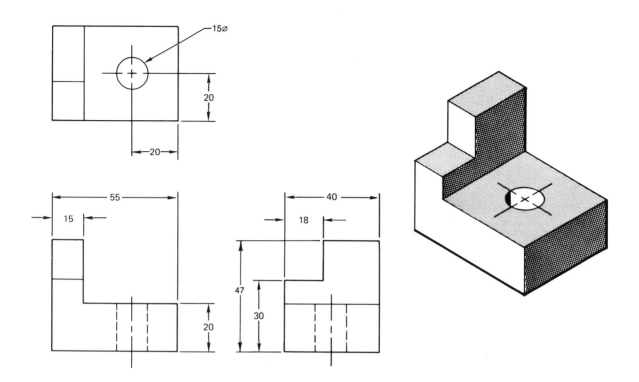

FIG. 23-15 Line shading examples

FIG. 23-16 Metric pens (J. S. Staedtler, Inc.)

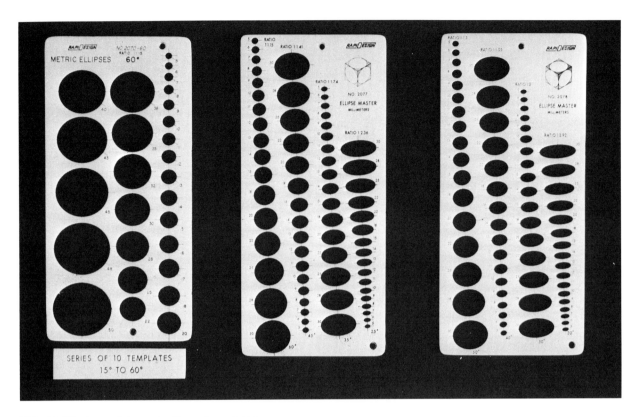

FIG. 23-17 Isometric and ellipse templates (RapiDesign, Inc.)

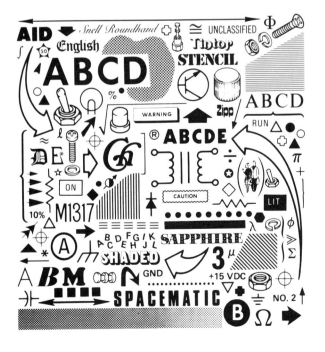

FIG. 23-18 Rub-on letters

FIG. 23-19 Airbrushes

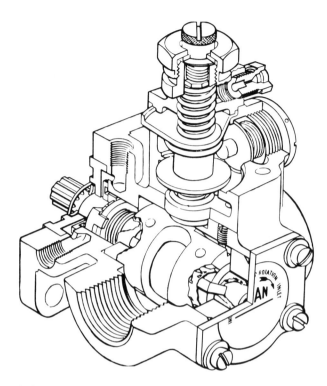

FIG. 23-20   Plain line drawing

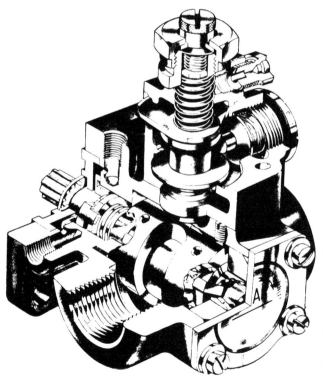

FIG. 23-21   Airbrushed line drawing

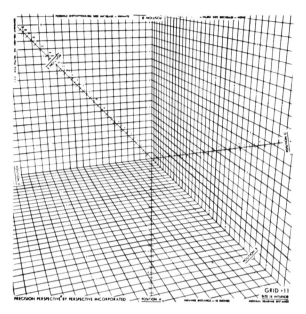

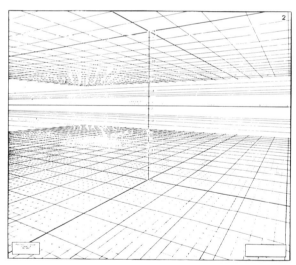

FIG. 23-22   Perspective grid sheets

ment the illustrator may use includes ink pens of various metric point sizes (Fig. 23-16); isometric and ellipse templates (Fig. 23-17); appliques in the form of rub-on letters, textures, and colors (Fig. 23-18); and, in some cases, the use of an airbrush (Fig. 23-19). Figure 23-20 shows a plain ink line drawing with no shading added. Figure 23-21 shows that same drawing fully airbrushed.

TYPES OF PICTORIAL DRAWINGS

There are three basic classifications of pictorial drawings: (1) oblique (Chapter 11), which includes cabinet and cavalier drawings; (2) axonometric (Chapter 11), which includes isometric, dimetric, and trimetric drawings; and (3) perspective (Chapter 22), which includes one-point, two-point, and three-point perspective drawings. Although perspective projection is the most visually correct method, isometric projection is much simpler and therefore is used most often. Aids used to make perspective projection easier are called perspective grid sheets (Fig. 23-22). If the object being illustrated is made up of a number of circles and arcs the illustrator will need a large number of ellipse templates to speed up the drawing process. Oblique drawings give the most distorted pictorial image and are rarely used by the illustrator. Because iso-metric projection is the most often used method of producing pictorial drawings, the rest of this chapter will deal with specific procedures for drawing and shading various isometric figures.

METHODS OF PRODUCING BASIC ISOMETRIC SHAPES

Chapter 11 illustrates the basic layout of isometric projection. Refer to that chapter for a review of blocking-in methods, isometric and nonisometric lines, the use of the isometric ellipse template, and the method of plotting irregular curves.

Many technical illustrations show a number of parts arranged in an order that indicates each part's relationship to the other. These drawings are called exploded assemblies (Fig. 23-23). The locations of various parts in an exploded assembly are usually indicated by center lines which connect one part to

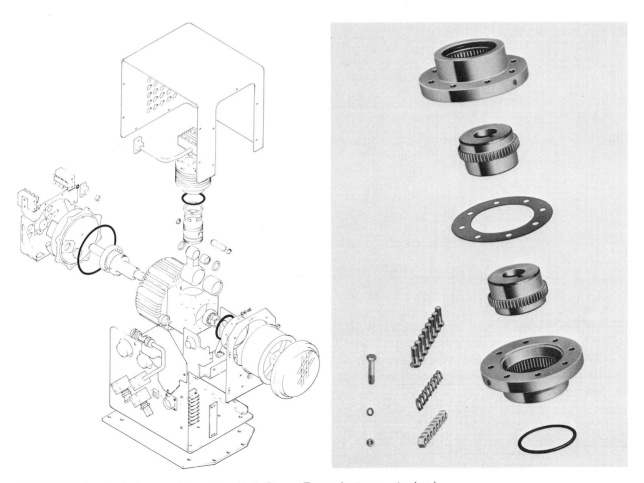

FIG. 23-23 Exploded assemblies (The Falk Corp., Texas Instruments, Inc.)

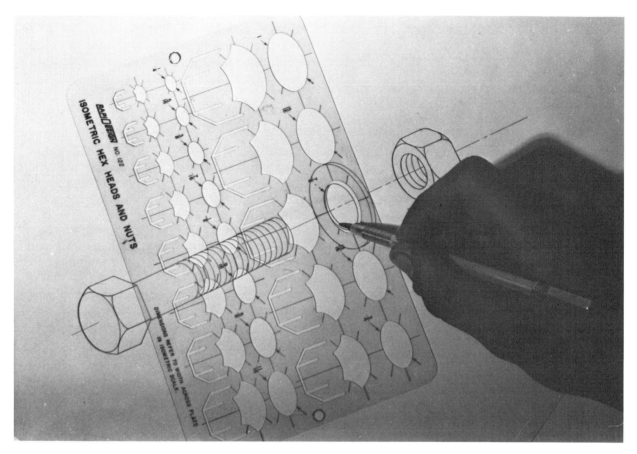

FIG. 23-24   Isometric hexagon template (Clark E. Smith)

another. In exploded assemblies of mechanical parts there are a number of shapes which often occur. In addition to the basic rectangles, circles, and arcs which were covered in Chapter 11, there are hexagonal bolts and nuts, spheres and portions of spheres, round head screws, fillister head screws, countersunk and counterbored holes, wheels or tori, and irregular pipe or wire curves.

Drawing Hexagon Bolts and Nuts

If an isometric hexagon template (Fig. 23-24) is available, it can be used the same as one would use an isometric ellipse template. If a template is not available or if the bolt size required is not on the template, follow the example illustrated in Figure 23-25.

The illustrator should refer to a chart like those found in the appendix for the distances across flats and corners. If only the distance across the flats is given, then the distance across the corners may be found as illustrated in Figure 23-26.

A similar procedure should be followed to draw a hexagon nut. Steps 1 through 7 in Figure 23-25 can be used to draw either a hexagon bolt or nut. For hex nuts step 8 should be modified as illustrated in Figure 23-27.

REFER TO APPENDIX TABLE FOR HEXAGON BOLTS AND NUTS
TO DETERMINE THE DISTANCE ACROSS THE FLATS.

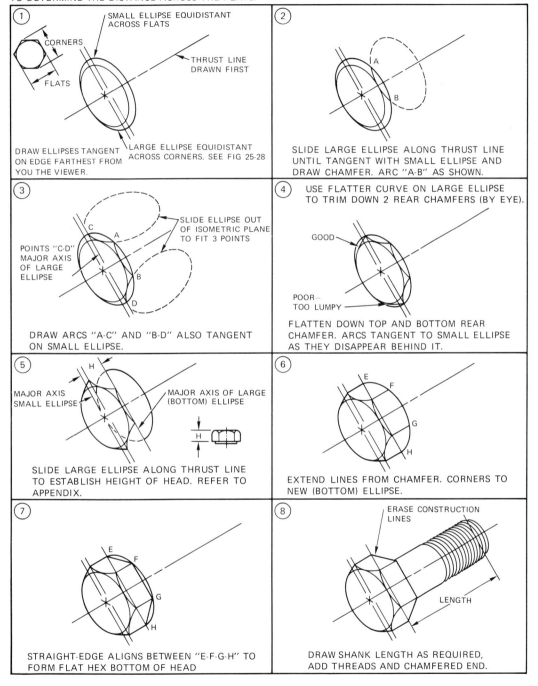

FIG. 23-25 Drawing a hex bolt head

FINDING THE DISTANCE ACROSS THE CORNERS WHEN THE DISTANCE
ACROSS THE FLATS IS KNOWN.

GIVEN: FROM TABLES A HEXAGON BOLT WITH A DISTANCE OF
         1.125" ACROSS THE FLATS.

PROCEDURE:

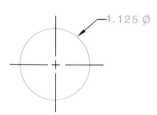

1. CONSTRUCT A CIRCLE
   WITH A DIAMETER OF
   1.125".

2. WITH A 30°-60° TRIANGLE
   SCRIBE LINES AS ILLUSTRATED
   TANGENT TO THE CIRCLE.
   MEASURE THE CORNER DISTANCE.

FIG. 23-26   Finding bolt head dimensions

DIAMETER EQUAL
TO MINOR DIAMETER
OF BOLT

OPPOSITE VIEW
OF SAME NUT

FIG. 23-27   Drawing a hex nut

DRAWING A SPHERE USING ISOMETRIC PROJECTION

1. DRAW AN ISOMETRIC
   AXIS.

2. DRAW 1 OR 2 ISOMETRIC
   CIRCLES AROUND THIS
   AXIS. THE ISOMETRIC
   CIRCLES ARE THE SIZE OF
   THE DESIRED SPHERE.

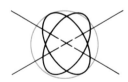

3. DRAW A CIRCLE TANGENT TO THE
   MAJOR DIAMETERS OF THE
   ISOMETRIC CIRCLES.

NOTE:

THE DIAMETER OF THE
ISOMETRIC SPHERE IS
GREATER THAN THE
DIAMETER OF THE
TRUE SPHERE.
REFER TO FIG. 23-29.

Drawing Spheres and Portions of Spheres

Spheres or portions of spheres are often required in
technical illustration. Figure 23-28 illustrates the
method of drawing an isometric sphere. A sphere's
outline will be a true circle but an isometric drawing of a 1-inch sphere will be larger than 1 inch.
This is because the major diameter of an isometric
circle (an ellipse) is larger than 1 inch (Fig. 23-29).

Drawing portions of spheres or flat surfaces on
spheres in isometric projection is a common technical illustration practice. Figure 23-30 illustrates
the method of producing these forms.

FIG. 23-28   Drawing an isometric sphere

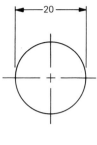

PLAN VIEW OF A
20 mm Ø SPHERE

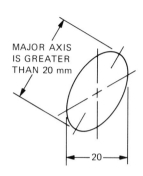

MAJOR AXIS
IS GREATER
THAN 20 mm

20 mm ISOMETRIC
CIRCLE IS TRUE
LENGTH ACROSS THE
ISOMETRIC AXIS

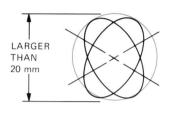

LARGER
THAN
20 mm

ISOMETRIC SPHERE
LARGER THAN
ORIGINAL SPHERE

FIG. 23-29 An isometric sphere is larger than the original sphere

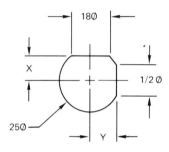

GIVEN: AN ORTHOGRAPHIC PROJECTION
OF A SPHERE WITH TWO SEGMENTS
REMOVED.

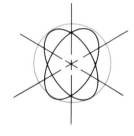

1. LAY OUT THE APPROPRIATE ISOMETRIC
SPHERE WITH ISOMETRIC AXES.

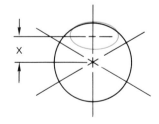

2. MEASURE DISTANCE X FROM THE CENTER
OF THE SPHERE. THIS IS THE CENTER OF
THE TOP SEGMENT. LAY OUT AN ISOMETRIC
CIRCLE AS INDICATED BY THE ORTHOGRAPHIC
PROJECTION.

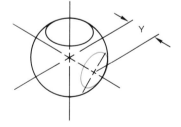

3. MEASURE DISTANCE Y FROM THE CENTER
OF THE SPHERE. THIS IS THE CENTER OF
THE SMALLER SEGMENT. LAY OUT AN
ISOMETRIC CIRCLE AS INDICATED BY THE
ORTHOGRAPHIC PROJECTION.

FIG. 23-30 Drawing flat surfaces on isometric spheres

Drawing Round Head Screws

Round head screws are common fasteners used in many assemblies. Drawing a round head screw in isometric is illustrated in Figure 23-31. Common dimensions for fasteners may be found in the appendix, in a machinery handbook, or in tables supplied by manufacturers of such fasteners.

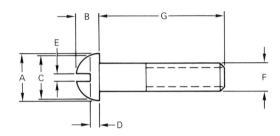

SCREW DIMENSIONS DERIVED FROM APPENDIX AND LAYOUT.

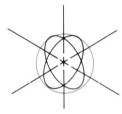

1. CONSTRUCT AN ISOMETRIC SPHERE EQUAL TO DIAMETER "A" FROM THE DIAGRAM ABOVE OR THE APPENDIX.

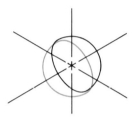

2. UTILIZE HALF THE SPHERE. (THE "B" DIMENSION IS VERY CLOSE TO HALF THE DIAMETER AND IS SUFFICIENT FOR ROUND HEAD SCREWS).

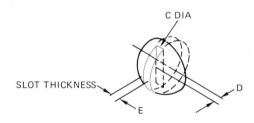

3. DIVIDE THE SLOT THICKNESS "E" ABOUT THE ISOMETRIC AXIS AS SHOWN AND DRAW IN THE TWO PARALLEL ARCS USING AN ISOMETRIC CIRCLE SLIGHTLY SMALLER THAN THAT CALLED FOR BY DIMENSION "A". MEASURE DISTANCE "C" AND "D" AND PLOT THEM AS SHOWN.

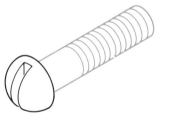

4. ADD THE SHAFT AND THREADS AS REQUIRED.

FIG. 23-31   Drawing a round head screw in isometric

Drawing Fillister Head Screws

Fillister head cap screws are drawn as shown in Figure 23-32. Standard dimensions for these screws can be found in the appendix, in a machinery handbook, or in tables supplied by manufacturers of these screws.

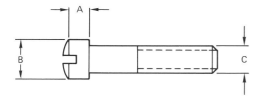

SEE APPENDIX FOR TYPICAL DIMENSIONS OF FILLISTER HEAD SCREWS.

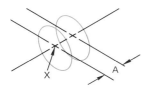

1. DRAW AN ISOMETRIC AXIS WITH TWO PARALLEL LINES "A" DISTANCE APART. DRAW TWO ISOMETRIC ELLIPSES OF DIAMETER "B" AS SHOWN.

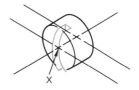

2. DRAW THE SLOT AS SHOWN USING THE SAME DIAMETER ELLIPSE AS GIVEN BY DIMENSION "B". THIS ELLIPSE WILL BE TANGENT AT POINT X AND ALIGNED ON A HORIZONTAL AND VERTICAL AXIS.

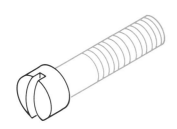

3. ADD THREADS AS REQUIRED.

FIG. 23-32 Drawing a fillister head screw in isometric

FLAT HEAD MACHINE SCREW

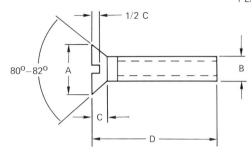

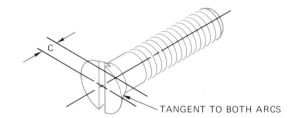

TANGENT TO BOTH ARCS

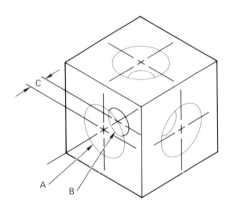

1. TO DRAW A COUNTERSUNK HOLE, LOOK UP DIMENSIONAL DATA FOR THE FLAT HEAD SCREW WHICH IS TO FIT THE HOLE. THIS DATA MAY BE FOUND IN THE APPENDIX.

2. LAY OUT DIMENSION "A" ON THE OUTSIDE SURFACE AND DIMENSION "B" A DISTANCE DOWN THE AXIS EQUAL TO "C".

FIG. 23-33   Drawing a countersunk hole

Drawing Countersunk and Counterbored Holes

Countersunk and counterbored holes are used to recess a bolt or screw head beneath the surface of a component. Methods for drawing these holes are illustrated in Figures 23-33 and 23-34.

Drawing a Torus

A torus is a donut shape which occurs in such common items as steering wheels, bicycle tires, and hand wheels. The method of producing a torus in isometric is illustrated in Figure 23-35.

Drawing Irregular Pipes or Wire Curves

Irregular pipes or wire curves are drawn as shown in Figure 23-36.

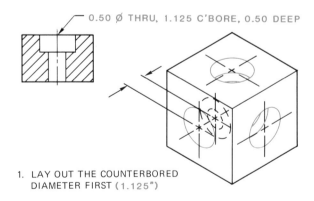

0.50 Ø THRU, 1.125 C'BORE, 0.50 DEEP

1. LAY OUT THE COUNTERBORED DIAMETER FIRST (1.125")

2. MEASURE THE DEPTH DOWN THE AXIS (0.50") DRAW THE HOLE BOTTOM (0.50") AT THIS DEPTH.

3. CHECK TO SEE IF THE SHAFT DIAMETER (0.50") WILL SHOW AS IT DOES IN THIS CASE.

FIG. 23-34   Drawing a counterbored hole

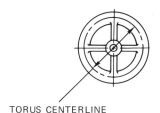

TORUS CROSS SECTION

TORUS CENTERLINE

1. DRAW AN ISOMETRIC AXIS AND AN ELLIPSE OF THE TORUS CENTERLINE.

2. DRAW FOUR ISOMETRIC ELLIPSES OF THE TORUS CROSS SECTIONAL DIAMETER IN THE POSITIONS INDICATED.

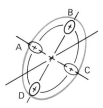

3. SELECT AN ISOMETRIC ELLIPSE LARGE ENOUGH TO ENCLOSE THE OUTER PORTIONS OF ELLIPSES A, B AND C. MOVE THE SAME ELLIPSE DOWN AND ENCLOSE ELLIPSES A, D AND C.

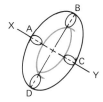

4. SELECT A SMALLER ISOMETRIC ELLIPSE TO CONNECT THE INNER PORTION OF ELLIPSES A, B AND D. MOVE THE SAME ELLIPSE ALONG AXIS X, Y TO CONNECT ELLIPSES B, C AND D. THIS COMPLETES THE WHEEL OR TORUS.

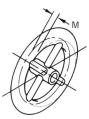

5. TO FIND THE CENTERLINE ON THE OUTSIDE SURFACE MOVE THE CENTERLINE ELLIPSE HALF THE DISTANCE "M" DOWN THE X, Y AXIS. ADD DETAIL AND FINISH THE WHEEL.

FIG. 23-35 Drawing a torus

1. DRAW THE IRREGULAR CURVE CENTER LINE IN AN APPROPRIATE POSITION.

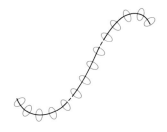

2. SELECT THE CORRECT ISOMETRIC ELLIPSE AND ALIGN THE MINOR AXIS ALONG THE IRREGULAR CURVE.

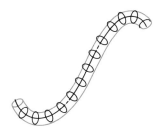

3. CONNECT THE ELLIPSES WITH A TANGENT LINE AS SHOWN.

FIG. 23-36 Drawing irregular curves

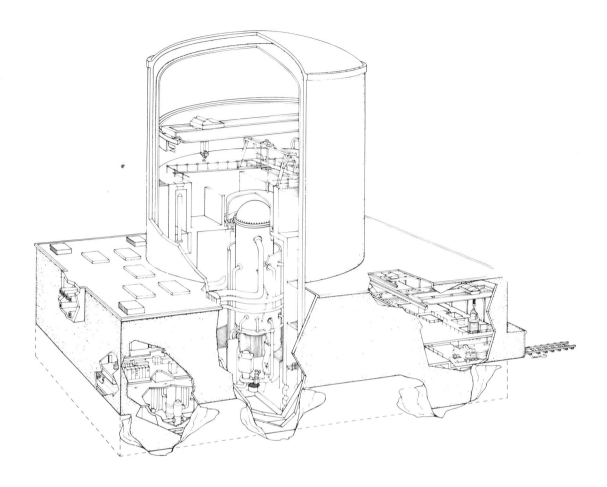

FIG. 23-37 An illustration without shading

Shading Technical Illustrations

The shading of technical illustrations may range from no shading at all (Fig. 23-37) to a completely rendered drawing (Fig. 23-38). Because drawing costs increase in proportion to the amount of shading done, most technical illustrations have a minimum of shading. The amount of shading is usually determined by how the illustration is to be used. If it is to be used as a sales device, more shading will be applied. If it is to be used in a repair manual or assembly sheet, a small amount of shading is usually applied. Figure 23-39 shows examples of shading on basic shapes. This method is called drop-line shading. Notice that the light source is from the upper left-hand corner and at an

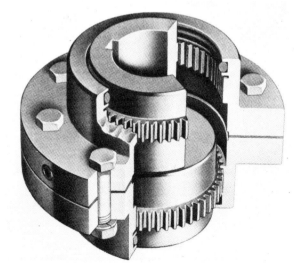

FIG. 23-38 A fully shaded illustration (The Falk Corp.)

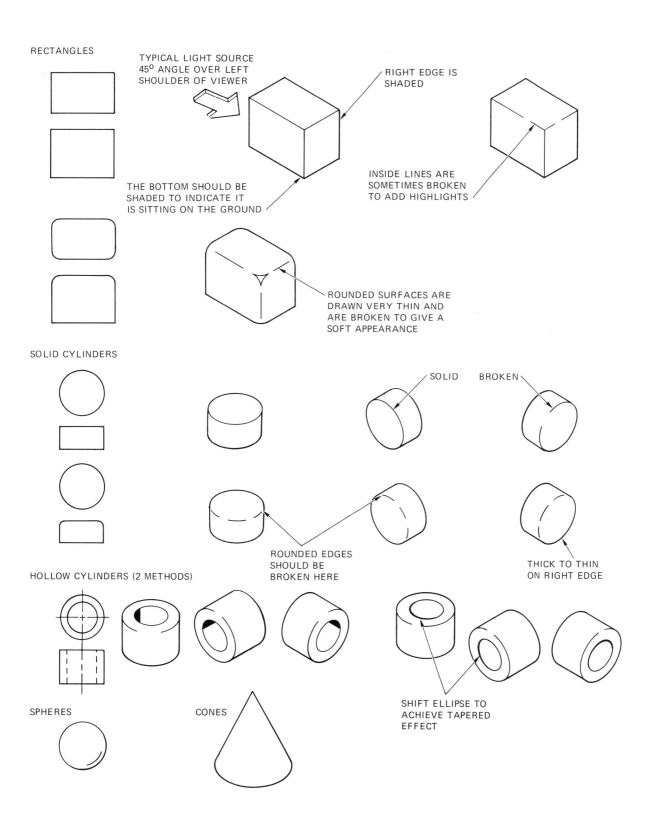

RECTANGLES

TYPICAL LIGHT SOURCE
45° ANGLE OVER LEFT
SHOULDER OF VIEWER

RIGHT EDGE IS
SHADED

THE BOTTOM SHOULD BE
SHADED TO INDICATE IT
IS SITTING ON THE GROUND

INSIDE LINES ARE
SOMETIMES BROKEN
TO ADD HIGHLIGHTS

ROUNDED SURFACES ARE
DRAWN VERY THIN AND
ARE BROKEN TO GIVE A
SOFT APPEARANCE

SOLID CYLINDERS

SOLID   BROKEN

THICK TO THIN
ON RIGHT EDGE

ROUNDED EDGES
SHOULD BE
BROKEN HERE

HOLLOW CYLINDERS (2 METHODS)

SHIFT ELLIPSE TO
ACHIEVE TAPERED
EFFECT

SPHERES   CONES

FIG. 23-39 Drop line shading

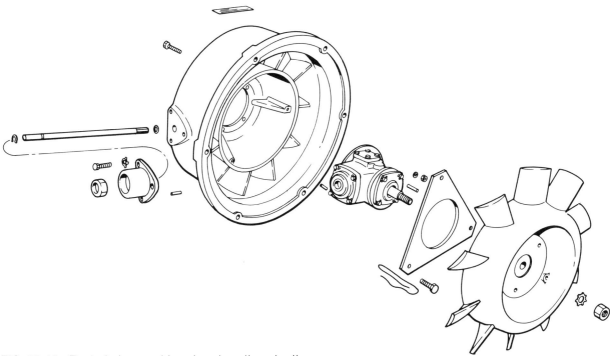

FIG. 23-40 Exploded assembly using drop line shading

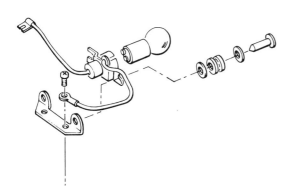

FIG. 23-41 Industrial example of drop line shading

angle of 45 degress. The light source is assumed to be coming over the viewer's shoulder. This is the standard that should be followed for most technical illustrations. Figures 23-40 and 23-41 show some typical industrial examples of this method of shading. Notice that the outline is always solid while two adjoining surfaces are often shown broken. Study these drawings and apply these procedures to your work.

PROBLEMS

1. Divide a piece of drawing paper into 12 equal parts. Draw Figure 23-42 with pencil. Overlay the pencil drawing with a sheet of vellum and trace it with ink.

2. Refer to Figure 23-25 and Figure 23-27 and the hexagon bolt and nut tables in the appendix. Select three different bolts and their corresponding nuts and draw them in isometric projection. All necessary dimensions except bolt length are given in the tables. Select a convenient length for the shafts of the bolts. Show threads continuing at least half the length of the shaft.

3. Draw the barbells in Figure 23-43 in orthographic projection and in isometric projection. Refer to Figure 23-28.

4. Draw the sphere in Figure 23-44 in isometric projection at twice the size indicated by the dimensions shown. Refer to Figure 23-30.

5. Select three different round head screws and draw them in isometric projection. Shaft lengths may vary but threads should extend

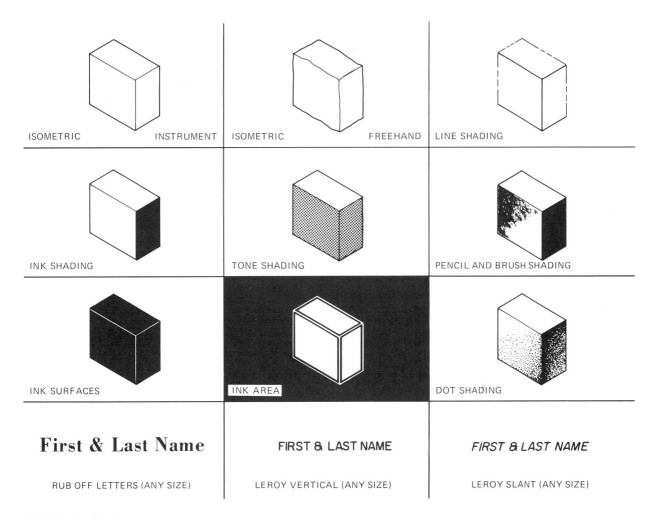

ISOMETRIC INSTRUMENT	ISOMETRIC FREEHAND	LINE SHADING
INK SHADING	TONE SHADING	PENCIL AND BRUSH SHADING
INK SURFACES	INK AREA	DOT SHADING
First & Last Name RUB OFF LETTERS (ANY SIZE)	**FIRST & LAST NAME** LEROY VERTICAL (ANY SIZE)	*FIRST & LAST NAME* LEROY SLANT (ANY SIZE)

FIG. 23-42 Problem

two-thirds of the distance up the shaft. Refer to Figure 23-31 and the Round Head Screw Table in the appendix.

6. Select three different fillister head screws and draw them in isometric projection. Shaft lengths may vary but threads should extend two-thirds of the distance up the shaft. Refer to Figure 23-32 and the Fillister Head Screw Table in the appendix.

7. Select three different flat head machine screws and draw them in isometric projection. Shaft lengths may vary but threads should extend two-thirds of the distance up the shaft. Refer to

Figure 23-33 and the Flat Head Machine Screw Table in the appendix.

8. Using the drawing from Problem 7 as a guide, draw an exploded assembly of a simple rectangle with three different countersunk holes to accept the three ílat head machine screws drawn in Problem 7 (Fig. 23-45). The rectangular box size will depend on the screws selected from Problem 7.

9. Using the drawing from Problem 6 as a guide, draw an exploded assembly of a simple rectangle with three counterbored holes to accept

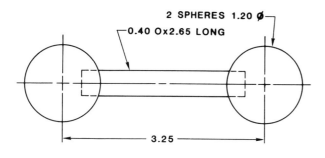

FIG. 23-43 Problem

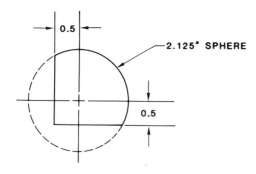

FIG. 23-44 Problem

the machine screws drawn in Exercise 6 (Fig. 23-46).

10. Select a series of drawings from Chapter 11 and draw them in isometric projection using the shading techniques illustrated in Figures 23-39 and 23-41. Refer to Figure 23-39 and Chapter 11.

11. Select a mechanical device such as a flashlight, pencil pointer, hand saw, plane, level, ball-point pen, fingernail clipper, lighter, or other small item with a number of parts. Using dividers and a scale draw the selected object to a suitable size (either enlarged or reduced) as an exploded isometric drawing. Refer to the illustrations given in this chapter for examples of shading and center line placement (Fig. 23-48).

12. Draw the problems in Figures 23-48 through 23-53. Transfer the measurements with a divider and double or triple each drawing's size.

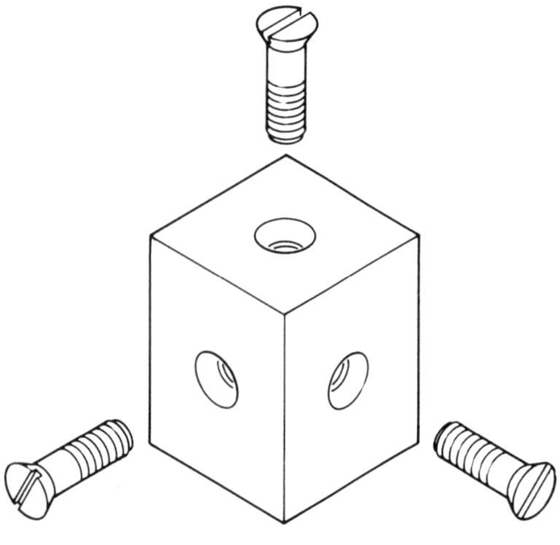

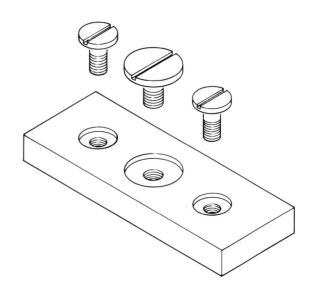

FIG. 23-45 Problem

FIG. 23-46 Problem

FIG. 23-47 Problem

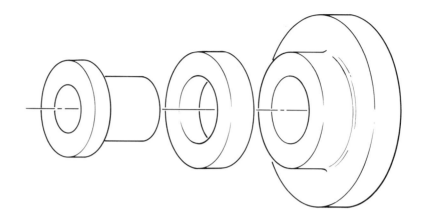

FIG. 23-48 Problem

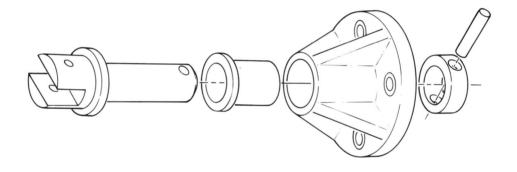

FIG. 23-49 Problem

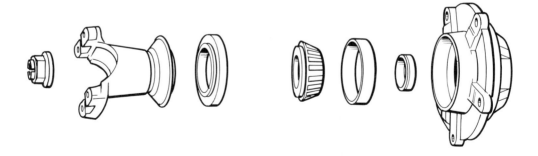

FIG. 23-50 Problem

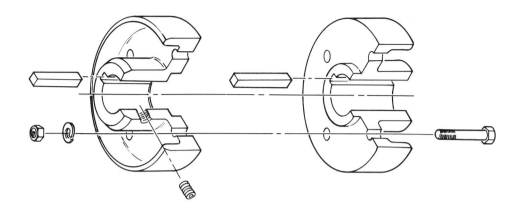

FIG. 23-51   Problem

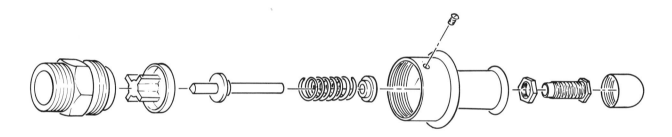

FIG. 23-52   Problem

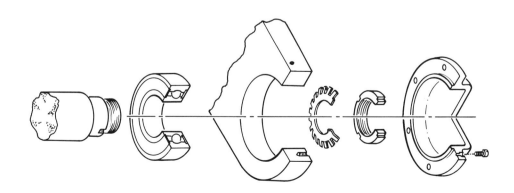

FIG. 23-53   Problem

1. WEATHER RADAR SCANNER
2. FORWARD RETRACTING NOSE GEAR; DUAL WHEELS WITH BRAKES
3. OUTWARD OPENING FORWARD PLUG-TYPE PASSENGER ENTRY DOOR
4. INTEGRAL PASSENGER STAIRS (TELESCOPE UNDER FLOOR)
5. COAT CLOSET
6. WING CENTER SECTION CONTAINING BLADDER-TYPE FUEL CELLS
7. KRUEGER FLAP
8. MAIN LANDING GEAR; DUAL WHEELS, INWARD RETRACTING
9. LEADING EDGE SLATS
10. AILERON CONTROL TAB
11. LOW SPEED OUTBOARD AILERON
12. FLIGHT SPEED BRAKES AND LATERAL CONTROL SPOILERS
13. HIGH SPEED INBOARD AILERON
14. GROUND SPEED BRAKES
15. PRATT & WHITNEY JT8D TURBOFAN ENGINE (14,000 LBS STATIC THRUST)
16. THRUST REVERSER UNIT WITH COVERING DOOR
17. INTAKE DUCT FOR CENTER ENGINE
18. CENTER ENGINE
19. DUAL, SEPARATELY POWERED RUDDER SEGMENTS
20. MOVABLE HORIZONTAL STABILIZER
21. ELEVATOR
22. ELEVATOR CONTROL TAB
23. REAR PLUG-TYPE PASSENGER ENTRY DOOR (INWARD OPENING)
24. CENTER ENGINE AIR INLET
25. TOURIST SECTION (SIX ABREAST SEATING)
26. TRIPLE SLOTTED, HIGH LIFT TRAILING EDGE FLAPS
27. DOUBLE UNIT MID-CABIN GALLEY
28. FIRST CLASS SECTION (FOUR ABREAST SEATING)

29. LAVATORY (ONE FORWARD; TWO AFT)
30. AFT INTEGRAL PASSENGER STAIRS
31. WINDOWS; THREE-PANE ACRYLIC PLASTIC ON 20'' SPACING
32. FAIL-SAFE FUSELAGE STRUCTURE; SEMI-MONOCOQUE, ALUMINUM ALLOY SKIN, STRINGERS Z-TYPE FRAMES
33. CONTROL CABIN
34. LIFE RAFT STOWAGE
35. WING LEADING EDGE AND ENGINE INLETS ANTI-ICED BY ENGINE BLEED AIR
36. RETRACTABLE TAIL SKID
37. FUEL DUMP CHUTE
38. OVERHEAD AIR DISTRIBUTION SYSTEM (SIDE WALL SYSTEM NOT SHOWN)

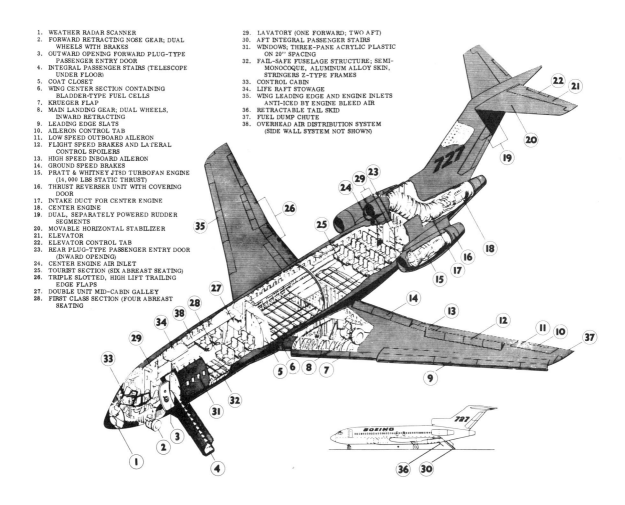

FIG. 23-54   Example of a shaded industrial illustration (Boeing Commercial Airplane Co.)

24

Chapter

Threads

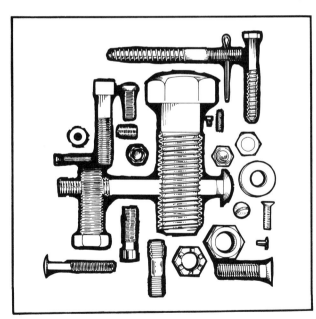

FIG. 24-1

INTRODUCTION

In the United States, over 300 billion fasteners are made each year. The automobile industry alone uses over 40 billion threaded fasteners annually. Their worth is $3 billion. The threaded fastener is a simple and efficient device to join objects together.

Examples of fasteners are seen in Figure 24-1.

Fasteners are made of regular steel, stainless steel, copper, monel, aluminum, brass, titanium, plastics, and wood. Metal fasteners with screw threads are the most common fasteners in industry.

Industrial fasteners are divided into two groups. (See Fig. 24-2.) The first is the permanent fasteners such as rivets and welds. The second is the removable fasteners such as bolts, screws, studs, nuts, keys, rings, and pins.

Specialized fasteners for thin materials, plastic, wood, and masonry can be designed and used to do any type of fastening job.

A draftsperson should know how to design, draw, and interpret blueprints using fasteners. This chapter will examine metal threaded fasteners. For additional information on the types, sizes, and functions of the different fasteners, refer to a machinist's handbook or engineering design texts.

THREAD SERIES

Threads are classified by systems or series. The Unified Thread System is one of the latest series. It has improved basic thread design. The Unified Thread System is interchangeable with the older American Standard Thread System and the British Standard

Whitworth series. The major differences among these are in the production methods of allowances, tolerances, and pitch diameters.

Common thread forms are shown in Figure 24-3. The V thread is used when a larger friction surface is required and for additional holding power. Square, acme, buttress, and worm threads are suitable for uses in which thrust or motion is transmitted. They are best for these uses because their thread shapes are the closest to being vertical to their axes. The knuckle thread is used for fast assembly. Two common examples are light bulbs and bottle caps.

Another major thread series is the metric threads. The ISO is the established metric thread series. (*ISO* stands for International Standards Organization.) There is also the newer optimum metric fastener system (OMFS), which was developed in the United States. The two metric thread series are interchangeable.

A "soft" conversion of U.S. inch fasteners to metric sizes is not practical because the fasteners themselves are not interchangeable with metric

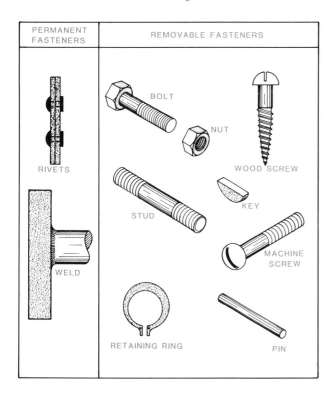

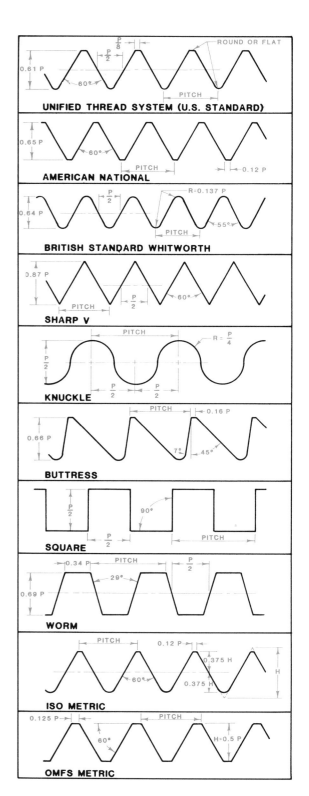

FIG. 24-2   Major types of industrial screws

FIG. 24-3   Profiles of ten types of threads

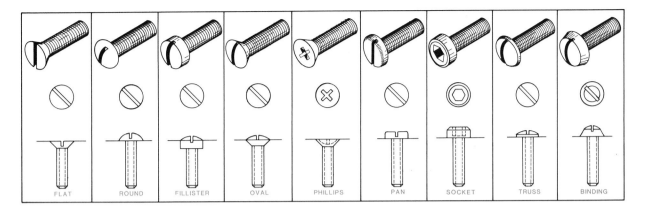

FIG. 24-4 Machine screws

threads. *Soft* conversion is the direct translation of existing inch dimensions into metric dimensions. For example, a direct translation of ⁵/₁₆-18 × 1 inch into metric reads M7.94 × 1.41 × 25.4 mm. There is no such metric size. *Hard* conversion, on the other hand, is conversion from an inch fastener to true metric fastener size of the recognized standard metric fastener systems.

The ISO has been working toward worldwide adoption of its metric standards. While the ISO threaded fasteners are adequate, U.S. threads have better performance qualities. Therefore, a completely new threaded fastener, OMFS, was developed to be compatible with ISO threads and yet superior to both U.S. inch fasteners and ISO fasteners.

COMMON TYPES OF THREADED FASTENERS

The draftsperson should be familiar with the common types of metal fasteners. Drawings must

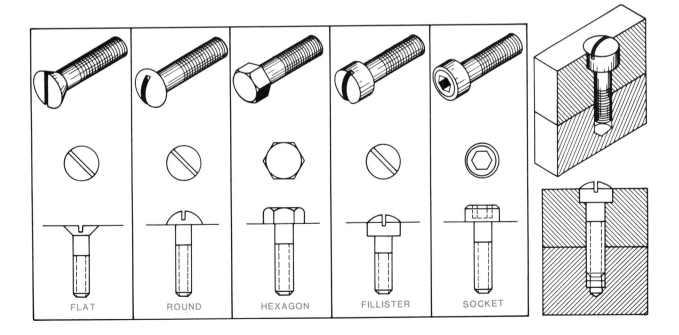

FIG. 24-5 Cap screws

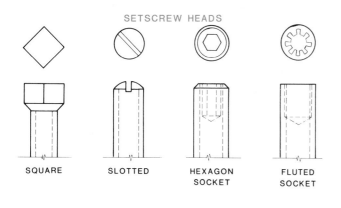

SETSCREW HEADS

SQUARE   SLOTTED   HEXAGON SOCKET   FLUTED SOCKET

SETSCREW POINTS

CONE POINT   FLAT POINT   DOG POINT   HALF DOG POINT   OVAL POINT   CUP POINT

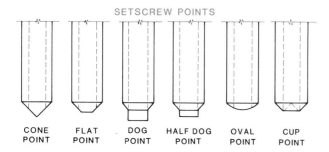

FIG. 24-6   Setscrews

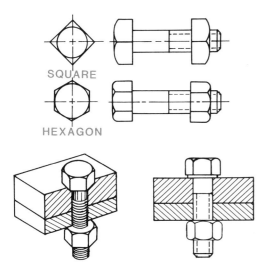

SQUARE

HEXAGON

FIG. 24-7   Square and hexagonal nuts and bolts

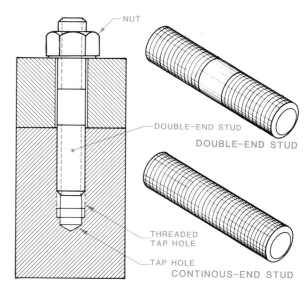

NUT

DOUBLE-END STUD

DOUBLE-END STUD

THREADED TAP HOLE

TAP HOLE

CONTINOUS-END STUD

FIG. 24-8   Studs

clearly communicate the type of fastener the designer wants. To do this, the draftsperson must know the physical properties of the fasteners and, in addition, the stresses and strains they will have to incur when in place.

Machine screws are the most common threaded fasteners in industry. They may be used with nuts or in a tapped hole. Figure 24-4 shows nine types of machine screws.

Cap screws are used in tapped holes. They join parts together by passing through a clearance hole in one part of an object and screwing into a tapped hole in another part (Fig. 24-5).

Setscrews are semipermanent fasteners that hold one part of an object from slipping when mated with another part. The setscrew has a clamping effect that resists motion between the assembled parts (Fig. 24-6).

Bolts are threaded fasteners that join parts by passing through a clearance hole in one part and screwing into a tapped hole in another part or into a nut (Fig. 24-7).

Studs are shafts threaded at both ends. They are used to assemble parts by screwing one end into a tapped hole and attaching the other end through the stud. The parts are then fastened together with a nut (Fig. 24-8).

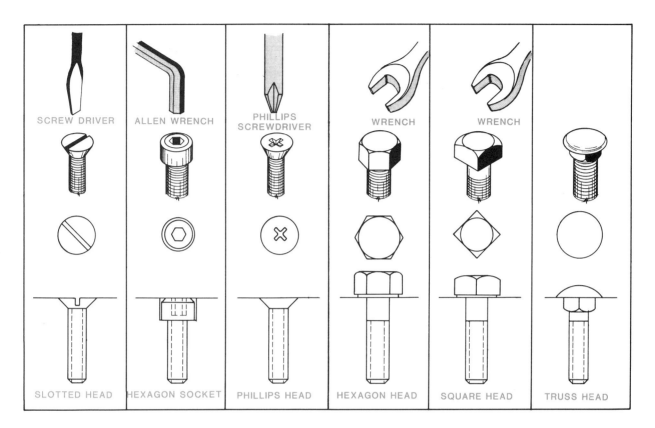

FIG. 24-9 Fastener heads

Six types of fastener *heads* are shown in Figure 24-9. These heads come in three weight sizes: regular, light, and heavy. Each head is designed to perform specific jobs. A general design rule is to keep the different styles to a minimum.

Different tools are needed to loosen the various fastener heads. Some of these are seen in the top of Figure 24-9.

Points, which are found at the end of the threaded shank, have several different shapes. Points are designed for ease of production or to perform a specific job. (Fig. 24-10).

Nuts secure threaded fasteners in place. Their design depends on their application. There are three weight sizes: regular, light, and heavy. (See Fig. 24-11.)

THREAD PARTS

The parts of a threaded fastener are identified in Figure 24-12. Most threads are graded in coarse, fine, extra-fine, and constant-pitch series (Fig. 24-13). Fasteners in the Unified Coarse (UNC) series can be assembled quickly. They are usually made

FIG. 24-10 Fastener points

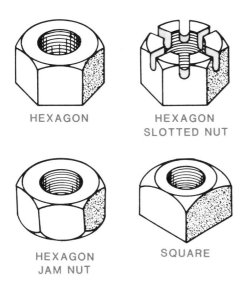

FIG. 24-11 Common types of nuts

8. *Minor diameter*. The smallest diameter of an internal or external screw thread. Also called root diameter or inside diameter for a nut.

9. *Pitch*. The distance from one point on a screw thread to a corresponding point on the next thread when measured parallel to the axis.

10. *Pitch diameter*. The diameter of an imaginary cylinder passing through the threads at points where the thread width and groove are equal.

11. *Right-hand thread*. A thread that will tighten when turned clockwise.

12. *Root*. The bottom of the thread cut into a cylinder. Another name for minor diameter.

13. *Thread class*. The closeness of fit between two threaded parts such as a nut and a bolt.

from cast iron or other soft metals. Those in the Unified Fine (UNF) series have a close and tight fit. The Unified Extra-Fine (UNEF) series of fasteners is used on thin materials and for very tight fits. Lastly, the constant-pitch series (8UN, 12UN, 16UN) has the same number of threads per inch. (*The* number before the UN in the thread designation is the number of threads.) Constant-pitch fasteners are used in and where large diameters are needed.

In drawing thread parts, the draftsperson must use correct terminology. Here is a short list of standard terms used for thread parts.

1. *Angles of thread*. The angle between the flanks of the thread that are created by the cutting tool.

2. *Crest*. The outermost surface of the thread.

3. *External thread*. The thread on the outside of a cylinder, as in a bolt.

4. *Internal thread*. The thread on the inside of a cylindrical hole, as in a nut.

5. *Lead*. The distance a screw thread will advance in one full revolution.

6. *Left-hand thread*. A thread that will tighten when turned counterclockwise.

7. *Major diameter*. The largest diameter of an internal or external screw thread.

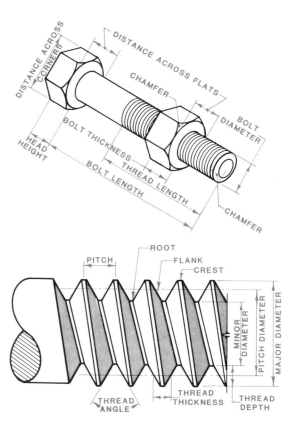

FIG. 24-12 Parts of a threaded fastener

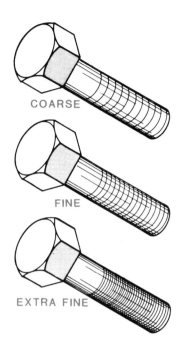

FIG. 24-13   Common types of thread series

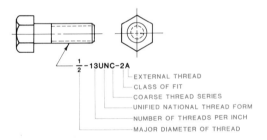

FIG. 24-14   Thread note for an external thread

Class 1 is a loose fit, Class 2 is a medium fit, and Class 3 a tight fit.

14. *Thread form.* The profile shape of a thread as viewed from the axial plane.

15. *Thread series.* The different types of threads that can be distinguished by the number of threads per inch (or millimetre) and by thread diameters.

THREAD DESIGNATION NOTES

The thread designation is indicated on a drawing by a note. An example of a note for an external thread

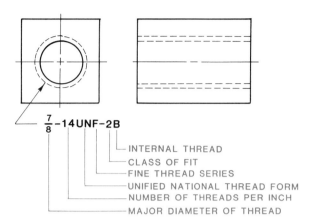

FIG. 24-15   Thread note for an internal thread

is shown in Figure 24-14. A note for an internal thread is shown in Figure 24-15.

There are other notations that can be added to a thread note. They are:

LH—left-hand thread

A—external thread

B—internal thread

DP—depth of thread

½ drill—tap drill size

CHMF—champer at end of threaded rod.

Information on the sizes of the parts of threaded fasteners will be found in the figures 24-16, 24-17, 24-18, and the appendix.

Figure 24-16 gives the threads per inch for commonly used bolts. Figure 24-17 gives the sizes for hexagonal bolt heads and nuts. Figure 24-18 compares the sizes of the Unified National (UN) series to the ISO metric thread series. These threads are not compatible. For additional information on threads, see the tables in the Appendix C.

DRAWING THREADED FASTENERS

When doing threaded fasteners in working drawings, be sure to draw the physical size of the fastener accurately (Fig. 24-19). The thread symbols can be estimated. The depth of the threads should be approximately ¹⁄₁₆ inch for small threads and ⅛ inch for the larger sizes. Threads can be drawn in

Major Diameter of Bolt	Number of Threads per Inch		Major Diameter of Bolt	Number of Threads per Inch	
	Coarse NC	Fine NF		Coarse NC	Fine NF
0 (.060)		80	5/8	11	18
1 (.073)	64	72	11/16		
2 (.086)	56	64	6/6	10	16
3 (.099)	48	56	13/16		
4 (.112)	40	48	7/8	9	14
5 (.125)	40	44	15/16		
6 (.138)	32	40	1	8	12
8 (.164)	32	36	1 1/16		
10 (.190)	24	32	1 1/8	7	12
12 (.216)	24	28	1 3/16		
1/4	20	28	1 1/4	7	12
5/16	18	24	1 5/16		
3/8	16	24	1 3/8	6	12
7/16	14	20	1 7/16		
1/2	13	20	1 1/2	6	12
9/16	12	18	1 9/16		

FIG. 24-16 Number of threads for commonly used bolts

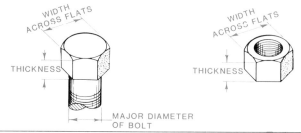

Major Diam. of Bolt	Bolt Head (in inches)				Nut (in inches)		
	Width Across Flats		Thickness of Head		Width Across Flats		Thickness of Nut
	Square Head	Hexagon Head	Square Head	Hexagon Head	Square Nut	Hexagon Nut	
1/4	3/8	7/16	11/64	11/64	7/16	7/16	7/32
5/16	1/2	1/2	13/64	7/32	9/16	9/16	17/64
3/8	9/16	9/16	1/4	1/4	5/8	5/8	21/64
7/16	5/8	5/8	19/64	19/64	3/4	3/4	3/8
1/2	3/4	3/4	21/64	11/32	13/16	13/16	7/16
5/8	15/16	15/16	27/64	27/64	1	1	35/64
3/4	1 1/8	1 1/8	1/2	1/2	1 1/8	1 1/8	21/32
7/8	1 5/16	1 5/16	19/32	37/64	1 5/16	1 5/16	49/64
1	1 1/2	1 1/2	21/32	43/64	1 1/2	1 1/2	7/8
1 1/8	1 11/16	1 11/16	3/4	3/4	1 11/16	1 11/16	1
1 1/4	1 7/8	1 7/8	27/32	27/32	1 7/8	1 7/8	1 3/32
1 3/8	2 1/16	2 1/16	29/32	29/32	2 1/16	2 1/16	1 13/64
1 1/2	2 1/4	2 1/4	1	1	2 1/4	2 1/4	1 5/16
1 5/8	2 7/16	2 7/16	1 3/32	1 1/16	2 7/16	2 7/16	1 27/64

FIG. 24-17

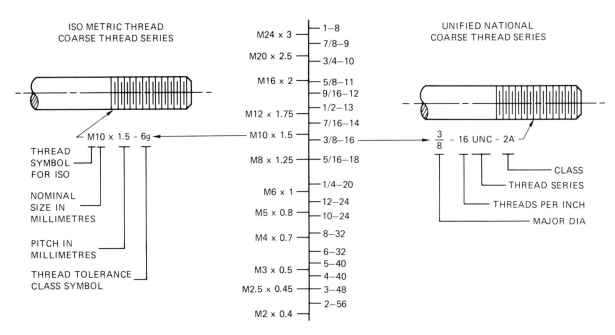

FIG. 24-18 Comparison of common threads for Unified National and Isometric thread series

pictorial, schematic, or simplified symbols (see Figs. 24-20 and 24-21). The simplified system is recommended because it reduces drawing time.

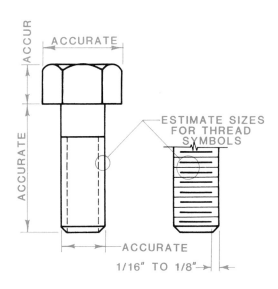

FIG. 24-19
The threaded fastener should be drawn accurately. The symbols for the threads can be estimated.

FASTENER TEMPLATES

Templates are a quick and accurate way to draw threaded fasteners. They come in many styles and are used in both working drawings and pictorial drawings. Figure 24-22 shows a fastener template. A nut and bolt template is seen in Figure 24-23. The steps in drawing the side view of a hexagonal bolt with a template are demonstrated in Figure 24-24. First, the straight edge of the head is drawn. Next, the top edge of the hexagonal head is filled in. Then the connecting lines are drawn between the top and bottom edges. Finally, the shaft is added in schematic or simplified symbols. Figure 24-25 shows the steps in drawing the end view of a hexagonal bolt with a template. Figure 24-26 shows the steps in drawing the side view of a nut. Figure 24-27 shows the steps in drawing the end view of a nut. An isometric template is used to draw an isometric bolt and nut, as shown in Figures 24-28 and 24-29. A completed assembly drawing is shown in Figure 24-30.

CUTTING THREADS

Small external threads can be cut with a threading die (Fig. 24-31) or cut on a lathe. Internal threads are cut with a tap (Fig. 24-32). Large internal

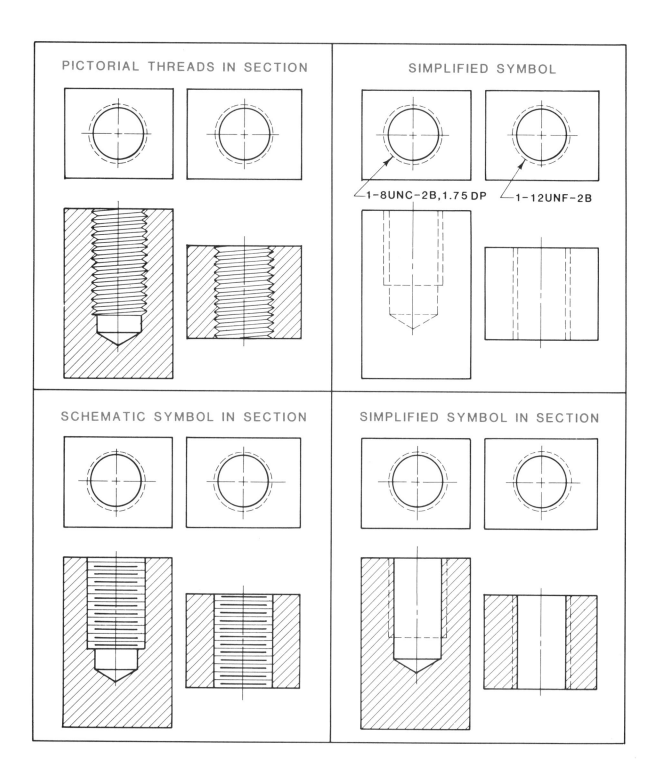

PICTORIAL THREADS IN SECTION

SIMPLIFIED SYMBOL

1–8UNC–2B,1.75 DP   1–12UNF–2B

SCHEMATIC SYMBOL IN SECTION

SIMPLIFIED SYMBOL IN SECTION

FIG. 24-20   Methods for drawing internal threads

threads can be cut with a boring bar in a lathe. When using a tap, always use the correct size tap drill (Fig. 24-33).

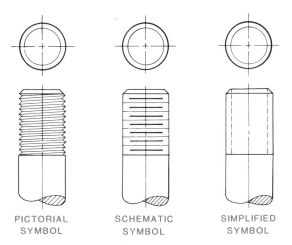

FIG. 24-21 Three types of thread symbols. The simplified symbol is preferred.

PICTORIAL SYMBOL   SCHEMATIC SYMBOL   SIMPLIFIED SYMBOL

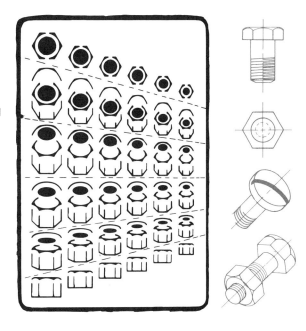

FIG. 24-23 Nut and bolt template

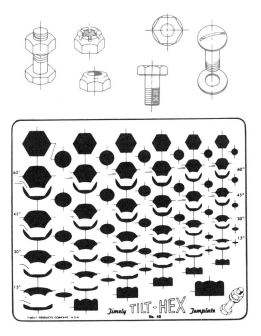

FIG. 24-22 Fastener template

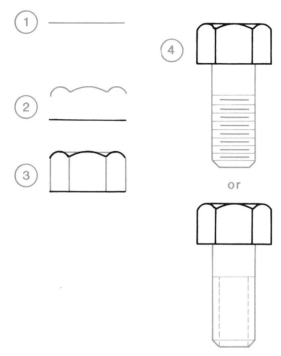

FIG. 24-24

Steps in drawing the side view of a bolt using a bolt template

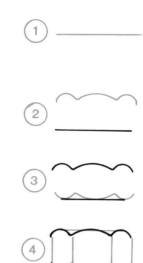

FIG. 24-26

Steps in drawing the side view of a jam nut using a bolt template

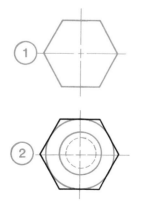

FIG. 24-25

Steps in drawing the end view of a bolt using a bolt template

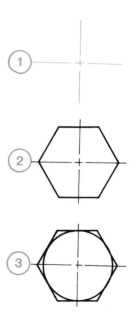

FIG. 24-27

Steps in drawing the end view of a nut using a bolt template

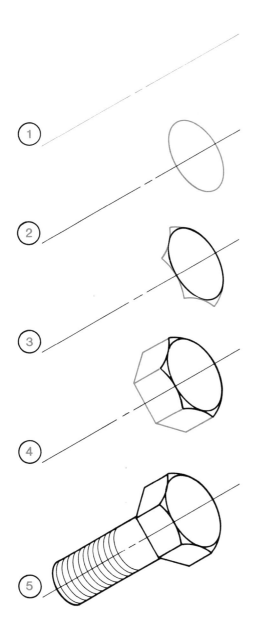

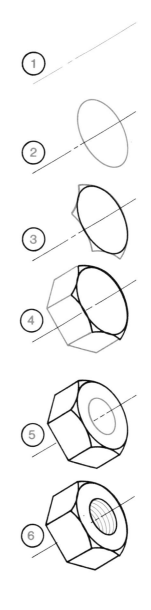

FIG. 24-28
Steps in drawing an isometric bolt using an isometric template

FIG. 24-29
Steps in drawing an isometric nut using an isometric template

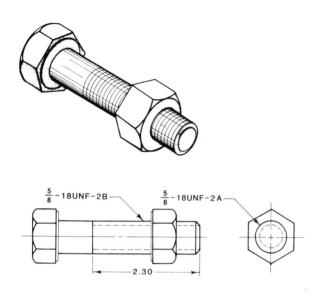

FIG. 24-30   Working drawing for a nut and bolt

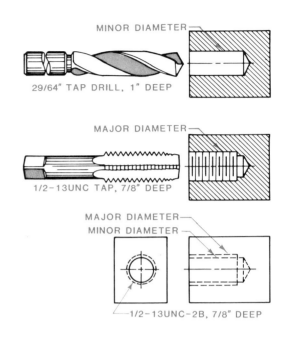

FIG. 24-32   Cutting internal threads

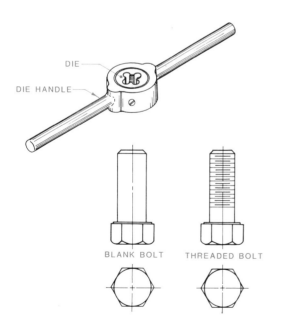

FIG. 24-31   Cutting external threads with a die

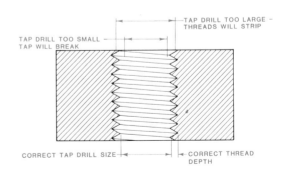

FIG. 24-33   Correct size for tap drill

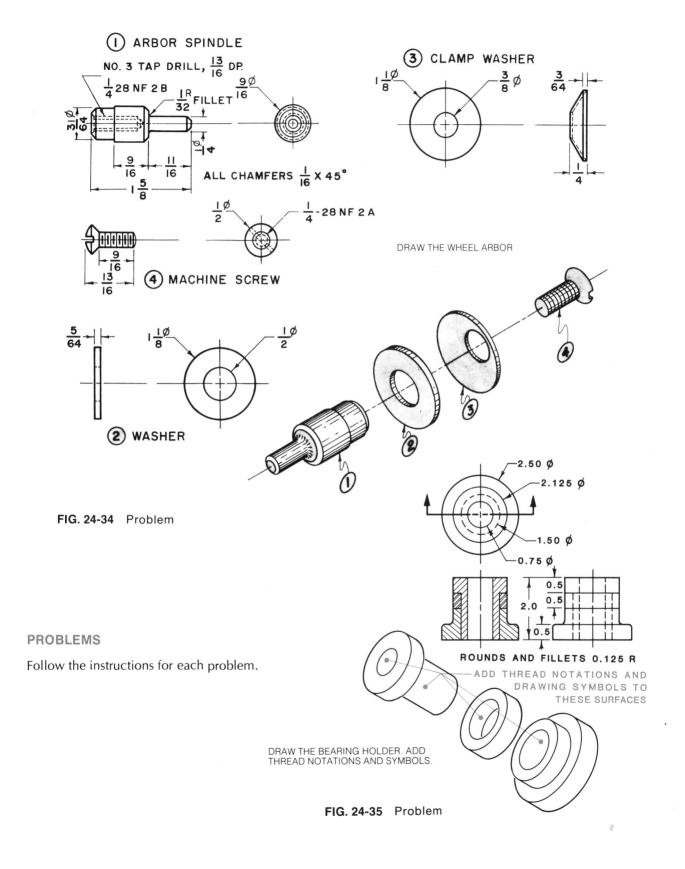

① ARBOR SPINDLE

NO. 3 TAP DRILL, $\frac{13}{16}$ DP.

$\frac{1}{4}$ 28 NF 2 B

$\frac{1}{32}$ R FILLET

$\frac{9}{16}$ Ø

$\frac{31}{64}$ Ø

$\frac{1}{4}$ Ø

$\frac{9}{16}$ $\frac{11}{16}$

$1\frac{5}{8}$

ALL CHAMFERS $\frac{1}{16}$ X 45°

③ CLAMP WASHER

$1\frac{1}{8}$ Ø $\frac{3}{8}$ Ø

$\frac{3}{64}$

$\frac{1}{4}$

DRAW THE WHEEL ARBOR

$\frac{1}{2}$ Ø $\frac{1}{4}$-28 NF 2 A

$\frac{9}{16}$

$\frac{13}{16}$

④ MACHINE SCREW

$\frac{5}{64}$ $1\frac{1}{8}$ Ø $\frac{1}{2}$ Ø

② WASHER

FIG. 24-34 Problem

PROBLEMS

Follow the instructions for each problem.

2.50 Ø
2.125 Ø
1.50 Ø
0.75 Ø

0.5
0.5
2.0
0.5

ROUNDS AND FILLETS 0.125 R

ADD THREAD NOTATIONS AND
DRAWING SYMBOLS TO
THESE SURFACES

DRAW THE BEARING HOLDER. ADD
THREAD NOTATIONS AND SYMBOLS.

FIG. 24-35 Problem

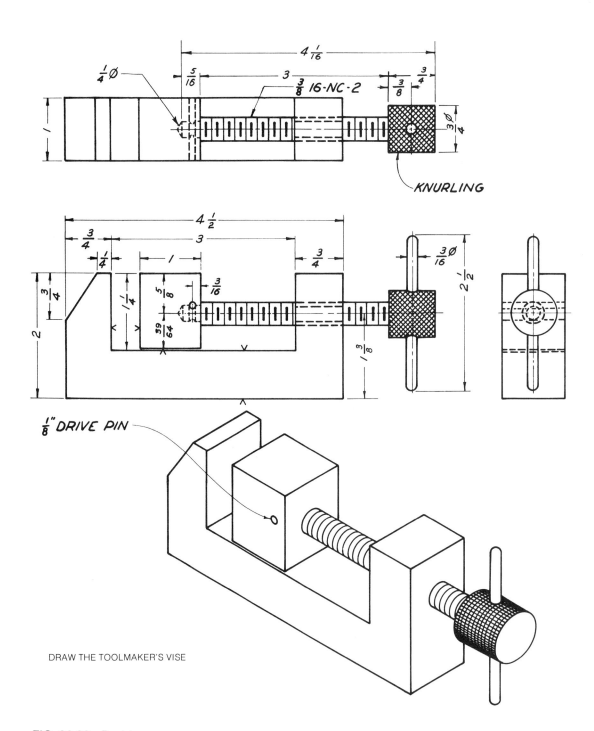

DRAW THE TOOLMAKER'S VISE

FIG. 24-36 Problem

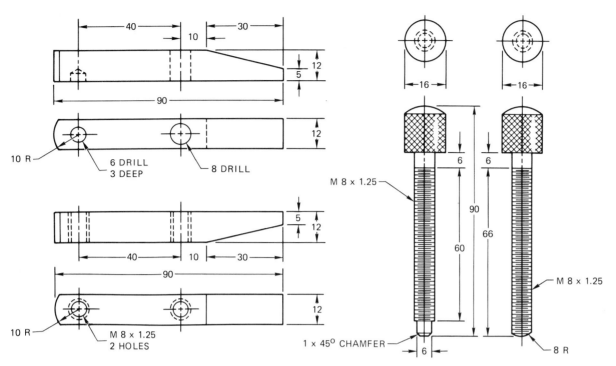

FIG. 24-37 Problem   DRAW THE METRIC CLAMP

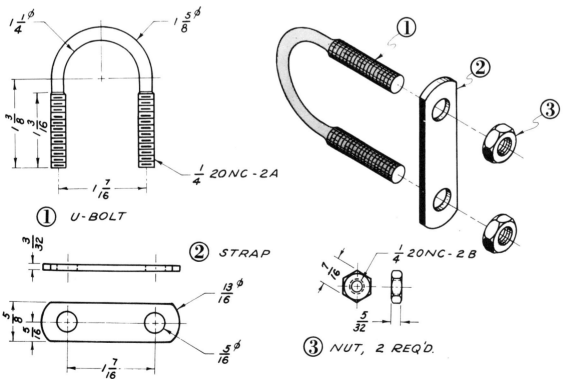

FIG. 24-38 Problem   DRAW THE U-BOLT AND STRAP

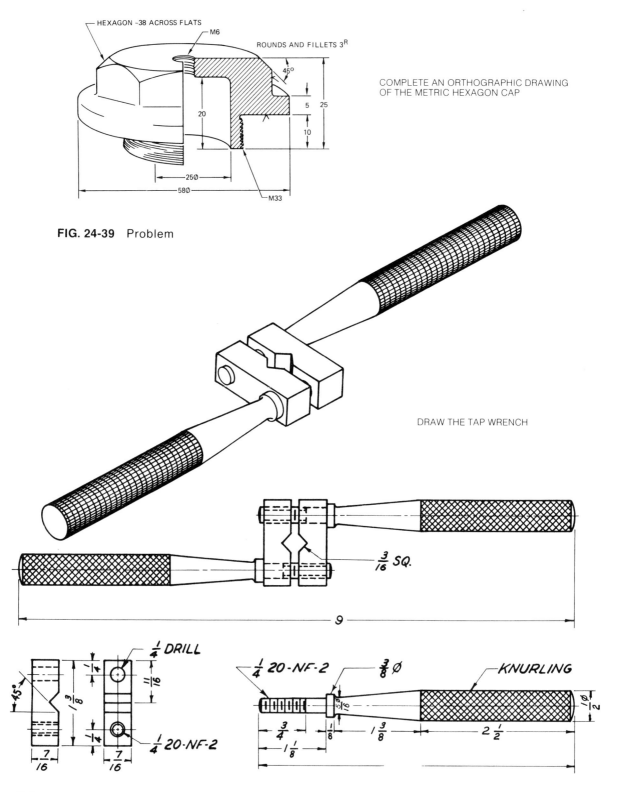

HEXAGON –38 ACROSS FLATS

M6

ROUNDS AND FILLETS 3^R

45°

COMPLETE AN ORTHOGRAPHIC DRAWING
OF THE METRIC HEXAGON CAP

25

20

5

10

25Ø

58Ø

M33

FIG. 24-39 Problem

DRAW THE TAP WRENCH

$\frac{3}{16}$ SQ.

9

$\frac{1}{4}$ DRILL

$\frac{1}{4}$

$\frac{11}{16}$

45°

$\frac{1}{3}{8}$

$\frac{1}{4}$

$\frac{1}{4}$ 20-NF-2

$\frac{7}{16}$

$\frac{7}{16}$

$\frac{1}{4}$ 20-NF-2

$\frac{3}{8}$ Ø

KNURLING

$\frac{3}{4}$

$\frac{1}{8}$

$1\frac{3}{8}$

$2\frac{1}{2}$

$1\frac{1}{8}$

$\frac{1}{2}$ Ø

FIG. 24-40 Problem

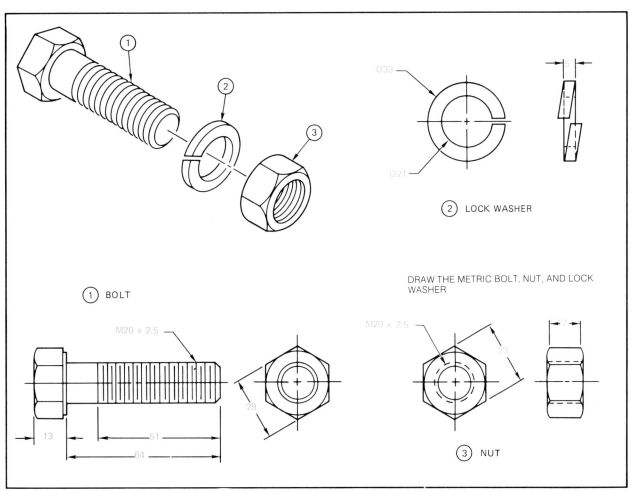

FIG. 24-41   Problem

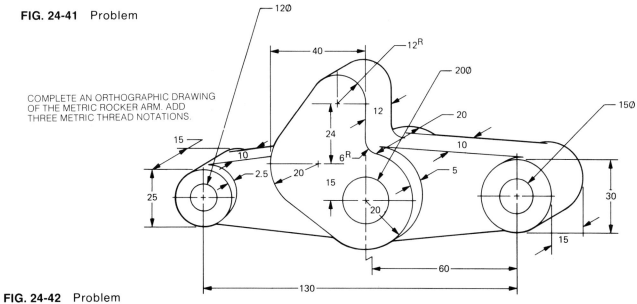

FIG. 24-42   Problem

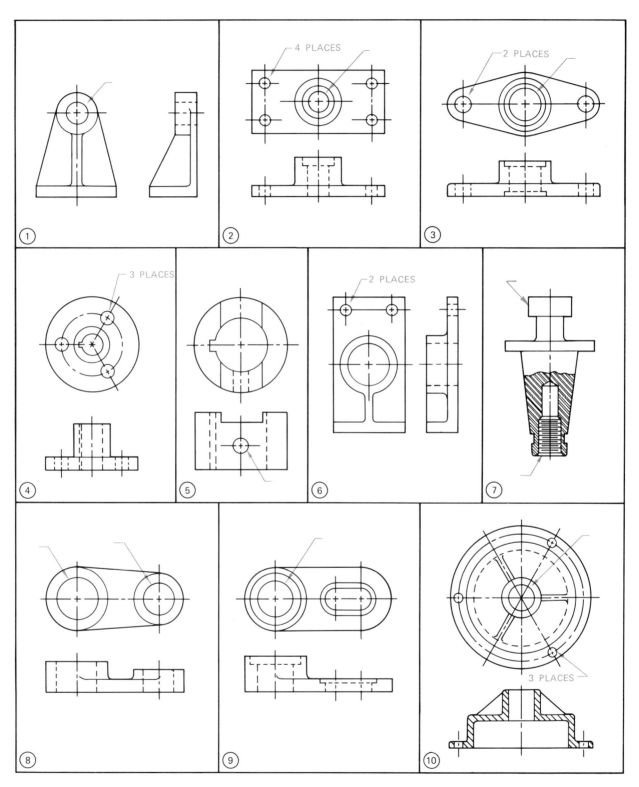

FIG. 24-43   Problem

MEASURE AND PRINT AN APPROPRIATE
THREAD NOTATION FOR EACH DRAWING.
DOUBLE THE SCALE (2:1) ON THE PAGE.

DRAW EACH PROBLEM AND COMPLETE
THE THREAD SYMBOLS. ADD DIMENSIONS
AND NOTATIONS.

Chapter

Welding

is being used in manufacturing in place of other methods of fabrication such as bolting and riveting. Welding is also replacing many castings or forgings. This is because it does not require dies or molds. As you study this chapter, you should gain the knowledge and skills necessary to draw and read welding prints.

TYPES OF WELDS

The two basic types of welding are the fusion process and the resistance process. The fusion process does not require that pressure be applied to the parts being joined. The resistance process does require pressure. The three principal welding processes are gas welding, arc welding, and resistance welding (Fig. 25-2).

Gas Welding

Gas welding joins metals by heating them with a gas flame to a temperature between 4000°F [2700°C] and 6500°F [3600°C]. The most common method is oxyacetylene welding. With this method, the flame is made by mixing and burning oxygen and acetylene gas. Parts can be joined with or without pressure. They can also be joined with or

INTRODUCTION

The process of permanently joining metals together by heating is called welding. Metal parts can be joined together with or without a filler metal. This can also be done with or without pressure. Welding

Fusion Welding					Resistance Welding	
Pressure Not Required					Pressure Required	
Gas Welding	Arc Welding		Thermit Welding	Brazing	Resistance Welding	Forge Welding
• Air acetylene • Oxyacetylene • Oxyhydrogen • Pressure gas	Carbon Electrode • Shielded carbon-arc • Shielded gas carbon-arc • Unshielded carbon-arc • Unshielded twin carbon-arc	Metal Electrode • Shielded arc—spot • Shielded arc—seam • Shielded metal—arc • Shielded atomic hydrogen • Shielded gas metal—arc • Shielded gas tungsten—arc • Shielded submerged arc • Gas shielded stud welding	• Pressure • Non-pressure	• Torch • Twin carbon arc • Furnace • Induction • Resistance • Dip • Block • Flow	• Spot • Seam • Projection • Flash • Upset • Percussion	• Roll • Die • Hammer

FIG. 25-2 Welding Processes

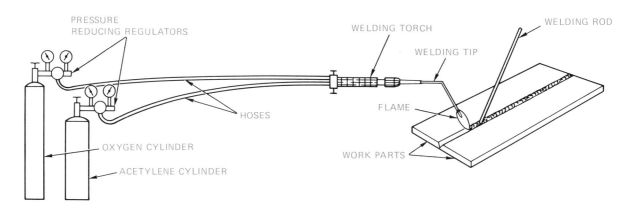

FIG. 25-3 Gas-welding principle

without a filler metal. Filler metal, if used, is in the form of a welding rod. The filler metal must combine with the parts being joined. The melting point of the filler must be the same or lower than the melting point of the metals being joined. The melting point of the filler metal is 800°F [425°C]. Figure 25-3 shows the operation of gas welding.

Arc Welding

Arc welding uses the heat of an electric arc to bring the metals to be fused to a molten state. Arc welding is divided into two groups. These two groups are carbon electrode and metal electrode. The carbon arc is used mainly in automatic equipment.

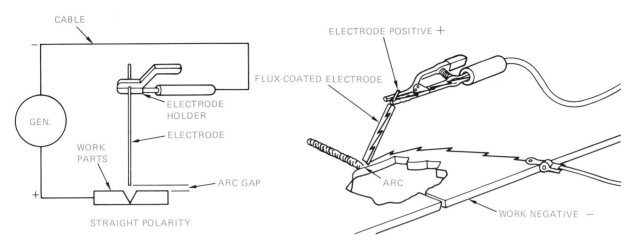

FIG. 25-4 D.C. arc welding principle

Arc-Welding Equipment

Arc-welding equipment consists of (Fig. 25-4):

1. A source of electric current (AC or DC)
2. An electrode holder
3. Cables
4. Electrodes

Shielded and Unshielded Electrodes

The arc must be shielded because, as it hardens, the molten metal combines with oxygen and nitrogen to form impurities that weaken the weld. Shielding can be obtained by a paste, powder, or fibrous flux that is added to the arc (Fig. 25-5). The shielded electrode is covered with a dry chemical coating. This coating forms a gaseous cloud that shields the molten metal from the atmosphere. The coating also forms a protective slag. The slag floats on the molten pool and hardens as the weld cools. This keeps impurities out of the weld.

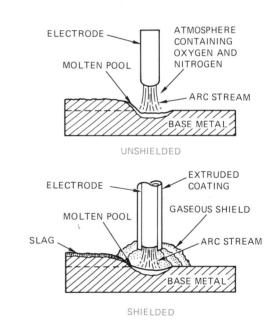

FIG. 25-5 Shielded and unshielded electrodes

Atomic-Hydrogen Welding

In atomic-hydrogen welding, the arc forms between two tungsten electrodes. The material is not part of the welding circuit. A stream of hydrogen passing through the arc is changed from molecular to atomic form. This gives off intense heat. This is an excellent process for welding materials that require a good finish weld.

Inert Gas Welding

Inert gas welding is a type of arc welding. An inert gas such as helium or argon is introduced through a small hole in the end of the electrode holder to shield the arc from impurities in the air. The tungsten electrode is not consumed. A filler metal can be used if extra metal is needed in the weld. Inert

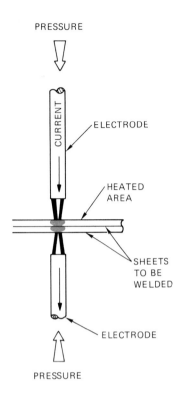

PRESSURE

CURRENT

ELECTRODE

HEATED AREA

SHEETS TO BE WELDED

ELECTRODE

PRESSURE

FIG. 25-6   Resistance welding joints by heat and pressure

gas welding is good for welding aluminum and magnesium.

Thermit Welding

Thermit welding uses the natural chemical reaction of aluminum with oxygen. A mixture of finely divided aluminum and iron oxide is ignited by a small quantity of ignition powder. The high temperature that results from the rapid burning melts metal. The metal then flows into a mold and fuses the parts being welded.

Brazing

Brazing is a type of welding process in which similar or different metals are joined by heating the brazing metals (usually copper, bronze, or silver) above 800°F [430°C] (the melting point of brazing metals) but below the melting point of the metals being joined.

Resistance Welding

In resistance welding, heat is generated by the resistance of the metal being welded with the passage of an electric current, while mechanical pressure is used to hold the heated parts together. No other materials such as filler metal, fluxes, or inert gases are needed. Resistance welds may have lap or butt type joints (Fig. 25-6).

The equipment for resistance welding is a transformer that produces high amperage and low voltage. Welding current flows into the metal to be joined through water-cooled tips, dies, wheels, or clamps. These are called electrodes. The electrodes are made from a high conductivity material like copper.

Resistance welding is popular because:

1. Welds are uniform

2. Production is fast

3. Parts do not warp from the heat

4. Skilled operators are not required

Spot Welds

Spot welds are individually formed resistance welds. They are the size of the electrodes. Current is passed through the metal parts by electrodes on opposite sides of the metals to be joined while pressure is applied (Fig. 25-7).

Seam Welds

Seam welds are made by heat created by the resistance to the flow of electric current in the metal parts being welded. The parts are held under pressure by circular electrodes. The seam weld is then formed by a series of overlapping spot welds (Fig. 25-8).

Projection Welds

Projection welding is similar to spot welding. However, projections or embossings on the materials to be welded supply the contact point for the path of the electric current. When these projections reach welding temperatures the electrodes force the parts to be welded together (Fig. 25-9).

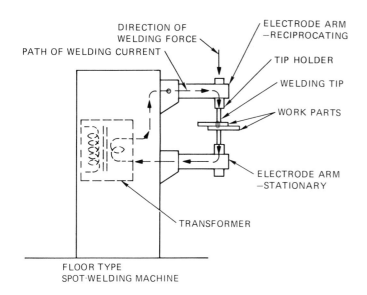

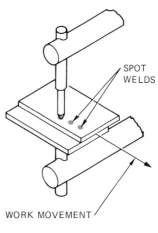

FIG. 25-7 Spot-welding principle

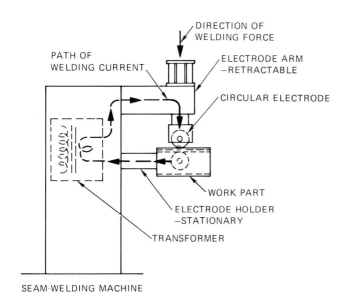

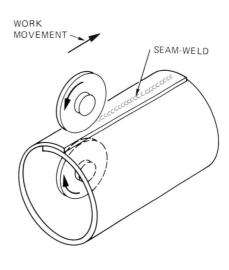

FIG. 25-8 Seam-welding principle

Flash Welds

Flash welds occur when an electric current is flashed over the end surfaces of parts to be welded. The heat produced by flashing and pressing fuses the two parts (Fig. 25-10).

Upset Welds

Upset welds are similar to flash welds. The difference is that the parts to be upset welded are butted together under pressure. Current is then passed through the electrode grip dies. Pressure is contin-

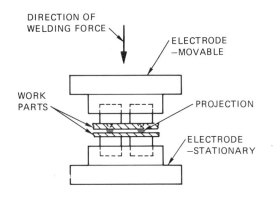

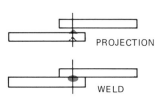

FIG. 25-9   Projection-welding principle

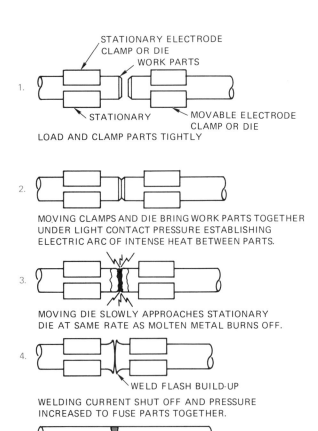

FIG. 25-10   Flash-welding principle

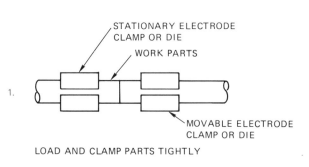

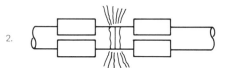

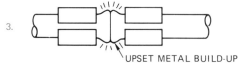

FIG. 25-11   Upset-welding principle

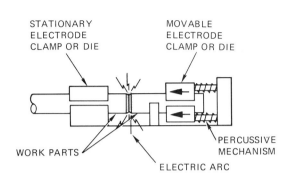

FIG. 25-12 Percussion-welding principle

ued after the current is shut off and an upsetting action takes place (Fig. 25-11).

Percussion Welds

Percussion welds occur when a spring loaded percussion mechanism propels two work parts toward each other at high speed. Just before the work parts contact each other, a high energy arc is discharged between them. Heat from the arc and the pressure from the collision cause the work parts to fuse at the weld joint (Fig. 25-12). Percussion welding is used to join metals that cannot be welded by other methods.

Forge Welding

Forge welding is the oldest type of welding. Pieces of metal are heated until the areas where they are to be joined are soft. When pressure is applied, the metal is fused together. Pressure can be applied by hammering, by rolling with a machine, or by stamping with dies.

Types of Welded Joints

There are five types of welded joints. They are based on the position of the parts being joined. Figure 25-13 shows the types of joints. It also lists the welding processes that can be applied to each joint.

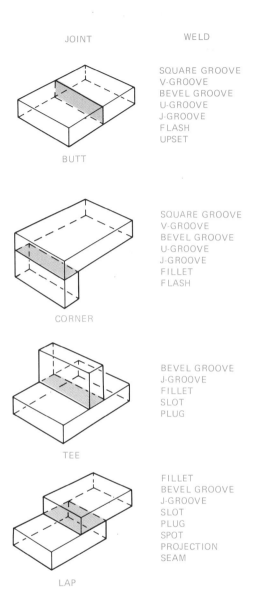

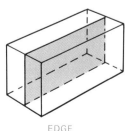

FIG. 25-13 Types of welded joints

1	2	3	4				
BACKING	FILLET	PLUG OR SLOT	GROOVE				
			SQUARE	V	BEVEL	U	J

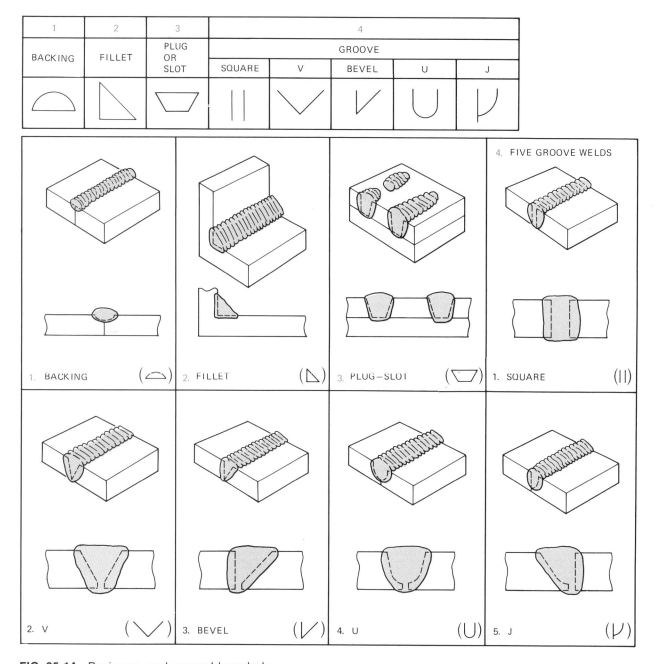

FIG. 25-14   Basic arc- and gas-weld symbols

Types of Fusion Welds

There are four types of welds that can be used to fuse joints together with the gas- and arc-welding processes. They are the bead, fillet, plug (slot), and groove welds. Figure 25-14 shows the welds and their symbols.

Bead welds are used to build up a surface. They are also used as a back-up weld on the opposite side of another weld.

Fillet welds are the most commonly used welds. This is because little joint preparation is needed.

Plug and slot welds are made through a sheet

of metal. They are used to fuse two metal sheets together. These welds are used when part of a surface cannot be reached. The plug weld is round. The slot weld, however, is a longer, continuous type of plug weld.

The various types of groove welds are used in butt welding.

Types of Resistance Welds

There are five basic types of resistance welds. These are the spot, projection, seam, flash, and upset welds. The symbols for resistance welds are shown in Figure 25-15. Examples of the symbols on welding drawings are shown later in this chapter.

Supplementary Weld Symbols

Supplementary symbols are needed to give the welder additional information about welds (Fig. 25-16). Examples of their use on drawings are shown later in this chapter.

Weld Symbol Designation

Before welding symbols were standardized, welds were indicated on drawings by heavy black areas and notes (Fig. 25-17). If the part had many welds it required extra drawing time and became cluttered.

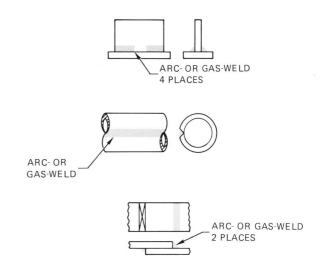

SPOT	PROJECTION	SEAM	FLASH— UPSET— PERCUSSION
◯	◯	⊖	||

FIG. 25-15 Resistance-weld symbols

WELD ALL AROUND	FIELD WELD	CONTOUR OF WELD		
		FLUSH	**CONVEX**	**CONCAVE**
		—	⌢	⌣

FIG. 25-16 Supplementary weld symbols

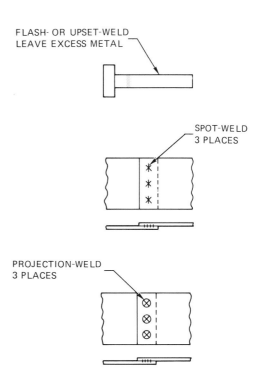

FIG. 25-17 Graphic and noted weld symbols

The American Welding Society symbols are recommended for use on all drawings requiring weld symbols. Under the AWS system, the weld is not shown on the drawing. It is indicated by a welding symbol. All the information required for a weld is given on the welding symbol or the bent

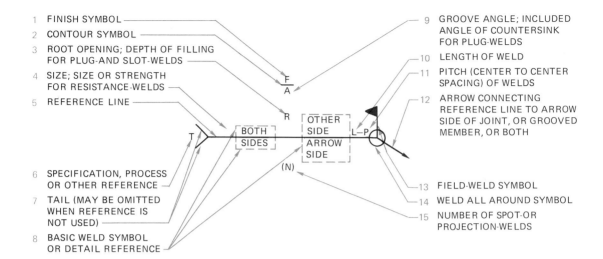

1 FINISH SYMBOL

2 CONTOUR SYMBOL

3 ROOT OPENING; DEPTH OF FILLING FOR PLUG-AND SLOT-WELDS

4 SIZE; SIZE OR STRENGTH FOR RESISTANCE-WELDS

5 REFERENCE LINE

6 SPECIFICATION, PROCESS OR OTHER REFERENCE

7 TAIL (MAY BE OMITTED WHEN REFERENCE IS NOT USED)

8 BASIC WELD SYMBOL OR DETAIL REFERENCE

9 GROOVE ANGLE; INCLUDED ANGLE OF COUNTERSINK FOR PLUG-WELDS

10 LENGTH OF WELD

11 PITCH (CENTER TO CENTER SPACING) OF WELDS

12 ARROW CONNECTING REFERENCE LINE TO ARROW SIDE OF JOINT, OR GROOVED MEMBER, OR BOTH

13 FIELD-WELD SYMBOL

14 WELD ALL AROUND SYMBOL

15 NUMBER OF SPOT-OR PROJECTION-WELDS

FIG. 25-18 Standard welding symbols

arrow (Fig. 25-18). Further detailed information for standard welding symbols follows:

1. The finish symbol specifies the finish of the weld. ''C'' is chip, ''M'' is machine, ''G'' is grind, ''R'' is rolling, and ''H'' is hammering.

2. The contour symbol specifies the final or finish contour of the weld. The weld will be finished flush or convex (Fig. 25-19).

3. The root opening specifies the size of plug and slot welds (Fig. 25-20).

4. The size of the weld is stated in inches and the strength in pounds per weld.

5. The reference line is the straight part of the ''bent arrow'' symbol that contains the weld data. These data are size, strength, type, position, length, and pitch (Fig. 25-21).

6. Additional information about the weld is placed inside the tail.

7. The tail of the symbol may be omitted if not used (Fig. 25-22).

8. The basic weld symbol indicates the type and position of the weld (Fig. 25-23).

9. This symbol specifies the included angle of the countersinking for plug welds (Fig. 25-24).

10. The length of the weld is specified in inches (Fig. 25-25).

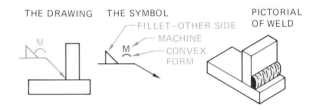

FIG. 25-19 The contour symbol

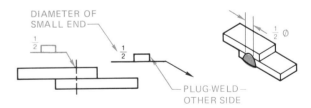

FIG. 25-20 The root opening

FIG. 25-21 The reference line

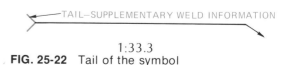

FIG. 25-22 Tail of the symbol

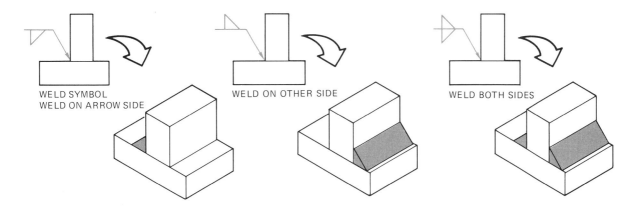

FIG. 25-23 Placement of weld symbols

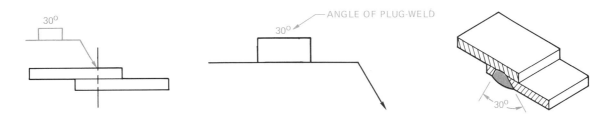

FIG. 25-24 Plug-weld symbol

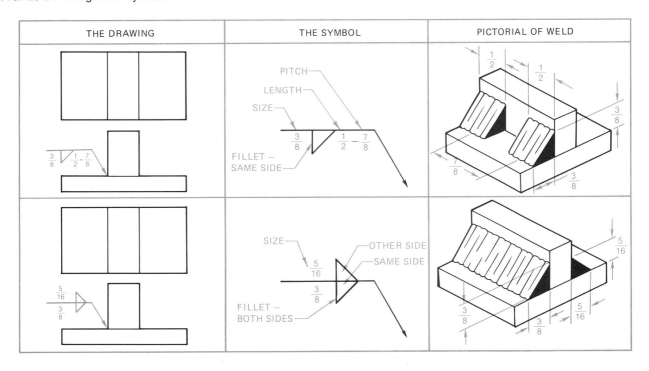

FIG. 25-25 Weld length specifications

THE WELD IS PREFORMED OUT OF THE
MANUFACTURING BUILDING IN A FIELD LOCATION.

FIELD-WELD

FIG. 25-26   The field-weld symbol

FILLET-WELD—ALL AROUND

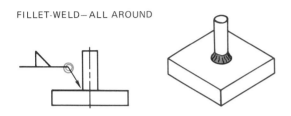

FIG. 25-27   Weld-all-around symbol

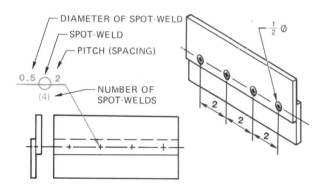

DIAMETER OF SPOT-WELD
SPOT-WELD
PITCH (SPACING)
0.5   2
(4)
NUMBER OF SPOT-WELDS
½ ∅

FIG. 25-28   Specification for number of welds

11. The pitch is the distance between centers of noncontinuous welds (Fig. 25-25).

12. The bent part of the reference line that points to the specific location of the weld is noted by this symbol.

13. A field-weld symbol specifies that the weld will not be made in the shop. It will be made in the field or where the final assembly will take place (Fig. 25-26).

14. This symbol specifies welds that extend all around the part (Fig. 25-27).

15. The number specifies the quantity of spot or projection welds (Fig. 25-28).

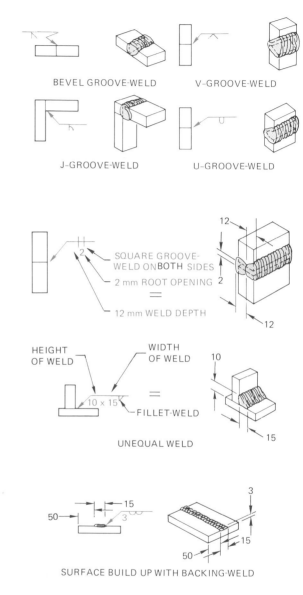

BEVEL GROOVE-WELD   V-GROOVE-WELD
J-GROOVE-WELD   U-GROOVE-WELD

SQUARE GROOVE-WELD ON BOTH SIDES
2 mm ROOT OPENING
12 mm WELD DEPTH

HEIGHT OF WELD   WIDTH OF WELD
10 x 15
FILLET-WELD
UNEQUAL WELD

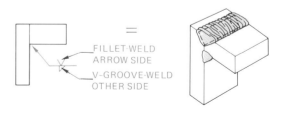

SURFACE BUILD UP WITH BACKING-WELD

FILLET-WELD ARROW SIDE
V-GROOVE-WELD OTHER SIDE

FIG. 25-29   Examples of arc and gas welds

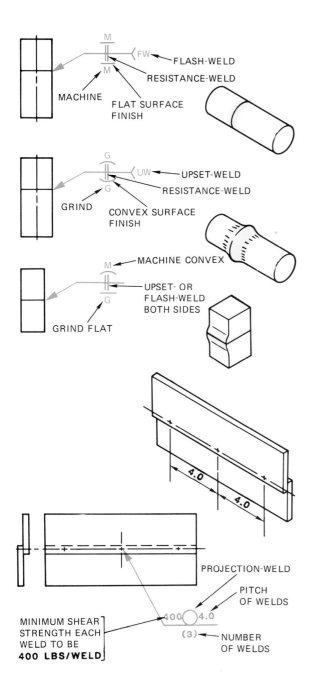

FIG. 25-30   Examples of resistance welds

The best way to learn welding symbols is to use them on welding drawings. The problems at the end of this chapter are a good starting point. The following examples show some additional uses of welding symbols. Figure 25-29 shows arc and gas welds. Figure 25-30 shows resistance welds.

Alternate symbols for some welds as they are used by different industries are shown in Figure 25-31. When symbols have a vertical leg, it is important to place it on the left side (Fig. 25-32).

Welding Positions

There are four different welding positions. These are flat, horizontal, overhead, and vertical (Fig. 25-33).

PLUG-WELD	▽	▭
SLOT-WELD	▽	⊔
SPOT-WELD	✕	○
PROJECTION-WELD	✕	○
SEAM-WELD	✕✕✕	⊖
FLASH-OR UPSET-WELD	|	||

FIG. 25-31   Alternate welding symbols

FIG. 25-32   The vertical leg of a welding symbol is placed on the left side

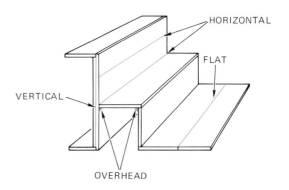

FIG. 25-33   Welding positions

PROBLEMS

Follow the instructions for each welding drawing problem in Figures 25-34 through 25-36.

Redraw the multiview welding drawings in Figure 25-37.

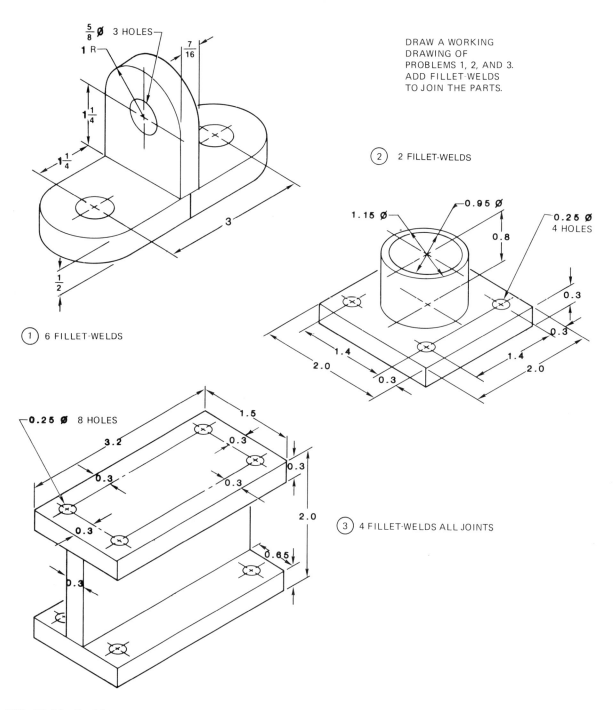

DRAW A WORKING DRAWING OF PROBLEMS 1, 2, AND 3. ADD FILLET-WELDS TO JOIN THE PARTS.

$\frac{5}{8}$ ∅ 3 HOLES

1 R

$1\frac{1}{4}$

$1\frac{1}{4}$

$\frac{7}{16}$

3

$\frac{1}{2}$

① 6 FILLET-WELDS

② 2 FILLET-WELDS

1.15 ∅

0.95 ∅

0.25 ∅ 4 HOLES

0.8

0.3

0.3

1.4

2.0

0.3

1.4

2.0

③ 4 FILLET-WELDS ALL JOINTS

0.25 ∅ 8 HOLES

3.2

1.5

0.3

0.3

0.3

0.3

0.3

2.0

0.3

0.65

0.3

FIG. 25-34 Problem

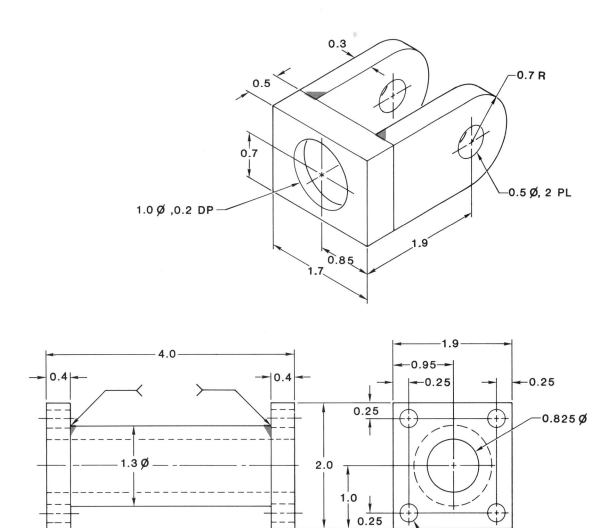

FIG. 25-35 Problem

DRAW AN ASSEMBLED
ORTHOGRAPHIC DRAWING FOR
THE TWO PROBLEMS ADD
RESISTANT WELDS NEEDED
TO ASSEMBLE ALL PARTS.
ALL THE STEEL PLATE IS
0.075'' THICK.

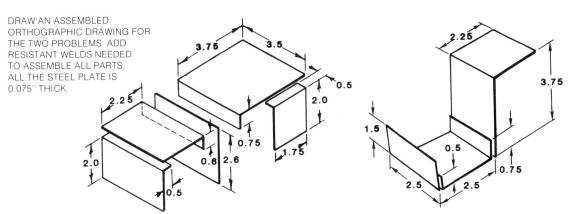

FIG. 25-36 Problem

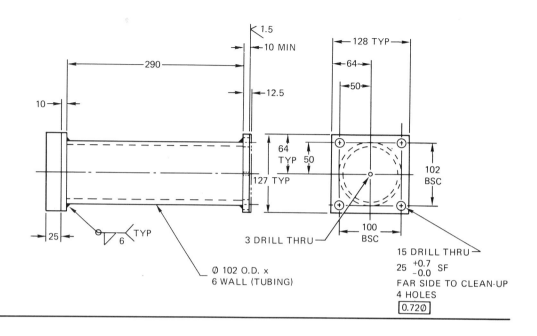

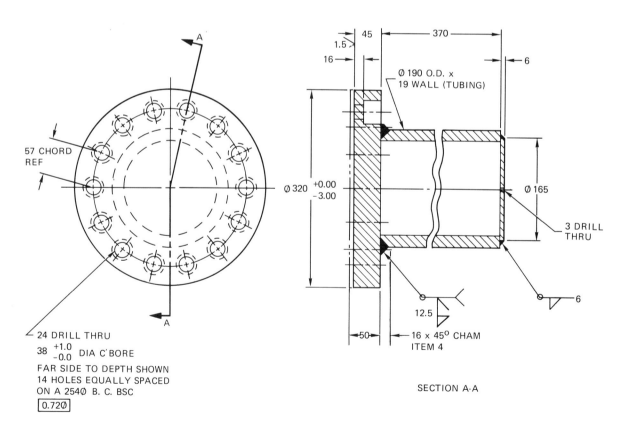

FIG. 25-37 Problems (Westinghouse Electric Corp., Marine Div., Sunnyvale, Ca.)

26

Chapter

Gearing

The Falk Corporation

INTRODUCTION

Gears are used to transmit rotary motion and power from one shaft to another. They come in standard types and sizes, which have a direct effect on their design and drawings. As you study this chapter, you should learn the functions of gears and how to draw gear teeth and gear symbols. Draftspersons have the task of making proper gear selections as the design requires. They do this by using standard gear design constants and formulas. Figure 26-2 illustrates some of the common gear types.

Figures 26-3 and 26-4 give the names of the various spur gear parts. Because the proportions of spur gears are standardized, one set of formulas has been developed. The chart in Figure 26-5 lists the formulas needed for external spur gears.

TOOTH SIZE

The term used for tooth size is *diametral pitch*. The diametral pitch is a ratio of the diameter of the gear to its number of teeth. Figure 26-6 shows several sizes of gear teeth and their corresponding diametral pitch. The larger the diametral pitch the larger the gear tooth. It is most important to keep in mind that the diametral pitch of mating gears must be the same. Determining the diametral pitch or size of gear tooth to be used is one of the first steps in gear design.

SPUR

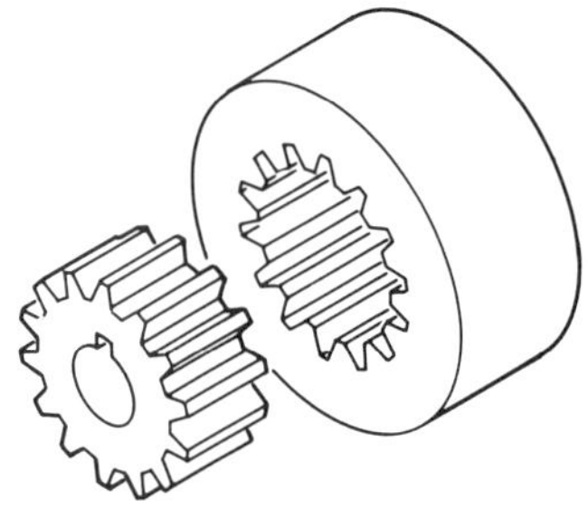

CYLINDRICAL SHAPED AND
OPERATE ON PARALLEL AXES.
THEY ARE THE SIMPLEST OF
THE GEAR TYPES.

HELICAL

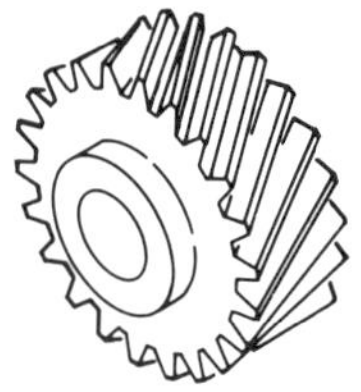

SIMILAR TO THE SPUR GEAR, BUT
SMOOTHER AND QUIETER DUE
TO A GREATER NUMBER OF
TEETH BEING IN MESH. MORE
EXPENSIVE THAN SPUR GEAR.

STRAIGHT
BEVEL

CONICAL IN FORM AND ACT ON
AXES THAT HAVE AN ANGULAR
RELATIONSHIP.

SPIRAL
BEVEL

DO THE SAME TASK AS A
STRAIGHT BEVEL GEAR, BUT
ARE SMOOTHER AND QUIETER.

WORM &
WORM GEAR

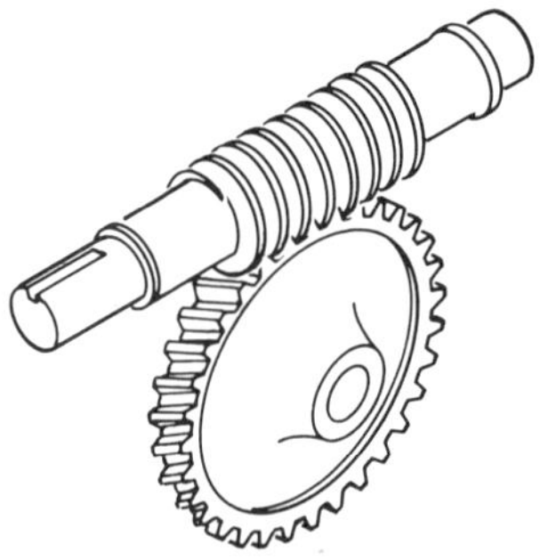

THE GEAR IS ALWAYS TURNED
BY THE MOTION OF THE WORM.
THE AXES ARE PERPENDICULAR
TO EACH OTHER. IT TAKES
MANY TURNS OF THE WORM
TO TURN THE GEAR ONE
REVOLUTION.

FIG. 26-2 Common standard gear types

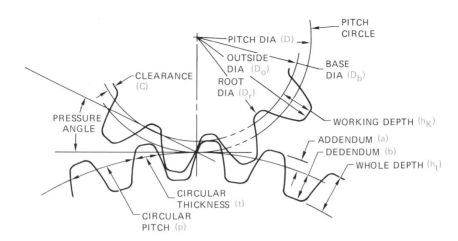

FIG. 26-3 Spur and helical gear nomenclature

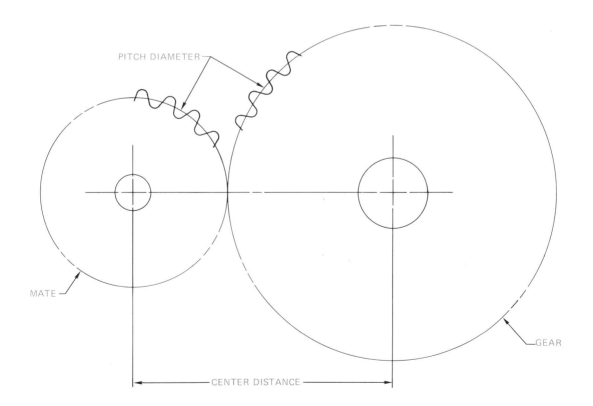

FIG. 26-4 Spur and helical gear nomenclature

Pitch Diameter (D)

Diametral Pitch (P) $P = \dfrac{N}{D}$

Number of Teeth (N)   N = P·D (Must be whole number)

Circular Pitch (p) $p = \dfrac{\pi D}{N}$ (π = 3.1416)

Addendum (a) $a = \dfrac{1}{P}$

Dedendum (b) $b = \dfrac{1.157}{P}$

Outside Diameter (Do)   Do = D + 2a

Root Diameter (D_R) D_R = D − 2b

Base Diameter (Db)   Db = D cos$\phi°$ (ϕ = pressure angle)

    Note: ϕ = pressure angle, common pressure angles are 20° and 14½°
        cos 20° = .93969
        cos 14½° = .96815

Circular Thickness (t)   t = π/2P

Working Depth (hk)   hk = 2a

Whole Depth (ht)   ht = a + b

Clearance (c)   c = b−a

Center Distance (C)   C = D

                 $\dfrac{\text{Gear + D Pinion}}{2}$

FIG. 26-5 Spur gear formulas

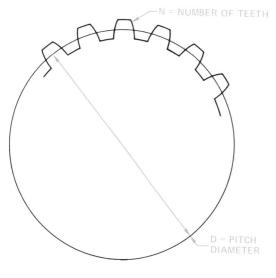

DIAMETRAL PITCH = $\dfrac{N}{D}$

EXAMPLE: DIAMETER = 10
             NUMBERED TEETH = 20 TEETH

$P = \dfrac{20}{10}$

DIAMETRAL PITCH = 2

FIG. 26-6 The gear module

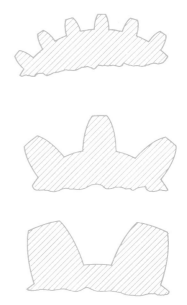

FIG. 26-7 Actual size diametral pitches

SAMPLES OF SPUR GEAR CALCULATIONS

These calculations make use of the formulas in Figure 26-5.

Given: Pitch Diameter Gear   = 6.00
            Pitch Diameter Mate   = 10.00
            Pressure Angle Mate   = 20°
            Diametral Pitch   = 6

Calculations:

Number of Teeth:

Gear: $N = P \times D$, $N = 6.0 \times 6$, $N = 36$

Mate: $N = P \times D$, $N = 10.0 \times 6$, $N = 60$

Circular Pitch:

$p = \dfrac{\pi D}{N}$,

$p = \dfrac{3.1416\,(6.0)}{36}$

$p = .5236$

Addendum:

$a = \dfrac{1}{P}$

$a = \dfrac{1}{6} = .1667$

Dedendum:

$b = \dfrac{1.157}{P}$

$b = \dfrac{1.157}{6}$

$b = .1928$

Outside Diameter:

Gear: $Do = D + 2a$,

$Do = 6.00 + 2(.1667)$,

$Do = 6.3334$

Mate: $Do = D + 2a$,

$Do = 10.00 + 2(.1667)$,

$Do = 10.3334$

Root Diameter:

Gear: $DR = D - 2b$,

$DR = 6.0 - 2(.1928)$,

$DR = 5.6144$

Mate: $DR = D - 2b$,

$DR = 10.0 - 2(.1928)$,

$DR = 9.6144$

Base Diameter:

Gear: $Db = D \cos \phi$,

$Db = 6.0\,(\cos 20°)$,

$Db = 6.0(.93969)$,

$Db = 5.6382$

Mate: $Db = D \cos \phi$,

$Db = 10.0\,(\cos 20°)$,

$Db = 10.0(.93969)$,

$Db = 9.3969$

Circular Thickness:

$t = \dfrac{\pi}{2P}$, $t = \dfrac{3.1416}{2 \times 6}$, $t = .2617$

Working Depth:

$hk = 2\,a$, $hk = 2(.1667)$, $hk = .3334$

Whole Depth:

$ht = a + b$, $ht = .1667 + .1928$, $ht = .3595$

Clearance:

$c = b - a$, $c = .1928 - .1667$,

$c = .0261$

Center Distance:

$C = \dfrac{D\text{ gear} + D\text{ mate}}{2}$

$C = \dfrac{6.0 + 10.6}{2}$

$C = \dfrac{16}{2}$ $C = 8.0$

USING GEAR TOOTH TEMPLATES

Since gear forms are standardized, it is unnecessary for the draftsperson to detail the gear tooth. The drawing of the gear is normally done with a template. Figure 26-8 illustrates the procedure.

DRAWING FORMAT FOR SPUR GEARS

A typical layout for spur gear detailing is illustrated in Figure 26-9. This drawing includes both the simplified drawing of the gear and the gear data chart. Note that only a few gear teeth are drawn as was shown in Figure 26-8. The phantom circles represent the outside diameter and the root diameter—the center line represents the pitch diameter. Dimensional values are indicated by X to show the number of decimals recommended for each dimension. On a real drawing the X would be replaced by the actual calculated quantity.

PROBLEMS

1. Using the formulas explained in this unit, do the following calculations.

Given: Pitch Diameter = 100 mm

Pressure Angle = 20°

Module = 5

1. DRAW PITCH DIAMETER AS A CENTER LINE.

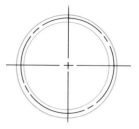

2. USING CONSTRUCTION LINES, DRAW IN THE OUTSIDE DIAMETER AND THE ROOT DIAMETER.

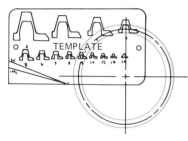

3. ALIGN THE TEMPLATE OF THE PROPER MODULE SIZE WITH THE OUTSIDE DIAMETER AND ROOT DIAMETER. DRAW IN ONE TOOTH AT A TIME.

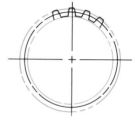

4. CONTINUE FOR AS MANY TEETH AS DESIRED (3 OR 4 IS NORMALLY SUFFICIENT). PUT OUTSIDE DIAMETER AND ROOT DIAMETER IN AS PHANTOM LINES.

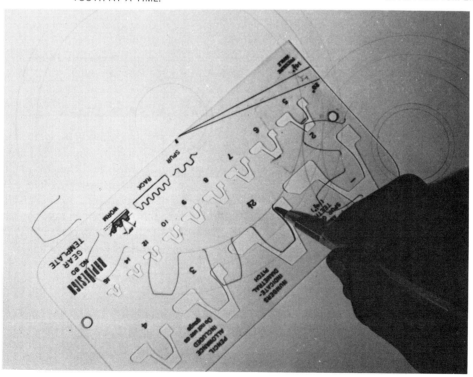

FIG. 26-8 Use of the gear tooth template (Clark E. Smith)

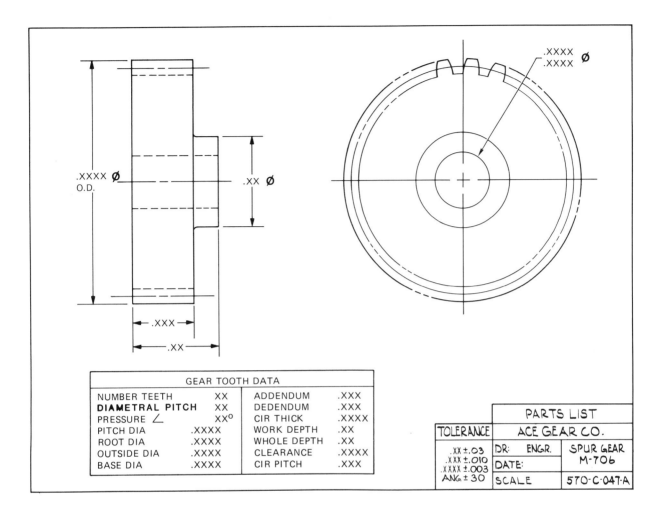

FIG. 26-9 A typical spur gear layout

PROBLEMS

1. Using the formulas explained in this unit, do the following calculations.

Given: Pitch Diameter = 12.00
        Pressure Angle = 20°
        Diametral Pitch = 5

Find: a. Addendum
        b. Dedendum
        c. Outside Diameter
        d. Number of Teeth
        e. Root Diameter
        f. Base Diameter
        g. Circular Pitch
        h. Working Depth
        i. Whole Depth
        j. Clearance

2. Given the data in Figure 26-10, draw a spur gear. Use Figure 26-9 as the format for the drawing.

3. Solve the calculations in Problem 1 using the gear blank shown in Problem 2. Draw a detail drawing of the spur gear using Figure 26-9 as the format for the drawing.

4. Practice drawing spur and rack gears with a gear template.

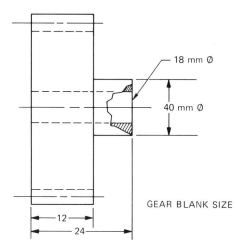

NUMBER OF TEETH	36
MODULE	2½
PRESSURE ANGLE	20°
PITCH DIAMETER	90.00
ROOT DIAMETER	83.75
OUTSIDE DIAMETER	95.00
BASE DIAMETER	84.572
ADDENDUM	2.50
DEDENDUM	3.125
CIRCULAR THICKNESS	3.927
WORKING DEPTH	5.00
WHOLE DEPTH	5.625
CLEARANCE	0.396
CIRCULAR PITCH	7.853

FIG. 26-10   Metric gear tooth data

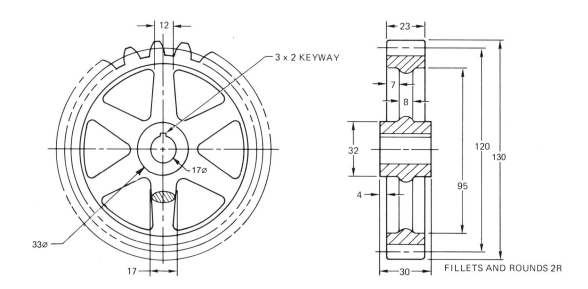

FIG. 26-11   Draw the two views of a spur gear.
Approximate the gear teeth with a gear
tooth template.

Cams

Chrysler Corporation

INTRODUCTION

Cams are mechanical devices that transform rotary motion to reciprocating (up and down or back and forth) motion. An example of this is a plate cam

mounted on a rotating shaft which causes a follower to go up and down (Fig. 27-2).

By studying Figure 27-2 you will note that the position of the follower is changed by the shape of the cam and its rotation. This is the basic principle of all cams. When designing a cam it is the movement of the follower that is most important. That is, the shape of the cam is determined by the desired reciprocating motion of the follower. In your study of this chapter, you should learn the different types of cams and followers (Fig. 27-3); how to draw them; and the definition of such cam terms as displacement, rise, fall, dwell, cam profile, and base circle.

TYPES OF CAMS

There are basically two types of cams. One is the plate cam and the other is the cylindrical or drum cam.

The plate cam is shaped like a disc with the contour of the cam on its circumference (Fig. 27-4). The follower, controlled by the plate cam, moves perpendicular to the cam's axis (the center line of the shaft). This is the typical cam used for opening and closing the valves of an automobile engine. It is the simplest cam to design and manufacture and is the one most used in industry.

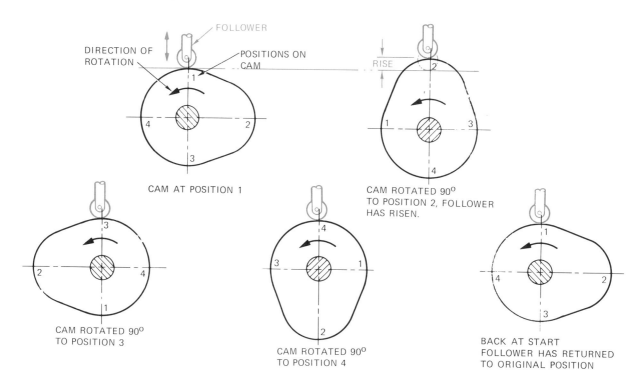

FOLLOWER

DIRECTION OF
ROTATION

POSITIONS ON
CAM

RISE

CAM AT POSITION 1

CAM ROTATED 90°
TO POSITION 2, FOLLOWER
HAS RISEN.

CAM ROTATED 90°
TO POSITION 3

CAM ROTATED 90°
TO POSITION 4

BACK AT START
FOLLOWER HAS RETURNED
TO ORIGINAL POSITION

FIG. 27-2   Cam and follower motion

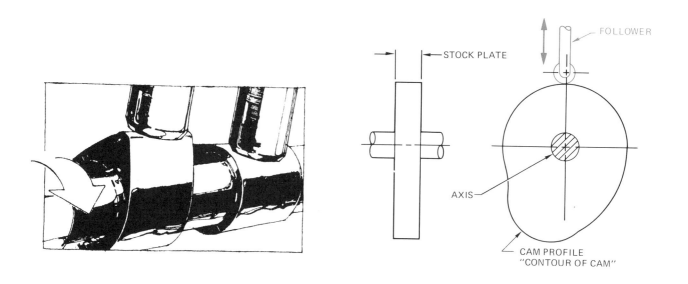

STOCK PLATE

FOLLOWER

AXIS

CAM PROFILE
"CONTOUR OF CAM"

FIG. 27-3   Cams in industry

FIG. 27-4   The plate cam

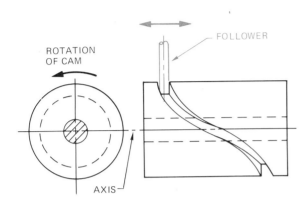

FIG. 27-5 The cylindrical cam

The cylindrical cam usually has a groove machined in its circumference (Fig. 27-5). The reciprocating motion of the follower is parallel to the cam's axis (the center line of the shaft).

TYPES OF FOLLOWERS

Followers are classified by the type of tip that comes in contact with the cam. The most common followers used for plate cams are shown in Figure 27-6. The type of followers used for a specific job depends on the speed of the cam, the pressure of the follower, the abruptness of the cam motion, and economic factors.

CAM TERMS

Displacement   The total movement of the follower measured in inches.
Rise   Upward movement of the follower.
Fall   Downward movement of the follower.
Dwell   No up or down movement of the follower. The cam rotates, but the follower remains stationary. Dwell is drawn as a horizontal line on the displacement diagram.
Cam Profile   The outside (circumference) shape of the cam face.
Base Circle   The circle at the lowest point of the cam face.

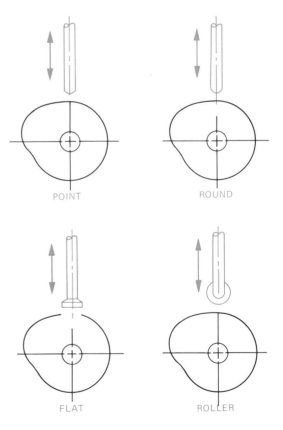

FIG. 27-6 Types of cam followers

DESIGNING AND DRAWING A PLATE CAM

Before attempting to draw a cam, the path of the follower must be determined and accurately plotted. As was stated at the beginning of this chapter, when designing a cam it is the movement of the follower that is most important. A chart called a displacement diagram is used to plot the desired motion of the follower. Figure 27-8 shows the relationship of the displacement diagram to the cam profile.

The displacement diagram is the first part to be drawn when laying out a cam. This layout will determine the *amount* of rise and fall of the follower, *when* it will be rising or falling, and the *rate* of rise and fall. The rate of rise and fall determines the speed and smoothness of the follower's motion. Some common motions that are used in plotting this displacement diagram are uniform motion, harmonic motion, constant acceleration, and cycloidal motion. The layouts of each of these motions are shown in Figure 27-9.

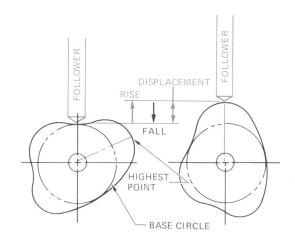

FOLLOWER AT LOWEST POINT ON CAM | CAM ROTATED SO FOLLOWER IS AT HIGHEST POINT ON CAM

FIG. 27-7   Cam terminology

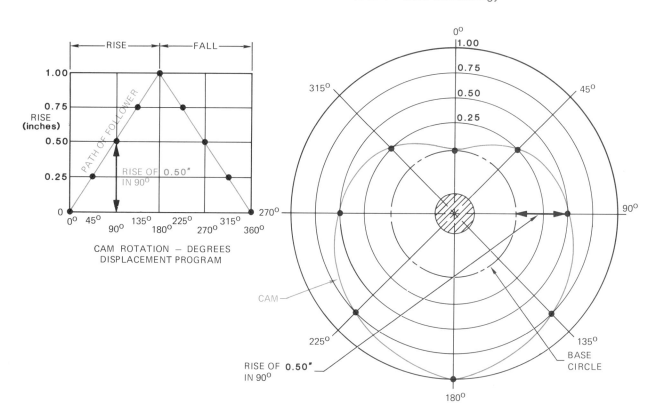

FIG. 27-8   A displacement diagram and cam profile

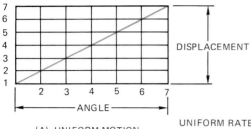

(A) UNIFORM MOTION

UNIFORM RATE OF SPEED.
A JERK AT START AND END.
USED FOR FEED CONTROL
ON MACHINE TOOLS.

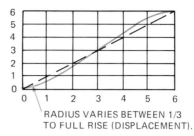

RADIUS VARIES BETWEEN 1/3
TO FULL RISE (DISPLACEMENT).

(B) MODIFIED UNIFORM MOTION
   — REDUCES JERK

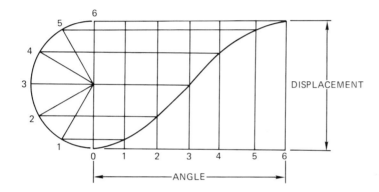

SMOOTH RISE AND FALL,
BUT PRODUCES A JERK IF
NEXT TO A UNIT OF DWELL.

(C) HARMONIC MOTION

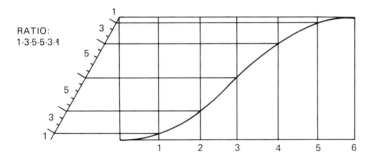

RATIO:
1-3-5-5-3-1

CONTINUOUS ACCELERATION,
SMOOTH, BUT ALSO PRODUCES
A JERK AT MEETING WITH A
DWELL UNIT.

(D) CONSTANT ACCELERATION

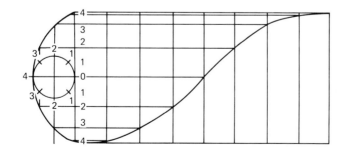

SMOOTHEST, NO JERK NEXT TO
DWELL. BEST FOR LIGHT LOADS
AT HIGH SPEEDS.

(E) CYCLOIDAL MOTION

FIG. 27-9 Common cam motions

A total displacement diagram is shown in Figure 27-10. This is a displacement diagram for a cam that:

1. Rises 2.0 in. at simple harmonic motion in 180°
2. Dwells for 45°
3. Falls 2.0 in. at modified constant motion to 360° (one revolution)

LAYOUT OF A PLATE CAM

Once the displacement diagram (Fig. 27-10) has been accurately laid out, the cam can be drawn. Figure 27-11 shows how to lay out a plate cam step by step. The basic steps are:

1. Draw the base circle.
2. Lay out degrees around it.
3. Plot points from the displacement diagram around the cam.
4. Connect the points.

PROBLEMS

1. Use Figure 27-9 to lay out the following displacement diagrams using the information in Figure 27-12.
2. Construct a displacement diagram using the following data:

   Displacement: 3.0 in.

   Circumference of base circle: 7.25 in.

   Rise: 3.0 in. at constant acceleration in 180°

   Fall: 3.0 in. at simple harmonic motion from 180° to 330°

   Dwell: 330° to 360°

3. Following the steps in Figure 27-11, lay out a plate cam using the displacement diagram in Figure 27-13.

4. Given the data in Figure 27-14, make a detail drawing of a place cam. This drawing will include the displacement diagram and the cam layout.

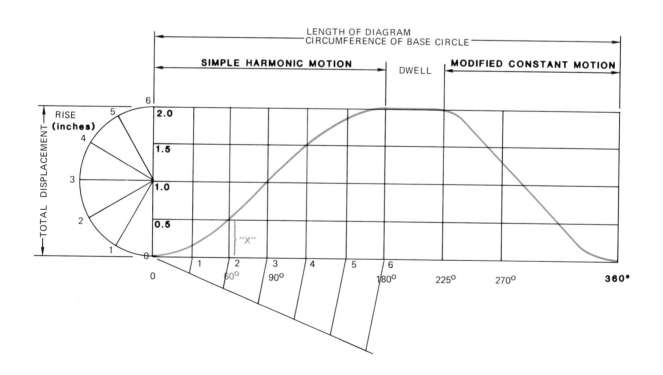

FIG. 27-10 Displacement diagram

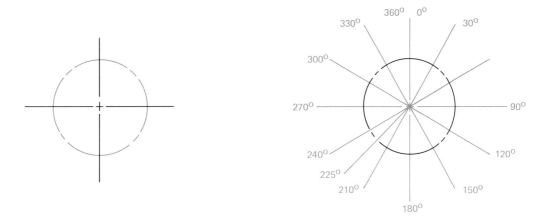

1. DRAW THE BASE CIRCLE.

2. LAY OUT ANGLES
    a)   AT LEAST EVERY 30°
    b)   AT ALL CRITICAL POINTS

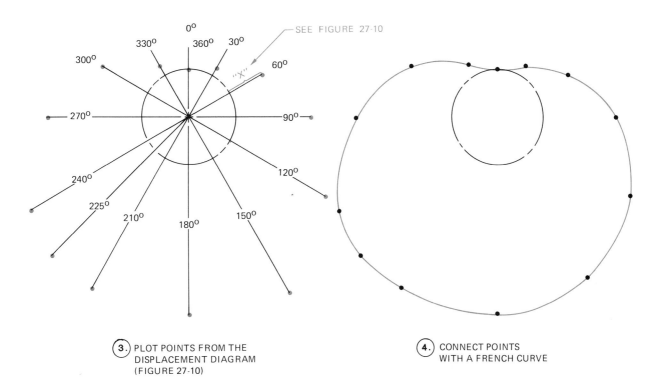

3. PLOT POINTS FROM THE
    DISPLACEMENT DIAGRAM
    (FIGURE 27-10)

4. CONNECT POINTS
    WITH A FRENCH CURVE

FIG. 27-11   Layout of a plate cam

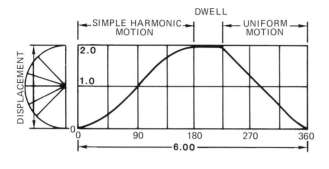

TYPES OF MOTION: RISES **2.0"** IN 180° BY SIMPLE
HARMONIC MOTION,

DWELLS FOR 30°

FALLS **2.0"** TO END OF CYCLE
BY UNIFORM MOTION

FIG. 27-12   Problem

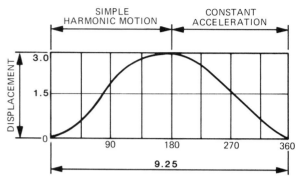

FIG. 27-13   Problem

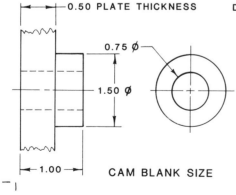

CAM BLANK SIZE

DISPLACEMENT DATA

   BASE CIRCLE: 3.00 DIAMETER
   CIRCUMFERENCE OF BASE
      CIRCLE: 9.42
   TOTAL DISPLACEMENT: 3.00
   RISE: 3.00 FOR 120° AT SIMPLE
      HARMONIC MOTION
   FALL: 1.50 FROM 120° TO 240°
      WITH CONSTANT ACCELERATION
   DWELL: FROM 240° TO 260°
   FALL: 1.50 FROM 260° TO 360° WITH
      UNIFORM MOTION

FIG. 27-14   Problem

28
Chapter

Electronics

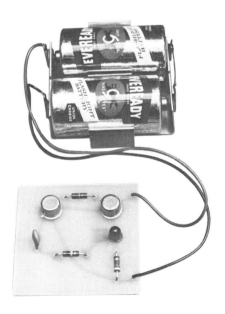

INTRODUCTION

Just as architects, sheet metal fabricators, and highway engineers have their own drawing requirements, so do those who work in the electronics industry. Every electronic device is composed of smaller bits and pieces. Each piece has its own name such as resistor, capacitor, or transistor, while together they are all called components. As you study this chapter, you should learn the principles of schematic symbols, schematic diagrams, printed circuit boards, and circuitry art.

The basic electronics drawing is called the schematic diagram. In it, each component is represented by a particular symbol. The symbols are part of an electronics alphabet which is recognized by most electronics workers throughout the world. The connections between the symbols form a road map of the many electronic paths or circuits. The schematic diagram is a complete description of how the components are interconnected electrically, but it does not describe the mechanical size or location of the parts. It is similar to airline route maps which really don't describe how to build the world but do show how to get from Los Angeles to Rome.

While an electronics technician needs only a schematic diagram and a handful of components to duplicate a very simple electronic device, a complete description of the device might require printed circuit layouts, parts placement diagrams, and even sheet metal drawings.

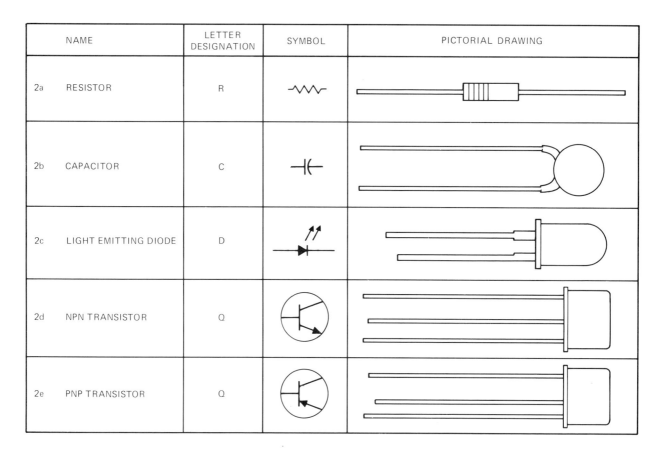

NAME		LETTER DESIGNATION	SYMBOL	PICTORIAL DRAWING
2a	RESISTOR	R		
2b	CAPACITOR	C		
2c	LIGHT EMITTING DIODE	D		
2d	NPN TRANSISTOR	Q		
2e	PNP TRANSISTOR	Q		

FIG. 28-2 Schematic symbols, renderings, and letter designations of the electronics parts used in the light flasher.

FLASHER

To illustrate the various techniques and electronics art forms, a simple device which can be easily duplicated is shown in Figure 28-1. This gadget is a solid-state light flasher built on a printed circuit board. The light blinks on for about 60 milliseconds every 1.2 seconds. The drain on the battery is so small that it can be left connected all the time.

THE COMPONENTS

The flasher has a total of eight components including the battery. Since this is a drafting text, there is no need to worry about how they work. The important things to know are what they look like, how

they are represented on a drawing, and how they are connected to each other.

Figure 28-2 shows a rendering of each component, its name, and its schematic symbol. Also, next to each name is the letter used to tie the schematic symbol to an item in the parts list. In a device with three resistors, like this one, the resistor symbols on the schematic diagram will be marked R1, R2, and R3. A more complete list of schematic symbols may be found in Figure 28-26.

The schematic diagram of the flasher is shown in Figure 28-3 and the parts list in Figure 28-4. A hand wired version of the flasher with the actual components in the same positions as the schematic symbols is shown in Figure 28-5. Because the flasher is a fairly simple circuit, it is easy to dupli-

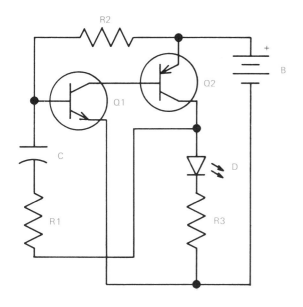

FIG. 28-3 The schematic diagram of the flasher is sort of a road map showing how the parts are connected together.

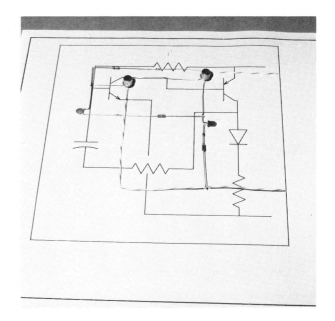

FIG. 28-5 A hand-wired flasher compared to its schematic diagram.

Parts List	
B	any 3 to 12 volt battery combination
C	0.01 μF, 10 volt ceramic capacitor
D	red light emitting diode
Q1	2N2219 (or equivalent) NPN silicon transistor
Q2	2N2905 (or eqivalent) PNP silicon transistor
R1	100 000 ohm, ¼ watt carbon resistor
R2	15 megohm, ¼ watt carbon resistor
R3	470 ohm, ¼ watt carbon resistor
Misc.	printed circuit board, battery holder, wire, and solder

FIG. 28-4 Parts list for the flasher gives the information required to purchase the parts.

FIG. 28-6 An electronic digital clock. (Caringella Electronics Incorporated)

cate the hand wired version. In more complicated devices, such as television receivers and computers, hand wiring is expensive and often requires troubleshooting to make it work.

CIRCUIT BOARDS

The electronic digital clock shown in Figure 28-6 is built on two printed circuit boards. The circuit boards shown in Figure 28-7 provide a mechanical

mounting for all of the components and contain most of the connections in a copper foil pattern. The following are some of the advantages of printed circuit board construction: (1) the wiring of each unit is identical to every other, (2) the components may often be inserted mechanically, and (3) all of the connections may be soldered in one operation.

The raw materials of printed circuit boards are large sheets of epoxy-glass (similar to the glass-fiber and resin combinations used to make auto bodies

FIG. 28-7 Interior view of the electronic digital clock showing the two printed circuit boards. (Caringella Electronics Incorporated)

the circuit board is a solid electrical conductor, similar to a piece of metal. In order to use it in an electronic circuit, copper foil must be removed so as to leave only the required "wires" or conductors. Holes must be punched or drilled so that the component leads may be inserted. The completed board must be cut to size.

The most common method of producing circuit boards uses chemical etching to remove the unwanted copper. A coating is applied to the surface of the copper by silk screening or by a photographic process that protects the wanted copper foil pattern from the action of the etching solution. After the etchant, as the solution is called, has done its job, the protective coating is removed with a solvent. The remaining copper is often plated with solder, tin, or gold or protected with a tarnish-resistant coating.

In order to produce the silk screen or photomask, an original drawing is required. This drawing, or master art, is usually produced two or four times the finished size of the circuit board. The combination of large artwork and photographic reduction makes it possible to produce well defined lines and finely detailed circuit boards.

There are no magic formulas to use in laying out a circuit board. Some points which make it easier are: (1) don't make it any smaller than it has to

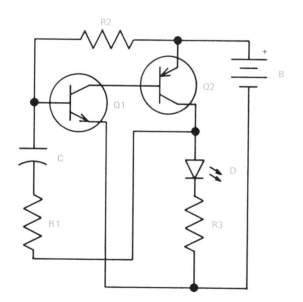

FIG. 28-8 The original flasher schematic diagram.

and boats), Teflon, paper-based phenolics, or similar materials. One or both sides of this insulating sheet are covered with thin copper foil. The foil is bonded to the sheet so that it cannot be removed except by milling or chemical etching. In this form,

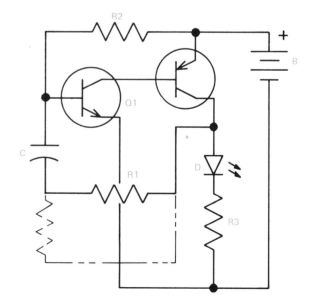

FIG. 28-9 One change eliminates the wire bridge.

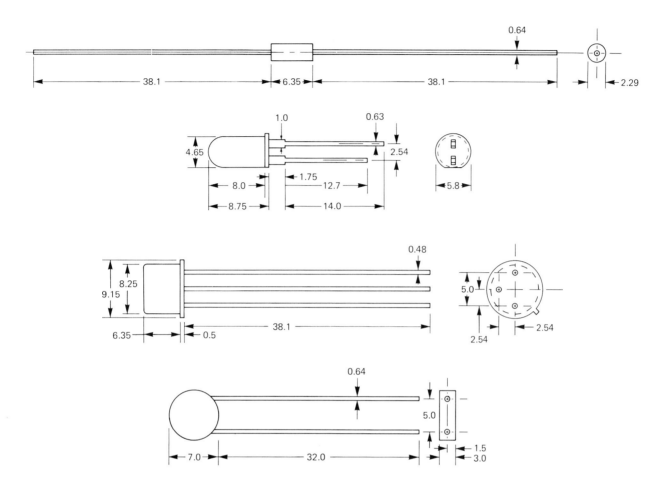

FIG. 28-10 Mechanical outlines and dimensions of the flasher parts.

be; (2) try to make the parts layout look like the schematic diagram; (3) if "wires" must cross, use the resistors or capacitors as the "bridges"; and (4) use round holes, if possible, instead of rectangular or odd shapes.

DRAWING A CIRCUIT BOARD

To get an idea of how the process works, let's go through it step by step. First examine the schematic diagram. In this case it is easy to get rid of the one wire bridge by redrawing the schematic diagram as shown in Figure 28-9.

Resistor R1 is used as a "bridge" to eliminate the crossed wires shown in the first schematic. The printed circuit layout can be made to follow the general lines of this new schematic diagram.

In order to place the components properly, it is necessary to know their mechanical dimensions. A dimensioned drawing of each component may usually be found in catalogs supplied by component manufacturers. Typical drawings are reproduced in Figure 28-10.

The schematic can be sketched on a sheet of ruled tracing paper or drafting film about four times as large as the finished circuit board will be. If the finished board is to be 2 inches on a side, start with a drawing at least 8 inches on a side. This step is shown in Figure 28-11.

Next, sketch in the actual mechanical outline of each component. These also must be four times larger than the real objects (Fig. 28-12).

The next step must be done with care, as it determines the actual location of each component.

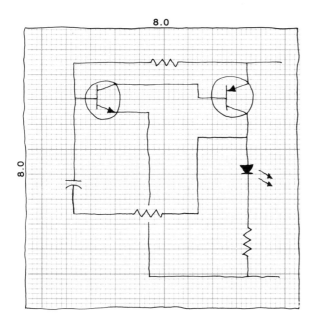

FIG. 28-11 The enlarged schematic is the first step in making the printed circuit board.

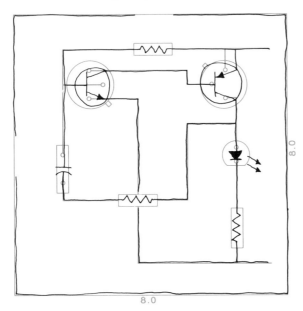

FIG. 28-12 The parts are sketched directly on Figure 28-11 or on a tracing paper overlay using Figure 28-11 as a guide.

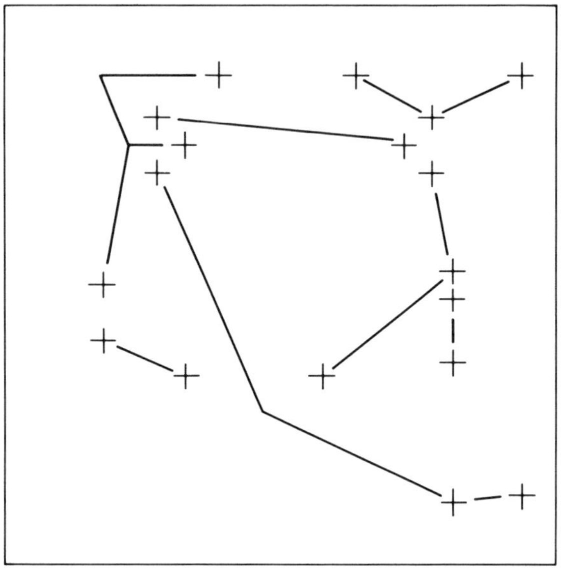

Locate and mark the holes for each component lead with fine crossed lines. Don't make the lines too thick as it will make the following steps less

FIG. 28-13 Another tracing paper overlay is used to accurately position the holes. As in Figure 28-12, this drawing may be done directly on the enlarged schematic.

precise (Fig. 28-13). Direct lines for connecting the holes may be added as guides. Untape the paper or film and turn it over. You are now looking at the "back" of the drawing. Tape it down so there are no wrinkles. Tape a larger sheet of clear Mylar film, clear acetate, or drafting film over the drawing.

The "master art" for the printed circuit board is made on this sheet of film. Mylar film is preferred to acetate or paper because it doesn't change size with changes in humidity. It is said to be non-hygroscopic; that is, it doesn't absorb moisture.

Instead of pencil, ink, or paint, the lines on the master art are made with die-cut pieces of matte-black crepe paper tape, similar to black masking tape. Many different die cut shapes are available for the hole pads as well as various widths of tape for the conductors. An assortment of shapes and pads is shown in Figure 28-14.

Pads for component lead holes are normally two to four times the diameter of the component lead hole itself. For example, a 0.04 in. hole is used for the components in the flasher. The pad on the printed circuit board should be 0.08 in. to 0.16 in. in diameter. The die cut pad on the drawing should be about 0.30 in. in diameter. The size of the hole in

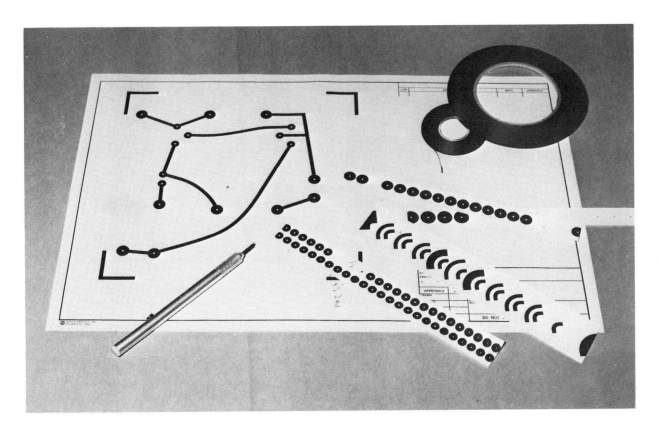

FIG. 28-14 Various shapes and sizes of precut pads and tapes are available for making printed circuit master art.

this donut pad should be as small as possible. Its only purpose is to help position the pad over the crossed lines on the drawing and to provide a guide for drilling the finished circuit board. Figure 28-15 shows the master art with only the pads in place.

The tape width used for the conductors can be scientifically determined if the thickness of the copper foil and the current flowing through the strip of foil are known. This is often required in more complex projects, but for very low current circuits such as the flasher, it is not important. The 1.5 and 3.0 mm wide tapes are easy to use and provide more than enough electrical capacity. The 1.5 mm tape is flexible enough to follow most curves without wrinkling. Try not to stretch the tape any more than necessary. The ends of the tape should overlap the donut pads almost to their centers.

Start by sticking the end of the tape down on the first pad. Carefully unwind the tape while applying it to the film. Cover the second pad, but don't rub it down yet. Using a small knife or single edge razor blade (an X-acto or other protected knife

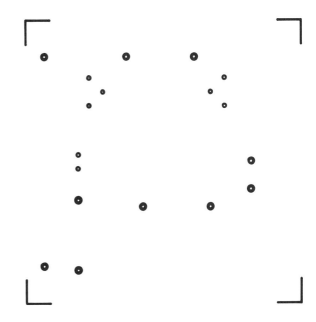

FIG. 28-15 Mark the corners and locate the hole pads first.

FIG. 28-16 The tape is cut by pulling it up against the knife blade so that the hole pads are not cut.

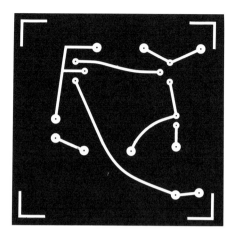

FIG. 28-18 Exact size reproduction of the negative used to make the circuit board.

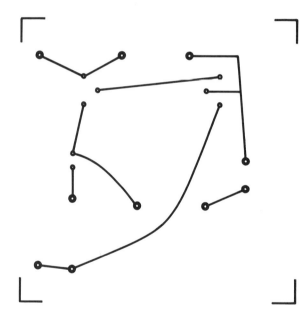

FIG. 28-17 Finished master art for the flasher.

It is practically impossible to describe exactly how to apply the tapes. After a little experimenting, you will develop a "feel" for it. Take your time and do it neatly. The finished master is shown in Figure 28-17. Notice the tape used to indicate the corners. Rub-down letters (four times size and bold) can be used to indicate part numbers, connections, or your initials.

PRODUCING A CIRCUIT BOARD

Converting the master art into a finished printed circuit board is a photographic process. First the master art is photographed using a precision graphic arts camera and litho type film. An exact size reproduction of the negative is shown in Figure 28-18. From this line negative, a silk screen can be made or a piece of presensitized circuit board can be printed under ultraviolet light. If only a few boards are to be made, presensitized board is usually used. If a large quantity is needed, the silk screen process is usually more economical.

In both cases an image is produced on the copper foil of the circuit board, permitting the etching of the unwanted copper while protecting the desired foil pattern.

A printed and etched circuit board is shown in Figure 28-19.

The final steps in the actual production of the board are drilling the holes and trimming to size. A

is best) press down lightly on the tape at the point where it should be cut. This point should be near the hole in the donut pad. Pull up gently on the tape so that it cuts against the knife edge (Fig. 28-16). Burnish the tape and pad with a burnishing tool or a fingernail.

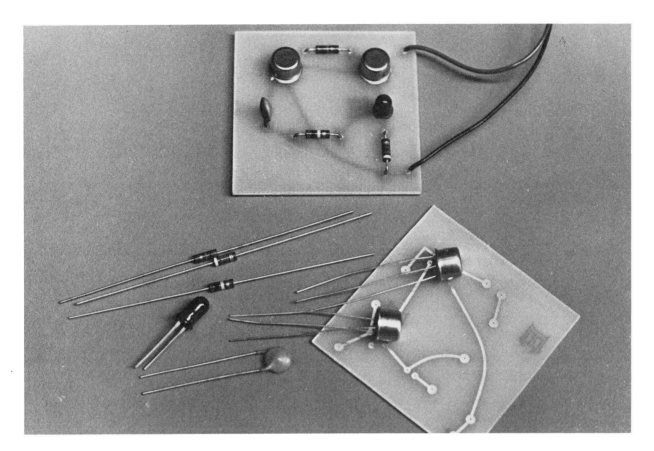

FIG. 28-19 The finished circuit board is shown with the completed flasher and all of the parts.

drilling guide drawing is often produced to document these operations. The easiest way to produce the drilling guide is to start with a brownline print made directly from the master art. This works quite well for a circuit as simple as the flasher. For more complex circuits, a screened print of the master art is made on a photosensitive drawing film.

The drilling guide should contain all of the mechanical data required to fabricate the circuit board. The drawing in Figure 28-20 shows the sizes and dimensions, as well as the specifications of the raw board material. Information on plating and protective coatings should also be given.

The hole sizes are indicated by letters keyed to a table. The table lists the key letter, hole diameter, number drill size (where applicable), and the total quantity of this size hole on the board.

ASSEMBLY DRAWINGS

A component placement drawing is required to detail the actual parts location on the circuit board. In addition, it is often necessary to silk screen a drawing of the parts location directly onto the plain side (no copper foil) of the circuit board. Manufacturers of electronic kits and manufacturers of products

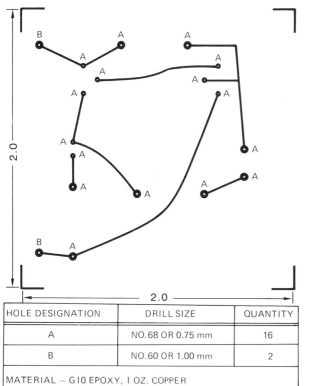

HOLE DESIGNATION	DRILL SIZE	QUANTITY
A	NO. 68 OR 0.75 mm	16
B	NO. 60 OR 1.00 mm	2
MATERIAL — G10 EPOXY, I OZ. COPPER		

FIG. 28-20 The drilling guide is made by adding information to a print of the master art.

which must be easy to repair usually use this technique.

Both drawings can be produced at the same time. Tape the master art, face down, to the drafting board along the bottom edge only. Cover the master art with a larger piece of tracing paper or drafting film and tape it to the drafting board along the top edge only. Carefully mark the location of each hole center and the four corners with a non-reproducing sky blue pencil. Remove the master art and tape down the bottom corners of the new drawing.

The outline of each component or its schematic symbol can now be drawn. Both methods are in common use, although outline drawings are usually preferred. The letter designation of each component is placed beside the schematic symbol or inside the outline. Rub-down letters make the job easy (see Figure 28-21). Since a silk screen may be made from the drawing, don't let any line go into the actual hole. Paint or heavy ink will be ap-

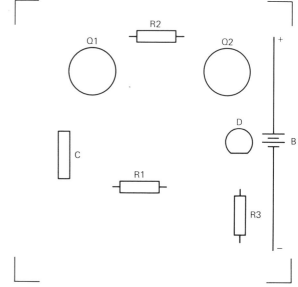

FIG. 28-21 A parts placement drawing may be used to produce a silk screen for the top of the circuit board and in making the composite print shown in Figure 23.

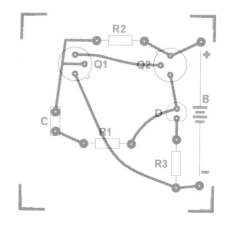

FIG. 28-22 Composite print of the master art and the parts placement drawing. This print is often called an X-ray view.

plied using the screen, and it might get messy when the paint is printed on the holes. The drawing itself may now be photographically reduced to produce the silk screen (usually called the top screen), and it

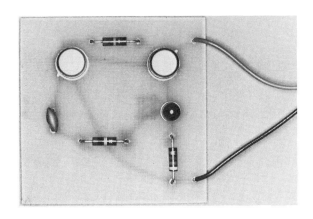

FIG. 28-23 The completed circuit board with all parts soldered in place.

may be used to produce blueline prints. Tape the drawing to the back side of the master art and use only one edge to keep it from wrinkling. Make sure that the corner marks register with each other. Run the combination through the printer together.

On more complex drawings, it is usually best to use a screened print of the master art instead of the actual solid master art to produce the component placement prints. An example of this type of print is shown in Figure 28-22.

The circuit board, with all of the components inserted and soldered, is shown in Figure 28-23. The two wires connect to the battery.

FIG. 28-24 One large and several small circuit boards are used in this stereo preamplifier. (BGW Systems)

FIG. 28-25   This key-to-disc computer terminal (a) has large complex circuit boards (b).

COMPLEX INDUSTRIAL ELECTRONICS

The same basic techniques used to produce the simple printed circuit board used in the flasher were used in designing the circuit board in the stereo preamplifier shown in Figure 28-24 and the computer key-to-disc terminal shown in Figure 28-25.

Figure 28-26 shows symbols for some of the electronic components commonly used in industry today.

Tiny microprocessor chips, like the one in the ceramic package shown in Figure 28-27 will be the heart of future control systems that will improve manufacturing operations, conserve heating and lighting energy, and increase gasoline mileage in automobiles. The ceramic package contains terminals that link the microprocessor to the rest of the control system. In the background, a photograph taken through a microscope shows the intricate details of a chip, which can contain as many as 6,000 transistors.

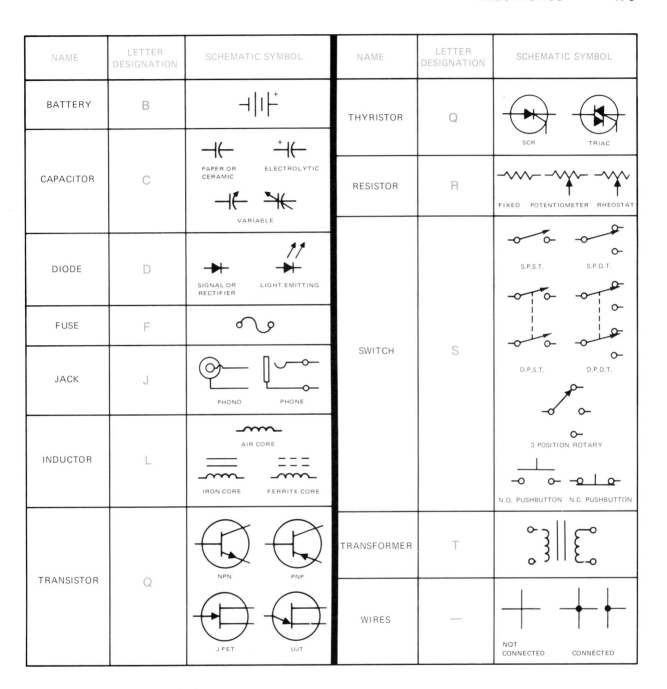

NAME	LETTER DESIGNATION	SCHEMATIC SYMBOL	NAME	LETTER DESIGNATION	SCHEMATIC SYMBOL
BATTERY	B		THYRISTOR	Q	SCR TRIAC
CAPACITOR	C	PAPER OR CERAMIC ELECTROLYTIC VARIABLE	RESISTOR	R	FIXED POTENTIOMETER RHEOSTAT
DIODE	D	SIGNAL OR RECTIFIER LIGHT EMITTING	SWITCH	S	S.P.S.T. S.P.D.T. D.P.S.T. D.P.D.T. 3 POSITION ROTARY N.O. PUSHBUTTON N.C. PUSHBUTTON
FUSE	F				
JACK	J	PHONO PHONE			
INDUCTOR	L	AIR CORE IRON CORE FERRITE CORE			
TRANSISTOR	Q	NPN PNP J FET UJT	TRANSFORMER	T	
			WIRES	—	NOT CONNECTED CONNECTED

FIG. 28-26 Schematic symbols.

PROBLEMS

1. Draw the electronic symbols in Figure 28-26.
2. Draw the schematic in Figure 28-3.
3. Draw the parts of the flasher in Figure 28-10.
4. Draw the layout in Figure 28-13. If tapes and hole pads are available, complete the artwork for the printed circuit board as in Figure 28-17.
5. Draw the amplifier in Figure 28-28.
6. Draw the schematic in Figure 28-29.
7. Draw the schematic in Figure 28-30.
8. Draw the schematic in Figure 28-31.
9. Draw the schematic in Figure 28-32.

FIG. 28-27   Miniaturized circuitry will be used in all
areas of electronics. (RCA)

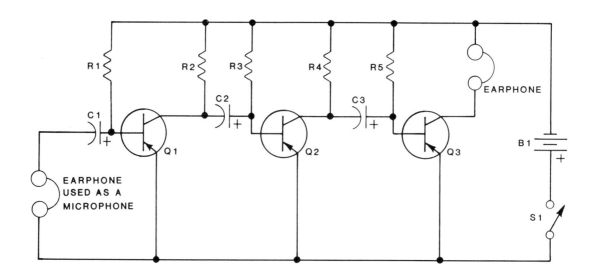

FIG. 28-28   Schematic diagram of a small audio amplifier

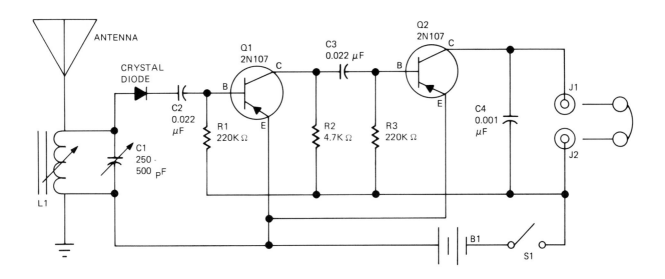

FIG. 28-29   Simple transistor radio

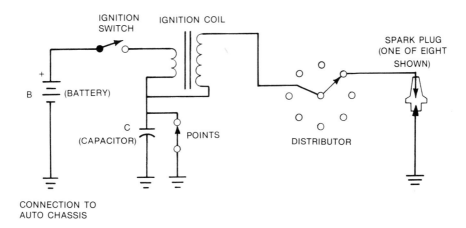

AUTOMOBILE IGNITION SYSTEM (SIMPLIFIED)

FIG. 28-30   Diagram of an automobile ignition system

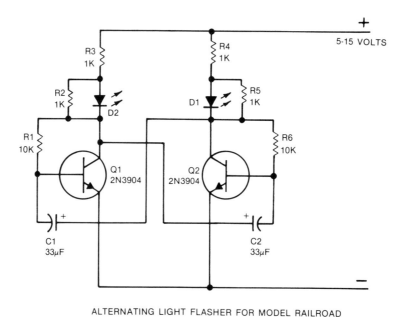

ALTERNATING LIGHT FLASHER FOR MODEL RAILROAD

FIG. 28-31   Diagram for an alternating light flasher

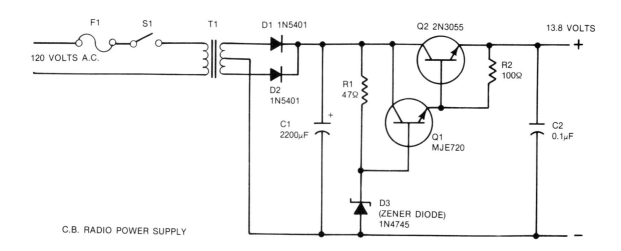

FIG. 28-32   Diagram for a c.b. radio power supply

Pipe Drafting

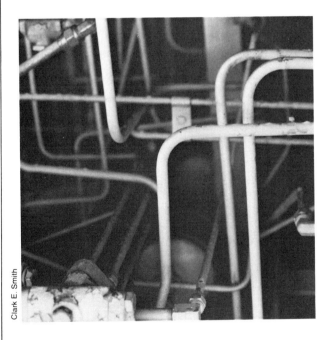

Clark E. Smith

INTRODUCTION

Pipe drafting is a type of assembly drawing which is used to represent piping layouts. Pipe drafting is used in any industry where it is necessary to transfer fluids, gases, and granular materials. As you

study this chapter, you should learn the different types of piping systems and piping materials, and the knowledge and skills necessary to draw piping systems.

STANDARDIZATION

Because piping is used almost everywhere, it has been standardized. There are standard sizes of pipes, fittings, and valves, which simplify the design, drawing, building, and repairing of pipe systems.

TYPES OF PIPE

Steel pipe is used for high temperature and/or pressure applications. Its size is specified by its nominal diameter, referred to as NPS (Nominal Pipe Size). This nominal size is roughly the inside diameter (I.D.) of the pipe. There are ten weights of pipes with different wall thicknesses and many sizes in each weight. The most common steel pipe is called Schedule 40 (Fig. 29-2).

Notice that while the inside diameter of the pipe varies from Schedule 40 to Schedule 80 the outside diameter (O.D.) remains the same. The inside diameter (I.D.) is smaller in Schedule 80 to allow for the increased wall thickness. This allows the external threads and fittings to remain the same.

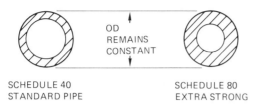

FIG. 29-2 Pipe wall thicknesses

Cast iron pipe is usually of large diameter, and used only for low pressure applications. A common application for cast iron pipe is large diameter underground sewer pipe.

Copper pipe and *tubing* is used where corrosion is a problem. Copper is one of the most expensive materials used for piping. It may be used for house plumbing and heating.

Plastic pipe is corrosion resistant, flexible, easily installed, and inexpensive. Sprinkler systems are a common application for plastic pipe.

PIPE CONNECTIONS

Pipes are connected by fittings. Fittings have been standardized to ensure that connections will not leak or vibrate loose. The different types of standard pipe connections are illustrated in Figures 29-3 through 29-8.

Screwed Pipe Threads

Pipe threads are tapered at a rate of 0.75 inch per foot to ensure a tight connection. Tapered pipe threads may be used for all types of pipe (Fig. 29-3).

Flanged Connections

Flanged connections are used to connect large diameter pipe. This type of connection allows rapid assembly and easy disassembly. The flange may be attached to the pipe by welding (shown in the example) or with a screw fitting (Fig. 29-4).

Welded Fittings

Welded fittings are excellent for high pressure systems. They are lightweight and less bulky than screwed or flanged fittings. Welded connections

should be used only where they are permanent and will not be disassembled. This is a common connection for steel pipe (Fig. 29-5).

Soldered Connections

Soldered connections are used with copper pipe and tubing. Fittings are made very precise and the

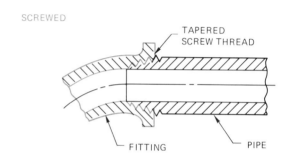

FIG. 29-3 Screwed pipe threads

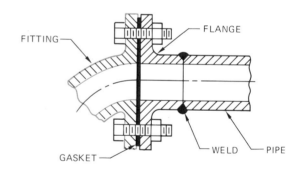

FIG. 29-4 Flanged connections

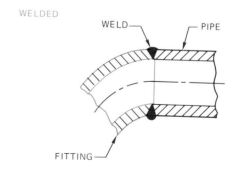

FIG. 29-5 Welded fittings

solder flows in to make a good, tight connection (Fig. 29-6).

Bell and Spigot Connection

The bell and spigot connection is used on clay and cast iron pipe. A common application is for domestic sewer lines. The pipe and fittings are manufactured with the bell and spigot on them (Fig. 29-7).

Cementing

Cementing (gluing) is a common method of connecting plastic pipe. It is easy and takes no special tools or skills. Because it is so easy to use, plastic pipe is popular with household do-it-yourselfers (Fig. 29-8).

PIPE FITTINGS

Pipe fittings are used to connect lengths of pipes, to change the direction of pipes, and to change pipe sizes. Some of the more common pipe fittings are shown in Figure 29-9. When a fitting is called out on a drawing, it is specified according to its nominal size, material, and strength factor.

VALVES

Valves play a major role in piping systems. They are used to stop or to control the flow of liquids through the pipe lines. There are many types of valves, but the three most common are gate valves, globe valves, and check valves. These valves and their functions are shown in Figures 29-10 through 29-12.

Gate Valves are the most common off/on valve. The gate valve is a simple, effective valve for stopping fluid flow, but it is unsuitable for regulating the rate of flow (Fig. 29-10).

Globe Valves are designed to regulate flow. As the valve is closed, it restricts and accurately regulates the flow of liquid. It may also be used to completely stop the flow (Fig. 29-11).

Check Valves permit liquids to flow in only one direction. If the fluid attempts to flow in the reverse direction, its force will close the check valve and stop the flow (Fig. 29-12).

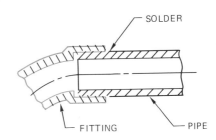

FIG. 29-6 Soldered connections

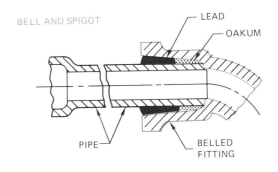

FIG. 29-7 Bell and spigot connections

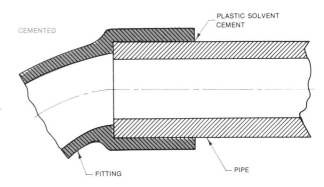

FIG. 29-8 Cementing

PIPING DRAWINGS

Piping drawings are system drawings. That is, they show how the pipe goes together to make up a system. The drawings show the location, size, and orientation of the pipe, fittings, and valves. Because

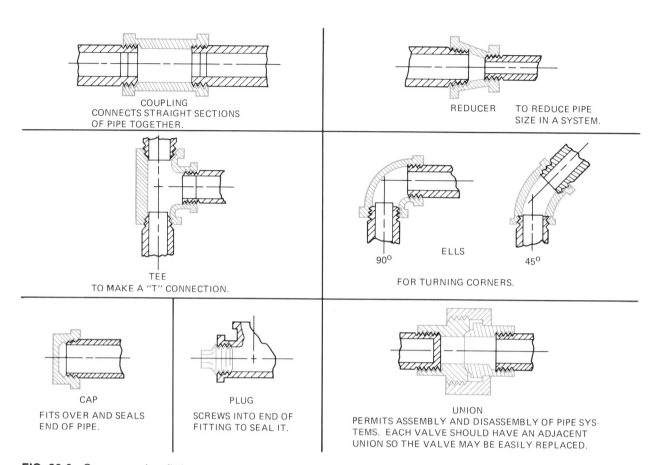

FIG. 29-9   Common pipe fittings

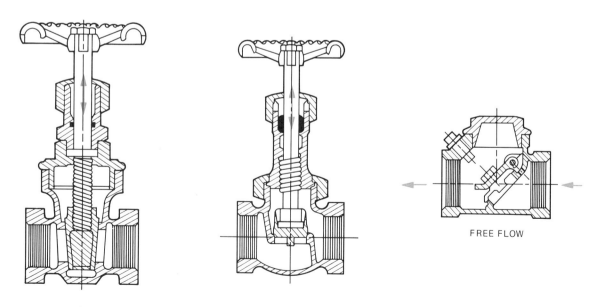

FIG. 29-10   Gate valve   **FIG. 29-11**   Globe valve   **FIG. 29-12**   Check valve

the components of a piping system are standard items, it is unnecessary to detail individual parts. To simplify the drawing, schematic symbols and simplified outlines of the components are used (Fig. 29-13).

Four types of piping drawings are commonly used. They are:

1. Single-line orthographic (Fig. 29-14)
2. Double-line orthographic (Fig. 29-15)
3. Single-line isometric (Fig. 29-16)
4. Double-line isometric (Fig. 29-17)

Isometric piping drawings are laid out and dimensioned in the same manner as any other isometric drawing (see Chapter 11). Figure 29-18 shows how some of the typical piping symbols are drawn on three isometric planes. Note how the symbols and dimensions are drawn on the isometric planes.

COMPONENT	SINGLE LINE	DOUBLE LINE
COUPLING		
CAP		
PLUG		
TEE		
90° ELL		
TURNED DOWN		
45° ELL		
REDUCER		
UNION		
GATE VALVE		
GLOBE VALVE		
CHECK VALVE		

FIG. 29-13 Symbols for screw connection fittings

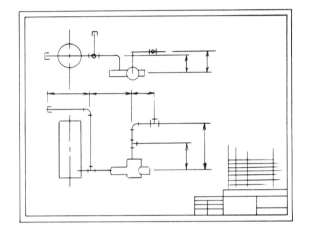

FIG. 29-14 Single-line orthographic

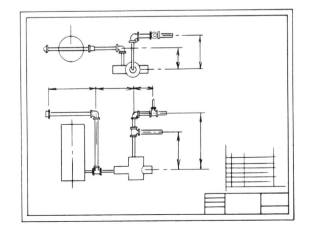

FIG. 29-15 Double-line orthographic

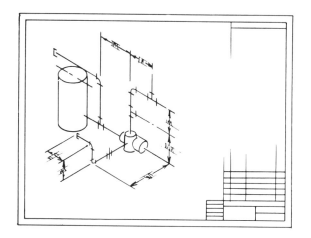

FIG. 29-16 Single-line isometric

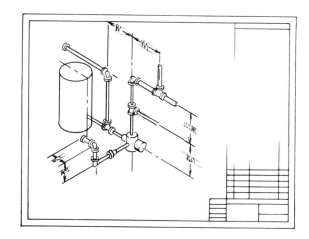

FIG. 29-17 Double-line isometric

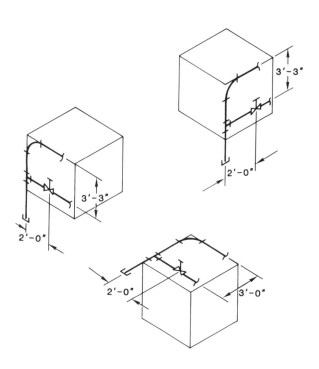

FIG. 29-18 Symbols and dimensions on an
isometric drawing

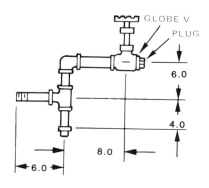

ALL PIPE 1.5 INCHES NOMINAL

FIG. 29-19 Pipe sizes may be shown in notes

DIMENSIONING

Piping drawings are dimensioned for locational purposes only. Locational dimensions are given to the center lines of pipes, fittings, and valves. The size of the pipe and fittings may be shown beside the components by a note or they may be called out in the parts list. Pipe lengths are usually not shown. They are left up to the pipe fitters (Fig. 29-19).

SIZES OF STANDARD PIPE

Figure 29-20 gives the nominal size, outside diameter, and inside diameter of steel pipe. It should be noted that the nominal size of the pipe is approximately the inside diameter size. Therefore a pipe referred to as a 1.5 inch pipe (its nominal size) has an inside diameter of approximately 1.5 inches (actually 1.610 inches).

PROBLEMS

1. Draw the three pipe segments in Figure 29-21. They are given as single-line orthographic drawings. Draw them as double-line orthographic drawings. The pipe is 1.5 inches nominal size.

2. Draw the three pipe segments in Figure 29-22. They are given as double-line orthographic drawings. Draw them as single-line orthographic drawings. The pipe is 1.0 inch nominal size.

Nominal Size (inches)	Outside Diameter (inches)	Inside Diameter (inches)
1/8	0.405	0.269
1/4	0.540	0.364
3/8	0.675	0.493
1/2	0.840	0.622
3/4	1.050	0.824
1	1.315	1.049
1 1/4	1.660	1.380
1 1/2	1.900	1.610
2	2.375	2.067
2 1/2	2.875	2.469
3	3.500	3.068
3 1/2	4.000	3.548
4	4.500	4.026
5	5.563	5.047
6	6.625	6.065
8	8.625	7.981
10	10.750	10.020
12	12.750	11.938
14	14.000	13.126
16	16.000	15.000
18	18.000	16.876
20	20.000	18.814
24	24.000	22.626

FIG. 29-20   Pipe sizes

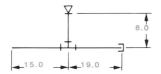
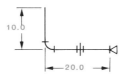

FIG. 29-21   Problem

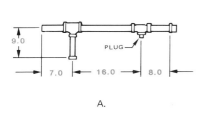
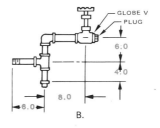
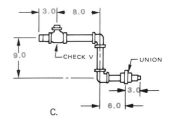

A.   B.   C.

FIG. 29-22   Problem

3. Figure 29-23 is a single-line isometric drawing of a simple piping system. Redraw it converting it to a single-line orthographic drawing. Show both the front and top projections. The pipe is 1.5" nominal, Schedule 40.

4. Figure 29-24 is a double-line orthographic drawing of a simple piping system. Redraw it converting it to a single-line isometric drawing. Refer to the examples in the chapter to aid in the proper orientation of the fittings on the drawing.

5. Draw a single-line isometric of the pipe segments in Figure 29-21.

6. Draw a double-line isometric of the pipe segments in Figure 29-21.

7. Draw a single-line isometric of the pipe segments in Figure 29-22.

8. Draw a double-line isometric of the pipe segments in Figure 29-22.

9. Draw a double-line isometric of the pipe segments in Figure 29-23.

10. Draw a double-line orthographic of the pipe segments in Figure 29-23.

11. Draw a double-line isometric of the pipe segments in Figure 29-24.

12. Draw a single-line orthographic of the pipe segments in Figure 29-24.

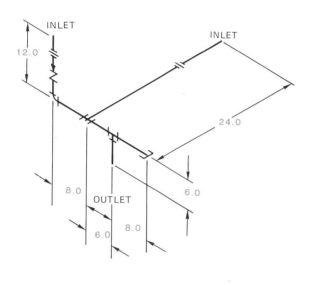

FIG. 29-23 Problem

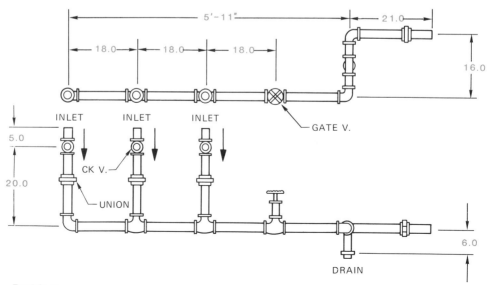

FIG. 29-24 Problem

Chapter

Design/Working Drawings

require a single drawing (Fig. 30-2) or the thousands of drawings required for the Northrop Corporation's F-5E supersonic aircraft shown in Figure 30-3. As you study this chapter, you should become familiar with the design and drawing processes and the different types of working drawings.

DESIGN

A finished product starts as an idea based on inventiveness and imagination. A well designed product such as the attack helicopter in Figure 30-4 will require a great deal of time and planning. Many factors will affect the design as it develops from ideas to sketches to drawings and is discussed by designers, engineers, fabricators, and sales and marketing personnel. The decisions made during this process are all a part of designing.

INTRODUCTION

The information in the preceding chapters can be used to produce a set of working drawings. To design and draw an object so it can be fabricated may

FIG. 30-2 A working drawing for a simple tool

FIG. 30-3 Thousands of drawings are required for a single aircraft (Northrop Corp.)

FIG. 30-4 Complex projects start as imaginative ideas (Hughes Helicopters, Culver City, CA)

The progressive steps from the concept of an idea to the finished product may be developed in the following manner:

I. Development of ideas and concepts with preliminary drawings (Fig. 30-5)
   A. Identification of needed item
   B. Sketching of preliminary ideas
   C. Modifications and refinements of sketches

D. Analysis of sketched designs (Fig. 30-6)
   1. Feasibility
   2. Functionality
   3. Economic Soundness
   4. Assembly traits
   5. Disassembly traits
   6. Serviceability
   7. Repairability
E. Final Decision—continue or stop

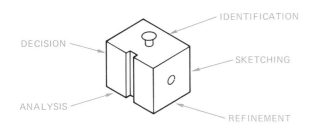

FIG. 30-5   Developing the creative idea

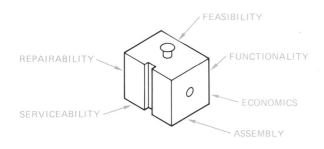

FIG. 30-6   Analyzing the idea

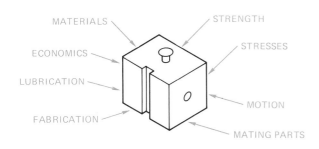

FIG. 30-7   Engineering data

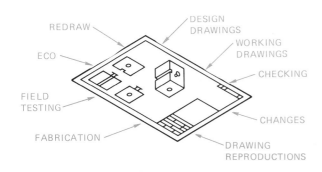

FIG. 30-8   Drawing and production

II.   Engineering data (Fig. 30-7)
   A.   Selection of materials
   B.   Calculation of strengths of materials
   C.   Calculation of stresses
   D.   Analysis of motions
   E.   Mating parts
   F.   Fabrication processes
   G.   Lubrication
   H.   Economical development
III.   Drafting production (Fig. 30-8)
   A.   Layout (design) drawing
   B.   Working drawing
   C.   Drawing checks
   D.   Engineering changes
   E.   Reproduction of drawings
   F.   Production of item
   G.   Field testing
   H.   Engineering change orders (ECO)
   I.   Redraw changes

The object of the complete design process is to achieve a sound design with maximum advantages and a minimum of disadvantages. A few examples

FIG. 30-9   Mass transit (Rohr industries, Inc.)

FIG. 30-10   Automotive products (General Motors Corp.)

FIG. 30-11   Commercial air transportation (Boeing)

FIG. 30-12 Defense products (General Dynamics)

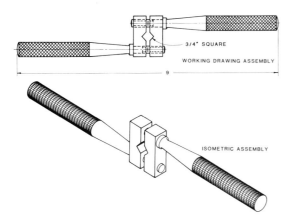

FIG. 30-13 Assembly drawings for a tap wrench

FIG. 30-14 Detail drawings for a tap wrench

of well designed end products are shown in:

1. Mass transit (Fig. 30-9)
2. Automotive transportation (Fig. 30-10)
3. Commercial air transportation (Fig. 30-11)
4. Defense (Fig. 30-12)

WORKING DRAWINGS

Working drawings can be put into two categories: assembly drawings and detail drawings. Figure 30-13 is an example of the assembly drawings for a tap wrench. Figure 30-14 is an example of a detail

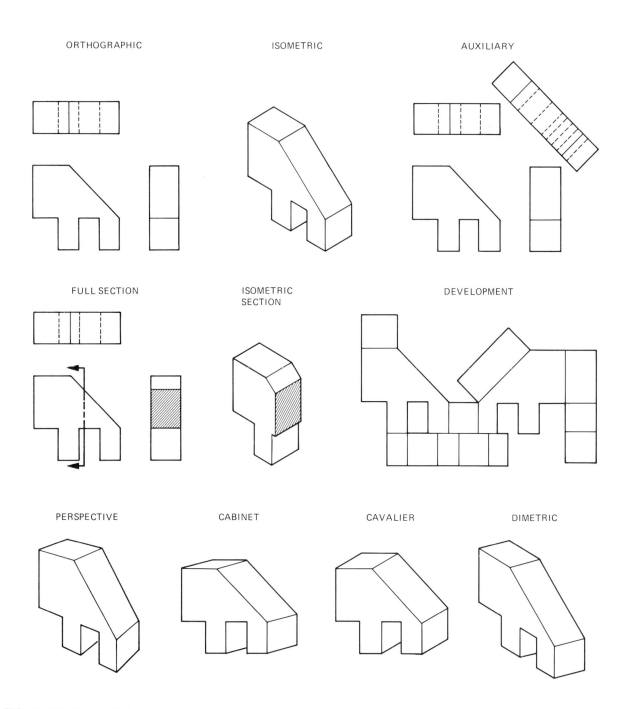

ORTHOGRAPHIC   ISOMETRIC   AUXILIARY

FULL SECTION   ISOMETRIC SECTION   DEVELOPMENT

PERSPECTIVE   CABINET   CAVALIER   DIMETRIC

FIG. 30-15 Types of drawings to select from

drawing for the same tap wrench. A working drawing should be the type of drawing which communicates most efficiently (Fig. 30-15). Figure 30-16 gives examples of several different types of working drawings for a small steam engine. Photographs can also be used. Figure 30-17 is an

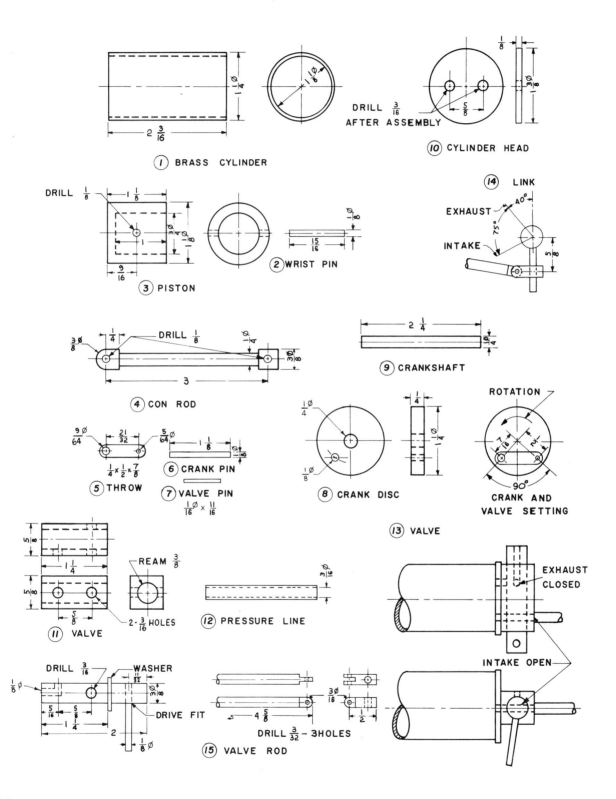

FIG. 30-16 A set of working drawings for a small steam engine

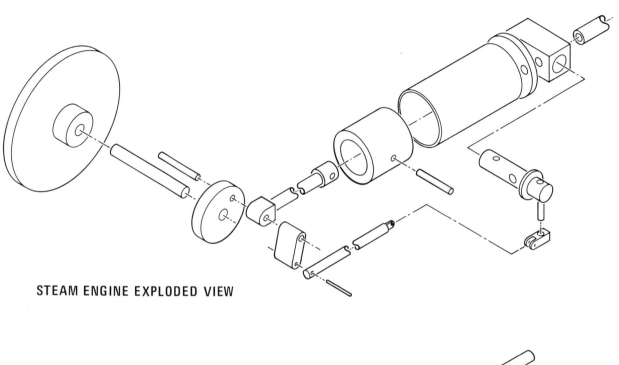

STEAM ENGINE EXPLODED VIEW

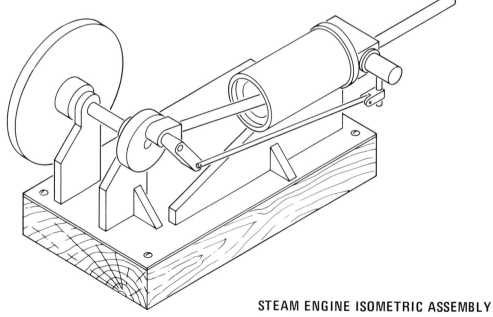

STEAM ENGINE ISOMETRIC ASSEMBLY

FIG. 30-16   A set of working drawings for a small
steam engine (Cont.)

FIG. 30-17 An assembly photograph (TRW)

FIG. 30-18 Detail photograph (TRW)

assembly photograph of a titanium propellant tank for spacecraft. Figure 30-18 is a detail photograph of the same tank.

There are several categories of working drawings for the draftsperson to choose from. It is not necessary to use all of them. The general rule is to use the minimum number of drawings that will communicate 100 percent of the information

needed to fabricate the item. The fabricator should not have to look elsewhere for additional information to manufacture the part.

The categories of working drawings are:

1. Layout assembly
2. Outline assembly
3. Operation assembly
4. Diagram assembly
5. Design assembly
6. Subassembly
7. Working assembly
8. Installation assembly
9. Detail drawing

A general rule that most draftspersons follow is to select multiview drawings for technical personnel and pictorial drawings for nontechnical personnel (Fig. 30-19).

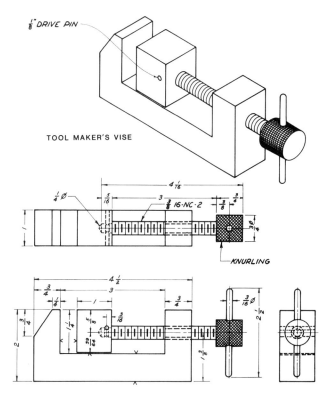

FIG. 30-19 Pictorial and multiview working drawings

LOCKHEED L-1011 TRISTAR

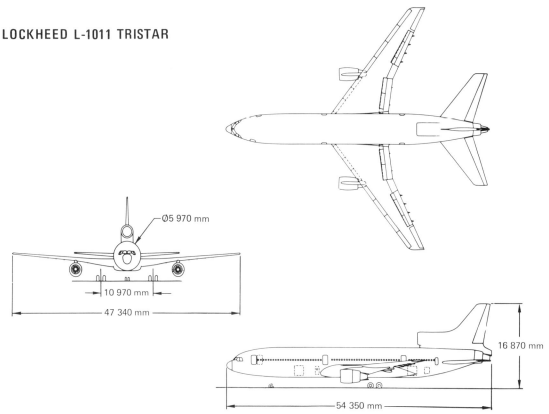

FIG. 30-20   Layout assembly drawing and product (Lockheed Co.)

1. The *layout assembly drawing* is the first pre-
    liminary step in developing a new design.
    Working from engineering sketches, the drafts-
    person makes the initial drawing of the full-
    size item to help in the development of the
    design. This initial sketch is called the layout
    assembly drawing. It shows all of the impor-
    tant features of the item to be assembled, leav-
    ing out the minor details such as nuts, bolts,
    and screws (Fig. 30-20).

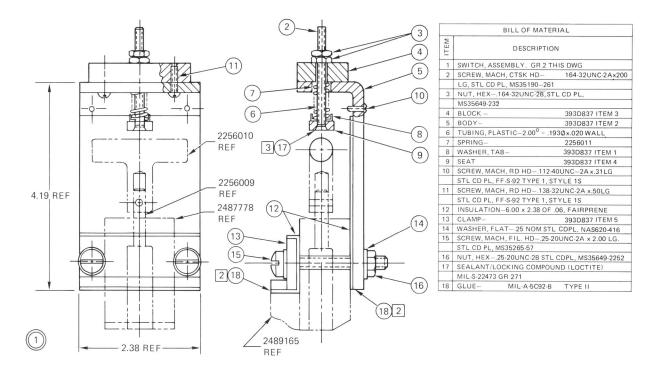

BILL OF MATERIAL		
ITEM	DESCRIPTION	
1	SWITCH, ASSEMBLY. GR.2 THIS DWG	
2	SCREW, MACH, CTSK HD—	164-32UNC-2A x200
	LG, STL CD PL, MS35190—261	
3	NUT, HEX—.164-32UNC-2B, STL CD PL,	
	MS35649-232	
4	BLOCK —	393D837 ITEM 3
5	BODY—	393D837 ITEM 2
6	TUBING, PLASTIC—2.00° = .193∅ x.020 WALL	
7	SPRING—	2256011
8	WASHER, TAB—	393D837 ITEM 1
9	SEAT	393D837 ITEM 4
10	SCREW, MACH, RD HD—.112-40UNC-2A x.31LG	
	STL CD PL, FF-S-92 TYPE 1, STYLE 1S	
11	SCREW, MACH, RD HD—.138-32UNC-2A x.50LG	
	STL CD PL, FF-S-92 TYPE 1, STYLE 1S	
12	INSULATION—6.00 x 2.38 OF .06, FAIRPRENE	
13	CLAMP—	393D837 ITEM 5
14	WASHER, FLAT—.25 NOM STL CDPL, NAS620-416	
15	SCREW, MACH, FIL. HD—.25-20UNC-2A x 2.00 LG.	
	STL CD PL, MS35265-57	
16	NUT, HEX—.25-20UNC-2B STL CDPL, MS35649-2252	
17	SEALANT/LOCKING COMPOUND (LOCTITE)	
	MIL-S-22473 GR 271	
18	GLUE— MIL-A-5C92-B TYPE II	

FIG. 30-21   The outline assembly

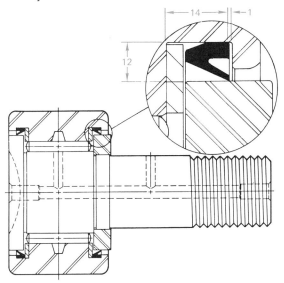

FIG. 30-22   The operation assembly drawing (McGill Mfg. Co.)

2. The *outline assembly drawing* shows the assembled product as an outline drawing. Few, if any, hidden lines are shown in an outline drawing. Some major dimensions are given to relate size, and each part is listed and identified on the drawing. This general appearance type of drawing can be used in catalogs and sales brochures (Fig. 30-21).

3. The *operation assembly drawing* is used to show the function of a single special item on the assembly drawing, or if more information is needed for fabrication (Fig. 30-22).

BILL	OF	MATERIALS
NUMBER	AMT	ITEM
CI	3	CAPACITOR 10 μF
BI	I	6 OR 9 VOLT BATTERY
RI	3	RESISTOR 100 KΩ
R2	I	RESISTOR 10 KΩ
R4	I	RESISTOR 4.7 KΩ
QI	3	TRANSISTOR 2N526
SI	I	SWITCH
AI	4	CONTACT CLIPS

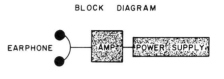

BLOCK DIAGRAM

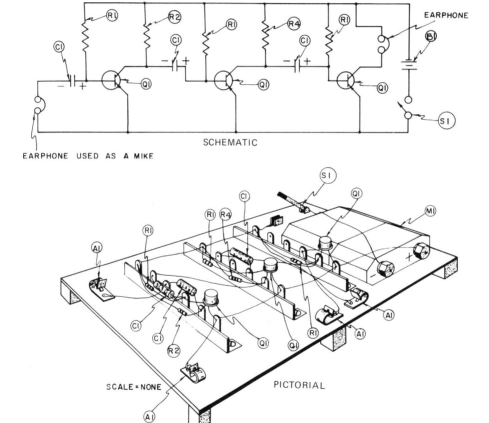

FIG. 30-23   Assembly diagram

4. The *diagram assembly drawing* is a single-line drawing showing the approximate location and form of parts in the assembly. Figure 30-23 is an example of an amplifier drawn in a diagram assembly.

5. *Design assembly drawings* are similar to layout assembly drawings. Figure 30-24 shows the design assembly drawing of a deep submersible submarine. Figure 30-25 shows the finished product, Lockheed's Deep Quest.

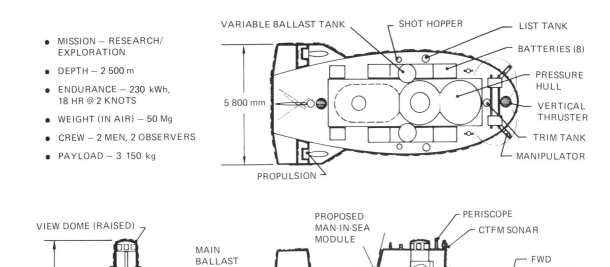

- MISSION – RESEARCH/ EXPLORATION
- DEPTH – 2 500 m
- ENDURANCE – 230 kWh, 18 HR @ 2 KNOTS
- WEIGHT (IN AIR) – 50 Mg
- CREW – 2 MEN, 2 OBSERVERS
- PAYLOAD – 3 150 kg

DEEP QUEST GENERAL ARRANGEMENT

FIG. 30-24   Design assembly drawings (Lockheed Missiles & Space Co.)

FIG. 30-25   The finished product, Lockheed's Deep Quest (Lockheed Missiles & Space Co.)

NOTES:

1. USE Ø1.2 PILOT HOLE
   IN ITEM 1 FOR LOCATION
   OF Ø1.6 HOLE

2. STAKE ITEM 1 TO ITEM 2

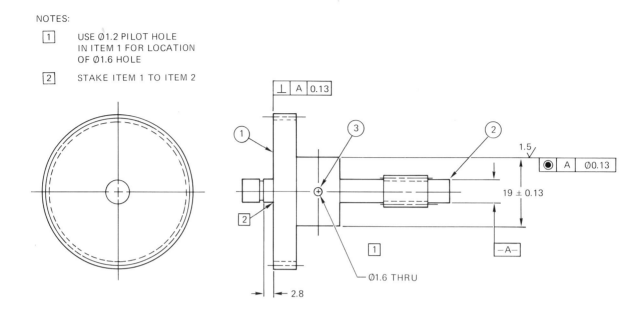

FIG. 30-26   Subassembly drawing

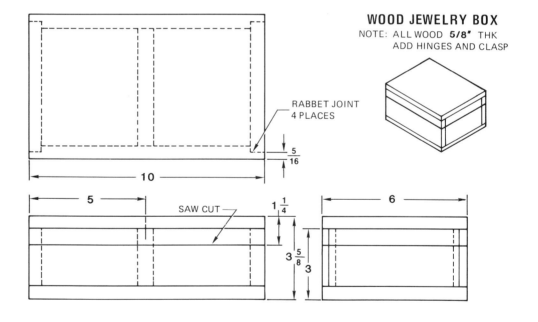

FIG. 30-27   The working assembly drawing

6. *Subassembly drawings* are used to show small parts of larger products. Figure 30-26 shows one gear assembly from a speed decreaser transmission.

7. The *working assembly drawing* is an assembly drawing with all the dimensions and notes required to fabricate the product. It is also called a detail assembly drawing. Each assembled part is completely detailed (Fig. 30-27).

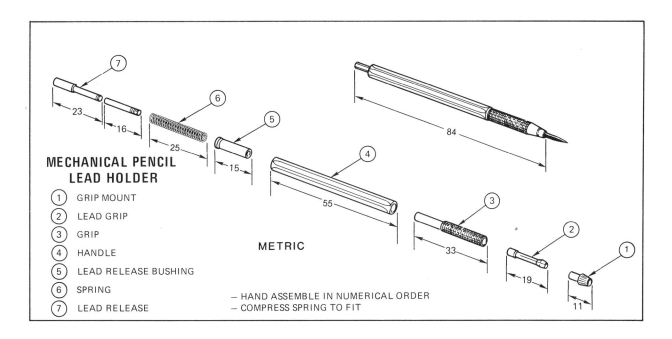

FIG. 30-28 The installation assembly drawing

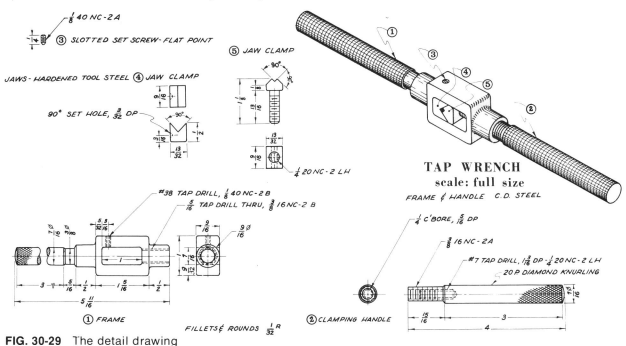

FIG. 30-29 The detail drawing

8. The *installation assembly drawing* is similar to the layout assembly drawing. The additional information it conveys includes dimensions and instructions on how the item is to be assembled, installed, and set into operation (Fig. 30-28).

9. The *detail drawing* is the most commonly used working drawing. It is a completely detailed drawing of a single item with dimensions and instructions for its fabrication (Fig. 30-29).

FIG. 30-30   Cartoon

designers and engineers make changes as the layout drawings are checked. Checkers will check prints and mark changes with color-coded pencils. The checker looks for correctness, completeness, quality, readability, and engineering changes.

It is easier for a checker to spot needed changes and errors than it is for the draftsperson who did the drawing.

Revisions and changes are usually made after the product is manufactured and field tested. A request to change the working drawing comes as an ECO (engineering change order). These changes are kept on record, usually on the original drawing, to eliminate the need for separate filing of related documents. Changes are recorded and identified in a revision strip near the change on the drawing (Fig. 30-31).

The drafting technician should choose the type of drawing that will communicate all the necessary information with the least chance of error, and the one that is quickest to draw (Fig. 30-30).

A series of detail and assembly drawings makes up the working drawings for a manufactured item (Fig. 30-16).

SIMPLIFIED DRAFTING

A few general drafting principles will help the draftsperson turn out faster and clearer working drawings:

1. Do not draw items that are not needed.
2. Draw only half of a symmetrical part.
3. Do not draw repetitive parts more than once.
4. Do not draw standard parts, list them.
5. Use labor saving devices such as templates and transfer letters.
6. Do not crowd drawings with too much line-work.

DRAWING CHANGES

Changes will often be necessary during the development of the design and even after the product is produced and tested. During the design process the

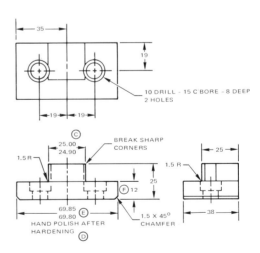

FIG. 30-31   Revisions and the revision strip

PROBLEMS

Freehand labeling is used for some of the drawings in this set of problems. This style resembles the usual style of working drawings.

1. Draw a full-size detail drawing of Figure 30-32. Draw the orthographic and isometric.

2. Draw a full-size detail drawing of Figure 30-33. Draw the orthographic and cabinet.

3. Draw a detail drawing of Figure 30-34. Draw an orthographic twice the original size (2:1).

4. Draw a full-size detail drawing of Figure 30-35. Draw the orthographic and isometric.

5. Draw the detail drawing and an assembly drawing of Figure 30-36.

6. Draw the detail drawing and an assembly drawing of Figure 30-37.

7. Draw the detail drawing and an assembly drawing of Figure 30-38.

8. Draw the assembly drawing of the Hughes 300C helicopter in Figure 30-39.

9. Draw the detail drawing of Figure 30-40.

10. Draw the detail drawing and isometric drawing of Figure 30-41.

11. Draw the detail drawing in Figure 30-42. Refer to the chapters on tolerances and positional tolerances.

12. Draw an outline assembly drawing for Figure 30-43.

13. Draw the orthographic and isometric drawing of Figure 30-44.

14. Draw Figure 30-45 complete isometric exploded and orthographic assembly drawings.

15. Draw Figure 30-46a. Note the changes made on Figure 30-46b as ECOs on the drawing.

16. Draw the c-clamp in Figure 30-47.

17. Draw the scribe holder in Figure 30-48.

18. Draw the tool holder in Figure 30-49.

19. Draw the wrench handle for sockets in Figure 30-50.

20. Draw the spoke wrench in Figure 30-51.

21. Draw the hand clamp in Figure 30-52.

22. Draw the door stop in Figure 30-53.

23. Draw the steel punch in Figure 30-54.

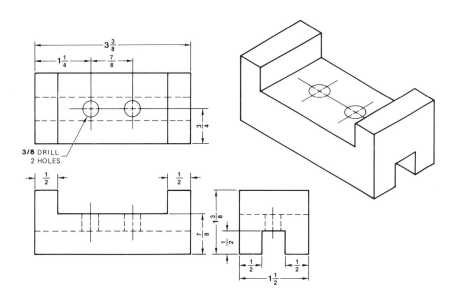

FIG. 30-32   Problem

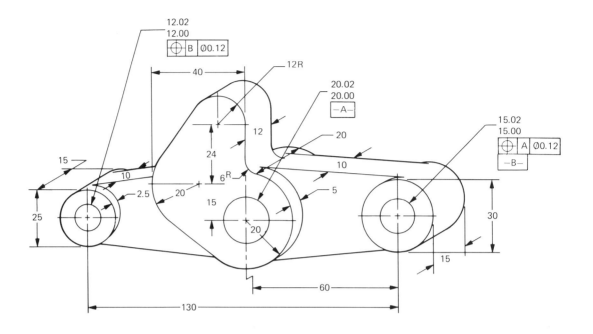

FIG. 30-33 Problem

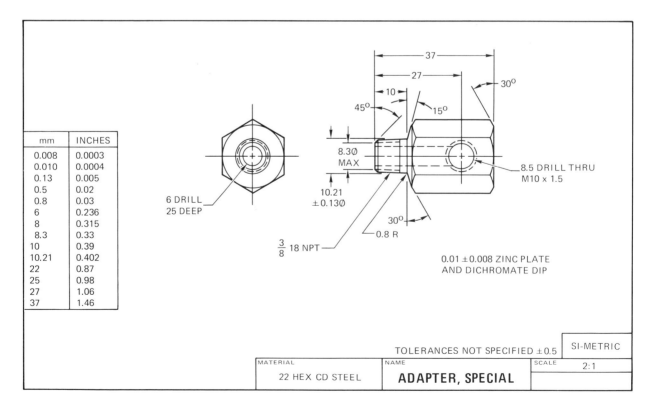

FIG. 30-34 Problem

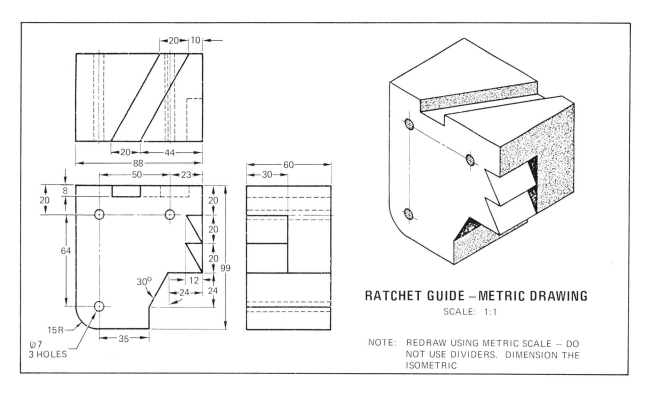

RATCHET GUIDE – METRIC DRAWING
SCALE: 1:1

NOTE: REDRAW USING METRIC SCALE — DO
       NOT USE DIVIDERS. DIMENSION THE
       ISOMETRIC

FIG. 30-35 Problem

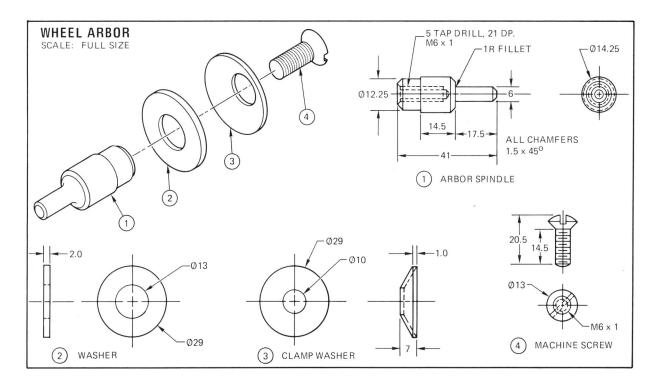

WHEEL ARBOR
SCALE: FULL SIZE

5 TAP DRILL, 21 DP.
M6 x 1
1R FILLET
Ø14.25
Ø12.25
14.5
17.5
41
6
ALL CHAMFERS
1.5 x 45°

① ARBOR SPINDLE

2.0
Ø13
Ø29

② WASHER

Ø29
Ø10
1.0
7

③ CLAMP WASHER

20.5
14.5
Ø13
M6 x 1

④ MACHINE SCREW

FIG. 30-36 Problem

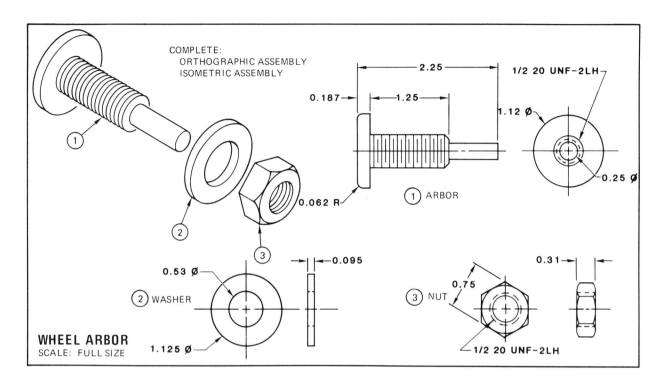

FIG. 30-37   Problem

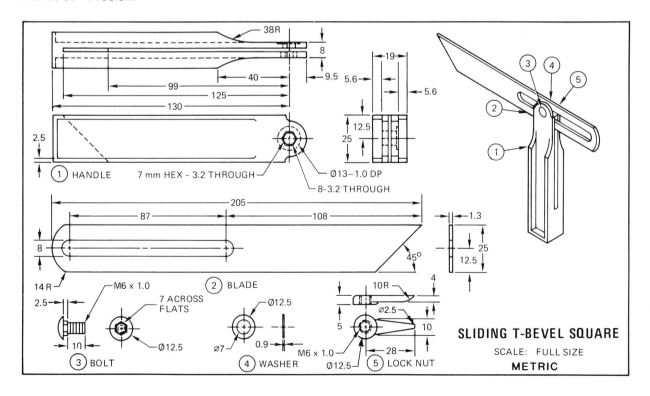

FIG. 30-38   Problem

Hughes 300C

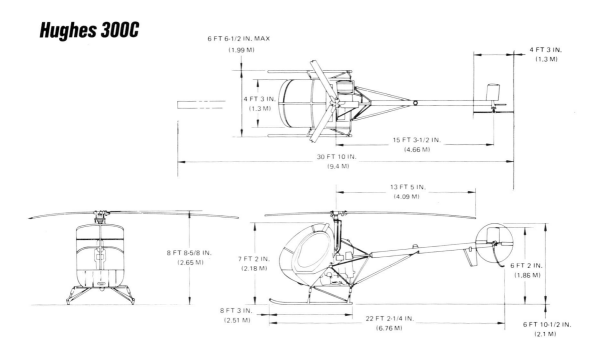

FIG. 30-39   Problem (Hughes Helicopter)

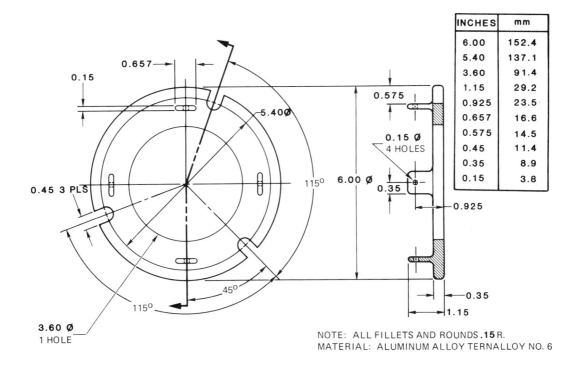

INCHES	mm
6.00	152.4
5.40	137.1
3.60	91.4
1.15	29.2
0.925	23.5
0.657	16.6
0.575	14.5
0.45	11.4
0.35	8.9
0.15	3.8

NOTE: ALL FILLETS AND ROUNDS **.15** R.
MATERIAL: ALUMINUM ALLOY TERNALLOY NO. 6

FIG. 30-40   Problem

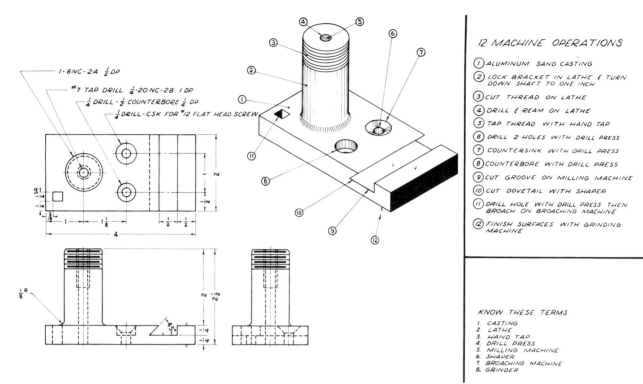

FIG. 30-41 Problem

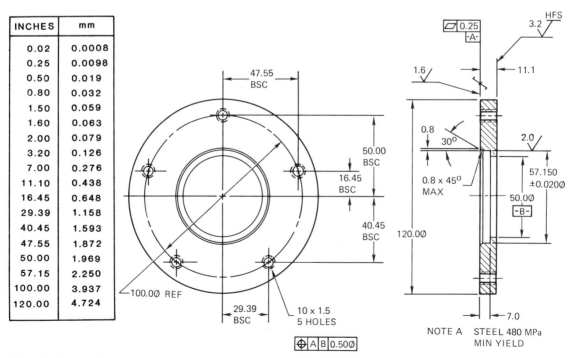

FIG. 30-42 Problem

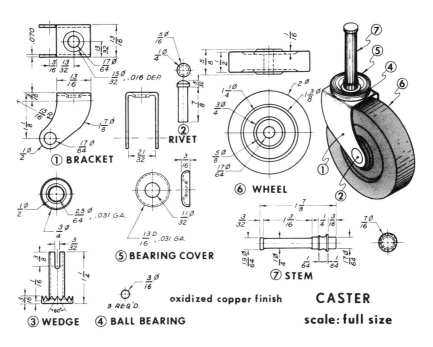

FIG. 30-43 Problem

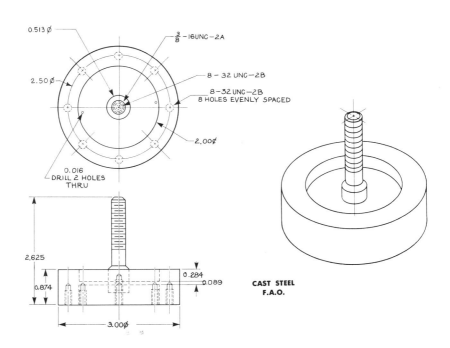

FIG. 30-44 Problem

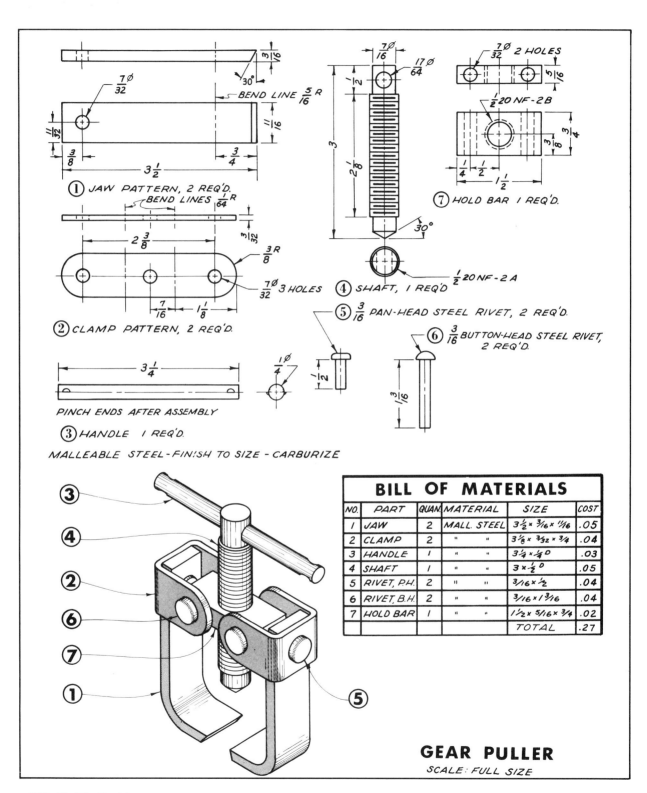

① JAW PATTERN, 2 REQ'D.

② CLAMP PATTERN, 2 REQ'D.

③ HANDLE 1 REQ'D.

MALLEABLE STEEL-FINISH TO SIZE - CARBURIZE

⑦ HOLD BAR 1 REQ'D.

④ SHAFT, 1 REQ'D

⑤ 3/16 PAN-HEAD STEEL RIVET, 2 REQ'D.

⑥ 3/16 BUTTON-HEAD STEEL RIVET, 2 REQ'D.

PINCH ENDS AFTER ASSEMBLY

BEND LINE 5/16 R

BEND LINES 1/64 R

BILL OF MATERIALS				
NO.	PART	QUAN	MATERIAL	SIZE	COST
1	JAW	2	MALL. STEEL	3½ × 3/16 × 11/16	.05
2	CLAMP	2	" "	3⅛ × 3/32 × 3/4	.04
3	HANDLE	1	" "	3¼ × ¼ D	.03
4	SHAFT	1	" "	3 × ½ D	.05
5	RIVET, P.H.	2	" "	3/16 × ½	.04
6	RIVET, B.H.	2	" "	3/16 × 1 3/16	.04
7	HOLD BAR	1	" "	1½ × 5/16 × 3/4	.02
				TOTAL	.27

GEAR PULLER

SCALE: FULL SIZE

FIG. 30-45 Problem

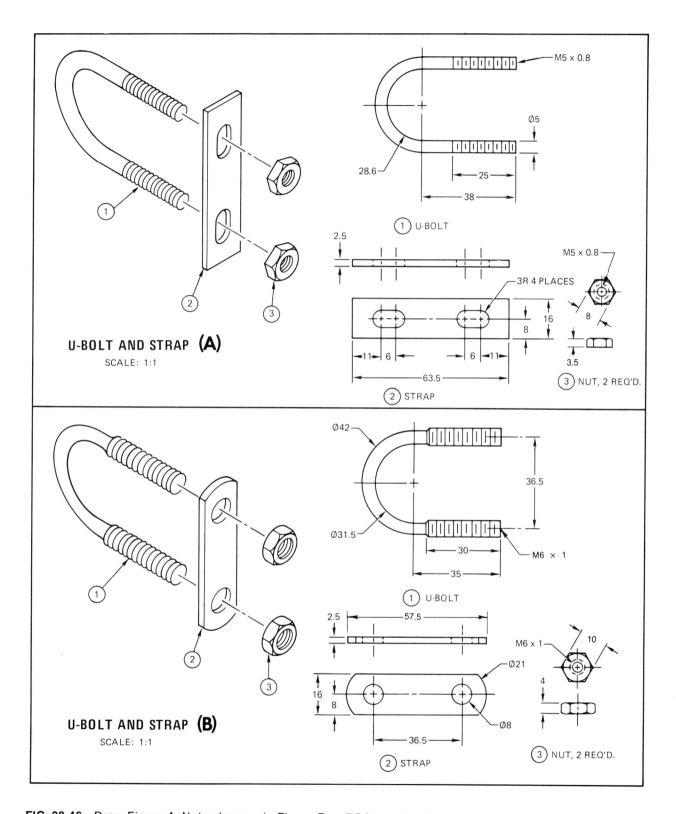

FIG. 30-46 Draw Figure A. Note changes in Figure B as ECOs on the drawing.

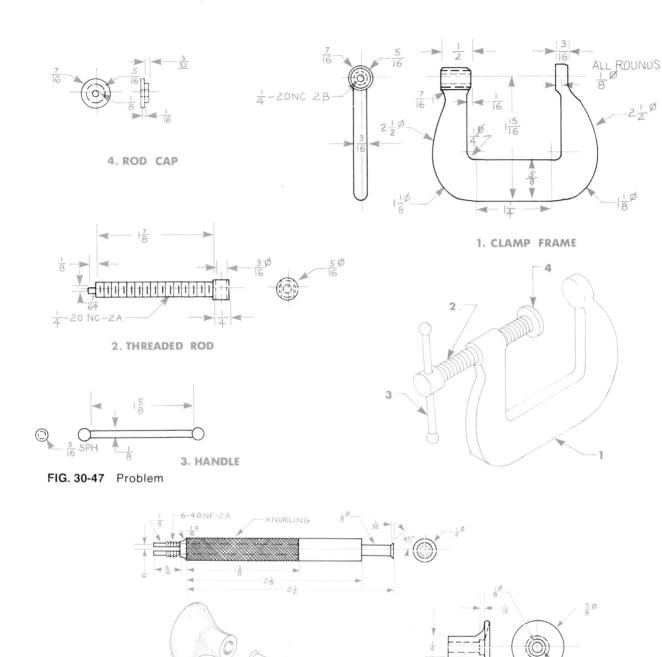

4. ROD CAP

1. CLAMP FRAME

2. THREADED ROD

3. HANDLE

FIG. 30-47   Problem

FIG. 30-48   Problem

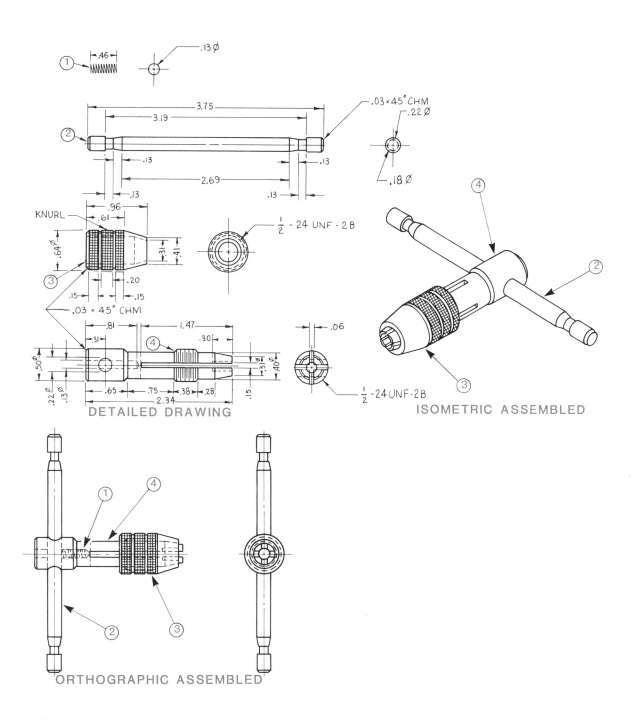

.46

.13 ∅

①

3.75

3.19

②

.13

.13

2.69

.13

.13

.96

KNURL

.61

.64 ∅

③

.31

.41

.20

.15

.15

.03 × 45° CHM

.03 × 45° CHM

.22 ∅

.18 ∅

$\frac{1}{2}$ - 24 UNF - 2 B

④

.81

1.47

.31

④

.30

.50 ∅

.31 ∅

.40

.22 ∅

.13 ∅

.65

.75

.38

.28

.15

2.34

DETAILED DRAWING

.06

$\frac{1}{2}$ - 24 UNF - 2B

ISOMETRIC ASSEMBLED

②

④

③

①

④

②

③

ORTHOGRAPHIC ASSEMBLED

FIG. 30-49 Problem

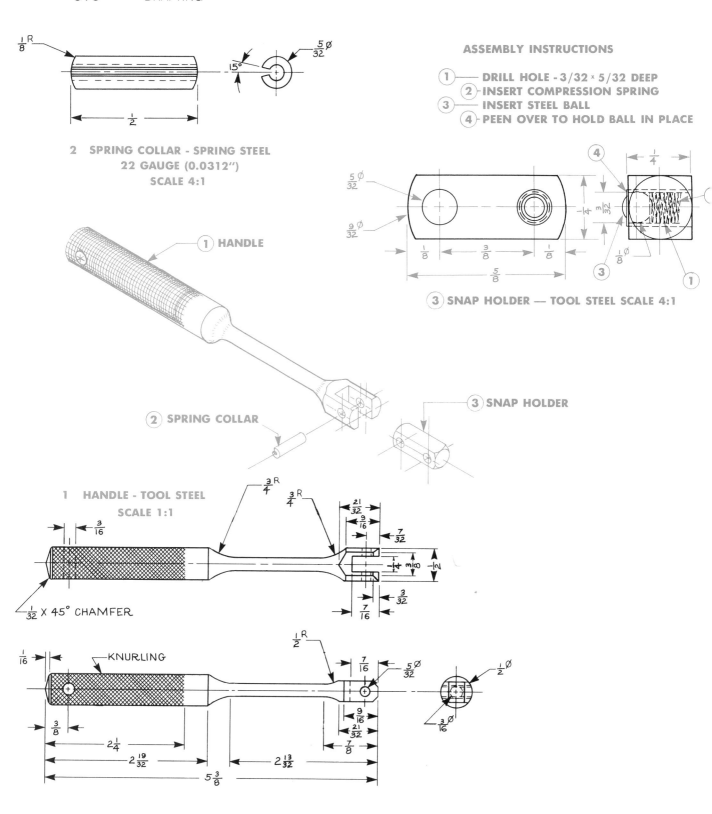

ASSEMBLY INSTRUCTIONS

1 — DRILL HOLE - 3/32 × 5/32 DEEP
2 - INSERT COMPRESSION SPRING
3 - INSERT STEEL BALL
4 - PEEN OVER TO HOLD BALL IN PLACE

2 SPRING COLLAR - SPRING STEEL
22 GAUGE (0.0312")
SCALE 4:1

1 HANDLE

3 SNAP HOLDER — TOOL STEEL SCALE 4:1

2 SPRING COLLAR

3 SNAP HOLDER

1 HANDLE - TOOL STEEL
SCALE 1:1

1/32 X 45° CHAMFER

KNURLING

FIG. 30-50   Problem

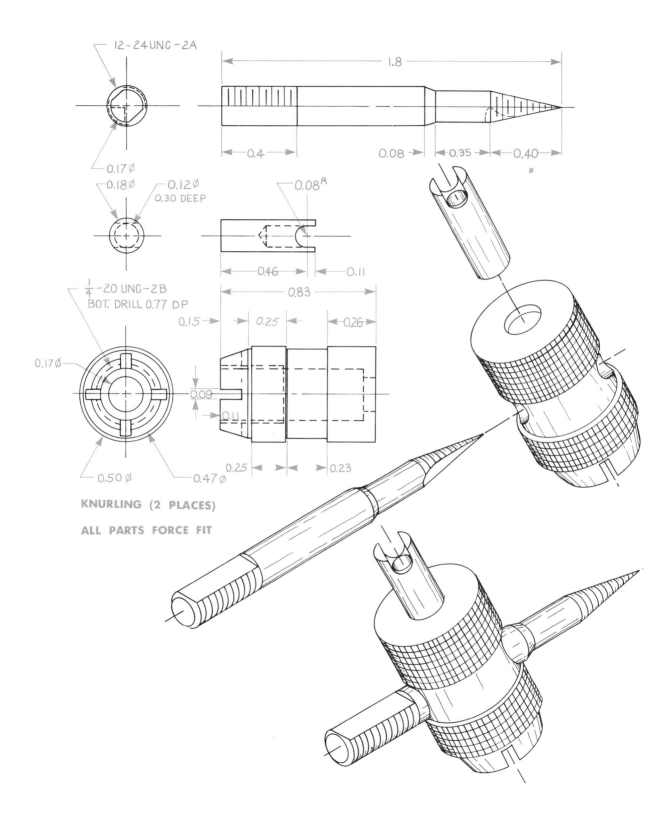

12-24UNC-2A

0.17⌀

0.18⌀

0.12⌀
0.30 DEEP

0.08^R

1.8

0.4   0.08   0.35   0.40

0.46   0.11

0.83

0.15   0.25   0.26

0.09

0.11

0.25   0.23

$\frac{1}{4}$-20 UNC-2B
BOT. DRILL 0.77 DP

0.17⌀

0.50 ⌀   0.47⌀

KNURLING (2 PLACES)

ALL PARTS FORCE FIT

FIG. 30-51 Problem

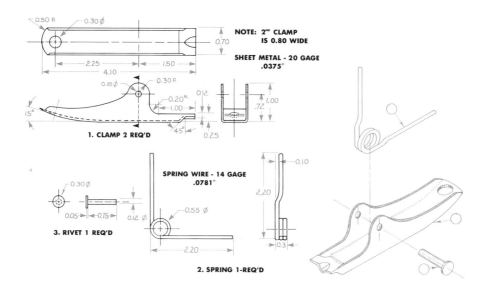

0.50 R   0.30 Ø

NOTE: 2⁰ CLAMP
IS 0.80 WIDE

0.70

2.25   1.50

4.10

SHEET METAL - 20 GAGE
.0375"

0.15 Ø   0.30 R

0.20 R   0.12

15°   1.00   1.00   .72

45°   0.25

1. CLAMP 2 REQ'D

SPRING WIRE - 14 GAGE
.0781"   0.10

0.30 Ø   2.20

0.05   0.75   0.12   0.55 Ø

3. RIVET 1 REQ'D

2.20   0.3

2. SPRING 1-REQ'D

0.40 HEX

FIG. 30-52 Problem

0.758 Ø

1.125 Ø

0.15 Ø, 3 HOLES EQ SP
0.20 Ø, 0.50 DP

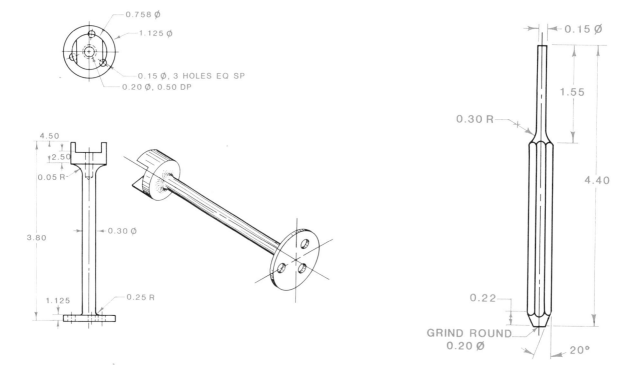

4.50

2.50

0.05 R

3.80   0.30 Ø

1.125   0.25 R

0.15 Ø

1.55

0.30 R

4.40

0.22

GRIND ROUND
0.20 Ø   20°

FIG. 30-53 Problem

FIG. 30-54 Problem

31
Chapter

The Design Process

Rohr Industries, Inc.

INTRODUCTION

This is a different kind of approach to designing. It is an attempt to show that all students have the ability to design and that with a background in drafting, you can put this design into practice.

As you study this chapter, you should develop confidence in your own creativity. You should also carry an idea through from its conception to a finished working drawing.

EUREKA

The problem of determining the amount of gold in the king's crown weighed heavily upon Archimedes as he stepped into his bath. Although bathing may have been a daily ritual, this particular morning when he immersed his body, he associated the rise in his bath water with the principle of specific gravity, a subject with which he was quite familiar. This discovery led to his famous exclamation of ''Eureka!'' (I have found it!). Since that time, ''Eureka'' has been associated with an experience of ecstatic joy and elation at the time of discovery.

With ''discovery,'' a creative act may occur; the experience need not be limited to humans. Take, for example, this experiment performed with a chimpanzee named Sultan.

Beyond some retaining bars in a room, out of arm's reach, lies a banana. In the background of the experiment room is placed a sawed-off castor oil bush from which branches can be easily broken off. It is impossible to squeeze the tree through the railings

FIG. 31-2 The creativity experiment

FIG. 31-3 Creativity and the design process

because of its awkward shape; besides, only one of the bigger apes could drag it as far as the bars. Sultan is led in, does not immediately see the banana, and, looking about him indifferently, sucks one of the branches of the tree. Soon, his attention is drawn to the fruit, and he approaches the bars and glances outside. The next moment he turns around, goes straight to the tree, seizes a thin slender branch, breaks it off with a sharp jerk, turns back to the bars, and attains his objective, the banana (Fig. 31-2).

At the particular moment of his discovery, had Sultan known how to speak, he would most cer-

tainly have shouted, "Eureka!" This chimpanzee used the creative process to design a rake.

THE CREATIVE PROCESS

Creating and designing may be considered part of the same process. Exactly what happens during the creative process is uncertain. However, it may be safe to say it exists at the time of a discovery or the budding of a new idea. The process may not stop there. It can go through various stages of development before it is seen in the form of a new product (Fig. 31-3). The design process may be seen as a

FIG. 31-5 Creativity in humans and in nature is similar (Bruning Div. of AM)

FIG. 31-5 Creativity in humans and in nature is similar (Bruning Div. of AM)

continuation of the creative process as it goes from the birth of an idea to its final stage—a finished product (Fig. 31-4).

The creative act cannot create something out of nothing. It involves uncovering, selecting, reshuffling, combining, and synthesizing already existing facts, ideas, and skills. Creativity in nature and in humans is similar. The analogy of a plant may help describe what happens during the preparation of a design. A plant assimilates certain ingredients from its environment: energy from the sun, various chemicals from the earth, and water and air. The plant "processes" these components by reshuffling, combining, and synthesizing them until it eventually creates a design in the form of a flower or fruit. But the flower or fruit in no way resembles the original components from which the design was created. The plant has a special system for selecting and converting the chemical components, as well as a genetic code to help the design process and to assure the outcome of a successful product (Fig. 31-6).

Humans have no special system or perfected code to guarantee the development of a perfect design. Furthermore, a design for utility or for

FIG. 31-6 A plant has special, built in systems to help the design process (Clark E. Smith)

FIG. 31-7 Design similarity in color T.V. tubes over a twenty year period (RCA)

FIG. 31-8 A creative design for a T.V. antenna (RCA)

aesthetics may or may not resemble the original model from which it was created. The design of a new television picture tube (Fig. 31-7) may have much in common with the original model, but the design of a new television antenna (Fig. 31-8) shows little resemblance to those seen on rooftops.

Creative designing is apparent when we consider the amount of selecting, reshuffling, and synthesizing of existing facts and ideas that had to take place between the original concept of a flying machine and our present day SST (Fig. 31-9). Also consider the transition from the first clumsy small-screen black and white television to today's small portable color model (Fig. 31-10).

The creative act may involve modifying old ideas and giving them a new dimension. Galileo turned telescopic toys, invented by Dutch opticians, to astronomical use. The steam engine, invented as a mechanical toy by Hero of Alexandria in the second century B.C., was put to practical use two thousand years later. Our present day rockets are actually based on the same concepts the Chinese used for gunpowder over four hundred years ago. Compare the original Chinese rockets to our present day models. Note their similarity in appearance. This is a good example of how the function of a design limits change (Fig. 31-11).

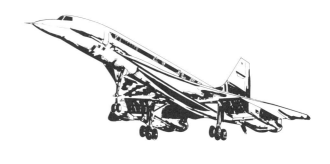

FIG. 31-9 Design progress in flying machines over seventy years

FIG. 31-10   Design progress in T.V. consoles over thirty years (RCA)

FIG. 31-11   Design progress in rockets over four hundred years (Rockwell international)

COMMUNICATING

Although a design or creative thought may be inspired, it will benefit no one else unless the inspiration can be practically communicated. Generally speaking, almost every product can be shown to involve some aspect of design. It is also assumed that the designer was familiar enough with the subject to be able to "communicate" the new idea. The importance of having the methods and tools for communicating an idea may be illustrated best by the following common experience:

One evening, while walking his dog, John Q. started to whistle a melodic tune. It seemed quite original and he felt that with a good arrangement it could become popular. Unfortunately, he didn't know the techniques for composing music so all he could do was to continue whistling. Needless to say, by the next morning he had forgotten the melody and the tune was lost forever.

Although he may have had some knowledge of the symbols involved in music composition, John Q. lacked the method of communicating them. He could not develop his initial inspiration through the design phase of the creative process.

Another example of the creative act will further illustrate the importance of communicating an idea. At the initial stage, an insight might give rise to a statement such as "the speed of light is constant." Anyone could have said that. However, without sufficient knowledge of physics the statement would have been much like the tune in John Q.'s head—uncommunicable. We could not have appreciated the importance of Einstein's creative experience until he could communicate it in the symbolic design $E = mc^2$.

An idea may be such an entirely new concept that every means of communicating (designing) would be helpful. Can you imagine the challenge that faced Felix Wankel when he dreamed about a rotary engine? Since it was so very different from what was known about engines, the end product of his Eureka experience had to be a design that would clearly communicate the engine's potential in order to be a success (Fig. 31-12).

Knowing the mechanics of design is only part of creating an idea for industrial production or scientific research. You should also be thoroughly familiar with the subject involved. Figure 31-13 il-

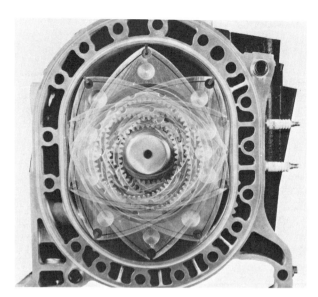

FIG. 31-12   The rotary engine, Wankel (Toyo Kogyo Company, Ltd./Mazda Motors of America.)

lustrates three different types of vehicles for which the designers' knowledge of aerodynamics was essential. The designers, however, had to draw on a reserve of knowledge greater than the basic principles of aerodynamics. All the vehicles function on similar principles, but each design is unique and serves a different purpose.

DEVELOPING CREATIVITY

Many aspects of the creative process cannot be explained. However, experiments have shown one thing for sure: everyone has the ability to be creative. Therefore, we should no longer view "creativity" as a characteristic found only in those so-called "gifted" people.

Just as children grow and develop at varying rates, so also do people develop creative abilities at

FIG. 31-13   Three vastly different flying machines that use the same aerodynamic principles (Hughes Helicopters)

FIG. 31-14   A collage of creativity development (Casey Gorman)

different rates. And because people mature at different rates and at different times, the creative process will have a "timely" nature. For example, you would not be disappointed if, at the age of four, you hadn't "created" in the fashion of a Mozart. In fact, you should never question your ability to be creative. Instead, you should expect, and be prepared for, many Eureka experiences.

"Being prepared" for the design phase of a creative act cannot be overemphasized. But how does one prepare for a process which is a subtle, timely, and uncertain occurrence? You will find that there are many elements of drafting that help develop the ability to design, and they should be practiced at every opportunity. These include the ability to draw, the use of drafting instruments, the knowledge of the language and symbols used to

FIG. 31-15   Practicing the art of design (Bruning Div. of AM)

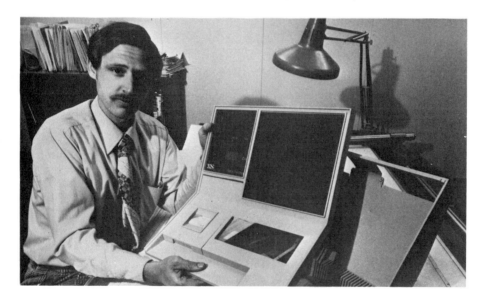

FIG. 31-16   Design of a modern check-out machine (National Semiconductor Corp.)

communicate specific information, and the types of engineering drawings and their uses (Fig. 31-15).

As you process these abilities and experiences you may eventually discover how to apply the basics of balance, ratio, motion, and distinctiveness in order to achieve good design. However, unlike a plant, the creation you design will not be the result of a predetermined, genetically structured code. It will be the result of a human process involving a good deal of study, perseverence and hard work.

Also, unlike a plant, each Eureka experience will result in a new creation (Fig. 31-16).

Some of the ways you can learn the basic concepts and structures of design which relate to your chosen field are sitting in classes, listening to lectures, getting helpful hints from teachers, and reading books. Another method which benefits learning is working closely with someone who has mastered the art of design. It is the human contact that helps to communicate the skills that cannot be

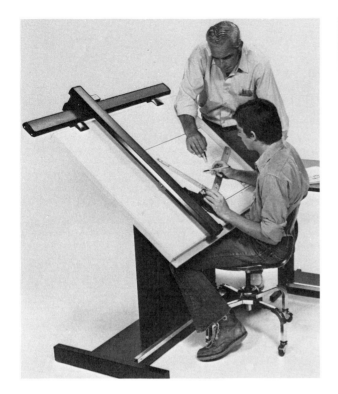

FIG. 31-17   Learning design through personal contact (Bruning Div. of AM)

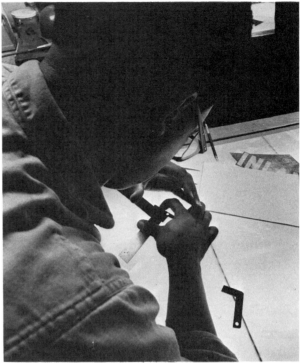

FIG. 31-18   Working at the creative act (Bob McCullough for Inland Steel Co.)

learned by any other means (Fig. 31-17). For example, efforts involving microscopy, chemistry, mathematics, and electronics could not produce a single violin such as the uneducated Stradivarius made more than 200 years ago. To learn by example we must respect an authority. We trust his way of doing things even when we cannot analyze and account in detail for his effectiveness. By watching a master and copying his efforts, an apprentice unconsciously picks up the rules of the art, some of which are not even known to the master himself. To learn from a master does not necessarily mean to imitate him, however. Remember that design refers to newness. With your own unique way of thinking and doing, the outcome of your creative act will be an original product (Fig. 31-18).

Learning to create is not like learning to play an instrument; nor is it like learning to distinguish between pieces of art. You can become a skilled pianist or the greatest art critic in the world, but neither can guarantee a Eureka experience.

How then does one go about preparing for the creative act? Louis Pasteur summed it up in the fol-

lowing way: "Saturate yourself through and through with your subject . . . and wait. Chance only favors invention for minds which are prepared for discovery by study and persevering efforts."

It is also possible to "practice" being creative so that it eventually becomes an art. All you need is the understanding of how a kaleidoscope works and a little imagination.

A kaleidoscope is shaped like a small telescope, with two mirrors set at an angle to the longitudinal axis. At one end there is an eyepiece attached to a tubular frame. At the other end there is a movable unit made up of two transparent discs that are secured to the tube. Between the discs there are a large number of colorful geometric figures which are free to move about as the movable unit is rotated. When looking through the eyepiece and holding the kaleidoscope up to the light we observe the geometric figures form a pattern in the mirrors. Rotating the discs changes the positions of the geometric figures so that one sees a variety of colorful designs. With many different sizes and colors of parts making up the contents of the unit, an unlimited number of structures may be observed.

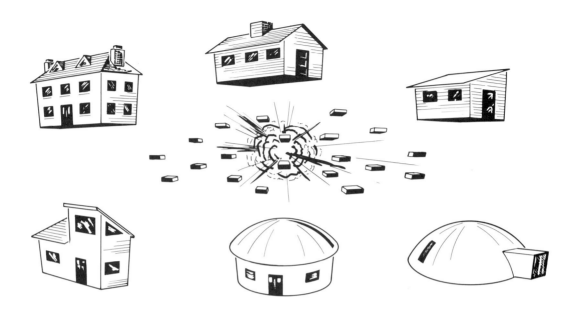

FIG. 31-19 Creating structures from a kaleidoscope of materials

Imagine your mind as a kaleidoscope. Look into it and envision a structure. Think of the components as bricks arranged in any fashion you like but without mortar to make them permanent. Next, imagine turning your mental kaleidoscope so that the bricks fall and tumble randomly. Now stop and envision a new structure. You can continue this activity as long as you like and the variety of structures (designs) that could result are limited only by your imagination (Fig. 31-19).

Of course, presuming that you have acquired some understanding of "good" design, you would probably quickly turn the kaleidoscope if you came up with the structure of a house that had no roof (Fig. 31-20).

Since you are not limited by an "enclosure" you can change the components of your kaleidoscope at will. Let's do so by examining the structure of Intelsat IV-A (Fig. 31-21). Since this exercise is for the purpose of practicing the art of creation, we research the areas within which we must work. We must consider a means of activating

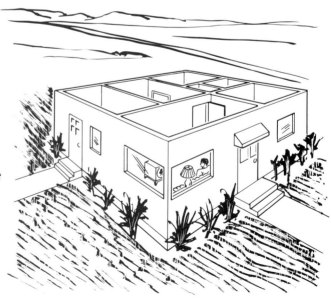

FIG. 31-20 A deficient possible design choice

FIG. 31-21   INTELSAT IV-A, a design problem (Hughes Aircraft Co., Los Angeles, CA)

a unique new antenna once in orbit, protection during blast-off, weight, positioning of electronic equipment, and the effects of space on the satellite. Assuming we are thoroughly familiar with our subject we can disassemble the structure into its basic parts. At this point we are at the "design phase" of the creative process. We can practice the art of creation by turning our mental kaleidoscope as frequently as we desire. Each time we stop we can envision a new structural design, recognizing the shortcomings of some designs and the merits of others. Hopefully, we will let the component parts and

their functions determine the nature of the final working design.

Practicing the art of design does not guarantee a creative act. It does, however, ensure that you will become proficient and comfortable at the "design phase" of the creative process. Given an idea, you will be able to skillfully design a successful product. If you were asked to design a hand-held measuring device you would know how to research the subject and come up with an appropriate design. (Would it look like the one in Figure 31-22? This is an excellent example of a structure which

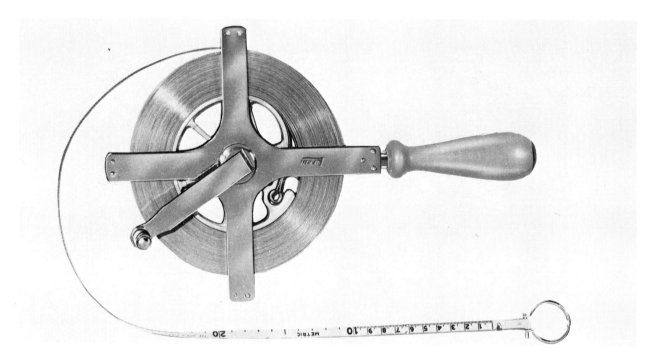

FIG. 31-22   A metric tape measure (The Cooper Group for Lufkin)

can be disassembled into component parts for your mental kaleidoscope.)

Practicing the art of design can be both exciting and rewarding. But the greatest benefits and satisfaction will be derived every time you turn your Eureka experience into a successful design.

PROBLEMS

Using the design steps in Chapter 31 and your creativity, design and communicate the following problems with drawings:

1. A canvas tent stake that will not pull out of sandy soil in a strong wind.

2. An ash tray that will not allow a cigarette to fall out.

3. A Christmas card for a draftsperson or designer.

4. A wall candle holder that will prevent the wall from getting burned.

5. A mailbox that signals there is mail in it.

6. A bathroom cabinet lock that a small child cannot open.

7. An apparatus that can be installed in a standard automobile to permit a legless person to drive.

8. A hand toothbrush with automatically fed toothpaste.

9. An interesting design for automobile hubcaps.

10. An automobile steering wheel with gauges and indicators in the wheel design.

11. A seat for a motorcycle.

12. A container to hold drafting instruments on the drafting table.

13. An adjustable lift (0″-2″) for the legs of a table.

14. A transistor radio that can be carried like a wallet.

Architectural Drafting

INTRODUCTION

This chapter will describe most of the design, construction, and drafting techniques needed to design and draw a family dwelling. Typical examples will

be shown. For further study, an architectural reference should be consulted.

A home may be designed and drawn by an architect, designer, or draftsperson. With experience and a sound background in architectural planning, almost anyone is capable of producing a set of architectural plans.

As you study this chapter, you should learn:

1. Dimensioning of architectural drawings.
2. How to design a family dwelling and orient that design to a lot.
3. The design functions of each room.
4. Architectural lettering.
5. The fundamentals of constructing a family dwelling.
6. How to draw a set of architectural working drawings.

ARCHITECTURAL DIMENSIONS

Without an accurate set of drawings, it would be impossible for the builder to build the structure. To put the design on paper accurately requires skill and knowledge in using an architect's scale. Before designing or drawing a structure, the dimensions of the structure must be understood.

The different values on the architect's scale (see Chapter 7, "Drafting Scales") are all read in the same manner. The feet are read forwards from the zero, and the inches are read backwards from the zero (Fig. 32-2). The number of divisions for feet and inches will vary with the size of the scale. Regardless of its size, however, each scale is read in the same manner.

The size of the drawings will be determined by the scale that is selected (Fig. 32-3). With some practice, one becomes used to choosing a scale by noting the overall dimensions of the architectural structure to be drawn. Then the final drawing will fit the sheet format. Once the drawings are accurate,

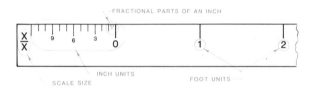

FIG. 32-2 Parts of an architectural scale

the dimensions are added so the builder will have no problems in reading them (Fig. 32-4). Examples of various architectural styles of dimension lines are shown in Figure 32-5.

Architectural drafting does not have standardized conventions as does engineering drawing. Each architect, designer, or architectural draftsperson will use drawing conventions to fit individual needs.

DESIGNING A FAMILY DWELLING

A good plan will not evolve unless the designer has an adequate knowledge and background in architectural design. The designer must suit the style and function of each room to serve the needs of the inhabitants with the greatest efficiency. The designer must plan, think, sketch, and resketch until the design goals are reached. A practical, well designed plan will be functional, safe, and attractive. Good design is sensible and does not have a lot of gimmicks. The building materials can be wood, metal, glass, stone, tile, plastics, or fabrics. These materials can be used in familiar or newly developed forms. An open mind will help create new and exciting designs.

Typical Scales for Architectural Drafting			
Type of Drawing	Architectural Scale	Engineering Scale	Metric Scale
Large plot plan	Not used	1:200	1" = 20'
Small plot plan	⅛" = 1'-0"	1:100	1" = 10'
Floor plan	¼" = 1'-0"	1:50	Not used
Detailed drawings	½" = 1'-0" ¾" = 1'-0" 1" = 1'-0"	1:20 1:10 1:5	Not used

FIG. 32-3 Scales and reductions for architectural drawings

The inhabitants of a home have the right to select their living environment. The selection of the style depends on the architectural tastes of the owners and the particular location of the home. Climate and economics also have a bearing on the development of a home. After the owners decide whether the home will have a formal plan, one with balanced design, dining room, and closed-off rooms (Fig. 32-6), or an informal plan with a rambling design, family room, and open rooms (Fig. 32-7), the designer is ready to begin planning the individual rooms.

An example of a two-story vacation home is shown in Figure 32-8. You should be familiar with the many different styles and their variations.

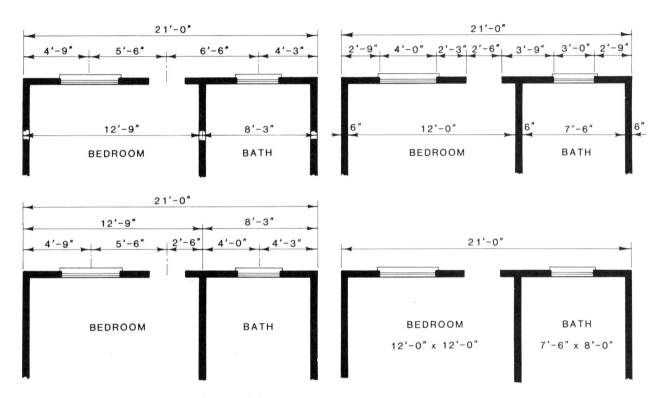

FIG. 32-4 Architectural dimensioning styles.

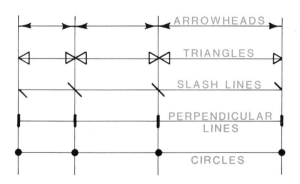

FIG. 32-5 Arrowheads and alternative forms of markings for dimensional lines

DESIGN #1833 © HOME PLANNERS, INC., DETROIT

FIG. 32-6 A formal plan will have a symmetrical design

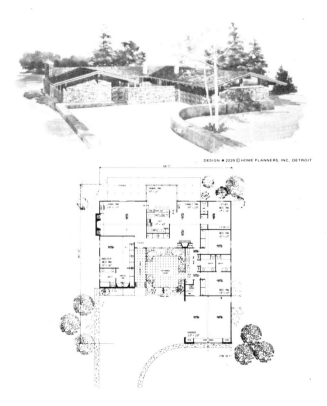

DESIGN # 2229 © HOME PLANNERS, INC., DETROIT

ROOM DESIGN

An efficient method of room design is to begin by making a list of all the activities that will take place in the room, and the equipment, furniture, and storage space that will be required (Fig. 32-9). Next, cut out scale templates of all the objects, and arrange them in the best order. Be sure to allow space for circulation and movement. Draw in the walls after you have found the most efficient arrangement. When all the rooms are completed, they must be fitted into a floor plan. Figure 32-10 shows the developmental steps and average room sizes.

FIG. 32-7 An informal plan will have a rambling design

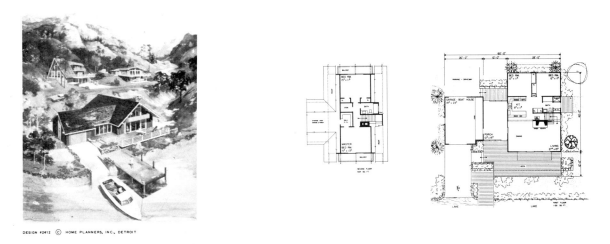

DESIGN #2412 © HOME PLANNERS, INC., DETROIT

FIG. 32-8   Two-story vacation home

FIG. 32-8   Two floor plans (Home Planners Inc.)

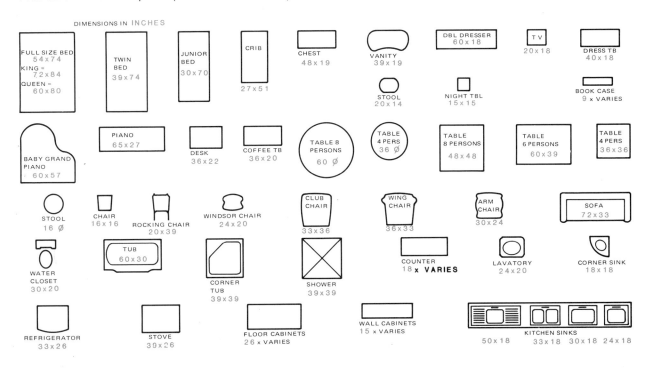

DIMENSIONS IN INCHES

FULL SIZE BED
54 x 74
KING =
7.2 x 84
QUEEN =
60 x 80

TWIN
BED
39 x 74

JUNIOR
BED
30 x 70

CRIB
27 x 51

CHEST
48 x 19

VANITY
39 x 19

DBL DRESSER
60 x 18

T V
20 x 18

DRESS TB
40 x 18

STOOL
20 x 14

NIGHT TBL
15 x 15

BOOK CASE
9 x VARIES

BABY GRAND
PIANO
60 x 57

PIANO
65 x 27

DESK
36 x 22

COFFEE TB
36 x 20

TABLE 8
PERSONS
60 Ø

TABLE
4 PERS
36 Ø

TABLE
8 PERSONS
48 x 48

TABLE
6 PERSONS
60 x 39

TABLE
4 PERS
36 x 36

STOOL
16 Ø

CHAIR
16 x 16

ROCKING CHAIR
20 x 39

WINDSOR CHAIR
24 x 20

CLUB
CHAIR
33 x 36

WING
CHAIR
36 x 33

ARM
CHAIR
30 x 24

SOFA
72 x 33

WATER
CLOSET
30 x 20

TUB
60 x 30

CORNER
TUB
39 x 39

SHOWER
39 x 39

COUNTER
18 x VARIES

LAVATORY
24 x 20

CORNER SINK
18 x 18

REFRIGERATOR
33 x 26

STOVE
39 x 26

FLOOR CABINETS
26 x VARIES

WALL CABINETS
15 x VARIES

KITCHEN SINKS
50 x 18   33 x 18   30 x 18   24 x 18

FIG. 32-9   Typical furniture and fixture dimensions.
Templates should be the same scale as
the floor plan.

1. MAKE A LIST OF ACTIVITIES, EQUIPMENT, AND FURNITURE.
2. DRAW TEMPLATES TO SCALE AND CUT OUT   SCALE: 1/4" 1'-0"
3. PLACE TEMPLATES.

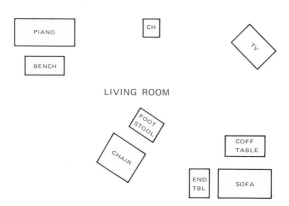

4. DRAW WALLS AROUND THE AREA.   23'-0"x16'-0"

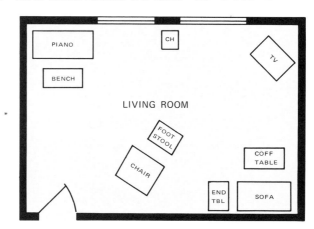

5. DESIGN ALL ROOMS USING STEPS 1 THROUGH 4

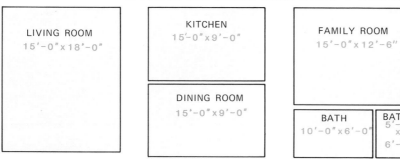

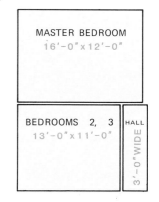

AVERAGE ROOM SIZES FOR A 3 BEDROOM HOME.

FIG. 32-10 Steps in planning a living room using furniture templates. The templates simplify the floor plan design process.

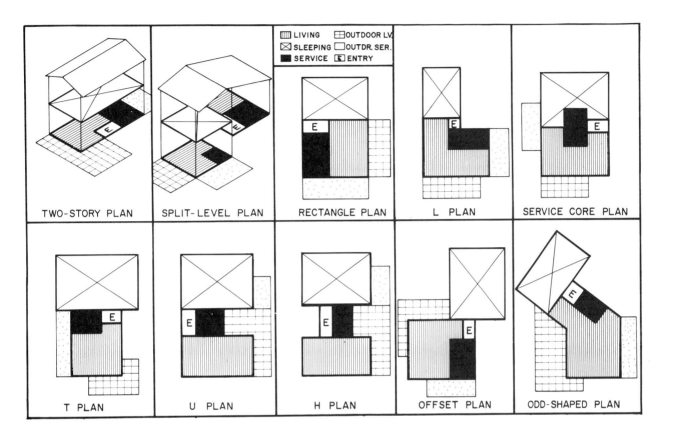

LIVING | **OUTDOOR LV.**
SLEEPING | **OUTDR. SER.**
SERVICE | **E ENTRY**

TWO-STORY PLAN   SPLIT-LEVEL PLAN   RECTANGLE PLAN   L PLAN   SERVICE CORE PLAN

T PLAN   U PLAN   H PLAN   OFFSET PLAN   ODD-SHAPED PLAN

FIG. 32-11   Possible entry locations

The basic areas of a family dwelling are:

1.   The entry area (close to the front door)
2.   The living area (living room, dining room, family room, den)
3.   The service area (kitchen, utility room, garage, workshop)
4.   The sleeping area (bedrooms, bathrooms)

ENTRY AREA

The best arrangement is one in which the owner can go directly from the front entry into any of the other areas (Fig. 32-11). The floor plan of a small home will call for an entry into one side of the living room. When space allows it, a separate area should be planned for the entry.

As noted in Figure 32-11, good traffic circulation in a home depends on the location of the entry area. The entrance should be recognized easily from the outside of the home and protected from the weather. Inside, the entry should have a closet and easy access to the living, sleeping, and service areas (Fig. 32-12).

LIVING ROOM

The living room is the center of the living area, where conversation, viewing television, listening to music, watching the fireplace, reading, and entertaining are enjoyed (Fig. 32-13). The living room size can be developed as shown in Figure 32-10.

FIG. 32-12 Proper utilization of the entry (Georgia-Pacific Corp.)

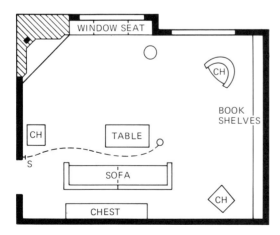

FIG. 32-13 The living room (Georgia-Pacific Corp.)

DINING ROOM

The dining area can be a separate formal dining room (Fig. 32-14), an alcove, or a table in the family room or kitchen (Fig. 32-15). It should be located as close to the kitchen appliances as possible. The following information should be considered when planning the dining area: the type of dining area wanted, the number of people to be seated, and the type of furniture to be used. It is important to allow space for circulation and chair movement (Fig. 32-16).

FAMILY ROOM

The family room is an informal living room, designed for many activities without damaging expensive furniture. Some activities for the family room are playing games, viewing television, listening to music, conversation, hobbies, and fireplace watching. The size of the family room will depend on the number of people in the family and their activities (Fig. 32-17).

FIG. 32-14 Formal dining room design

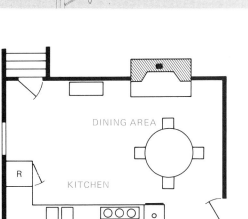

FIG. 32-15 Informal dining room area (Georgia-Pacific Corp.)

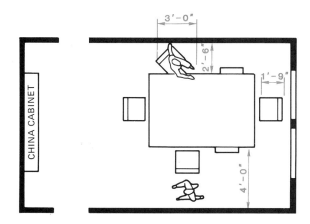

FIG. 32-16 Space allowed for circulation and chair movement

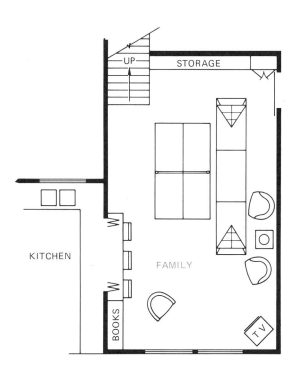

FIG. 32-17 The family room (Georgia-Pacific Corp.)

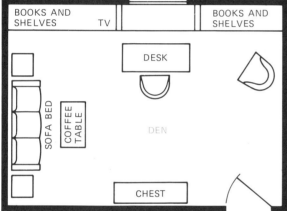

FIG. 32-18 The den (Georgia-Pacific Corp.)

DINING·COOKING·PREPARATION·STORAGE·ENTRY

FIG. 32-19 ''Production line'' for the efficient kitchen

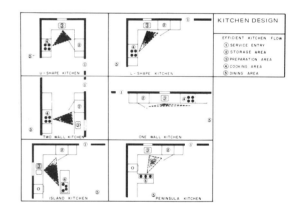

FIG. 32-20 Six basic kitchen designs

DEN

Although it is usually considered as a part of the living area, the den may be designed for any area of the house. And it can serve many purposes: a reading room, an office, a hobby area, a television viewing area, or a guest room. The size of the den will depend on the functions it is to serve (Fig. 32-18).

KITCHEN

The kitchen must be efficiently planned. Placement of the refrigerator/freezer, sink, and cooking appliances in relation to the service entry and dining area is key to efficient kitchen design. The first stop made by anyone entering the service door with a load of groceries should be at the refrigerator and

food storage areas (note that the refrigerator door should open toward the sink). The second stop is at the sink or counter where the food is prepared. The third and final stop is at the cooking appliances, which should be located as near the dining area as possible (Fig. 32-19).

Points placed near the refrigerator, sink, and stove form what is referred to as the ''work triangle.'' If the perimeter of this ''work triangle'' is between 12 and 21 feet, the size of the kitchen work area is usually well planned.

Figure 32-20 illustrates six basic kitchen designs, and each has many variations that will allow for proper circulation while maintaining an efficient work triangle design. Figure 32-21 gives the average dimensions of floor and wall cabinets used in the kitchen.

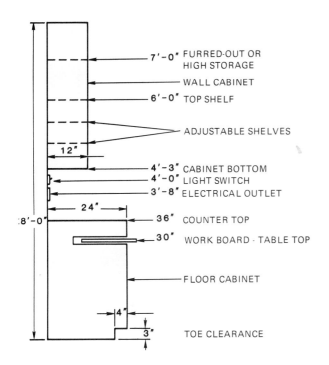

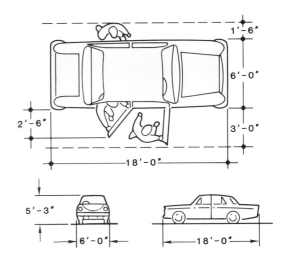

FIG. 32-23 Average automobile sizes

FIG. 32-21 Wall and floor cabinet dimensions

UTILITY ROOM

The utility room is also called the laundry room, service porch, or mud-room. It functions as a place for laundry, storage, ironing, and sewing. The water heater, heating unit, and pantry are often placed in this area. The utility room can be small or large, depending on the services needed (Fig. 32-22).

GARAGE

The garage is primarily used to shelter the family automobile. It is also used for bikes, freezers, laundry, workshop, and storage of garden tools and boats. Plan the garage size carefully, allowing for both automobiles and storage items. Figure 32-23 shows the most important dimensions of a typical automobile.

Figure 32-24 shows typical garage sizes. Exterior design of a garage, whether it is attached or detached, should conform to the design of the house.

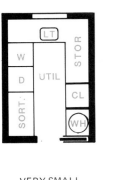

VERY SMALL
7'-0" x 10'-0"

SMALL
9'-0" x 10'-0"

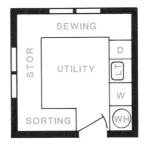

AVERAGE
10'-0" x 10'-0"

LARGE
10'-0" x 12'-0"

FIG. 32-22 Utility room sizes

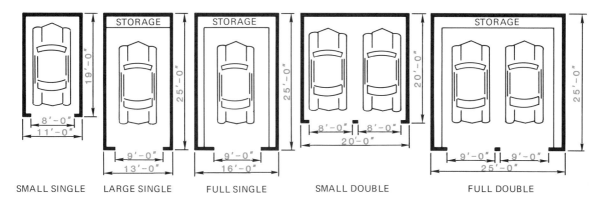

FIG. 32-24   Average garage sizes

SMALL SINGLE   LARGE SINGLE   FULL SINGLE   SMALL DOUBLE   FULL DOUBLE

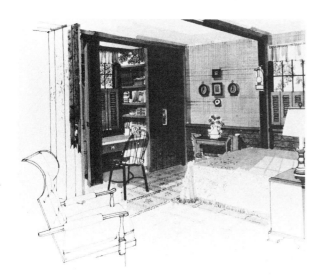

FIG. 32-25   The bedroom (Georgia-Pacific Corp.)

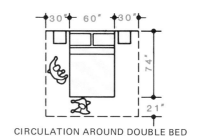

CIRCULATION AROUND DOUBLE BED

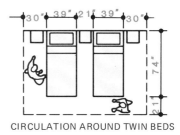

CIRCULATION AROUND TWIN BEDS

FIG. 32-26   Circulation around bedroom furniture

WORKSHOP

A home workshop can be a simple small table in one corner of a room or garage, or it can be a large room complete with power tools and special ventilation systems.

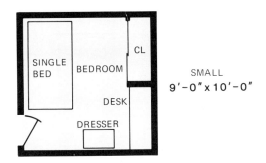

SMALL
9'-0" x 10'-0"

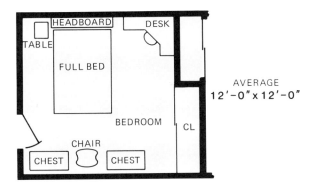

AVERAGE
12'-0" x 12'-0"

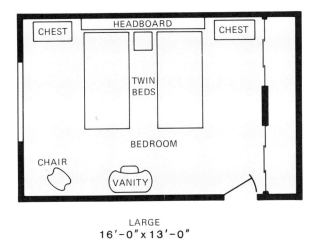

LARGE
16'-0" x 13'-0"

FIG. 32-27 Typical bedroom sizes

BEDROOM

Bedrooms are designed for sleeping, dressing, relaxing, and reading (Fig. 32-25). Bedrooms should be located in the quietest areas of the site. Their size depends on the number of people who will use them and the amount of furniture required (Fig. 32-26). Typical bedroom sizes are shown in Figure 32-27.

The decor of the bedroom should be comfortable and pleasing to the individuals using it (Fig. 32-28). If the privacy of a second bedroom for children is needed, the plan shown in Figure 32-29 would be an excellent compromise.

FIG. 32-28 Bedroom decor (Georgia-Pacific Corp.)

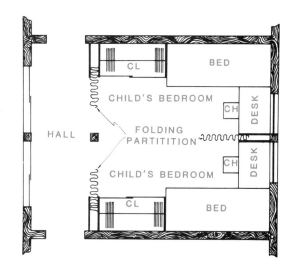

FIG. 32-29 A bedroom for children

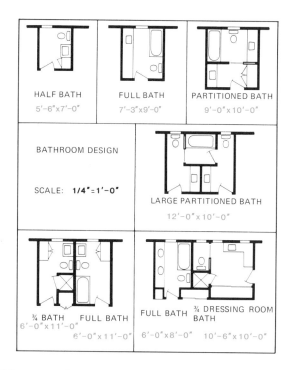

FIG. 32-30 Typical bathroom designs

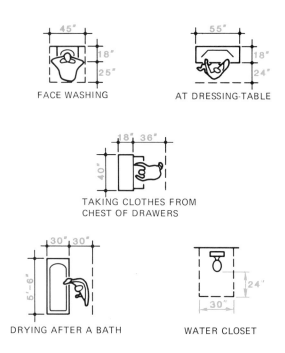

FIG. 32-32 Bathroom and dressing room space requirements

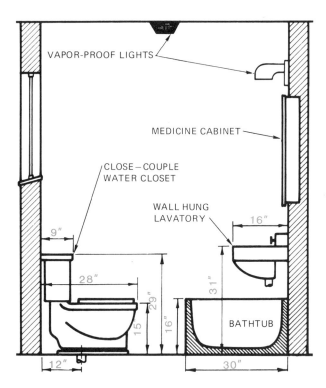

FIG. 32-31 Typical fixture sizes

BATHROOMS

Bathrooms must be accessible from all the bedrooms. With a small building budget, the architect will plan for one bathroom located near the sleeping area. A larger budget will allow the architect to plan a master bathroom off the master bedroom, and an adjacent bathroom (with common wall plumbing) for the other bedrooms and guests. With unlimited funds an architect could plan an entry bath (powder room), a bath by the service entry (mud-room), and a bath in each bedroom.

Figure 32-30 shows some typical bathroom designs. Figure 32-31 shows the typical fixture sizes, and Figure 32-32 shows their space requirements. Consider the following factors when planning a bathroom: style, size, fixture placement, lighting, ventilation, heating, and efficient layout of plumbing lines.

STORAGE

A well-planned home should have a minimum of 7 percent of its floor space for storage (Fig. 32-33). Storage areas consist of walk-in closets, wardrobe

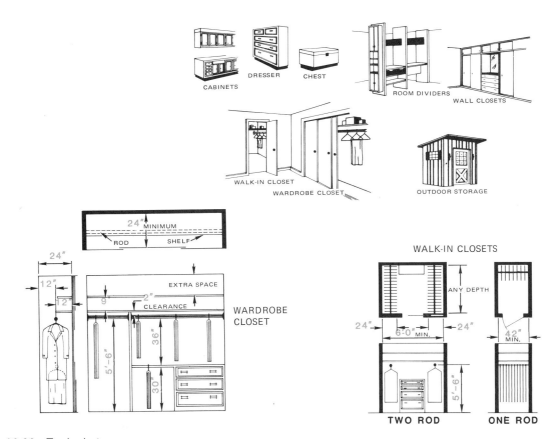

FIG. 32-33 Typical storage areas

closets, wall closets, room dividers, free standing storage, kitchen cabinets, and outdoor storage.

DRAWING THE ARCHITECTURAL PLANS

When all the planning and sketching is finished, a complete set of plans must be drawn. Plans are needed to apply for a building permit, to get bids, and for the actual building of the home.

The remainder of this chapter will show how to prepare a set of architectural working drawings.

ARCHITECTURAL LETTERING

It is important for the draftsperson to accurately communicate the information set forth on a set of architectural drawings. Clear, precise lettering, as shown in Chapter 4, is most important. Many architectural draftspersons prefer to develop their own flowing style of architectural lettering. However, the single stroke Gothic letter style is recommended by most architects. If you do develop your own architectural lettering style, do not make it so elaborate that it will not be readable (Fig. 32-34).

ARCHITECTURAL SYMBOLS

Architectural symbols are the shorthand of architectural drawing. Their simplicity of form reduces drawing time and line work. Typical symbols that are used on architectural floor plans are shown in Figure 32-35. Templates can be used to further reduce drawing time (Fig. 32-36).

Because lettering is very time consuming, an-

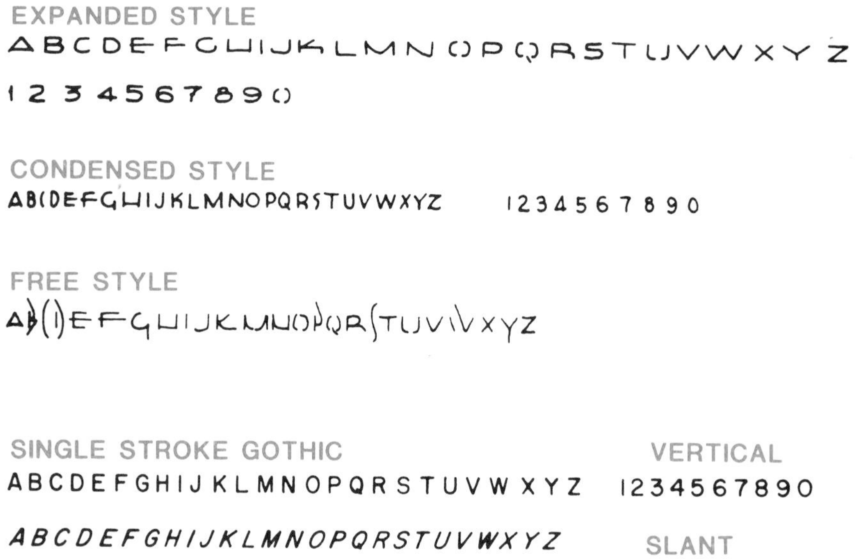

EXPANDED STYLE
A B C D E F G H I J K L M N O P Q R S T U V W X Y Z

1 2 3 4 5 6 7 8 9 ( )

CONDENSED STYLE
A B C D E F G H I J K L M N O P Q R S T U V W X Y Z   1 2 3 4 5 6 7 8 9 0

FREE STYLE
A B C D E F G H I J K L M N O P Q R S T U V W X Y Z

SINGLE STROKE GOTHIC   VERTICAL
A B C D E F G H I J K L M N O P Q R S T U V W X Y Z   1 2 3 4 5 6 7 8 9 0

A B C D E F G H I J K L M N O P Q R S T U V W X Y Z   SLANT

FIG. 32-34 Architectural lettering styles

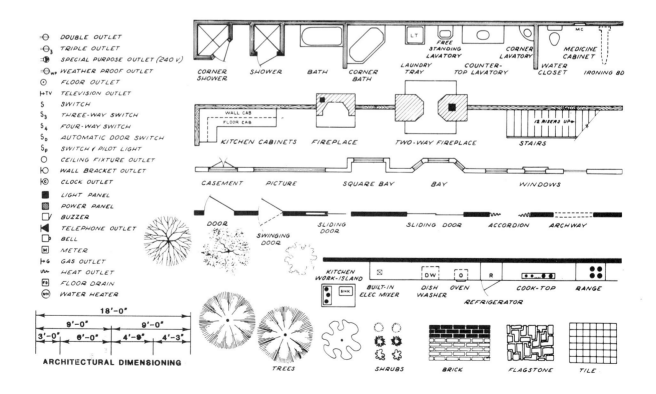

FIG. 32-35 Architectural symbols

other good method of reducing drawing time is to use abbreviations. Architectural abbreviations have not been standardized; however, some of the most typical abbreviations are:

apartment	APT	footing	FTG	riser	RIS
asphalt	ASPH	foundation	FND	roofing	RFG
basement	BSMT	garage	GAR	room	RM
bathroom	B	gas	G	screen	SCR
bathtub	BT	glass	GL	section	SECT
beam	BM	grade	GR	sheathing	SHTHG
bearing	BRG	hall	H	shower	SHWR
bedroom	BR	height	HT	siding	SDG
building	BLDG	I-beam	I	sink	S
brick	BRK	interior	INT	specifications	SPEC
ceiling	CLG	joist	JST	square metre	m²
cleanout	CO	kitchen	K	steel	STL
closet	CL	lath	LTH	switch	SW
concrete	CONC	light	LT	telephone	TEL
counter	CTR	linoleum	LINO	terra cotta	TC
detail	DET	living room	LR	thermostat	T
diameter	DIA	louver	LV	thickness	THK
dining room	DR	maximum	MAX	tread	TR
dishwasher	DW	minimum	MIN	vent	V
door	DR	on center	OC	vertical	VERT
dryer	D	partition	PARTN	vinyl tile	VTILE
entrance	ENT	plaster	PLAS	washing machine	WM
exterior	EXT	plate	PL	water closet	WC
firebrick	FBRK	plumbing	PLMB	water heater	WH
fireplace	FP	recessed	REC	waterproof	WP
flashing	FL	refrigerator	REF	weatherstrip	WS
floor	FLR	reinforced	REINF	wrought iron	WI

Do not use a period after abbreviations unless the abbreviation spells a word. An example is mixture—"MIX."

ORIENTATION

Orientation is the placement of a home, and the arrangement of its rooms, on a building site. While there are many factors involved in the orientation of a home, we will study only the most basic—sun, view, and noise (Fig. 32-37).

The sun is the most important orientation factor. In the Northern Hemisphere, the sun rises in the east and travels across the southern sky, giving those rooms that face east and south exposure to morning sun. The south side of a home in the Northern Hemisphere is exposed to the sun all day long (the north side is never exposed to sunshine). The sun sets in the west, giving those rooms that face west the late afternoon sun. The summer sun is higher in the sky than the winter sun (Fig. 32-38). Therefore the roof overhang on the south side of a home must be planned so the walls will be shaded in the heat of the summer and the sun will be allowed to shine through in the winter.

Sources of noise must be considered in planning the quiet areas of the home like the bedrooms and den, which should be placed as far away from external and internal sources of noise as possible. An experienced architect will use landscaping,

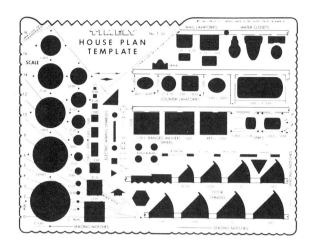

FIG. 32-36 Architectural template (Timely Products Co., Baltimore, OH)

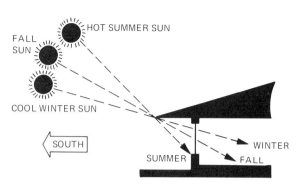

FIG. 32-38 Seasonal sun angles

double pane windows, insulation, draperies, fences, and parts of the structure itself to stop exterior noise. To stop internal noise an architect will use walls, doors, carpeting, soundproofing materials, heavy insulation, draperies, closet walls, and rubber pads under appliances.

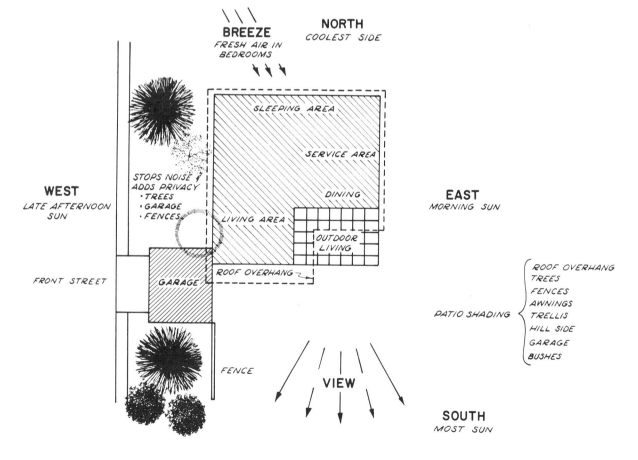

FIG. 32-37 Orientation of a home

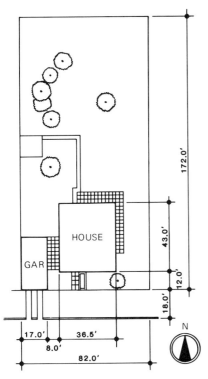

SCALE 1:100

FIG. 32-39 A plot plan with a scale of 1:100

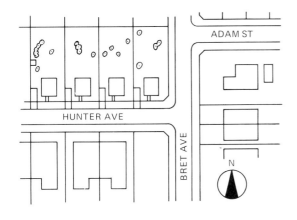

FIG. 32-40 Plat plan

PLOT PLAN

A plot plan is a drawing that shows the location of all the structures on a lot (Fig. 32-39). Another type of plot plan shows an entire block or subdivision

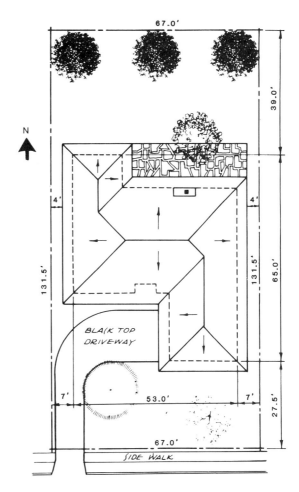

FIG. 32-41 A plot plan with a scale of ⅛″ = 1′-0″

and its streets. This is called a plat plan (Fig. 32-40). A typical plot plan, drawn at a scale of ⅛″ = 1′-0″, is shown in Figure 32-41.

SETBACKS

A setback is the distance from the property line to the structure. This distance is usually controlled by local building codes. As a general rule, half of the area within a property line can be built upon. Figure 32-42 is an example of the step-by-step process of planning, orienting, and drawing a plot plan. Figure 32-43 shows a few examples of different types of setbacks for different types of lots.

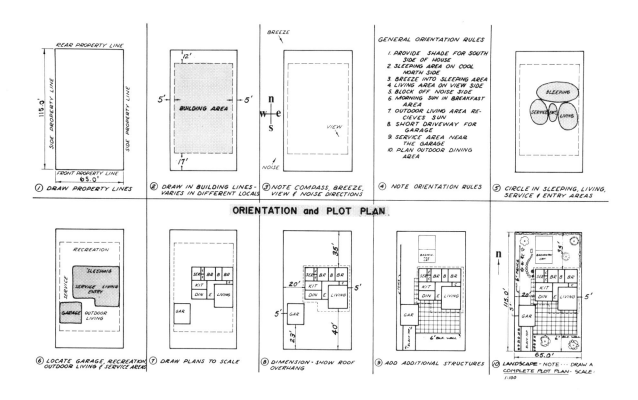

FIG. 32-42   The process of drawing a plot plan

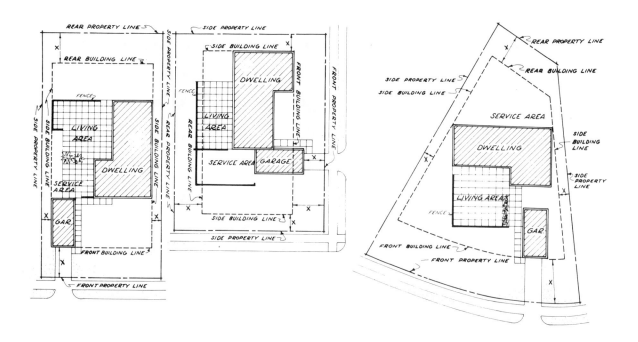

FIG. 32-43   Property lines, setbacks, and buildable area. (Note: X denotes setbacks.)

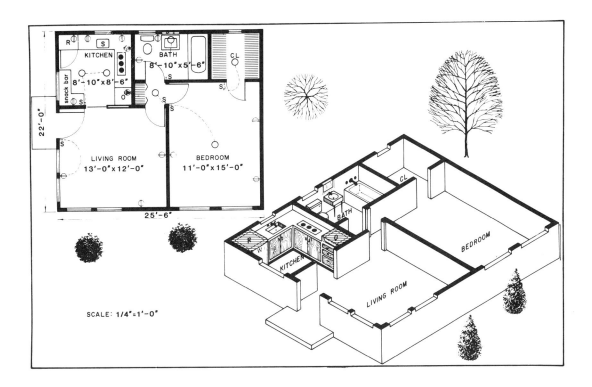

FIG. 32-44 A floor plan and isometric floor plan

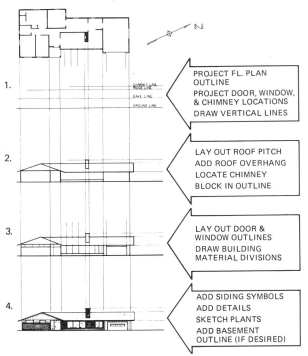

FIG. 32-45 Steps for drawing an elevation

ISOMETRIC FLOOR PLAN

The isometric floor plan is used by people who cannot visualize how a house will look from the standard floor plan. An isometric floor plan should be drawn as shown in Figure 32-44. Refer back to Chapter 11 for a review of isometric drawings.

EXTERIOR ELEVATIONS

Exterior elevations of a house are projected from the floor plan (Fig. 32-45). All sides of a house should be drawn in elevation drawings. Standard dimensions for ceilings, windows, and doors are given in Figure 32-46. Exterior material symbols are shown in Figure 32-47.

INTERIOR ELEVATION

Interior elevations are drawn when it is necessary for the builders to have additional information on items such as cabinets and stonework for interior construction and finish work. Drawing an interior elevation is similar to the process of projecting an

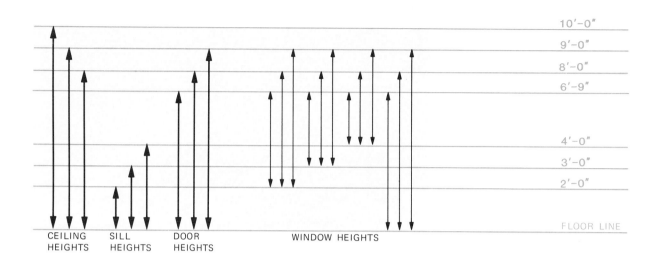

FIG. 32-46 Typical ceiling, door, and window heights

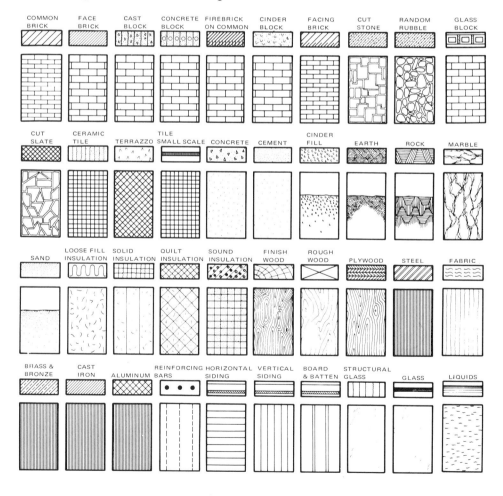

FIG. 32-47 Building materials in elevation and section

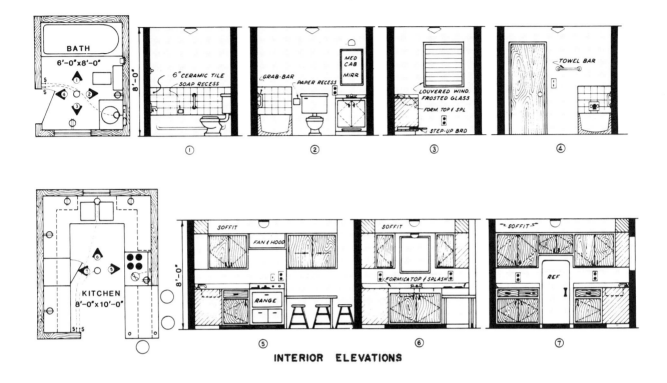

FIG. 32-48 Interior elevations

exterior elevation. Figure 32-48 shows the process used to draw interior elevations.

ROOFS

There are many roofing styles. A good exterior design requires thorough planning of both the style and the pitch (angle) of the roof (Fig. 32-49). The entire roof should be shown on each exterior elevation. A roof plan should also be drawn (Fig. 32-50). The roof is usually drawn in the elevation and projected to the roof plan (Fig. 32-51). The down slope of the roof should be shown with an arrow.

EXTERIOR PERSPECTIVE

Before beginning an architectural exterior perspective drawing, review Chapter 22.

A two-point perspective can be drawn by following the steps in Figure 32-52 and estimating sizes. Architectural perspectives can be drawn in from many different positions relative to the horizon (Fig. 32-53). A perspective drawing and a floor plan will give most people a complete description of a home's design (Fig. 32-54).

Figure 32-55 shows how to draw some of the symbols used to represent the textures of building materials.

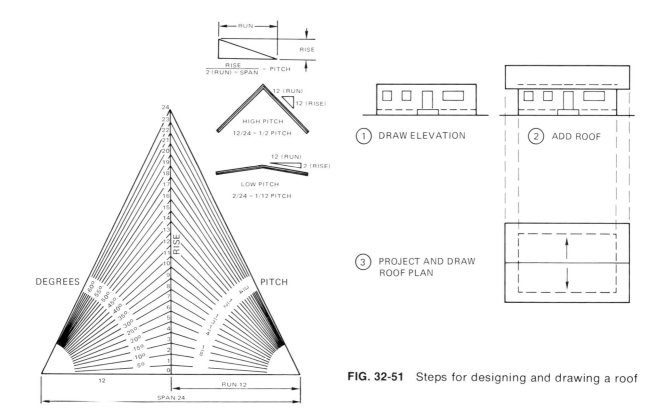

FIG. 32-51   Steps for designing and drawing a roof

FIG. 32-49   Roof pitches

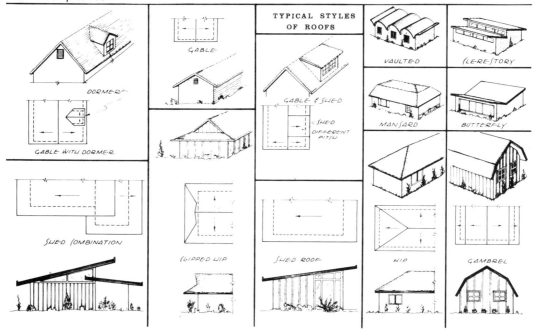

FIG. 32-50   Typical styles of roofs

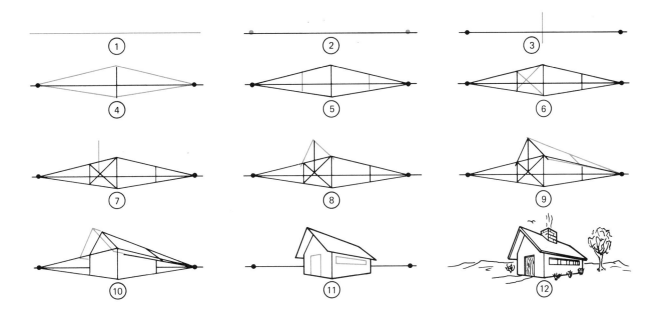

FIG. 32-52 Steps for drawing a two-point perspective exterior

ON HORIZON

ABOVE HORIZON

BELOW HORIZON

FIG. 32-53 Levels of exterior perspective

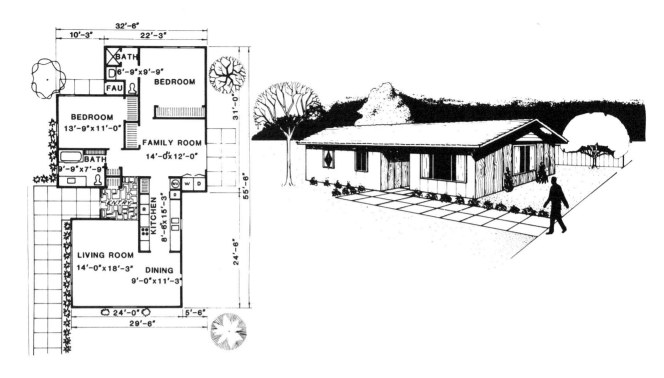

FIG. 32-54 A perspective with floor plan

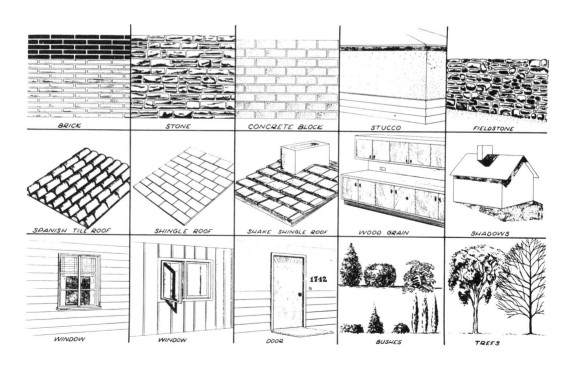

FIG. 32-55 Typical textures for building materials

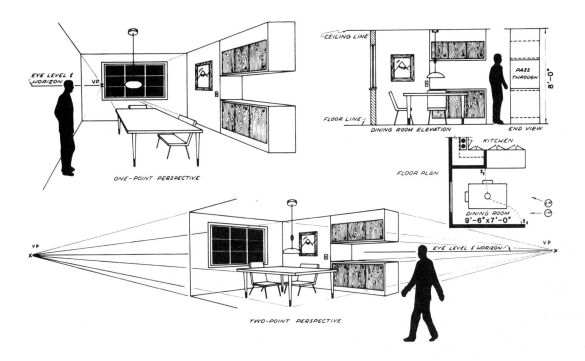

FIG. 32-56   Interior perspectives

INTERIOR PERSPECTIVE

Interiors can be illustrated with either one-point or two-point perspective drawings. Figure 32-56 shows the same dining room plan drawn as an interior elevation, and with one-point and two-point perspectives. The steps for making one-point and two-point interior perspective drawings are shown in Figure 32-57.

CONSTRUCTION DETAILS

Many different methods of construction may be used to build the same structure. Depending on weather, costs of building materials, and the local building codes, building procedures will vary in different areas of the country. This text cannot cover all of the construction details an architect must know to complete the plans of a typical family dwelling. Additional references should be consulted if you need them. The sizes for building materials are shown in Figure 32-58. These are standard

sizes in today's building industry. The lumber is measured and priced in units of board feet.

SLAB FOUNDATION

The slab foundation is nothing more than a large slab of concrete poured into wood forms to hold it in place on the ground (Fig. 32-59). A waterproof membrane is placed between the ground and the slab to keep moisture out, and a footing is placed under walls. Steel reinforcing bars can be placed in the footing, and welded wire mesh can be placed in the slab for additional strength.

T-FOUNDATION

The T-foundation is an inverted concrete T which is poured around the perimeter of the house (Fig. 32-60). The floor framing is added to the top of the foundation. Engineering tables must be used to determine the span, spacing, and size of the structural

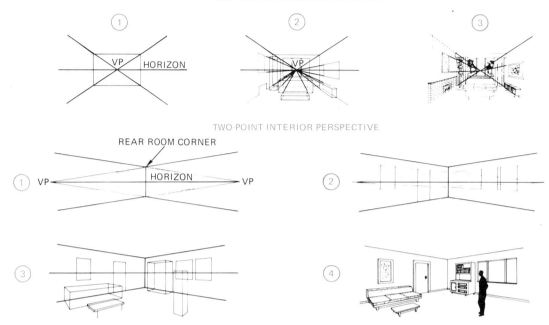

FIG. 32-57 Steps for drawing one- and two-point interior perspectives

Thickness (inches)		Width (inches)	
Nominal Size	**Dressed Size**	**Nominal Size**	**Dressed Size**
1	¾	2	1½
1¼	1	3	2½
1½	1¼	4	3½
2	1½	5	4½
2½	2	6	5½
3	2½	7	6½
3½	3	8	7¼
4	3½	9	8¼
4½	4	10	9¼
5 and larger	½ off	11	10¼
		12	11¼
		14	13¼
		16	15¼

FIG. 32-58a Cross-sectional sizes of constructional lumber

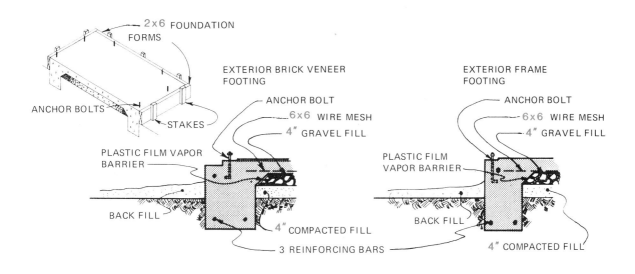

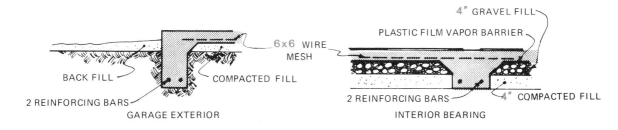

FIG. 32-59 The slab foundation

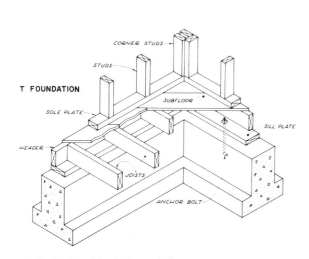

FIG. 32-60 The T foundation

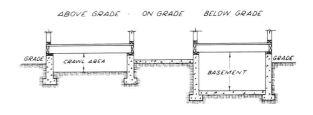

FIG. 32-61 Levels of foundations

members. This information is available from local building departments, licensed architects, or licensed structural engineers. Examples of the T and slab foundation levels are shown in Figure 32-61.

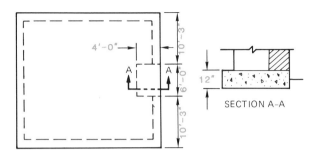

FIG. 32-62 The fireplace footing

FIREPLACES

The information needed to build a fireplace is size and location of footing (Fig. 32-62), size of opening, and interior finish (Fig. 32-63). No further construction details are needed for an experienced builder.

FRAMING

House framing is the same almost everywhere. Unless the local building department requires a framing plan for the building permit, an experienced builder will not need one. If the home is to be built by an inexperienced builder, all framing

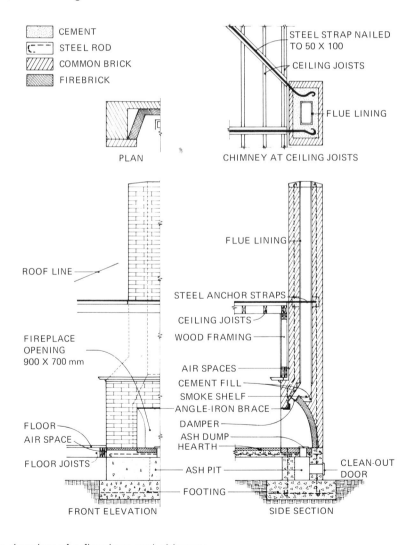

FIG. 32-63 Section drawing of a fireplace and chimney

plans should be drawn to scale to eliminate building errors and help in the purchase of building materials. Typical floor, wall, and roof framing plans are shown in Figures 32-64, 32-65, and 32-66. Structural sizes, spans, and spacing should be obtained from engineering tables or codes. A complete framing plan is shown in the set of plans contained in Figures 32-67a through h.

SPECIFICATIONS

The specifications list is a typed list with descriptions of the construction details, fixtures, appliances, and finish of a construction job. Since the architectural drawings serve as part of the legal

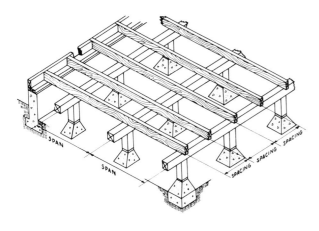

FIG. 32-64 Floor framing

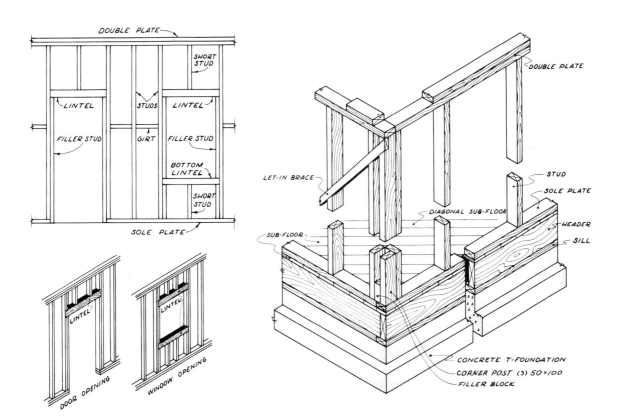

FIG. 32-65 Wall framing

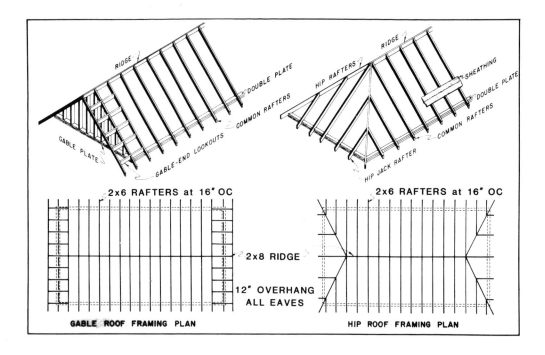

FIG. 32-66   Roof framing

contract between the owner and the contractor, it is necessary for all information pertaining to the building of a structure to be on either the drawings or on the specifications list. The toothbrush holder in the bathroom is a small but significant example. Does the owner want a built-in toothbrush holder installed by the contractor, or will it be installed by the family?

SCHEDULES

All fixtures, appliances, and finishes that are not in the specifications or on the drawings must be placed on a schedule list.

PLUMBING PLANS

An experienced plumber will not need a detailed plumbing plan to install the plumbing in a family dwelling. The plumber only needs to know the types of plumbing fixtures and their locations. A separate, detailed plumbing plan will help an inexperienced person purchase the plumbing lines and fixtures, and aid in their installation. Commercial buildings require a separate plumbing plan.

THE COMPLETE SET OF PLANS

The final design step in designing a home is to draw the complete set of plans. These plans are called the working drawings. The number of working drawings will vary according to the requirements of the local building codes and the contractor. Accuracy speeds construction and eliminates costly errors.

A full set of plans is shown in Figure 32-67a, b, c, d, e, f, g, h. They include the:

Plot plan
Roof plan
Floor plan
Slab or T-foundation and details
Garage plan
Garage foundation and details
Interior elevations
Roof framing plan
Exterior elevations
Window and door schedules
Perspective
Eave detail
Floor framing plan
Exterior wall framing plan
House section

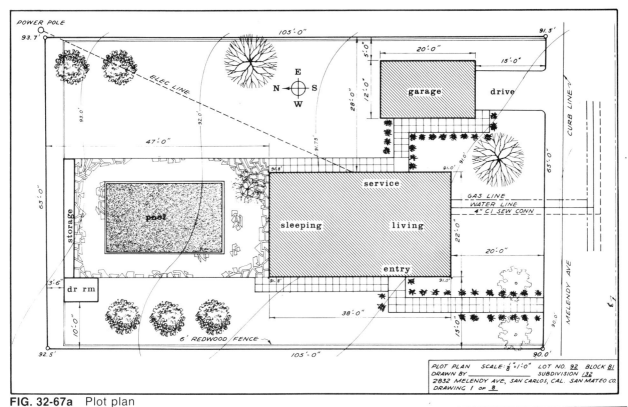

FIG. 32-67a Plot plan

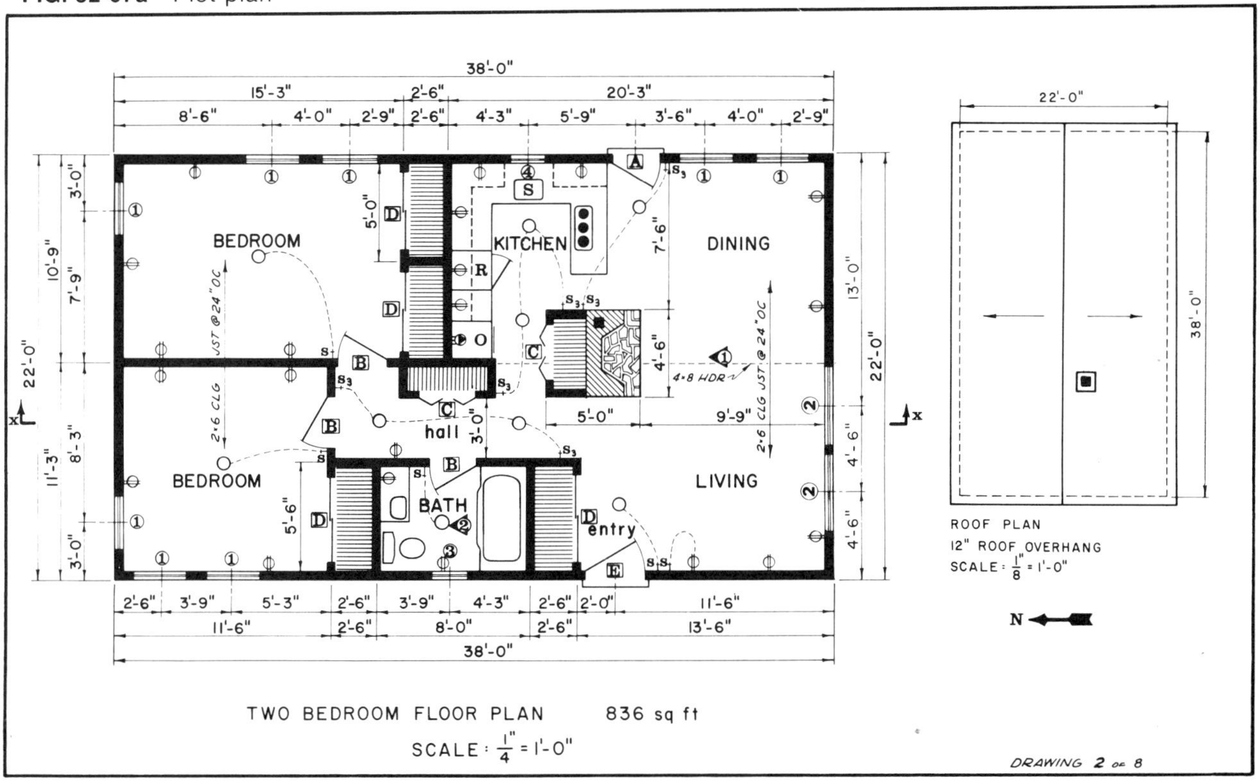

POWER POLE

ELEC LINE

PLOT PLAN   SCALE: $\frac{1}{8}$"=1'-0"   LOT NO. 92   BLOCK 81
DRAWN BY _____   SUBDIVISION 132
2832 MELENDY AVE, SAN CARLOS, CAL. SAN MATEO CO.
DRAWING 1 OF 8

TWO BEDROOM FLOOR PLAN   836 sq ft

SCALE: $\frac{1}{4}$" = 1'-0"

ROOF PLAN
12" ROOF OVERHANG
SCALE: $\frac{1}{8}$" = 1'-0"

DRAWING 2 OF 8

FIG. 32-67b Floor and roof plans

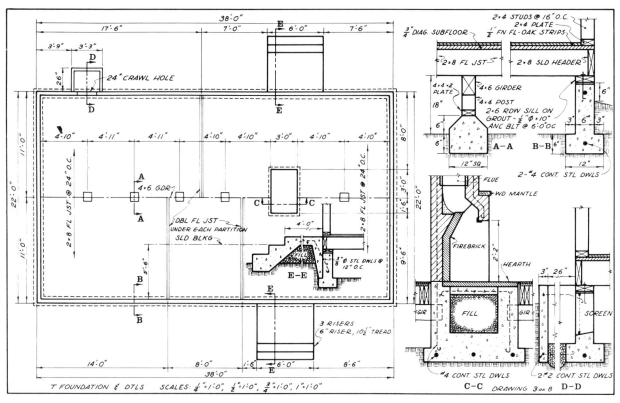

FIG. 32-67c T-foundation plan

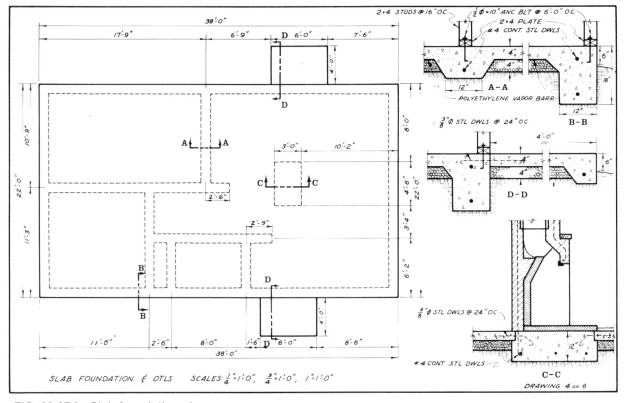

FIG. 32-67d Slab foundation plan

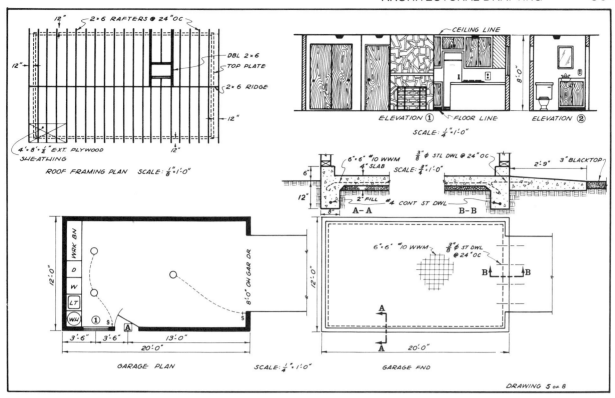

FIG. 32-67e Garage, garage foundation, interior elevation, and roof framing plans

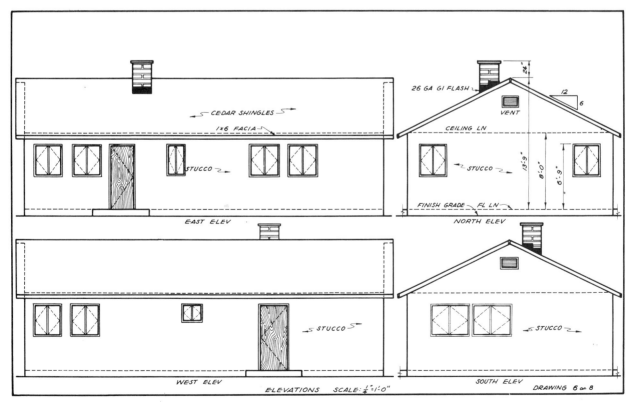

FIG. 32-67f Exterior elevation plan

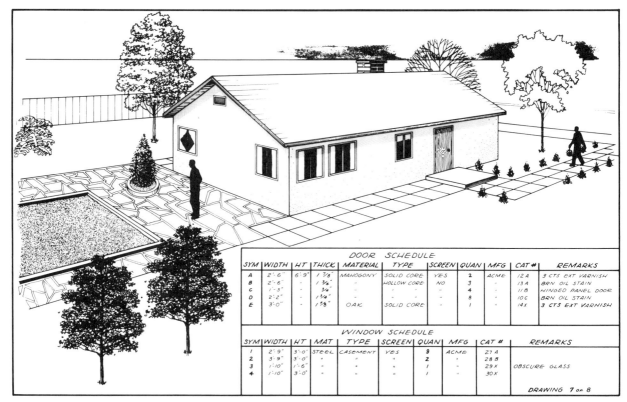

DOOR SCHEDULE									
SYM	WIDTH	HT	THICK	MATERIAL	TYPE	SCREEN	QUAN	MFG	CAT #	REMARKS
A	2'-6"	6'-9"	1 3/8"	MAHOGANY	SOLID CORE	YES	2	ACME	12 A	3 CTS EXT VARNISH
B	2'-6"	"	1 3/8"	"	HOLLOW CORE	NO	3	"	13 A	BRN OIL STAIN
C	1'-5"	"	3/4"	"	"		4	"	11 B	HINGED PANEL DOOR
D	2'-2"	"	1 3/4"	"	"		8	"	10 C	BRN OIL STAIN
E	3'-0"	"	1 3/8"	OAK	SOLID CORE		1	"	14 X	3 CTS EXT VARNISH

WINDOW SCHEDULE									
SYM	WIDTH	HT	MAT	TYPE	SCREEN	QUAN	MFG	CAT #	REMARKS
1	2'-9"	3'-0"	STEEL	CASEMENT	YES	9	ACME	27 A	
2	3'-9"	3'-0"	"	"	"	2	"	28 B	
3	1'-10"	1'-6"	"	"	"	1	"	29 X	OBSCURE GLASS
4	1'-10"	3'-0"	"	"	"	1	"	30 X	

DRAWING 7 OF 8

FIG. 32-67g Window and door schedules and perspective

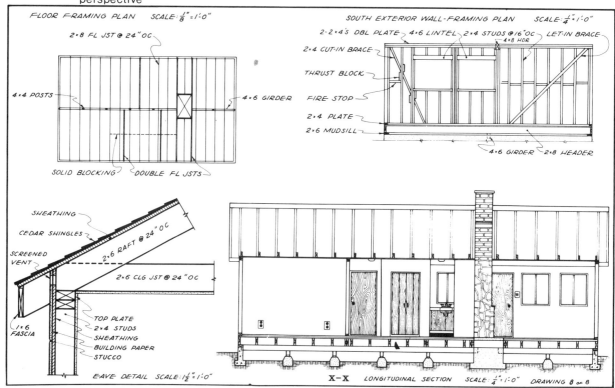

FIG. 32-67h Eave detail, floor framing, and house section plans

PROBLEMS

1. Practice drawing line work and architectural symbols freehand and with drawing instruments.

2. Draw the floor plan in Figure 32-68. Add the electrical symbols, built-in equipment, and furniture.

3. Orient and design a home for the property in Figure 32-69.

4. Orient and design a small cabin for Figure 32-70.

5. Draw the east and west elevations in Figure 32-68.

6. Draw the north and south elevations for Figure 32-68.

7. Redesign the roof style for the house in Figure 32-68.

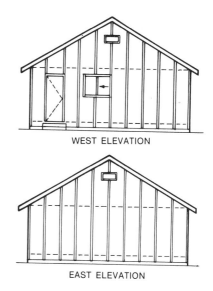

WEST ELEVATION

EAST ELEVATION

ASSIGNMENT:
(1) Complete ALL electrical
(2) Complete ALL built-in equipment
(3) Add special equipment
(4) Draw furniture

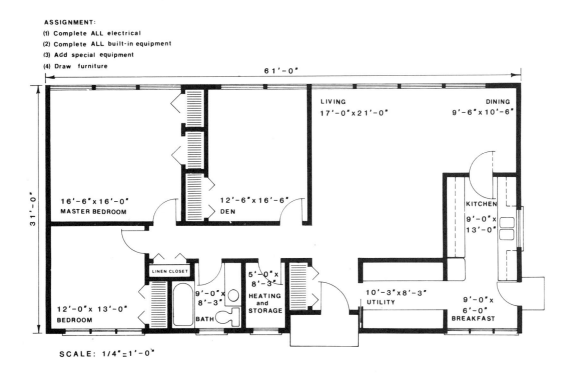

FIG. 32-68 Problem

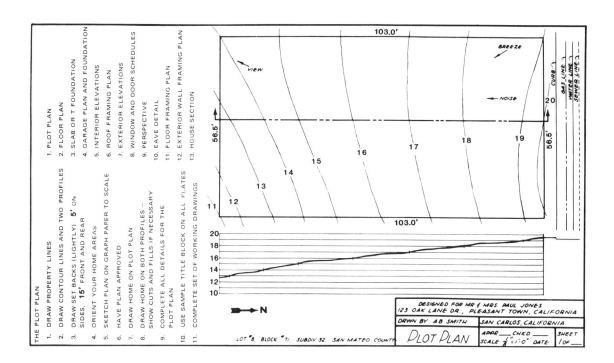

THE PLOT PLAN

1. DRAW PROPERTY LINES
2. DRAW CONTOUR LINES AND TWO PROFILES
3. DRAW SET BACKS (LIGHTLY) **5'** ON SIDES, **15'** FRONT AND REAR
4. ORIENT YOUR HOME AREAS
5. SKETCH PLAN ON GRAPH PAPER TO SCALE
6. HAVE PLAN APPROVED
7. DRAW HOME ON PLOT PLAN
8. DRAW HOME ON BOTH PROFILES – SHOW CUTS AND FILLS IF NECESSARY
9. COMPLETE ALL DETAILS FOR THE PLOT PLAN
10. USE SAMPLE TITLE BLOCK ON ALL PLATES
11. COMPLETE SET OF WORKING DRAWINGS

1. PLOT PLAN
2. FLOOR PLAN
3. SLAB OR T FOUNDATION
4. GARAGE PLAN AND FOUNDATION
5. INTERIOR ELEVATIONS
6. ROOF FRAMING PLAN
7. EXTERIOR ELEVATIONS
8. WINDOW AND DOOR SCHEDULES
9. PERSPECTIVE
10. EAVE DETAIL
11. FLOOR FRAMING PLAN
12. EXTERIOR WALL FRAMING PLAN
13. HOUSE SECTION

DESIGNED FOR MR & MRS PAUL JONES
123 OAK LANE DR., PLEASANT TOWN, CALIFORNIA

DRWN BY A B SMITH	SAN CARLOS, CALIFORNIA	
	PLOT PLAN	
LOT #8, BLOCK #71, SUBDIV 32, SAN MATEO COUNTY	APPR ___ CHKD ___ SCALE ⅛"=1'-0" DATE ___	SHEET 1 OF ___

FIG. 32-69   Problem

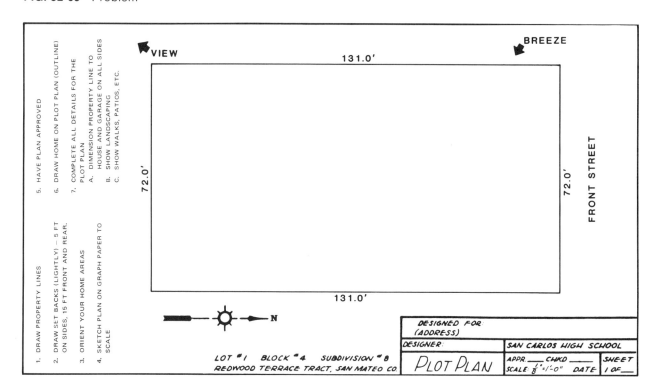

1. DRAW PROPERTY LINES
2. DRAW SET BACKS (LIGHTLY) – 5 FT ON SIDES, 15 FT FRONT AND REAR.
3. ORIENT YOUR HOME AREAS
4. SKETCH PLAN ON GRAPH PAPER TO SCALE
5. HAVE PLAN APPROVED
6. DRAW HOME ON PLOT PLAN (OUTLINE)
7. COMPLETE ALL DETAILS FOR THE PLOT PLAN
   A. DIMENSION PROPERTY LINE TO HOUSE AND GARAGE ON ALL SIDES
   B. SHOW LANDSCAPING
   C. SHOW WALKS, PATIOS, ETC.

DESIGNED FOR: (ADDRESS)		
DESIGNER:	SAN CARLOS HIGH SCHOOL	
LOT #1 BLOCK #4 SUBDIVISION #8 REDWOOD TERRACE TRACT, SAN MATEO CO.	PLOT PLAN	APPR ___ CHKD ___ SCALE ⅛"=1'-0" DATE ___	SHEET 1 OF ___

FIG. 32-70   Problem

Chapter

Computer Graphics

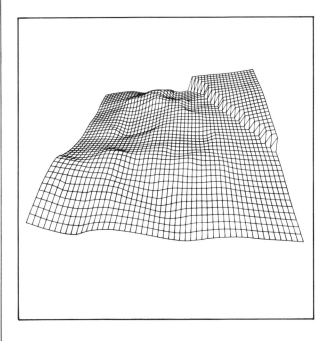

picture turning it around to view it from various positions in order to formulate all of its features. With computer graphics there is no need to do this design process mentally (Fig. 33-2). All the information necessary to produce the series of positions can be fed into the computer, processed, and displayed on the CRT (Cathode Ray Tube). You can then visually study the bracket and at the same time retain in the memory of the computer the capability to return to any particular view or views that require your further attention.

Most modern industries are finding new ways to use the computer. This chapter will help you understand how computer graphics works. It will also help you understand how it is used, and how it will affect you. You should learn the functions of the digitizer, the plotter, the CRT video display, and how information is stored in data banks. You should also learn to graphically compute from a data table on graph paper.

INTRODUCTION

Imagine that you are an industrial designer creating a specialized bracket for a piece of machinery. As you visualize the bracket in your mind, you

THE IMPORTANCE OF COMPUTER GRAPHICS

Working drawings consist of lines, dimensions, and notes giving the nature of the materials, finish, and perhaps processing steps. Designers, engineers, and

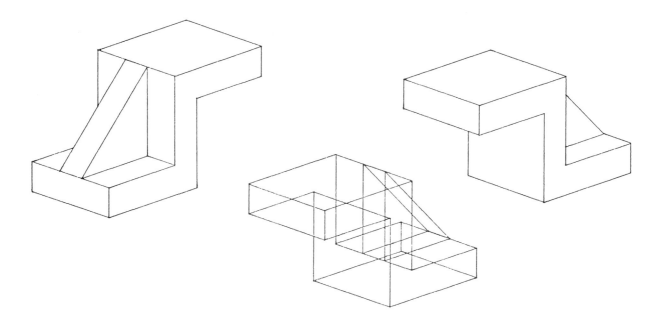

FIG. 33-2 Computer-revolved part drawings

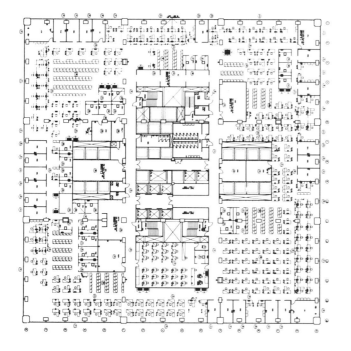

FIG. 33-3 Computer-generated architectural
drawing

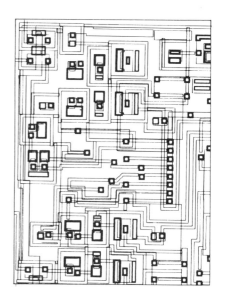

FIG. 33-4 Computer-generated integrated
circuit drawing

draftspersons follow standard procedures for economically putting these ideas on paper.

As our physical world becomes more technical
and sophisticated, structures, tools, and machines

become more complicated. Working drawings for a
110-story building (Fig. 33-3), a computer involving
hundreds of thousands of solid-state components,
an integrated circuit mask calling for thousands of

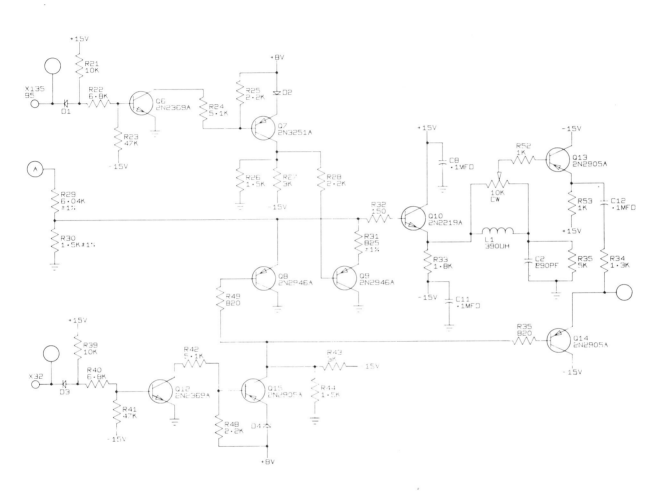

FIG. 33-5 Computer-generated electronic schematic

connections to be specified within a space the size of a postage stamp (Fig. 33-4), and a colored cross-sectional X ray of the human brain are typical problems that would take too many man hours and dollars to produce by traditional methods. Electrical schematics (Fig. 33-5) are another example of computer-generated drawings.

There are two types of computer graphics. These are computer-aided and computer-produced. The term *computer-aided* graphics refers to the generation of information that designers, engineers, and draftspersons need to produce working drawings. In *computer-produced* graphics the computer controls a device that automatically makes the drawing or

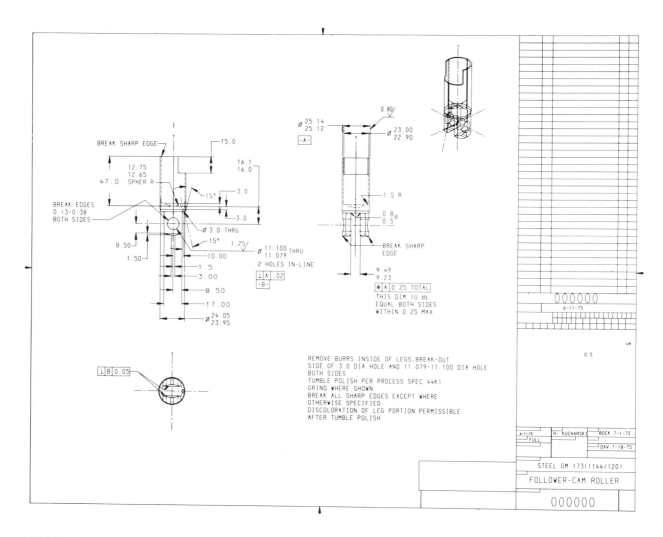

FIG. 33-6 Examples of computer-produced
drawings

picture (Fig. 33-6). The equipment and program-
ming necessary to produce engineering drawings is
expensive. It ordinarily costs less to produce small
and simple drawings by hand.

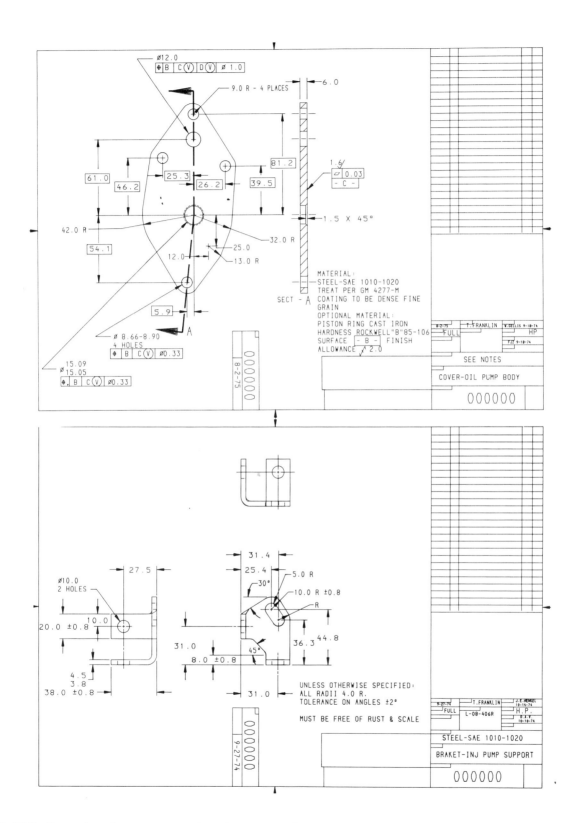

FIG. 33-6   Examples of computer-produced drawings (Cont.)

FIG. 33-7 A computer storage bank (Armstrong Cork Co.)

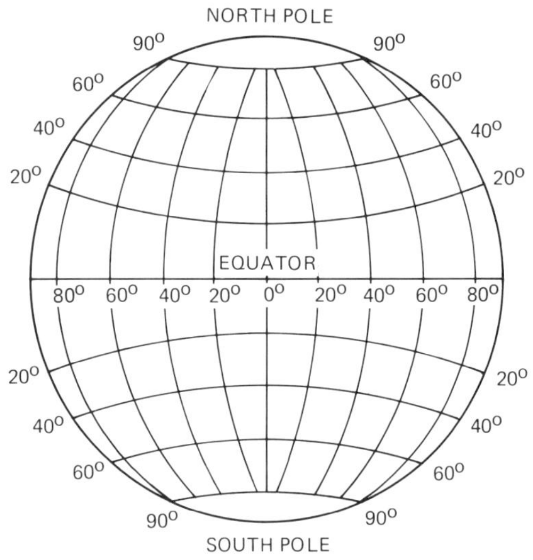

FIG. 33-8 Digitizing a ship's position by latitude and longitude

THE HARDWARE

The computer is the heart of a computer graphics system. We will take a brief look at what the computer does (Fig. 33-7).

A digital computer can receive data (information) from different sources (inputs). It can store it (in memory banks), process it (perform mathematical computations, make logical decisions, etc.), and retain the answers for future use. It may also graphically display the results (output). The physical equipment in computer systems is called ''hardware.'' The instructions and programs that tell the computer what to do and how to do it are called ''software.''

OPERATION

Almost any computer can handle the simple processing needed for most computer graphics problems. Our main concern here is with the input and output devices.

Digital computers work with *discrete* (individual) *digits* (units). These are similar to those you get when you push buttons on a pocket calculator, typewriter, or push-button telephone. Digitizing means using numbers to tell where something is. In navigation, a ship at sea is located by its latitude and longitude (Fig. 33-8). Figure 33-9 is a digitizer that can locate and plot these navigational points.

Figure 33-10 shows how a simple drawing can be digitized on graph paper (grid-size controls scale). The horizontal axis is called the x axis. The vertical axis is called the y axis. The x axis location is generally given first. Each point is located by giving both the x and y location. Line (1) is described by (1) 5 60—75 60. This means it starts at x5 y60 and continues to x75 y60. The data table gives the digitizing information for all five lines. Note that wherever possible one line continues

FIG. 33-9 A complete electronic digitizer (Auto-trol Corp., Denver, CO)

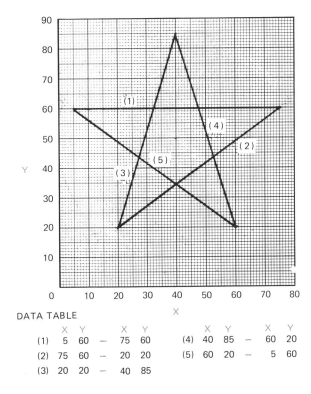

DATA TABLE

	X	Y		X	Y		X	Y		X	Y
(1)	5	60	—	75	60	(4)	40	85	—	60	20
(2)	75	60	—	20	20	(5)	60	20	—	5	60
(3)	20	20	—	40	85						

FIG. 33-10 Digitizing a drawing

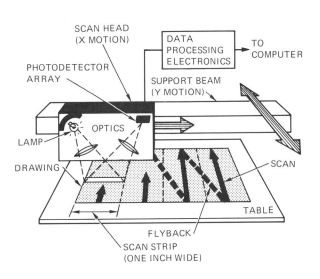

FIG. 33-11 A light-sensitive automatic digitizer

from where the previous line finished. This saves machine operation time. In this case the digitizing was done by hand. Automatic digitizers can produce a list of these points by light-sensitive scanning methods (Fig. 33-11). The more versatile

FIG. 33-12   Digitizing with a cursor (Bill Butlerfield)

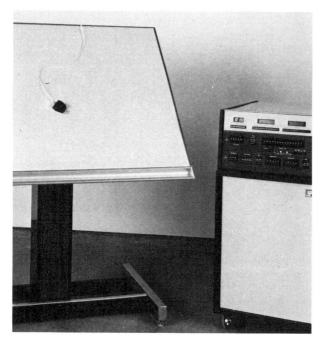

FIG. 33-13   Digitizing with a probe (Auto-trol Corp.)

FIG. 33-14   A probe being used for architectural design (Tektronix)

FIG. 33-15 A CRT displaying a computer-generated drawing (Tektronix)

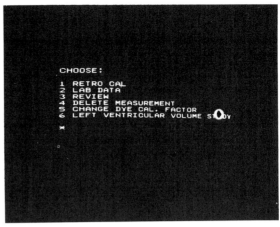

FIG. 33-16 A "menu" being displayed on a screen (Burroughs Corp.)

digitizers use a cursor (Fig. 33-12) or a probe (Fig. 33-13). The cursor is a glass handle with cross hairs. It is placed over the desired points. Its position is noted by the digitizer and translated to digits for the computer. Probes or pens may also be used to pinpoint the locations on a drawing. These are then translated to a computer and CRT in x and y data. The location may be sensed electrically by horizontal and vertical wires in a grid under the drawing board surface, or by acoustic or electromagnetic sensors. Figure 33-14 shows a probe being used in architectural design.

Another instrument for inputting information to the computer is the light pen (Fig. 33-15). Wherever the light pen touches the face of the CRT it triggers circuits in the computer that produce light. The designer can sketch an idea, erase it, change it, or add to it. He can do this until he has the best solution to his problem. Any part or all of his plan can be stored in the computer for analysis or later use. The computer can also produce a copy printed on paper.

Symbols and instructions from the computer can be displayed on the screen in a form called a "menu" (Fig. 33-16). The designer can point to any item on the menu with the light pen. The computer will then follow those programs.

Input information for computer graphics may also come from a telewriter (Fig. 33-17), magnetic tape, punched tape, or punched cards. It can also come directly from the computer as a result of data analysis from many different sources.

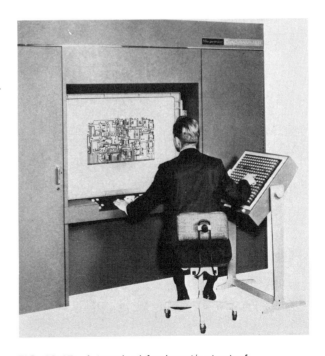

FIG. 33-17 A terminal for input/output of information (Mergenthaler Linotype Co.)

FIG. 33-18 A plotter drawing a printed circuit (Gerber Scientific Instrument Company)

OUTPUT

When the designer and the computer have the necessary data, it can be used to help the draftsperson make the actual drawing. This is a good example of how computer-aided graphics works. The computer may also control an output device and produce the desired graphics itself.

However, because the development of the software (the program and the instructions to operate output devices) is so expensive, most drawings are still done manually. It is economical to go to this expense only if the program can be used many times with programmed modifications.

On a small scale, a CRT can be used as an output device to display a solution. A photostatic or printed permanent copy can be made from the screen.

For regular or large size drawings, a plotter (Fig. 33-18) can be used. This machine is the opposite of the digitizer. The plotter has a cursor, which may have a pen, cutter, or electrostatic point. This cursor will move to x and y positions according to the data table. Programs in the computer can cause the cursor to make straight lines, circles, or curves. It can also make letters or numbers, characters, symbols, or any other imaginable shape. Plotters may be flatbed (Fig. 33-19) or the drum type (Fig. 33-20). In the latter type a continu-

FIG. 33-19 A computer-controlled flatbed plotter (Auto-trol Corp., Denver, CO)

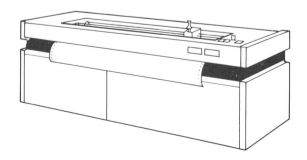

FIG. 33-20 A high speed drum plotter (Broomall Industries)

ous roll of paper passes under the cursor point. Both the paper and the cursor point movements are controlled by the computer. This is so that any shape may be drawn.

Computer graphics is also used for visual displays that are difficult to do manually. These displays show a machined part, or any form, turning in space. This is so it can be viewed from any angle. This is relatively easy to do once the computer programs have been developed (Fig. 33-21, 33-22). Graphic display enables a scientist to analyze scientific research data at a glance (Fig. 33-23).

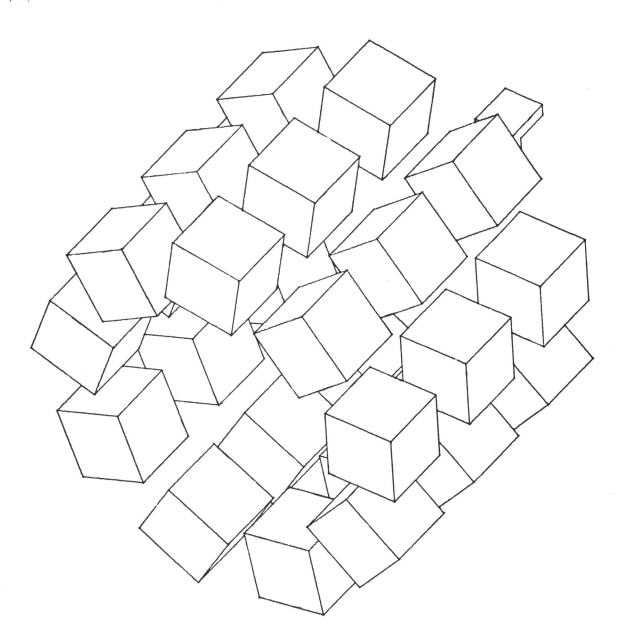

FIG. 33-21 A cube rotated to nine different angles by a computer (JPL)

The unusual distortion of letters and forms seen on ordinary TV programs is another example of computer graphics at work (Fig. 33-24).

An industrial application which is closely re-lated to drafting is numerically controlled machining. Figure 33-25 shows the steps required to produce a finished machine part from a drawing.

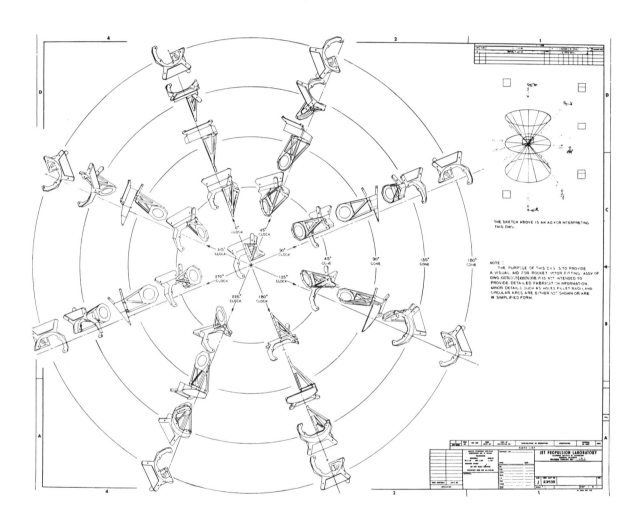

FIG. 33-22   A complex part rotated in many planes (JPL)

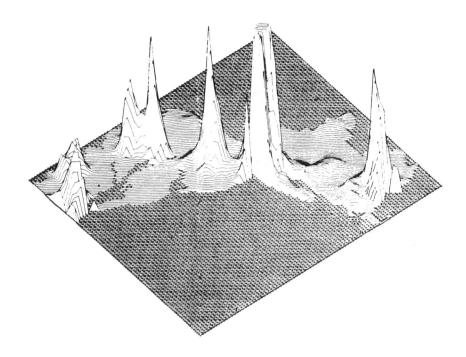

FIG. 33-23 Computer-generated 3-D representation of scientific data

FIG. 33-24 Computer graphics distorted on a TV tube

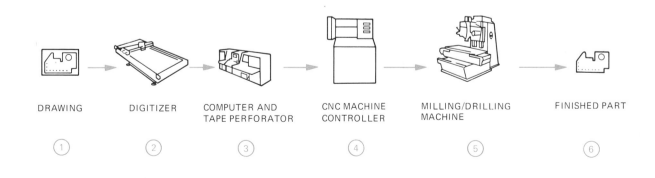

DRAWING	DIGITIZER	COMPUTER AND TAPE PERFORATOR	CNC MACHINE CONTROLLER	MILLING/DRILLING MACHINE	FINISHED PART
①	②	③	④	⑤	⑥

FIG. 33-25 Numerically controlled machining is computer controlled

PROBLEMS

1. Write a Data Table for Figure 33-26.

2. Use this Data Table to draw a figure on graph paper. Use any size grids.

	x	y	–	x	y
(1)	43	27	–	55	14
(2)	55	14	–	30	10
(3)	30	10	–	43	27
(4)	43	27	–	32	88
(5)	32	88	–	10	10
(6)	10	10	–	30	10
(7)	55	14	–	70	20
(8)	70	20	–	32	88

3. Use this Data Table to draw a figure on graph paper. Use any size grids.

	x	y	–	x	y
(1)	10	50	–	10	28
(2)	10	28	–	20	22
(3)	20	22	–	20	86
(4)	20	86	–	15	89
(5)	15	89	–	15	47
(6)	14	47	–	10	50
(7)	10	50	–	15	53
(8)	15	89	–	35	100
(9)	35	100	–	40	97
(10)	40	97	–	20	86
(11)	40	96	–	49	101
(12)	49	101	–	64	92
(13)	64	92	–	35	75
(14)	35	75	–	30	78
(15)	40	97	–	40	84
(16)	40	84	–	30	78
(17)	30	78	–	30	28
(18)	30	28	–	20	22
(19)	30	28	–	35	25
(20)	35	25	–	35	75
(21)	64	92	–	64	53
(22)	64	53	–	54	47
(23)	54	47	–	54	36
(24)	54	36	–	35	25
(25)	54	36	–	70	27
(26)	70	27	–	70	38
(27)	70	38	–	54	47
(28)	64	65	–	90	50
(29)	90	50	–	70	38
(30)	70	27	–	90	39
(31)	90	39	–	90	50

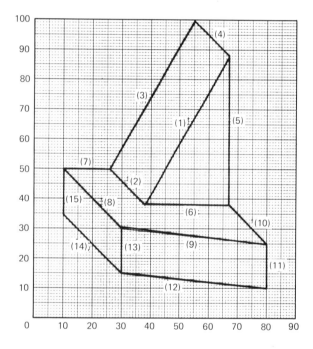

FIG. 33-26 Problem

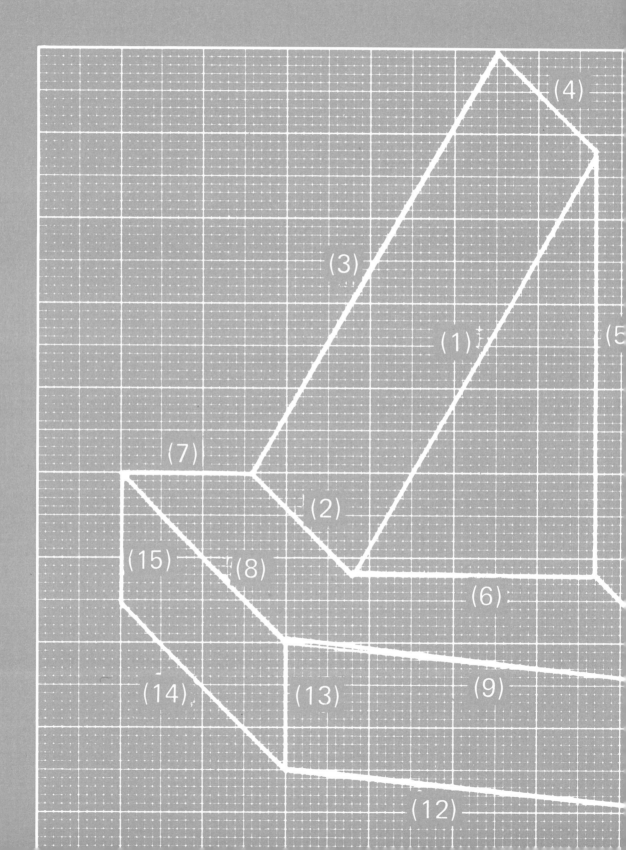

Part

Reproduction

A finished mechanical drawing represents time, effort, and materials—all are costly. Drawings are valuable documents which must be handled with care. They are used in all phases of design, manufacturing, and sales. For these reasons, they must be readily reproducible.

Part VI of this test introduces the methods and processes used to reproduce, copy, store, and retrieve an original drawing.

34

Chapter

Microfilming

Bell & Howel

Engineering documentation is important at every phase of design and production, including:
Research and development
Drafting and design
Cost control
Scheduling
Purchasing
Quality control
Production
Inspection
Engineering changes
Sales

Microfilming supports these areas by providing the capacity for filing millions of engineering drawings and documents so that any one of them can be rapidly retrieved and printed.

The concept of microfilming is simple. Documents are photographed and stored on a small strip of film. Each picture can be enlarged and viewed, or a print can be made from the film. Microfilming is developing into a large new industry called micrographics.

As you study this chapter, you should become familiar with the different types of microfilm, and the storage and retrieval systems for microfilm. You should also develop the skills needed to make drawings for microfilming.

INTRODUCTION

Drawings and papers containing information for production and operations in a company or industry are called documents. The handling of these documents is called documentation.

TYPES OF MICROFILM

Strips of microfilm come in widths of 16 mm, 35 mm, and 105 mm. This film can be processed into rolls, aperture cards, or microfiche (Fig. 34-2), and can be duplicated quickly. All types of film can be retrieved manually or by fully automated equipment.

An aperture card is a file card (approximately 80 mm X 150 mm) with a hole for the microfilm. Identification for manual or automatic retrieval is placed on the card.

Microfiche is a sheet of transparent film (approximately 100 mm X 150 mm) that contains many small microfilm strips set in rows.

MICROFILMING EQUIPMENT

Microfilming equipment is basically simple: A microfilm camera is used to take the pictures on a specially lighted table (Fig. 34-3). Automatic microfilm developers and duplicators are used to process the film (Fig. 34-4). Microfilm viewers are used to

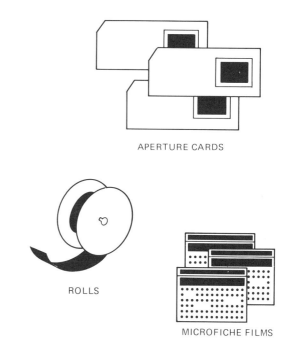

APERTURE CARDS

ROLLS

MICROFICHE FILMS

FIG. 34-2   Types of microfilm

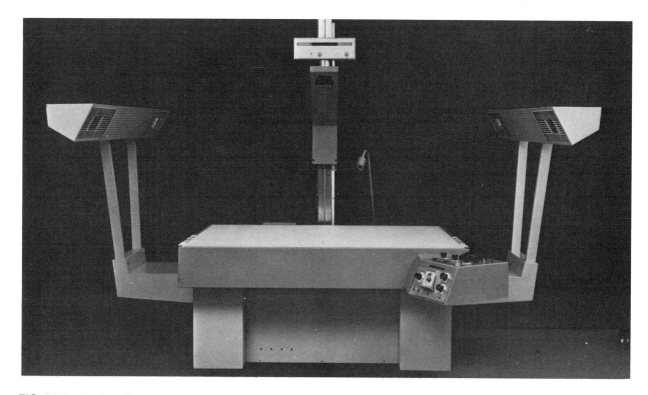

FIG. 34-3   A microfilm camera (Eastman Kodak Co.)

enlarge the image (Fig. 34-5). A "blowback" printer may be used to make a print from the film (Fig. 34-6). An automated retrieval system may be used as a storage file (Fig. 34-7).

FIG. 34-4   A microfilm duplicator (GAF Corp.)

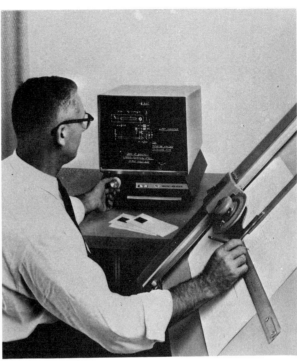

FIG. 34-5   A microfilm viewer (Teledyne Post)

FIG. 34-6   A microfilm reader-printer (GAF Corp.)

FIG. 34-7   An automated retrieval system (GAF Corp.)

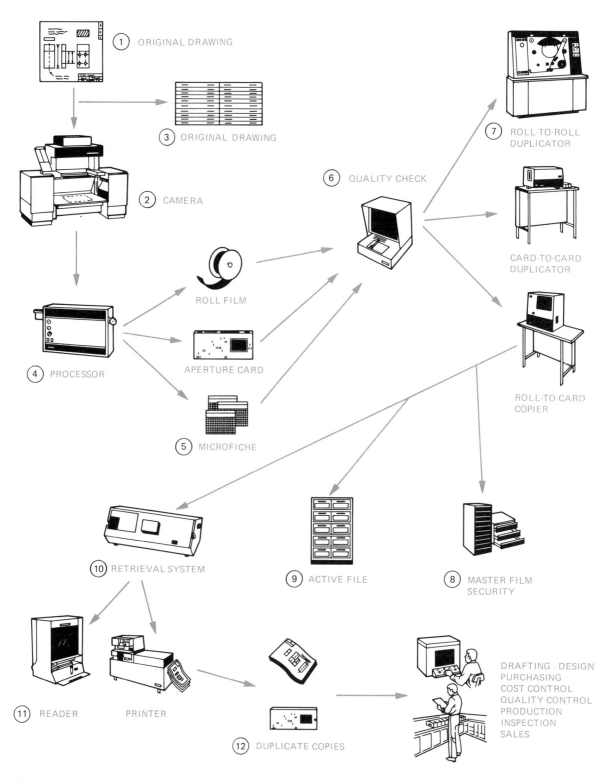

FIG. 34-8 Typical microfilming operations

MICROFILMING OPERATIONS

Each company that uses microfilm has equipment to fit its needs. Each type of microfilming equipment has a wide range of uses. A typical microfilming flow pattern is shown in Figure 34-8.

1. The original drawing or document is completed and checked for microfilming quality.

2. A camera photographs the document on microfilm.

3. The original document is stored.

4. The microfilm is processed.

5. The microfilm is placed on rolls, aperture cards, or microfiche, and identified for retrieval.

6. The microfilm is quality checked.

7. The microfilm is duplicated and quality checked.

8. The master microfilm is placed in a security file.

9. A duplicate microfilm is placed in the active file.

10. Upon retrieval, the microfilm can be viewed directly, duplicated in microfilm, or a print can be made. Blowback prints are either 250 × 300 mm or 500 × 600 mm in size.

Drawing Preparation for Microfilming

The legibility and quality of a drawing are very important for microfilming. A drawing will usually be reduced in size thirty times; then the print-out or blowback will be enlarged to about 60% of the drawing's original size. Poor quality in the original drawing will not produce readable blowbacks.

Pencil or ink can be used for the original drawing. Ink makes a better microfilm image than pencil because of the density and consistency of line width. All lines must be smudgefree with a strong white background for effective contrast. There should be a minimum distance of $1/16$ inch between parallel lines. All drawing on vellum must be on the front side. The camera will not pick up work on the back side. Any blurred or fuzzy lines will appear very fuzzy on the blowbacks.

The minimum width of lines for A, B, and C paper sizes is 0.015 inch. The minimum width of lines for D paper size is 0.02 inch.

The rules for clarity of lettering are the same as those for any other line work. Letters should be legible, dense, and not too small or spaced too closely, to prevent them from bleeding together and blurring. The minimum height of letters on A, B, and C paper sizes is $1/8$ inch. The minimum height of letters on D or larger paper sizes is $3/16$ inch. The spacing between lines of letters should be a minimum of $1/8$ inch. The microfont style of lettering (Fig. 34-9) produces fewer reading errors in the blowbacks and should be used on all microfilm drawings.

Revisions

Revisions of engineering changes or original drawings should be made with the same style of line work and lettering as the original work. A quick, simple, and efficient method used to make an engineering change is to make a print on a wash-off film. Wash-off film has lines that are wet-erasable for easy changes. After the changes are made, a new microfilm copy of the revised drawing must be made.

PROBLEMS

Complete an orthographic and an isometric drawing of the following problems using microfont lettering and the general rules for preparing drawings for microfilming (Fig. 34-10 through 34-15).

FIG. 34-9 Microfont lettering for microfilmed drawings

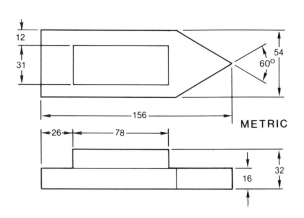

FIG. 34-10   Problem

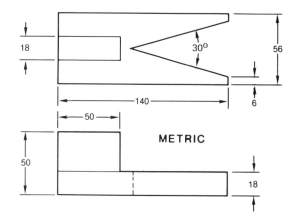

FIG. 34-11   Problem

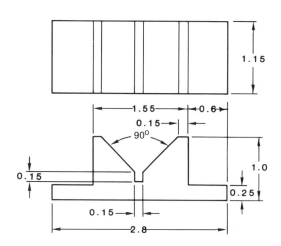

FIG. 34-12   Problem

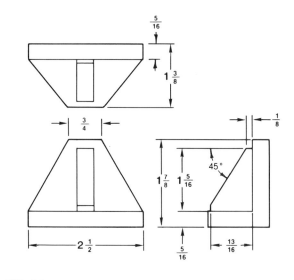

FIG. 34-13   Problem

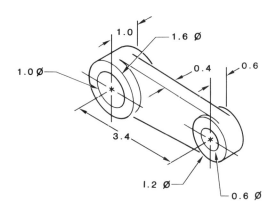

FIG. 34-14   Problem

35
Chapter

Reproduction Processes

Gaf

INTRODUCTION

There are many different processes for duplication drawings and documents. The most commonly used processes for drafting are:

Microfilming
Blueprinting

Diazo printing
Printing
Xerography
Photography
Thermography

The quality of the original drawing is important to any form of reproduction. You must have good originals to obtain good reproductions. As you study this chapter, you should practice reproducing your drawings with as many different processes as possible.

BLUEPRINTING

A blueprint is a full-size negative copy of an original drawing. It has a dark blue background and white lines. Blueprints are made by exposing paper coated with ultraviolet iron salts. Figure 36-2 illustrates the steps for making a blueprint.

The original drawing should be on transparent paper (tracing paper, vellum, or film) and have dark, dense lines. All surfaces that are exposed to the ultraviolet light will remain blue when developed. If the lines are not dense enough, the light will seep through and expose the chemical. Partially exposed lines will not be white when

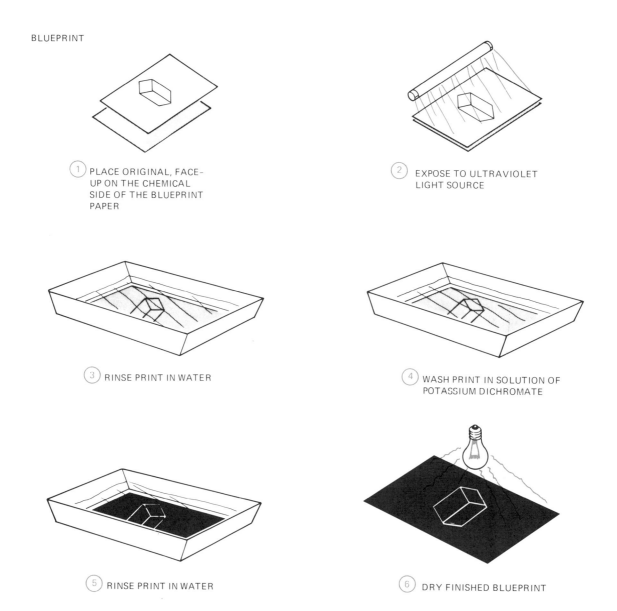

BLUEPRINT

① PLACE ORIGINAL, FACE-
UP ON THE CHEMICAL
SIDE OF THE BLUEPRINT
PAPER

② EXPOSE TO ULTRAVIOLET
LIGHT SOURCE

③ RINSE PRINT IN WATER

④ WASH PRINT IN SOLUTION OF
POTASSIUM DICHROMATE

⑤ RINSE PRINT IN WATER

⑥ DRY FINISHED BLUEPRINT

FIG. 35-2   The blueprint process

developed. Instead, they will be a shade of blue. Such blueprints are difficult to read.

Blueprinting is one of the oldest methods of reproducing engineering drawings. It is inexpensive, but because of the wet process it is slow.

DIAZO PRINTING

A diazo print, also called a whiteprint, is also a full-size copy of the original drawing. It is a positive reproduction with dark lines and a white background. The advantages of a whiteprint are usability of the white background for additional drawing, low cost, and quick reproduction in a printer-developer (Fig. 35-3). Whiteprints can be developed with either a dry (ammonia vapor) or a wet chemical process. The dry method is the fastest and most commonly used process (Fig. 35-4).

This reproduction method is based on light-sensitive diazonium salts from which a colored,

FIG. 35-3 Diazo printer-developers (Teledyne, GAF Corp.)

DIAZO WHITEPRINT

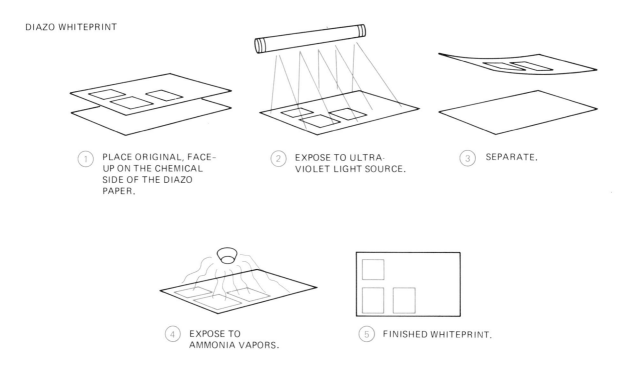

1. PLACE ORIGINAL, FACE-UP ON THE CHEMICAL SIDE OF THE DIAZO PAPER.

2. EXPOSE TO ULTRA-VIOLET LIGHT SOURCE.

3. SEPARATE.

4. EXPOSE TO AMMONIA VAPORS.

5. FINISHED WHITEPRINT.

FIG. 35-4   The diazo whiteprint process

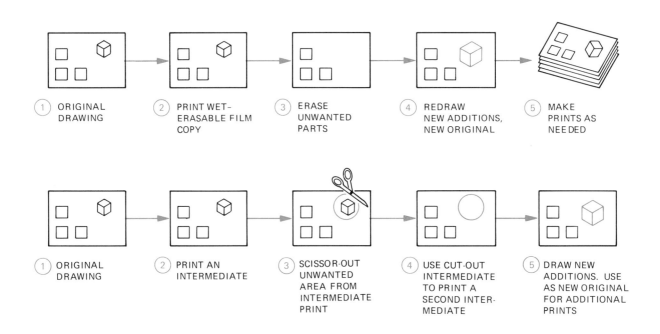

1. ORIGINAL DRAWING

2. PRINT WET-ERASABLE FILM COPY

3. ERASE UNWANTED PARTS

4. REDRAW NEW ADDITIONS, NEW ORIGINAL

5. MAKE PRINTS AS NEEDED

1. ORIGINAL DRAWING

2. PRINT AN INTERMEDIATE

3. SCISSOR-OUT UNWANTED AREA FROM INTERMEDIATE PRINT

4. USE CUT-OUT INTERMEDIATE TO PRINT A SECOND INTER-MEDIATE

5. DRAW NEW ADDITIONS. USE AS NEW ORIGINAL FOR ADDITIONAL PRINTS

FIG. 35-5   Methods of making revisions

positive azo dye image is made. Light causes the chemical coating of diazo paper to disappear, leaving a totally white surface. Ammonia fumes will turn the chemical dark where the lines blocked the light of the printer. Undeveloped paper must be stored in a dark place away from ammonia fumes.

The whiteprinter has a speed control which is used to vary the exposure time and a developing chamber which contains the ammonia fumes. The variables which determine exposure time are:

1. The developing speed of the whiteprint paper. Speeds vary from very slow to fast.

2. The density of the line work on the original drawing. Dark lines can be run through at slower speeds. Lightly lined drawings must have less exposure time, leaving more background on the print.

3. The transparency of the original drawing paper. A good transparency allows faster speeds and better prints.

Overexposure will allow light to burn out the chemical behind the line work thereby producing a print with very light lines.

Underexposure will prevent sufficient burning off of the background chemical thereby producing a dark background.

The best print is the one with the most contrast between its lines and the background.

The whiteprinter can also make different types of copies. Some of these copies are:

1. Intermediates: Intermediates are transparent prints made from the original drawing. The intermediate print is used as a master to make additional prints, saving wear and tear on the original drawing.

2. Brown Prints (also called Van Dyke prints): A brown print is an intermediate. The brown print is a negative copy. The background is solid and the lines are clear.

3. Transparencies: A transparency is a clear plastic print that can be used on an overhead projector or as an intermediate.

PRINTING

If many copies are required they can be run on a printing press. Printing is an inexpensive way to duplicate large numbers of copies.

XEROGRAPHY

Xerography is the trade name of an electrostatic type of reproduction process. It is a dry process that uses an electrostatic force to deposit dry powder on a sheet of paper. The powder is then heat fused to the paper in order to produce a black line copy. Copies are made quickly and inexpensively on ordinary untreated paper. Some copy machines can also enlarge and reduce copies of the original drawings.

Recently Xerographic equipment has been developed that can be connected to computers. The drawing information can be stored and sent anywhere in the world by microwave or telephone.

The disadvantage of the Xerography reproduction process is that the machines are too expensive for low-volume users and do not copy large, solid areas very well.

PHOTOGRAPHY

A photographic copy camera can take photographs of drawings with good control of enlargements and reductions. High contrast films and print papers produce a sharp, dark line. This type of reproduction is used where artwork must be of excellent quality.

THERMOGRAPHY

A thermographic copy can be made from any type of drawing that is done with a marker containing either carbon or a metallic substance that will absorb more heat than the background. The print paper is sensitive to heat, not light, and a developer is not required.

REVISIONS

The main concern when making revisions to a drawing is to save man-hours. Figure 35-5 shows two ways to eliminate redrawing the entire drawing. Be certain to match new additions to the existing drafting style.

DRAWING PAPERS

Either translucent or opaque drawings papers are used for reproduction. The translucent papers include tracing paper, vellum, tracing cloth, and polyester film.

Blueprinting and diazo printing must be done with translucent papers. The other processes (microfilming, printing, Xerography, photography, and thermography) can use any type of paper.

PROBLEMS

1. As shown in Figure 35-6, make a series of very heavy and very light lines on a number of different types and qualities of translucent paper. Make prints and compare the line density and background of prints exposed at varying speeds.

2. Make an orthographic and an isometric drawing of Figure 35-7 on translucent paper. Make a duplicate copy on as many different types of repdoduction machines as possible.

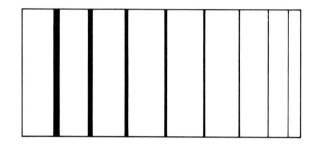

FIG. 35-6 Problem

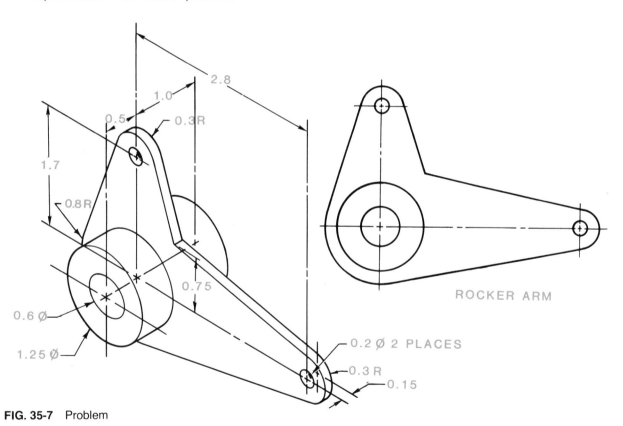

FIG. 35-7 Problem

Appendix A

Abrasive A material that cuts materials softer than itself. It may be used in loose form, mounted on cloth, paper, or bonded on a wheel.

Accurate Made within the tolerance allowed.

Acoustic tile Tile made of sound-absorbing materials.

Acute angle An angle of less than 90 degrees.

Addendum The radial distance between the pitch circle and the top of a gear tooth.

Airbrush A device used to spray ink or paint by means of compressed air.

Allen screw A special set screw or cap screw with a hexagon socket in its head.

Allowance Intentional difference in the dimensions of mating parts to provide for different classes of fits.

Alloy A mixture of two or more metals combined together to form a new metal.

Anchor bolt A bolt used to secure frameworks to piers or foundations.

Angle The figure formed by two lines coming together at a point.

Angle iron A bar of structural iron in the form of a right angle.

Anneal To soften metals or remove internal stresses by heating and slowly cooling.

Anodize To protect aluminum by oxidizing it in an acid bath with a DC current.

Aperture card A data-processing card containing a cut-out space into which a piece of microfilm is mounted.

Arc A portion of a circle.

Assembly A unit fitted together from manufactured parts.

Assembly drawing A drawing showing the working relationship of various parts of a machine or structure as they fit together.

Axis The central line of a drawing or of a geometrical figure.

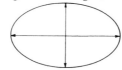

Babbitt or babbitt metal A metal used as an antifriction lining for bearings that is composed of antimony, tin, and copper.

Backlash The play (lost motion) between moving parts, such as a threaded shaft and nut or the teeth of meshing gears.

Bar Square or round rod; also flat steel up to 150 mm in width.

Bar stock Bars or other material to be worked on.

Basic dimension A theoretically exact value used to describe the size, shape, or location of a feature.

Basic size That size from which the limits of size are derived by the application of allowances and tolerances.

Beam compass A compass used to draw large circles and arcs.

Bearing Any part that bears up or supports another part.

Bearing plate A steel plate, usually at the base of a column, used to distribute a load over a larger area.

Bearing wall A wall that supports loads other than its own weight.

Bend allowance The amount of sheet metal required to make a bend over a specific radius.

Bevel Any surface that is not at right angles to an adjacent surface; called a miter when the angle is 45°.

Bill of material A list of the pieces and hardware needed for a particular project.

Blueprint A drawing that has been transferred to a sensitized paper by exposure to light and some type of developer, such as water or ammonia.

Bolt circle A circular centerline locating the centers of holes positioned about a common center point.

Bond The joining together of building materials to ensure solidity.

Bore To make a hole.

Braze To join metals by the fusion of nonferrous alloys, such as brass or zinc, that have melting temperatures above 430 °C but lower than the metals being joined.

Brick veneer A brickwork facing applied as exterior finish.

Brinell A scale for designating the hardness of a piece of metal.

Burr The ragged edge of a piece of metal that results from cutting or punching.

Bushing A hollow cylindrical sleeve which acts as a bearing between rotating parts, such as a shaft and pulley.

Butt joint A joint in which the ends of two pieces of material come together without overlapping.

Butt weld A weld, usually electric, in which the two pieces do not overlap but are joined directly at their ends.

Caliper A measuring device with two adjustable legs used for measuring thicknesses or diameters.

Callout A note on the blueprint giving a dimension, specification, or machine process.

Cam A machine part mounted on a revolving shaft, used for changing rotary motion into an alternating back and forth motion.

Cantilever A beam, girder, or truss overhanging one or both supports.

Casting A part formed by pouring molten metal into a mold of the desired shape and allowing the metal to harden.

Caulk To fill cracks and seams with caulking material.

Ceiling joist A joist to which the finishing materials of a ceiling are fixed.

Center drill A special drill used to produce bearing holes in the ends of a workpiece that is to be mounted between centers.

Center gauge A flat gauge used for measuring angles and for setting a threading tool at right angles to the stock to be threaded in the lathe.

Chamfer To bevel a sharp external edge; also, a beveled edge.

Channel iron A rolled bar that consists of a web and two flanges.

Circle A closed curve with all of its points equidistant from its center.

Circular pitch The distance from a point on one gear tooth to the same point on the next tooth measured along the pitch circle.

Circumference The perimeter of a circle.

Clearance The distance by which one part clears another part.

Cold-rolled steel (CRS) Open-hearth or Bessemer steel which has been rolled while cold to produce smooth, accurate stock, containing 0.12% to 0.20% carbon.

Collar A round flange or ring fitted on a shaft to prevent sliding.

Concave A curved inward; a hollowed out portion; a portion of the inside of a hollow cylinder.

Concentric Having a common center.

Cone A solid figure generated by a line fixed at one end while the other end moves in a circular path.

Contour The outline of an object.

Conventional Customary or traditional; not original.

Convex Curved outward; surface of a cylinder.

Counterbore To enlarge a hole to a specified diameter and depth.

Counterclockwise In a circular direction from right to left.

Countersink To form a conical depression to fit the head of a screw so that the face of the screw will be level with the surface.

Cross-hatching A series of closely spaced parallel lines drawn obliquely to indicate sectional views.

Curtain wall A wall that carries no building load other than its own weight.

Cylinder A geometric figure with a uniform circular cross-section through its entire length.

Datum A point, line, surface, or plane assumed to be exact for purposes of computation, from which the locations of other features are established.

Dedendum Distance from the pitch circle to the bottom of the tooth space.

Design size The size of a feature after allowances for clearance and tolerances have been assigned.

Detail drawing A drawing of a single part that provides all the information necessary for the production of that part.

Development The pattern of the surface of an object drawn as a flat plane.

Diagonal A line joining opposite corners of a rectangle; diagonal members used for stiffening and wind bracing.

Diameter The length of a straight line running through the center of a circle and terminating at the circumference.

Diametral pitch The number of gear teeth per mm of pitch diameter.

Diazo microfilm A film using ammonia-developed diazo emulsion for making duplicate copies of the original silver microfilm.

Die One of a pair of hardened metal blocks for forming, impressing, or cutting out a desired shape.

Die casting A very accurate and smooth casting made by pouring a molten alloy into a metal mold or die. Distinguished from a casting made in sand.

Die stamping The process of cutting or forming a piece of sheet metal with a die.

Dividers A tool used for dividing spaced equally, for transferring measures, for scribing arcs on wood and metal surfaces, and for general layout work on metal.

Draft The clearance on a pattern or mold that allows easy withdrawal of the pattern from the mold.

Driftpin A tapered pin used to line

up holes before riveting or bolting two pieces together.

Drill To make a cylindrical hole using a revolving tool with cutting edges.

Duct A sheet metal pipe or passageway used to move air.

Eccentric Not having a common center. A device that changes rotary motion into reciprocating (back and forth) motion.

Elevation The drawing of a house showing how it looks from the front or from any side.

Ellipse A closed curve in the form of a symmetrical oval.

Engineering drawing Technical drawing or drafting; the graphic language of the engineer.

Equilateral A figure having sides of equal length.

Exploded view Separate parts of a single assembly projected away from each other or separated to show relationships among the parts of one drawing.

Eyebolt A bolt provided with a loop or eye instead of the customary head.

FAO An abbreviation used on detail drawings to indicate that the piece is to be finished all over.

Fabricate To cut, punch, and assemble members in the shop.

Face To machine (finish) a flat surface on a lathe with the surface perpendicular to the axis of rotation.

Feature A portion of a part, such as a diameter, hole, keyway, or flat surface.

Ferrous Metals that have iron as their base material.

Filler metal Metal to be added in making a weld.

Fillet A concave intersection used to strengthen the area between two surfaces.

Fillister The cylindrical head of a screw, slotted to fit a screw driver.

Finish General finish requirements such as paint, chemical, or electroplating rather than surface texture or roughness.

Fit The relationship between two mating parts with respect to the amount of clearance or interference present when they are assembled.

Fixture A device used to position and hold a part in a machine tool. It does not guide the cutting tool.

Flange A projecting rim used to add strength, to provide for an attachment to another part, or to act as a guide (as on a railroad car wheel).

Flashing The sheet metal built into the joints of a well or covering the valleys, ridges, and hips of a roof for the purpose of preventing leakage.

Floor plan The drawing that shows the exact shape, dimensions, and arrangements of the rooms of a building.

Footing A concrete form projecting at the base of a wall for the purpose of distributing the load over a greater area, preventing excessive settlement.

Force fit An interference between two mating parts sufficient to require force to press the pieces together. The joined pieces are considered permanently assembled.

Foreshorten To show lines or objects shorter than their true lengths. Foreshortened lines are not perpendicular to the line of sight.

Forge To form material using heat and pressure.

Free fit The fit between two mating parts used when tolerances are liberal. Clearance is sufficient to permit a shaft to turn freely without binding or overheating when properly lubricated.

Frustum The bottom figure formed by cutting off a portion of a cone or pyramid parallel to its base.

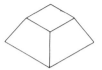

Functional drawing A drawing using the fewest number of views and the fewest number of lines to provide the exact information required.

Furring The application of thin wood, metal, or other building material to a wall, beam, ceiling, or the like to level a surface for lathing, boarding, etc., or to make an air space within a wall.

Fusion The melting together of filler metal and base metal, or of base metal only, that results in two materials uniting.

Gauge A device for determining whether a dimension on an object is within specified limits.

Gage line The centerline for rivet holes.

Gasket A thin piece of rubber, metal, or some other material, placed between surfaces to make a tight joint.

Gear A toothed wheel used to transmit power or motion from one shaft to another. A machine element used to transmit motion or force.

Geometric dimensioning and tolerancing A means of dimensioning and tolerancing a drawing with respect to the actual function or relationship of part features so that it can be most economically produced. It includes positional and form dimensioning and tolerancing.

Girder A horizontal member, either single or built up, acting as a principal beam.

Glazing The act of furnishing or fitting with glass.

Graduate To divide a scale or dial into regular spaces.

Groove weld A weld made in the groove between two members to be joined.

Gusset plate A plate used to connect various members, such as in a truss.

Harden To make carbon steel stronger by heating and quenching in water or oil.

Hardness test Technique used to measure the degree of hardness of heat-treated materials.

Head The horizontal piece forming the top of a wall opening, such as a door or window.

Headers In masonry, those stones or bricks extending over the thickness of a wall; in carpentry, the large beam into which the common joists are framed.

Heat-treat To change the properties of metal by carefully controlled heating and cooling.

Helix The curve formed on a cylinder by an ordinary screw thread.

Hex nut A nut with six sides or faces.

Hexagon A six-sided figure with each corner at a 120-degree angle. Each center angle is 60 degrees.

Hip The intersection between two sloping surfaces forming an exterior angle.

Hone A method of finishing a hole or other surface to a precise tolerance by using an abrasive block and rotary motion.

Hub The central portion of a wheel or gear.

Incline Making an angle with another line or plane.

Index To divide a cylindrical piece of work into equal parts in order to cut gear teeth or to drill equally spaced holes.

Indicator A precision measuring instrument for checking the trueness of work.

Inscribe To draw one figure within another figure.

Insulator A nonconductor, such as the glass or porcelain holders that are used to carry electric wires.

Joist A horizontal beam used as a support for a floor, ceiling, or flat roof.

Journal Portion of a rotation shaft supported by a bearing.

Intermittent welding Welding in which the continuity is broken by recurring unwelded spaces.

Jig A device that holds a work in position and guides the cutting tool.

Knurl To form a series of regular dents in a cylindrical surface so that it can be held or turned by hand.

Laminate To form an object by fastening together several layers or thin sheets of material.

Lap joint A joint between two overlapping members.

Laterals Members used to prevent lateral deflection.

Lay out To locate and scribe points for machining and forming operations.

Legend The title or a brief description of a drawing; a special mark which appears on maps and drawings together with its definition.

Limits The extreme permissible dimensions of a part resulting from the application of a tolerance.

Line conventions Symbols that furnish a means of representing or describing some part of an object; the symbols are expressed by a combination of line weight and style.

Lintel The horizontal structural member that supports the wall over an opening.

Longitudinal section A lengthwise section of the object that is being drawn.

Major diameter The largest diameter of a thread measured perpendicular to the axis.

Maximum material condition The state that exists when two mating objects both contain the maximum amount of material; for example, minimum hole diameter and maximum shaft diameter.

Mesh To engage gears to a working contact.

Microfilm A high-resolution photographic film, usually 35 mm size for engineering drawings and documents.

Micrometer caliper A measuring device used to determine the exact measurements of thicknesses.

Microprint A 450 mm x 610 mm print made from microfilm.

Mill To remove material by means of a rotating cutter on a milling machine.

Minor diameter The smallest diameter on a screw thread measured across the root of the thread and perpendicular to the axis. Also known as the *root diameter*.

Miter To match together, as two pieces of molding, on a line bisecting the angle of junction.

Miter joint A joint made by cutting the ends of two pieces at the same angle, usually 45 degrees.

Mold The material that forms the cavity into which molten metal is poured.

National Coarse (NC) The coarse thread series of the American Standard screw threads.

National Fine (NF) The fine thread series of the American Standard screw threads.

Neck A groove cut near the end of a shaft.

Nonferrous Metals such as aluminum, magnesium, and copper, which are not derived from an iron base or an iron alloy base.

Object line The heavy, full line on a drawing which describes the outline of an object.

Obtuse angle   An angle greater than 90 degrees.

Octagon   An eight-sided geometric figure with each corner forming a 135-degree angle. Each center angle is 45 degrees.

Ogee   An S-shaped curve.

Orthographic drawing   A multiview drawing that shows every feature of an object in its true size and shape.

Overall dimensions   Dimensions that give the entire length or width of an object, as contrasted with dimensions that show such minor details of an object as the location of a hole.

Parallel lines   Lines which lie side by side and are the same distance apart at all points.

Pattern   In sheet metal, a drawing or a template the exact size and shape of the workpiece; in molding, the model from which castings are made.

Pentagon   A five-sided geometric figure with each corner forming a 118-degree angle. Each center angle is 72 degrees.

Perimeter   The boundary of a geometric figure.

Perpendicular   A line at right angles to another line.

Perspective drawing   A method of pictorial drawing that can represent an object on a single plane as it appears to the eye.

Photo drawing   A drawing prepared using a photograph on which dimensions, notes, and specifications have been added.

Pinion   The smaller of two mating gear wheels.

Pitch   The distance from a point on one thread to a corresponding point on the next thread.

Pitch circle   An imaginary circle corresponding to the circumference of the friction gear from which the spur gear was derived.

Plan view   The view obtained by looking directly down on an object; the top view.

Plate   A horizontal member that carries other structural members—usually the top timber of a wall that carries the roof trusses or rafters directly; to electrochemically coat a metal object with another metal; another name for a drawing.

Positional tolerancing   The permitted variation of a feature from the exact or true position indicated on the drawing.

Process specifications   A description of the exact procedures, materials, and equipment to be used in performing a particular operation such as milling.

Profile   The outline of an object; to machine an outline with a rotary cutter.

Project   To extend from.

Projection   A method of representing three-dimensional objects on a plane having only length and breadth.

Rack   A flat strip with teeth designed to mesh with teeth on a gear. Used to change rotary motion to linear motion.

Radius   The length of a straight line running from the center of a circle to the perimeter of the circle.

Rafter   A member in a roof framework usually extending from the ridge to the eaves.

Ratchet   A gear with triangular-shaped teeth which are engaged by a pawl.

Ream   To finish an extremely accurate hole with a fluted cutting tool.

Rectangle   A geometric figure with opposite sides equal in length and each corner forming a 90 degree angle.

Reference dimension   A dimension that is used only for information purposes and does not govern production or inspection operations.

Relief   An offset of surfaces to provide clearance for machining.

Rib   A relatively thin flat member acting as a brace or support.

Right angle   A 90 degree angle. The angle that is formed by a line that is perpendicular to another line.

Rivet   To fasten with rivets; to batter or upset the headless end of a pin used as a permanent fastening.

Root opening   The separation between members to be joined.

Rotate   To turn or revolve around a point.

Rough layout   A rough plan that arranges lines and symbols in relation to one another.

Round   The rounded corner of two surfaces.

Scale   A measuring instrument used by the draftsperson; the outside coating of a casting.

Scale drawing   A drawing made smaller or larger than the piece it represents, but to a definite proportion that is usually specified on the drawing.

Screw thread   A thread cut on the outside of a cylindrical piece such as a bolt.

Seam   The line formed where two edges are joined together.

Section   A cross-sectional view at a specified point of a part or assembly.

Segment   Any part of a divided line.

Series circuit   An electric circuit in which all parts are connected in such a way that a current can pass through in a single path.

Shaft   A cylindrical piece of steel used to carry pulleys or to transmit power by rotation.

Shim   A thin plate of metal used to adjust the distance between two surfaces.

Short circuit   A path of low resistance across an electric circuit. It usually results in an excessive flow of current.

Shoulder A plane surface on a shaft, perpendicular to the axis and formed by a difference in diameter.

Shrink rule A rule used by pattern makers in which the graduations are proportionately larger than standard to allow for shrinkage of the cast metal.

Sketch To draw without the aid of drafting instruments.

Solder To join with solder that is composed of lead and tin.

Span Distance between centers of supports of a truss, beam, or girder.

Specification A detailed description of a part or material giving all information such as quality, size, and quantity, not shown on the graphic part of a blueprint.

Spin To form a rotating piece of sheet metal into a desired shape by pressing it against a rotating form with a smooth tool.

Splice A longitudinal connection between the parts of a continuous member.

Spline A shaft or arbor fitting with a groove or keyway.

Square To machine or cut at right angles. A geometric figure with four equal length sides and four right (90 degree) angles.

Stair landing A platform between flights of stairs or at the termination of a flight of stairs.

Stair riser The vertical board under the tread in stairs.

Stair tread The horizontal part of a step of a stair.

Stiffener Angle, plate, or channel riveted to a member to prevent buckling.

Stretch-out A full size drawing or pattern of a sheet metal object.

Stringer A longitudinal member used to support loads directly.

Strut A compression member in a framework.

Surface texture The lay, roughness, waviness, and flaws of a surface.

Symbol A figure or character used in place of a word or group of words.

Tabular dimension A type of rectangular datum dimensioning in which dimensions from mutually perpendicular datum planes are listed in a table on the drawing instead of on the pictorial portion.

Tangent A line drawn to the surface of an arc or circle so that it contacts the arc or circle at only one point.

Tap To cut threads in a hole with a rotating tool called a *tap* which has threads and flutes to give cutting edges.

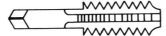

Taper A piece that increases or decreases in size at a uniform rate to assume a wedge or conical shape.

Technical illustration A pictorial drawing used to simplify and interpret technical information.

Template or templet A flat form or pattern of full size, used to lay out a shape and to locate holes or other features.

Tensile strength The maximum load or pull a piece can support without failure.

Thread To cut a screw thread; an internal or external form that can be screwed together as a fastener.

Threshold The stone, wood, or metal piece which lies directly under a door.

Tolerance The total amount of variation permitted from the design size of a part.

Torque The rotational or twisting force in a turning shaft.

Train A series of meshed gears.

Triangle A three-sided figure.

Truss A rigid framework formed by a series of triangles for carrying loads.

Typical (TYP) A term used to indicate that the dimension or feature applies to all locations which appear to be identical in size and configuration.

Undercut To cut leaving an overhanging edge; a cut having inwardly sloping sides.

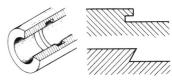

Upset To forge a larger diameter or shoulder on a bar.

Vernier A small auxiliary scale used to obtain fractional parts of a major scale.

Vertical At right angles to a horizontal line or plane.

Web A thin, flat part joining larger parts. Also known as a rib.

Woodruff key A semicircular flat key.

Working drawing A drawing that gives the craftsman the necessary information to make and assemble a product.

Appendix B

ANMC American National Metric Council. Established by ANSI as a coordinating center for metric activities in the United States.

ANSI American National Standards Institute. Voluntary federation to coordinate the development of national standards. Called ANSI.

AO ISO size system for paper.

Ampere SI unit of electric current.

BIPM International Bureau of Weights and Measures.

Base unit One of the seven units that form the SI Metric System.

CGS The metric system based on centimetre-gram-second units.

Candela (cd) SI unit of luminous intensity.

Celsius (°C) An alternate scale to the SI kelvin scale of temperature.

Centi- (c) A prefix meaning 1/100 (10^{-2}).

Centigrade The common name given to the Celsius temperature scale.

Conversion The process of finding equivalent values. Intended here to signify the adoption of SI units.

Coulomb (C) SI unit of electrical charge.

Customary measurement system The predominant measurement system in any country. In the United States the units are the foot-pound-second system.

Deci- Metric prefix meaning one tenth (10^{-1}).

Decimal system Number system based on multiples and subdivisions of the number 10. The metric system is one such decimal system.

Degree (°) A unit of angular measure.

Deka- Metric prefix meaning ten times (10).

Derived units Compound units formed by the algebraic combination of base units, supplementary units, or other derived units. Several of the units in this class have been assigned special names and symbols. The remaining derived units are identified by the names and symbols of the given algebraic expression.

Energy The SI unit of energy is the joule (J). The electrical unit of energy is the kilowatt hour (kWh). In physics, the unit of energy is the electron volt (eV).

Farad (F) SI unit of capacitance.

Force The acceleration of a mass. It is measured in newtons.

Geometric progression A numerical series in which each number is multiplied by a constant value to determine the next larger number.

Giga- (G) A metric prefix of 10^9.

Gram (g) A metric unit of mass or weight equal to 1/1 000 kilogram, nearly equal to the weight of one cubic centimetre of water at its maximum density.

Hard conversion The original data or design sizes changed to the SI metric modules.

Hectare (ha) A unit of area equal to 10 000 square metres.

Hecto- (h) A prefix meaning one hundred (10^2).

Henry (H) SI unit of inductance.

Hertz (Hz) SI unit of frequency.

Hour (h) A unit of 3 600 seconds of time.

ISO International Organization for Standardization. A nongovernmental group which works on a global basis and whose sole purpose is coordination, approval, and issuance of voluntary international standards.

International System of Units (SI) The modern metric system defined and adopted by the General Conference of Weights and Measures.

Joule (J) SI unit of energy.

Kelvin (K) SI unit of thermodynamic temperature.

Kilo- (k) SI prefix meaning one thousand (10^3).

Kilogram The basic SI unit of mass; equal to 1 000 grams.

Krypton 86 A colorless inert gaseous element; its light wavelength is used as the basis for defining the metre.

Litre (1) Unit of volume; equal to one cubic decimetre (dm^3).

Lumen (lm) SI unit of luminous flux (amount of light). A 100 watt light bulb emits about 1 700 lumens (170 lm/10W).

Lux (lx) SI unit of illumination. It is the density or radiant flux of one lumen per square metre.

Mega- (M) SI prefix meaning one million (10^6).

Metre (m) Basic SI unit of length; approximately equal to 1.1 yards.

Metric system Measurement system developed in France at the time of the French Revolution, based primarily on the metre, a length defined at that time as one ten millionth of the distance from the North Pole to the equator.

Metric ton Measure of weight equal to 1 000 kilograms, or about 2 200 pounds.

Metricize To convert any other unit to its metric (SI) equivalent. This may be an exact, rounded, or rationalized equivalent.

Micro- (μ) SI prefix meaning one millionth (10^{-6}).

Milli- (m) SI prefix meaning one thousandth (10^{-3}).

Mole (mol) The basic SI unit for amount of substance.

Nano- (n) SI prefix meaning one billionth (10^{-9}).

Newton (N) SI unit for that force which when applied to a body having a mass of 1 kg gives it an acceleration of one metre per second. It is independent of the earth's gravitation.

OMFS Optimum Metric Fastener System. OMFS fasteners are compatible with ISO metric series. Thread profile differs from ISO only in a slight reduction of thread height to permit a larger root radius in the external thread for improved fatigue resistance.

Ohm (Ω) SI electrical unit of resistance. An ohm is equal to the voltage divided by the amperage ($\Omega = V/A$).

Pascal (Pa) SI unit of pressure which is a force delivered over a given area.

Pico- (p) SI prefix meaning one trillionth (10^{-12}).

Power The time rate of doing work, measured in units of joules per second. The SI unit of power is the watt (W). See watt.

Preferred numbers A certain series of numbers (often called *Renard numbers*) increasing in a geometric progression. When these numbers are used for sizes of parts, they provide a logical and orderly assortment of sizes, with maximum incremental efficiency for a given range.

Prefixes Multiples and submultiples described in increments of 1 000 and applied to SI units.

Rounding off Manipulation of numerical data applied to values that are the result of conversion of a quantity (customary to SI units, or vice versa). The object is to rewrite the value by rounding within the desired accuracy.

SI The International Metric System of Units (from the French "Système International d'Unités"). Composed of seven base units: length, mass, time, electric current, temperature, amount of substance, and luminous intensity; two supplementary units of plane and solid angles; and derived units of compound units formed by algebraic combination of base units, supplementary units, and other derived units.

Second (s) Smallest basic unit of time in the SI and customary systems.

Siemens (S) SI unit of conductance. Conductance is the opposite of resistance. A passive circuit element which has a conductance of one siemens will allow a current flow of one ampere when a potential difference of one volt is applied to it.

Soft conversion The original data or

design is based upon the customary modules. The conversion to metric values consists of multiplying the customary value by the appropriate conversion factor and, where desired, rounded off.

Square metres (m²) SI measurement of area.

Tera- SI prefix meaning one trillion (10^{12}).

Tesla (T) SI unit of flux density.

Tonne (t) 1 000 kilograms; also called a metric ton.

U.S. Customary Units Units based upon the yard and pound that are commonly used in the United States of America and defined by the National Bureau of Standards. Some of these units have the same name as similar units in the United Kingdom (British, United Kingdom, or Imperial units) but are not necessarily equal to them.

USMA United States Metric Association, Inc. A volunteer organization promoting the adoption of SI in America.

Volt (V) SI unit of potential difference. It is also called voltage or electromotive force.

Watt (W) SI unit of power.

Weber (Wb) SI unit of magnetic flux.

Appendix C

REFERENCE TABLES

The following tables may be used for engineering drawings. They cover only the most commonly used information. For more detailed engineering data use, references from a machinist's handbook or engineering tables. Up-to-date references should be used to keep up with rapidly changing industrial standards.

UNIFIED AND AMERICAN SCREW THREADS

Size	Coarse Thread Series UNC & NC Threads Per Inch	Tap Drill	Fine Thread Series UNF & NF Threads Per Inch	Tap Drill	Extra Fine Thread Series UNEF & NEF Threads Per Inch	Tap Drill	8-Pitch Thread Series 8 N Threads Per Inch	Tap Drill	12-Pitch Thread Series 12 N Threads Per Inch	Tap Drill	16-Pitch Thread Series 16 N Threads Per Inch	Tap Drill
.060		80	80	3/64								
1 .073	64	No. 53	72	No. 53								
2 .086	56	No. 50	64	No. 50								
3 .099	48	No. 47	56	No. 45								
4 .112	40	No. 43	48	No. 42								
5 .125	40	No. 38	44	No. 37								
6 .138	32	No. 36	40	No. 33								
8 .164	32	No. 29	36	No. 29								
10 .190	24	No. 25	32	No. 21								
12 .216	24	No. 16	28	No. 14	32	No. 13						
1/4	20	No. 7	28	No. 3	32	7/32						
5/16	18	F	24	I	32	9/32						
3/8	16	5/16	24	Q	32	11/32						
7/16	14	U	20	25/64	32	13/32						
1/2	13	27/64										
1/2	12		20	29/64	28	15/32			12	27/64		
9/16	12	31/64	18	33/64	24	33/64			12	31/64		
5/8	11	17/32	18	37/64	24	37/64			12	35/64		
3/4	10	21/32	16	11/16	20	45/64			12	43/64	16	11/16
7/8	9	49/64	14	13/16	20	53/64			12	51/64	16	13/16
1	8	7/8	12	59/64	20	61/64	8	7/8	12	59/64	16	15/16
1 1/8	7	63/64	12	1 3/64	18	1 5/64	8	1	12	1 3/64	16	1 1/16
1 1/4	7	1 7/64	12	1 11/64	18	1 3/16	8	1 1/8	12	1 11/64	16	1 3/16
1 3/8	6	1 7/32	12	1 19/64	18	1 5/16	8	1 1/4	12	1 19/64	16	1 5/16
1 1/2	6	1 11/32	12	1 27/64	18	1 7/16	8	1 3/8	12	1 27/64	16	1 7/16
1 3/4	5	1 9/16			16	1 11/16	8	1 5/8	12	1 43/64	16	1 11/16
2	4 1/2	1 25/32			16	1 15/16	8	1 7/8	12	1 51/64	16	1 15/16
2 1/4	4 1/2	2 1/32					8	2 1/8	12	2 11/64	16	2 3/16
2 1/2	4	2 1/4					8	2 3/8	12	2 27/64	16	2 7/16
2 3/4	4	2 1/2					8	2 5/8	12	2 43/64	16	2 11/16
3	4	2 3/4					8	2 7/8	12	2 59/64	16	2 15/16
3 1/4	4	3					8	3 1/8	12	3 11/64	16	3 3/16
3 1/2	4	3 1/4					8	3 3/8	12	3 27/64	16	3 7/16
3 3/4	4	3 1/2					8	3 5/8	12	3 43/64	16	3 11/16
4	4	3 3/4					8	3 7/8	12	3 59/64	16	3 15/16
4 1/4							8	4 1/8	12	4 11/64	16	4 3/16
4 1/4							8	4 3/8	12	4 27/64	16	4 7/16
4 3/4							8	4 5/8	12	4 43/64	16	4 11/16
5							8	4 7/8	12	4 59/64	16	4 15/16
5 1/4							8	5 1/8	12	5 11/64	16	5 3/16
5 1/2							8	5 3/8	12	5 27/64	16	5 7/16
5 3/4							8	5 5/8	12	5 43/64	16	5 11/16
6							8	5 7/8	12	5 59/64	16	5 15/16

HEXAGON BOLTS

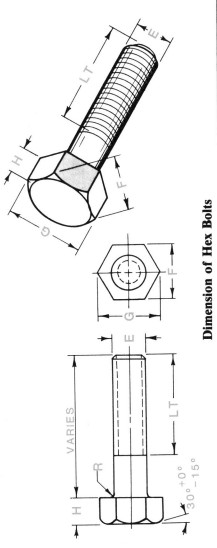

Dimension of Hex Bolts

Nominal Size or Basic Product Diameter		E Body Diameter	F Width Across Flats			G Width Across Corners		H Height			R Radius of Fillet		LT Thread Length for Bolt Lengths	
		Max	Basic	Max	Min	Max	Min	Basic	Max	Min	Max	Min	6 in. and Less Basic	Over 6 in. Basic
1/4	0.2500	0.260	7/16	0.438	0.425	0.505	0.484	11/64	0.188	0.150	0.03	0.01	0.750	1.000
5/16	0.3125	0.324	1/2	0.500	0.484	0.577	0.552	7/32	0.235	0.195	0.03	0.01	0.875	1.125
3/8	0.3750	0.388	9/16	0.562	0.544	0.650	0.620	1/4	0.268	0.226	0.03	0.01	1.000	1.250
7/16	0.4375	0.452	5/8	0.625	0.603	0.722	0.687	19/64	0.316	0.272	0.03	0.01	1.125	1.375
1/2	0.5000	0.515	3/4	0.750	0.725	0.866	0.826	11/32	0.364	0.302	0.03	0.01	1.250	1.500
5/8	0.6250	0.642	15/16	0.938	0.906	1.083	1.033	27/64	0.444	0.378	0.06	0.02	1.500	1.750
3/4	0.7500	0.768	1 1/8	1.125	1.088	1.299	1.240	1/2	0.524	0.455	0.06	0.02	1.750	2.000
7/8	0.8750	0.895	1 5/16	1.312	1.269	1.516	1.447	37/64	0.604	0.531	0.06	0.02	2.000	2.250
1	1.0000	1.022	1 1/2	1.500	1.450	1.732	1.653	43/64	0.700	0.591	0.09	0.03	2.250	2.500
1 1/8	1.1250	1.149	1 11/16	1.688	1.631	1.949	1.859	3/4	0.780	0.658	0.09	0.03	2.500	2.750
1 1/4	1.2500	1.277	1 7/8	1.875	1.812	2.165	2.066	27/32	0.876	0.749	0.09	0.03	2.750	3.000
1 3/8	1.3750	1.404	2 1/16	2.062	1.994	2.382	2.273	29/32	0.940	0.810	0.09	0.03	3.000	3.250
1 1/2	1.5000	1.531	2 1/4	2.250	2.175	2.598	2.480	1	1.036	0.902	0.09	0.03	3.250	3.500
1 3/4	1.7500	1.785	2 5/8	2.625	2.538	3.031	2.893	1 5/32	1.196	1.054	0.12	0.04	3.750	4.000
2	2.0000	2.039	3	3.000	2.900	3.464	3.306	1 11/32	1.388	1.175	0.12	0.04	4.250	4.500
2 1/4	2.2500	2.305	3 3/8	3.375	3.262	3.897	3.719	1 1/2	1.548	1.327	0.19	0.06	4.750	5.000
2 1/2	2.5000	2.559	3 3/4	3.750	3.625	4.330	4.133	1 21/32	1.708	1.479	0.19	0.06	5.250	5.500
2 3/4	2.7500	2.827	4 1/8	4.125	3.988	4.763	4.546	1 13/16	1.869	1.632	0.19	0.06	5.750	6.000
3	3.0000	3.081	4 1/2	4.500	4.350	5.196	4.959	2	2.060	1.815	0.19	0.06	6.250	6.500
3 1/4	3.2500	3.335	4 7/8	4.875	4.712	5.629	5.372	2 3/16	2.251	1.936	0.19	0.06	6.750	7.000
3 1/2	3.5000	3.589	5 1/4	5.250	5.075	6.062	5.786	2 5/16	2.380	2.057	0.19	0.06	7.250	7.500
3 3/4	3.7500	3.858	5 5/8	5.625	5.437	6.495	6.198	2 1/2	2.572	2.241	0.19	0.06	7.750	8.000
4	4.0000	4.111	6	6.000	5.800	6.928	6.612	2 11/16	2.764	2.424	0.19	0.06	8.250	8.500

30° +0° −15°

VARIES

FINISHED HEXAGON BOLTS

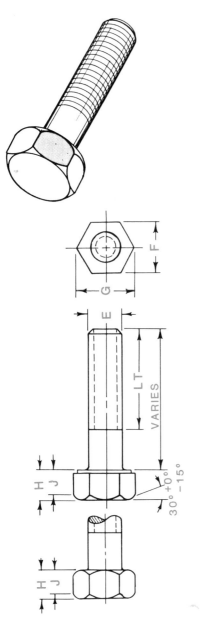

Dimensions of Hex Cap Screws (Finished Hex Bolts)

Nominal Size or Basic Product Diameter		E Body Diameter		F Width Across Flats			G Width Across Corners		H Height			J Wrenching Height	LT Thread Length for Screw Lengths	
		Max	Min	Basic	Max	Min	Max	Min	Basic	Max	Min	Min	6 in. and Less Basic	Over 6 in. Basic
1/4	0.2500	0.2500	0.2450	7/16	0.438	0.428	0.505	0.488	5/32	0.163	0.150	0.106	0.750	1.000
5/16	0.3125	0.3125	0.3065	1/2	0.500	0.489	0.577	0.557	13/64	0.211	0.195	0.140	0.875	1.125
3/8	0.3750	0.3750	0.3690	9/16	0.562	0.551	0.650	0.628	15/64	0.243	0.226	0.160	1.000	1.250
7/16	0.4375	0.4375	0.4305	5/8	0.625	0.612	0.722	0.698	9/32	0.291	0.272	0.195	1.125	1.375
1/2	0.5000	0.5000	0.4930	3/4	0.750	0.736	0.866	0.840	5/16	0.323	0.302	0.215	1.250	1.500
9/16	0.5625	0.5625	0.5545	13/16	0.812	0.798	0.938	0.910	23/64	0.371	0.348	0.250	1.375	1.625
5/8	0.6250	0.6250	0.6170	15/16	0.938	0.922	1.083	1.051	25/64	0.403	0.378	0.269	1.500	1.750
3/4	0.7500	0.7500	0.7410	1 1/8	1.125	1.100	1.299	1.254	15/32	0.483	0.455	0.324	1.750	2.000
7/8	0.8750	0.8750	0.8660	1 5/16	1.312	1.285	1.516	1.465	35/64	0.563	0.531	0.378	2.000	2.250
1	1.0000	1.0000	0.9900	1 1/2	1.500	1.469	1.732	1.675	39/64	0.627	0.591	0.416	2.250	2.500
1 1/8	1.1250	1.1250	1.1140	1 11/16	1.688	1.631	1.949	1.859	11/16	0.718	0.658	0.461	2.500	2.750
1 1/4	1.2500	1.2500	1.2390	1 7/8	1.875	1.812	2.165	2.066	25/32	0.813	0.749	0.530	2.750	3.000
1 3/8	1.3750	1.3750	1.3630	2 1/16	2.062	1.994	2.382	2.273	27/32	0.878	0.810	0.569	3.000	3.250
1 1/2	1.5000	1.5000	1.4880	2 1/4	2.250	2.175	2.598	2.480	15/16	0.974	0.902	0.640	3.250	3.500
1 3/4	1.7500	1.7500	1.7380	2 5/8	2.625	2.538	3.031	2.893	1 3/32	1.134	1.054	0.748	3.750	4.000
2	2.0000	2.0000	1.9880	3	3.000	2.900	3.464	3.306	1 7/32	1.263	1.175	0.825	4.250	4.500
2 1/4	2.2500	2.2500	2.2380	3 3/8	3.375	3.262	3.897	3.719	1 3/8	1.423	1.327	0.933	4.750	5.000
2 1/2	2.5000	2.5000	2.4880	3 3/4	3.750	3.625	4.330	4.133	1 17/32	1.583	1.479	1.042	5.250	5.500
2 3/4	2.7500	2.7500	2.7380	4 1/8	4.125	3.988	4.763	4.546	1 11/16	1.744	1.632	1.151	5.750	6.000
3	3.0000	3.0000	2.9880	4 1/2	4.500	4.350	5.196	4.959	1 7/8	1.935	1.815	1.290	6.250	6.500

30° +0°
−15°

VARIES

LT

E G F

H J

SQUARE BOLTS

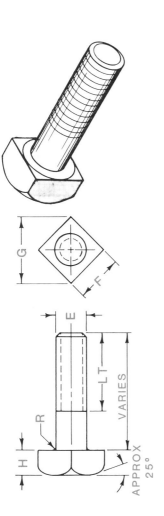

Dimensions of Square Bolts

Nominal Size or Basic Product Diameter		E Body Diameter Max	F Width Across Flats Basic	F Width Across Flats Max	F Width Across Flats Min	G Width Across Corners Max	G Width Across Corners Min	H Height Basic	H Height Max	H Height Min	R Radius of Fillet Max	R Radius of Fillet Min	LT Thread Length for Bolt Lengths — 6 in. and Less Basic	LT Thread Length for Bolt Lengths — Over 6 in. Basic
1/4	0.2500	0.260	3/8	0.375	0.362	0.530	0.498	11/64	0.188	0.156	0.03	0.01	0.750	1.000
5/16	0.3125	0.324	1/2	0.500	0.484	0.707	0.665	13/64	0.220	0.186	0.03	0.01	0.875	1.125
3/8	0.3750	0.388	9/16	0.562	0.544	0.795	0.747	1/4	0.268	0.232	0.03	0.01	1.000	1.250
7/16	0.4375	0.452	5/8	0.625	0.603	0.884	0.828	19/64	0.316	0.278	0.03	0.01	1.125	1.375
1/2	0.5000	0.515	3/4	0.750	0.725	1.061	0.995	21/64	0.348	0.308	0.03	0.01	1.250	1.500
5/8	0.6250	0.642	15/16	0.938	0.906	1.326	1.244	27/64	0.444	0.400	0.06	0.02	1.500	1.750
3/4	0.7500	0.768	1 1/8	1.125	1.088	1.591	1.494	1/2	0.524	0.476	0.06	0.02	1.750	2.000
7/8	0.8750	0.895	1 5/16	1.312	1.269	1.856	1.742	19/32	0.620	0.568	0.06	0.02	2.000	2.250
1	1.0000	1.022	1 1/2	1.500	1.450	2.121	1.991	21/32	0.684	0.628	0.09	0.03	2.250	2.500
1 1/8	1.1250	1.149	1 11/16	1.688	1.631	2.386	2.239	3/4	0.780	0.720	0.09	0.03	2.500	2.750
1 1/4	1.2500	1.277	1 7/8	1.875	1.812	2.652	2.489	27/32	0.876	0.812	0.09	0.03	2.750	3.000
1 3/8	1.3750	1.404	2 1/16	2.062	1.994	2.917	2.738	29/32	0.940	0.872	0.09	0.03	3.000	3.250
1 1/2	1.5000	1.531	2 1/4	2.250	2.175	3.182	2.986	1	1.036	0.964	0.09	0.03	3.250	3.500

HEXAGON NUTS AND JAM NUTS

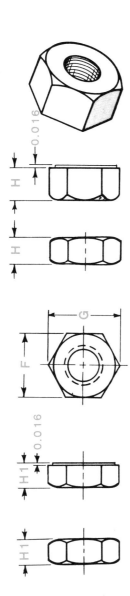

Dimensions of Hex Nuts and Hex Jam Nuts

Nominal Size or Basic Major Diameter of Thread		F Width Across Flats			G Width Across Corners		H Thickness Hex Nuts			H1 Thickness Hex Jam Nuts		
		Basic	Max	Min	Max	Min	Basic	Max	Min	Basic	Max	Min
1/4	0.2500	7/16	0.438	0.428	0.505	0.488	7/32	0.226	0.212	5/32	0.163	0.150
5/16	0.3125	1/2	0.500	0.489	0.577	0.557	17/64	0.273	0.258	3/16	0.195	0.180
3/8	0.3750	9/16	0.562	0.551	0.650	0.628	21/64	0.337	0.320	7/32	0.227	0.210
7/16	0.4375	11/16	0.688	0.675	0.794	0.768	3/8	0.385	0.365	1/4	0.260	0.240
1/2	0.5000	3/4	0.750	0.736	0.866	0.840	7/16	0.448	0.427	5/16	0.323	0.302
9/16	0.5625	7/8	0.875	0.861	1.010	0.982	31/64	0.496	0.473	5/16	0.324	0.301
5/8	0.6250	15/16	0.938	0.922	1.083	1.051	35/64	0.559	0.535	3/8	0.387	0.363
3/4	0.7500	1 1/8	1.125	1.088	1.299	1.240	41/64	0.665	0.617	27/64	0.446	0.398
7/8	0.8750	1 5/16	1.312	1.269	1.516	1.447	3/4	0.776	0.724	31/64	0.510	0.458
1	1.0000	1 1/2	1.500	1.450	1.732	1.653	55/64	0.887	0.831	35/64	0.575	0.519
1 1/8	1.1250	1 11/16	1.688	1.631	1.949	1.859	31/32	0.999	0.939	39/64	0.639	0.579
1 1/4	1.2500	1 7/8	1.875	1.812	2.165	2.066	1 1/16	1.094	1.030	23/32	0.751	0.687
1 3/8	1.3750	2 1/16	2.062	1.994	2.382	2.273	1.11/64	1.206	1.138	25/32	0.815	0.747
1 1/2	1.5000	2 1/4	2.250	2.175	2.598	2.480	1 9/32	1.317	1.245	27/32	0.880	0.808

SQUARE NUTS

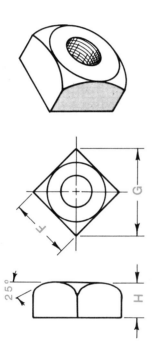

Dimensions of Square Nuts

Nominal Size or Basic Major Diameter of Thread		F Width Across Flats			G Width Across Corners		H Thickness		
		Basic	Max	Min	Max	Min	Basic	Max	Min
1/4	0.2500	7/16	0.438	0.425	0.619	0.584	7/32	0.235	0.203
5/16	0.3125	9/16	0.562	0.547	0.795	0.751	17/64	0.283	0.249
3/8	0.3750	5/8	0.625	0.606	0.884	0.832	21/64	0.346	0.310
7/16	0.4375	3/4	0.750	0.728	1.061	1.000	3/8	0.394	0.356
1/2	0.5000	13/16	0.812	0.788	1.149	1.082	7/16	0.458	0.418
5/8	0.6250	1	1.000	0.969	1.414	1.330	35/64	0.569	0.525
3/4	0.7500	1 1/8	1.125	1.088	1.591	1.494	21/32	0.680	0.632
7/8	0.8750	1 5/16	1.312	1.269	1.856	1.742	49/64	0.792	0.740
1	1.0000	1 1/2	1.500	1.450	2.121	1.991	7/8	0.903	0.847
1 1/8	1.1250	1 11/16	1.688	1.631	2.386	2.239	1	1.030	0.970
1 1/4	1.2500	1 7/8	1.875	1.812	2.652	2.489	1 3/32	1.126	1.062
1 3/8	1.3750	2 1/16	2.062	1.994	2.917	2.738	1 13/64	1.237	1.169
1 1/2	1.5000	2 1/4	2.250	2.175	3.182	2.986	1 5/16	1.348	1.276

25°

MACHINE SCREW HEADS AND NUTS

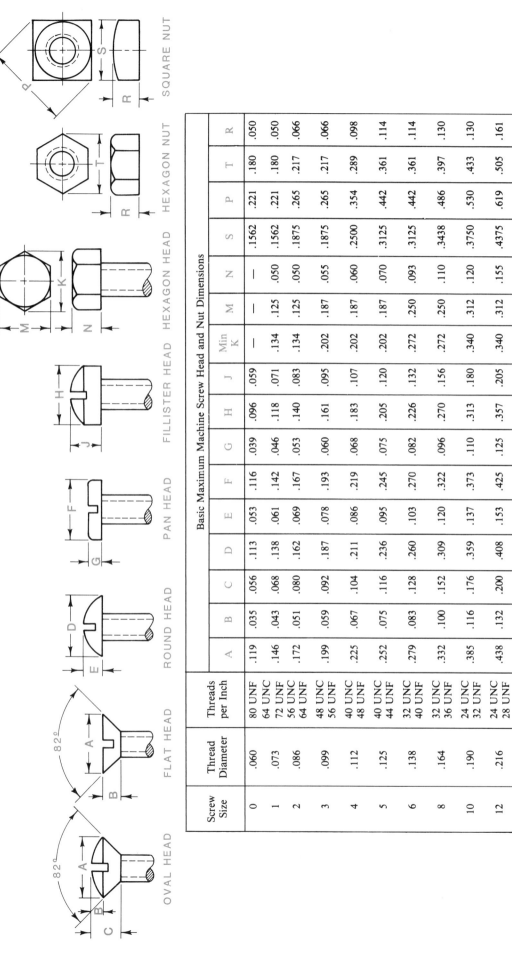

SQUARE NUT   HEXAGON NUT   HEXAGON HEAD   FILLISTER HEAD   PAN HEAD   ROUND HEAD   FLAT HEAD   OVAL HEAD

Basic Maximum Machine Screw Head and Nut Dimensions

Screw Size	Thread Diameter	Threads per Inch	A	B	C	D	E	F	G	H	J	Min K	M	N	S	P	T	R
0	.060	80 UNF	.119	.035	.056	.113	.053	.116	.039	.096	.059	—	—	—	.1562	.221	.180	.050
1	.073	64 UNC / 72 UNF	.146	.043	.068	.138	.061	.142	.046	.118	.071	.134	.125	.050	.1562	.221	.180	.050
2	.086	56 UNC / 64 UNF	.172	.051	.080	.162	.069	.167	.053	.140	.083	.134	.125	.050	.1875	.265	.217	.066
3	.099	48 UNC / 56 UNF	.199	.059	.092	.187	.078	.193	.060	.161	.095	.202	.187	.055	.1875	.265	.217	.066
4	.112	40 UNC / 48 UNF	.225	.067	.104	.211	.086	.219	.068	.183	.107	.202	.187	.060	.2500	.354	.289	.098
5	.125	40 UNC / 44 UNF	.252	.075	.116	.236	.095	.245	.075	.205	.120	.202	.187	.070	.3125	.442	.361	.114
6	.138	32 UNC / 40 UNF	.279	.083	.128	.260	.103	.270	.082	.226	.132	.272	.250	.093	.3125	.442	.361	.114
8	.164	32 UNC / 36 UNF	.332	.100	.152	.309	.120	.322	.096	.270	.156	.272	.250	.110	.3438	.486	.397	.130
10	.190	24 UNC / 32 UNF	.385	.116	.176	.359	.137	.373	.110	.313	.180	.340	.312	.120	.3750	.530	.433	.130
12	.216	24 UNC / 28 UNF	.438	.132	.200	.408	.153	.425	.125	.357	.205	.340	.312	.155	.4375	.619	.505	.161
1/4	.250	20 UNC / 28 UNF	.507	.153	.232	.472	.175	.492	.144	.414	.237	.409	.375	.190	.4375	.619	.505	.193
5/16	.3125	18 UNC / 24 UNF	.635	.191	.290	.590	.216	.615	.178	.518	.295	.545	.500	.230	.5625	.795	.650	.225
3/8	.375	16 UNC / 24 UNF	.762	.230	.347	.708	.256	.740	.212	.622	.355	.614	.562	.295	.6250	.884	.722	.257

CAP SCREW HEADS

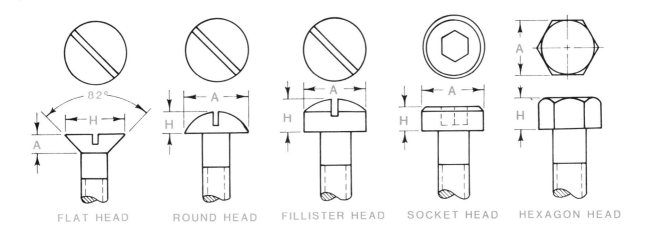

FLAT HEAD   ROUND HEAD   FILLISTER HEAD   SOCKET HEAD   HEXAGON HEAD

Nominal Diameter		Flat Head		Round Head		Fillister Head		Socket Head		Hexagon Head	
		A	H	A	H	A	H	A	H	A	H
0	.060							.096	.060		
1	.073							.118	.073		
2	.086							.140	.086		
3	.099							.161	.099		
4	.112							.183	.112		
5	.125							.205	.125		
6	.138							.226	.138		
8	.164							.270	.164		
10	.190							5/16	.190		
1/4	.250	.500	.140	.437	.191	.375	.172	3/8	1/4	7/16	11/64
5/16	.312	.625	.177	.562	.245	.437	.203	15/32	5/16	1/2	7/32
3/8	.375	.750	.210	.625	.273	.562	.250	9/16	3/8	9/16	1/4
7/16	.437	.812	.210	.750	.328	.625	.297	21/32	7/16	5/8	19/64
1/2	.500	.875	.210	.812	.354	.750	.328	3/4	1/2	3/4	11/32
9/16	.562	1.000	.244	.937	.409	.812	.375				
5/8	.625	1.125	.281	1.000	.437	.875	.422	15/16	5/8	15/16	27/64
3/4	.750	1.375	.352	1.250	.546	1.000	.500	1 1/8	3/4	1 1/8	1/2
7/8	.875	1.625	.423			1.125	.594	1 5/16	7/8	1 5/16	37/64
1	1.000	1.875	.494			1.312	.656	1 1/2	1	1 1/2	43/64
1 1/8	1.125	2.062	.529					1 11/16	1 1/8	1 11/16	3/4
1 1/4	1.250	2.312	.600					1 7/8	1 1/4	1 7/8	27/32
1 3/8	1.375	2.562	.665					2 1/16	1 3/8	2 1/16	29/32
1 1/2	1.500	2.812	.742					2 1/4	1 1/2	2 1/4	1

PLAIN WASHERS

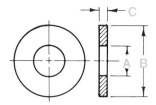

Inside Diameter A	Outside Diameter B	Nominal Thickness C	Inside Diameter A	Outside Diameter B	Nominal Thickness C	Inside Diameter A	Outside Diameter B	Nominal Thickness C
5/64	3/16	0.020	13/32	13/16	0.065	15/16	2	0.165
3/32	7/32	0.020	7/16	7/8	0.083	15/16	2 1/4	0.165
3/32	1/4	0.020	7/16	1	0.083	15/16	3 3/8	0.180
1/8	1/4	0.022	7/16	1 3/8	0.083	1 1/16	2	0.134
1/8	5/16	0.032	15/32	59/64	0.065	1 1/16	2 1/4	0.165
5/32	5/16	0.035	1/2	1 1/8	0.083	1 1/16	2 1/2	0.165
5/32	3/8	0.049	1/2	1 1/4	0.083	1 1/16	3 7/8	0.238
11/64	13/32	0.049	1/2	1 5/8	0.083	1 3/16	2 1/2	0.165
3/16	3/8	0.049	17/32	1 1/16	0.095	1 1/4	2 3/4	0.165
3/16	7/16	0.049	9/16	1 1/4	0.109	1 5/16	2 3/4	0.165
13/64	15/32	0.049	9/16	1 3/8	0.109	1 3/8	3	0.165
7/32	7/16	0.049	9/16	1 7/8	0.109	1 7/16	3	0.180
7/32	1/2	0.049	19/32	1 3/16	0.095	1 1/2	3 1/4	0.180
15/64	17/32	0.049	5/8	1 3/8	0.109	1 9/16	3 1/4	0.180
1/4	1/2	0.049	5/8	1 1/2	0.109	1 5/8	3 1/2	0.180
1/4	9/16	0.049	5/8	2 1/8	0.134	1 11/16	3 1/2	0.180
1/4	9/16	0.065	21/32	1 5/16	0.095	1 3/4	3 3/4	0.180
17/64	5/8	0.049	11/16	1 1/2	0.134	1 13/16	3 3/4	0.180
9/32	5/8	0.065	11/16	1 3/4	0.134	1 7/8	4	0.180
5/16	3/4	0.065	11/16	2 3/8	0.165	1 15/16	4	0.180
5/16	7/8	0.065	13/16	1 1/2	0.134	2	4 1/4	0.180
11/32	11/16	0.065	13/16	1 3/4	0.148	2 1/16	4 1/4	0.180
3/8	3/4	0.065	13/16	2	0.148	2 1/8	4 1/2	0.180
3/8	7/8	0.083	13/16	2 7/8	0.165	2 3/8	4 3/4	0.220
3/8	1 1/8	0.065	15/16	1 3/4	0.134	2 5/8	5	0.238
							2 7/8	5 1/4	0.259
							3 1/8	5 1/2	0.284

LOCK WASHERS

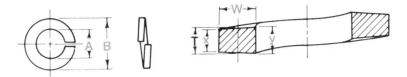

Dimensions of Regular Helical Spring Lock Washers

Nominal Washer Size		A		B	T	W
		Inside Diameter		Outside Diameter	Mean Section Thickness $\left(\frac{X + Y}{2}\right)$	Section Width
		Max	Min	Max	Min	Min
No. 2	0.086	0.094	0.088	0.172	0.020	0.035
No. 3	0.099	0.107	0.101	0.195	0.025	0.040
No. 4	0.112	0.120	0.114	0.209	0.025	0.040
No. 5	0.125	0.133	0.127	0.236	0.031	0.047
No. 6	0.138	0.148	0.141	0.250	0.031	0.047
No. 8	0.164	0.174	0.167	0.293	0.040	0.055
No. 10	0.190	0.200	0.193	0.334	0.047	0.062
No. 12	0.2;16	0.227	0.220	0.377	0.056	0.070
1/4	0.250	0.262	0.254	0.489	0.062	0.109
5/16	0.312	0.326	0.317	0.586	0.078	0.125
3/8	0.375	0.390	0.380	0.683	0.094	0.141
7/16	0.438	0.455	0.443	0.779	0.109	0.156
1/2	0.500	0.518	0.506	0.873	0.125	0.171
9/16	0.562	0.582	0.570	0.971	0.141	0.188
5/8	0.625	0.650	0.635	1.079	0.156	0.203
11/16	0.688	0.713	0.698	1.176	0.172	0.219
3/4	0.750	0.775	0.760	1.271	0.188	0.234
13/16	0.812	0.843	0.824	1.367	0.203	0.250
7/8	0.875	0.905	0.887	1.464	0.219	0.266
15/16	0.938	0.970	0.950	1.560	0.234	0.281
1	1.000	1.042	1.017	1.661	0.250	0.297
1 1/16	1.062	1.107	1.080	1.756	0.266	0.312
1 1/8	1.125	1.172	1.144	1.853	0.281	0.328
1 3/16	1.188	1.237	1.208	1.950	0.297	0.344
1 1/4	1.250	1.302	1.271	2.045	0.312	0.359
1 5/16	1.312	1.366	1.334	2.141	0.328	0.375
1 3/8	1.375	1.432	1.398	2.239	0.344	0.391
1 7/16	1.438	1.497	1.462	2.334	0.359	0.406
1 1/2	1.5(0	1.561	1.525	2.430	0.375	0.422

INCH-METRIC THREAD COMPARISON

Inch Series			Metric			
Size	Diameter (inches)	TPI	Size	Diameter (inches)	Pitch (mm)	TPI (Approximate)
			M1.4	.055	.3 .2	85 127
#0	.060	80				
			M1.6	.063	.35 .2	74 127
#1	.073	64 72				
			M2	.079	.4 .25	64 101
#2	.086	56 64				
			M2.5	.098	.45 .35	56 74
#3	.099	48 56				
#4	.112	40 48				
			M3	.118	.5 .35	51 74
#5	.125	40 44				
#6	.138	32 40				
			M4	.157	.7 .5	36 51
#8	.164	32 36				
#10	.190	24 32				
			M5	.196	.8 .5	32 51
			M6	.236	1.0 .75	25 34
1/4	.250	20 28				
5/16	.312	18 24				
			M8	.315	1.25 1.0	20 25
3/8	.375	16 24				
			M10	.393	1.5 1.25	17 20
7/16	.437	14 20				
			M12	.472	1.75 1.25	14.5 20
1/2	.500	13 20				
			M14	.551	2 1.5	12.5 17
5/8	.625	11 18				
			M16	.630	2 1.5	12.5 17
			M18	.709	2.5 1.5	10 17
3/4	.750	10 16				
			M20	.787	2.5 1.5	10 17
			M22	.866	2.5 1.5	10 17
7/8	.875	9 14				
			M24	.945	3 2	8.5 12.5
1"	1.000	8 12				
			M27	1.063	3 2	8.5 12.5

METRIC/INCH DRILL SIZES

Standard Inch Drills	Metric Drills mm	Equiv. Inch	Standard Inch Drills	Metric Drills mm	Equiv. Inch
0.0156	0.40	0.0157		1.75	0.0689
0.018	0.45	0.0177	0.070	1.8	0.0709
0.020	0.50	0.0197	0.073	1.85	0.0728
0.0025	0.58	0.0228		1.9	0.0748
0.024	0.60	0.0236	0.076	1.95	0.0768
0.026	0.65	0.0256	0.0781	2.0	0.0787
0.028	0.70	0.0276	0.081	2.05	0.0807
0.0295	0.75	0.0295	0.0827	2.1	0.0827
0.0312	0.80	0.0315		2.15	0.0846
0.033	0.85	0.0335	0.086	2.2	0.0866
0.035	0.90	0.0354	0.089	2.25	0.0886
0.037	0.95	0.0374	0.0906	2.3	0.0906
0.039	1.00	0.0394		2.35	0.0925
0.041	1.05	0.0413	0.0938	2.4	0.0945
0.043	1.10	0.0433	0.096	2.45	0.0965
0.0453	1.15	0.0453	0.098	2.5	0.0984
0.0469	1.20	0.0472	0.0995	2.55	0.1004
0.0492	1.25	0.0492	0.1015	2.55	0.1004
0.0512	1.30	0.0512	0.1024	2.6	0.1024
0.0531	1.35	0.0531	0.104	2.65	0.1043
0.055	1.4	0.0551	0.1065	2.7	0.1063
0.0571	1.45	0.0571	0.1094	2.75	0.1083
0.0591	1.5	0.0591	0.1110	2.8	0.1102
0.061	1.55	0.0610	0.113	2.85	0.1122
0.0625	1.6	0.0630		2.9	0.1142
0.0635	1.6	0.0630	0.116	2.95	0.1161
0.065	1.65	0.0650	0.120	3.0	0.1181
0.067	1.7	0.0669	0.122	3.1	0.1220

METRIC/INCH DRILL SIZES (CONT.)

Standard Inch Drills	Metric Drills		Standard Inch Drills	Metric Drills	
	mm	Equiv. Inch		mm	Equiv. Inch
0.125	3.2	0.1260	0.252	6.4	0.2520
0.1285	3.25	0.1280	0.257	6.5	0.2559
0.1299	3.3	0.1299	0.261	6.6	0.2598
0.1339	3.4	0.1339	0.2656	6.7	0.2638
0.136	3.5	0.1378		6.8	0.2677
0.1378	3.5	0.1378	0.272	6.9	0.2717
0.1406	3.6	0.1417	0.277	7.0	0.2756
0.144	3.7	0.1457		7.1	0.2795
0.147	3.75	0.1476	0.2812	7.2	0.2835
0.1495	3.8	0.1496	0.2854	7.25	0.2854
0.152	3.9	0.1535		7.3	0.2874
0.154	3.9	0.1535	0.290	7.4	0.2913
0.1562	4.0	0.1575	0.2913	7.4	0.2913
0.159	4.0	0.1575	0.295	7.5	0.2953
0.161	4.1	0.1614	0.2969	7.6	0.2992
0.166	4.2	0.1654	0.302	7.7	0.3031
0.1695	4.3	0.1693	0.3071	7.8	0.3071
0.1719	4.4	0.1732	0.3125	7.9	0.3110
0.173	4.4	0.1723	0.316	8.0	0.315
0.177	4.5	0.1772		8.1	0.3189
0.180	4.6	0.1811	0.323	8.2	0.3228
0.182	4.6	0.1811	0.3281	8.3	0.3268
0.185	4.7	0.1850	0.332	8.4	0.3307
0.189	4.8	0.1890		8.5	0.3346
0.191	4.9	0.1929	0.339	8.6	0.3386
0.1935	4.9	0.1929	0.3438	8.7	0.3425
0.196	5.0	0.1968	0.348	8.8	0.3465
0.199	5.1	0.2008		8.9	0.3504
0.201	5.1	0.2008	0.3543	9.0	0.3543
0.2031	5.2	0.2047	0.358	9.1	0.3583
0.204	5.2	0.2047	0.3594	9.2	0.3622
0.2055	5.2	0.2047	0.368	9.3	0.3661
0.209	5.3	0.2087		9.4	0.3701
0.213	5.4	0.2126	0.375	9.5	0.3740
0.2188	5.5	0.2165	0.377	9.6	0.378
0.221	5.6	0.2205		9.7	0.3819
0.2244	5.7	0.2244	0.386	9.8	0.3868
0.228	5.8	0.2283	0.3906	9.9	0.3898
0.2344	5.9	0.2323		10.0	0.3937
0.238	6.0	0.2362	0.397	10.1	0.3976
0.2402	6.1	0.2402		10.2	0.4016
0.242	6.1	0.2402	0.404	10.3	0.4055
0.246	6.2	0.2441	0.4062	10.4	0.4094
0.250	6.3	0.2480	0.413	10.5	0.4134

WIRE GAUGE SIZES

American Wire Gauge		Metric Wire Sizes		American Wire Gauge		Metric Wire Sizes	
No.	Inch	mm	Inch Equiv.	No.	Inch	mm	Inch Equiv.
4/0	0.4600	11.8	0.4646	22	0.0254	0.63	0.0248
3/0	0.4096	10.0	0.3937	23	0.0226	0.56	0.0220
2/0	0.3648	9.0	0.3543	24	0.0201	0.50	0.0197
0	0.3249	8.0	0.3150	25	0.0179	0.45	0.0177
1	0.2893	7.1	0.2795	26	0.0159	0.40	0.0158
2	0.2576	6.3	0.2480	27	0.0142	0.355	0.0140
3	0.2294	5.6	0.2205	28	0.0126	0.315	0.0124
4	0.2043	5.0	0.1969	29	0.0113	0.280	0.0010
5	0.1819	4.5	0.1772	30	0.0100	0.250	0.00984
6	0.1620	4.0	0.1575	31	0.00893	0.224	0.00882
7	0.1443	3.55	0.1398	32	0.00795	0.200	0.00787
8	0.1285	3.15	0.1240	33	0.00708	0.180	0.00709
9	0.1144	2.80	0.1102	34	0.00631	0.160	0.00630
10	0.1019	2.50	0.0984	35	0.00562	0.140	0.00552
11	0.0907	2.24	0.0882	36	0.00500	0.125	0.00492
12	0.0808	2.00	0.0787	37	0.00445	0.112	0.00441
13	0.0702	1.80	0.0709	38	0.00397	0.100	0.00394
14	0.0641	1.60	0.0630	39	0.00353	0.090	0.00354
15	0.0571	1.40	0.0552	40	0.00315	0.080	0.00315
16	0.0508	1.25	0.0492	41	0.00280	0.071	0.00280
17	0.0453	1.12	0.0441	42	0.00249	0.063	0.00248
18	0.0403	1.00	0.0403	44	0.00198	0.050	0.00197
19	0.0359	0.90	0.0359	46	0.00157	0.040	0.00158
20	0.0320	0.80	0.0320	48	0.00124	0.0315	0.00124
21	0.0285	0.71	0.0285	50	0.00099	0.025	0.00098

SHEET METAL THICKNESS

Standard Inch Thickness	Metric ISO Series		Standard Inch Thickness	Metric ISO Series	
	mm	Inch		mm	Inch
0.004	0.100	0.0039	0.040	1.00	0.0394
0.006	0.160	0.0063	0.050	1.25	0.0492
0.008	0.200	0.0079	0.063	1.60	0.0630
0.010	0.250	0.0098	0.080	2.00	0.0787
0.012	0.315	0.0124	0.100	2.50	0.0984
0.016	0.400	0.0158	0.125	3.15	0.1240
0.020	0.50	0.0197	0.160	4.00	0.1575
0.025	0.63	0.0248	0.200	5.00	0.1969
0.032	0.80	0.0315	0.250	6.30	0.2480

Nominal Size		Outside Diameter			Inside Diameter		
USA[1]	ISO[2]	USA[1]		ISO[2]	USA[1]		ISO[2]	R10[3]
inch	mm	inch	mm	mm	inch	mm	mm	mm
1/8	6	0.405	10.2	10.2	0.269	6.8	6.2	6.5
1/4	8	.540	13.7	13.6	.364	9.2	8.9	8.0
								10.0
3/8	10	.675	17.1	17.1	.493	12.5	12.4	12.5
1/2	15	.840	21.3	21.4	.622	15.8	16.1	16.0
3/4	20	1.050	26.7	26.9	.824	20.9	21.6	20.0
1	25	1.315	33.4	33.8	1.049	26.6	27.3	25.0
1 1/4	32	1.660	42.2	42.4	1.380	35.1	35.9	31.5
1 1/2	40	1.900	48.3	48.4	1.610	40.9	41.9	40.0
2	50	2.375	60.3	60.2	2.067	52.5	52.9	50.0
2 1/2	65	2.875	73.0	76.0	2.469	62.7	68.7	63.0
3	80	3.500	88.9	88.8	3.068	77.9	80.7	80.0
3 1/2	90	4.000	101.6	101.2	3.548	90.1	93.1	
4	100	4.500	114.3	114.0	4.026	102.3	105.0	100.0
5	125	5.563	141.3	139.6	5.047	128.2	129.9	125.0
6	150	6.625	168.3	165.2	6.065	154.1	155.5	160.0
8	200	8.625	219.1	(4)	7.981	202.7		200.0
10	250	10.750	273.0	(4)	9.970	253.2		250.0
12	300	12.750	323.9	(4)	12.000	304.8		315.0

[1] ASTM A53-68, standard pipe.
[2] ISO Recommendation R65, Medium series.
[3] R10 Series of preferred number.
[4] USA dimensions are used.

Seven SI Base Units

Quantity	Name of Unit	Symbol
1 Length	Metre	m
2 Mass	Kilogram	kg
3 Electric current	Ampere	A
4 Thermodynamic temperature	Kelvin	K
5 Luminous intensity	Candela	cd

Quantity	Name of Unit	Symbol
6 Time	Second	s
7 Amount of substance	Mole	mol

SI Supplementary Units

Plane angle	Radian	rad
Solid angle	Steradian	sr

Special Units Used with SI

Quantity	Name of Unit	Unit Symbol	Expressed in Other Units
Energy	Electron volt	eV	1 eV = 1.602 19 $\times$ 10^{-19} J (approximately)
Mass of an atom	Atomic mass unit	u	1 u = 1.660 53 $\times$ 10^{-27} kg (approximately)
Length	Astronomical unit	AU	1 AU = 149 800 $\times$ 10^6 m (system of astronomical constants, 1964)
	Parsec	pc	1 pc = 206 265 AU = 30 857 $\times$ 10^{12} m (approximately)
Pressure of fluid	Bar	bar	1 bar = 10^5 Pa
Navigational distance	Nautical mile	$\cdots$	1 852 m
Nautical speed	Knot	$\cdots$	(1 852/3 600) m/s 0.514 44 m/s (approximately)

Other Units Used with SI

Quantity	Name of Unit	Unit Symbol	Expressed in Other Units
Time	Minute	min	1 min = 60 s
	Hour	h	1 h = 60 min
	Day	d	1 d = 24 h
Plane angle	Degree	°	1° = (π/180) rad
	Minute	′	1′ = (1/60)°
	Second	″	1″ = (1/60)′
Volume	Litre	l	1 l = 1 dm^3
Mass	Tonne	t	1 t = 10^3 kg

Spelling and Symbols for Metric Units

Unit	Symbol	Unit	Symbol	Unit	Symbol
Acre	acre	Cubic centimetre	cm³	Decilitre	dl
Are	a	Cubic decimetre	dm³	Decimetre	dm
Barrel	bbl	Cubic dekametre	dam³	Dekagram	dag
Board foot	fbm	Cubic foot	ft³	Dekalitre	dal
Bushel	bu	Cubic hectometre	hm³	Dekametre	dam
Carat	c	Cubic inch	in³	Dram, avoirdupois	dr avdp
Celsius, degree	°C	Cubic kilometre	km³	Fathom	fath
Centare	ca	Cubic metre	m³	Foot	ft
Centigram	cg	Cubic mile	mi³	Furlong	furlong
Centilitre	cl	Cubic millimetre	mm³	Gallon	gal
Centimetre	cm	Cubic yard	yd³	Grain	grain
Chain	ch	Decigram	dg	Gram	g
Hectare	ha	Microinch	μin	Rod	rod
Hectogram	hg	Microlitre	μl	Second	s
Hectolitre	hl	Mile	mi	Square centimetre	cm²
Hectometre	hm	Milligram	mg	Square decimetre	dm²
Hogshead	hhd	Millilitre	ml	Square dekametre	dam²
Hundredweight	cwt	Millimetre	mm	Square foot	ft²
Inch	in	Minim	minim	Square hectometre	hm²
International Nautical Mile	INM	Ounce	oz	Square inch	in²
		Ounce, avoirdupois	oz avdp	Square kilometre	km²
Kelvin	K	Ounce, liquid	liq oz	Square metre	m²
Kilogram	kg	Ounce, troy	oz tr	Square mile	mi²
Kilolitre	kl	Peck	peck	Square millimetre	mm²
Kilometre	km	Pennyweight	dwt	Square yard	yd²
Link	link	Pint, liquid	liq pt	Stere	stere
Liquid	liq	Pound	lb	Ton, long	long ton
Litre	litre	Pound, avoirdupois	lb avdp	Ton, metric	t
Metre	m	Pound, troy	lb tr	Ton, short	short ton
Microgram	μg	Quart, liquid	liq qt	Yard	yd

No period is used with symbols for units.
The same symbol is used for both singular and plural.
The exponents "2" and "3" are used to signify "square" and "cubic," respectively, instead of the symbols "sq" or "cu."

DERIVED SI UNITS

Quantity	Unit	SI Symbol	Formula
Acceleration	Metre per second squared	m/s²	· · ·
Angular acceleration	Radian per second squared	rad/s²	· · ·
Angular velocity	Radian per second	rad/s	
Area	Square metre	m²	· · ·
Density	Kilogram per cubic metre	kg/m³	· · ·
Electric capacitance	Farad	F	A·s/V
Electric charge	Coulomb	C	A·s
Electric field strength	Volt per metre	V/m	· · ·
Electric resistance	Ohm	Ω	V/A
Electromotive force	Volt	V	W/A
Energy	Joule	J	N·m
Force	Newton	N	kg·m/s²
Frequency	Hertz	Hz	s⁻¹
Illumination	Lux	lx	lm/m²
Inductance	Henry	H	V·s/A
Kinematic viscosity	Square metre per second	m²/s	· · ·
Luminance	Candela per square metre	cd/m²	· · ·
Luminous flux	Lumen	lm	cd·sr
Magnetic field strength	Ampere per metre	A/m	· · ·
Magnetic flux	Weber	Wb	V·s
Magnetic flux density	Tesla	T	Wb/m²
Magnetomotive force	Ampere	A	· · ·
Potential difference	Volt	V	W/A
Power	Watt	W	J/s
Pressure	Newton per square metre	N/m²	· · ·
Quantity of heat	Joule	J	N·m
Stress	Newton per square metre	N/m²	· · ·
Velocity	Metre per second	m/s	· · ·
Viscosity	Newton second per square metre	N·s/m²	· · ·
Voltage	Volt	V	W/A
Volume	Cubic metre	m³	· · ·
Work	Joule	J	N·m

PREFERRED ISO METRIC BOLT LENGTHS AND DIAMETERS (B.S. 3692:1963 APPENDIX)

Nominal Length mm

Coarse Thread

Nominal Size D	16	20	25	30	35	40	45	50	55	60	65	70	75	80	90	100	110	120	130	140	150	160	170	180	190	200	220	240
M3	X	X	X	X																								
M4		X	X	X	X	X	X	X																				
M5		X	X	X	X	X	X	X	X	X	X	X																
M6			X	X	X	X	X	X	X	X	X	X	X	X														
M8				X	X	X	X	X	X	X	X	X	X	X	X	X	X											
M10				X	X	X	X	X	X	X	X	X	X	X	X	X	X	X	X	X	X	X						
M12					X	X	X	X	X	X	X	X	X	X	X	X	X	X	X	X	X	X	X	X				
M16						X	X	X	X	X	X	X	X	X	X	X	X	X	X	X	X	X	X	X	X	X		
M20								X	X	X	X	X	X	X	X	X	X	X	X	X	X	X	X	X	X	X	X	
M24										X	X	X	X	X	X	X	X	X	X	X	X	X	X	X	X	X	X	X
M30													X	X	X	X	X	X	X	X	X	X	X	X	X	X	X	X
M36															X	X	X	X	X	X	X	X	X	X	X	X	X	X

Fine Thread

Nominal Size D	16	20	25	30	35	40	45	50	55	60	65	70	75	80	90	100	110	120	130	140	150	160	170	180	190	200	220	240
M8				X	X	X	X	X	X	X	X	X	X	X	X	X	X	X	X	X	X							
M10					X	X	X	X	X	X	X	X	X	X	X	X	X	X	X	X	X	X						
M12					X	X	X	X	X	X	X	X	X	X	X	X	X	X	X	X	X	X	X	X				
M16							X	X	X	X	X	X	X	X	X	X	X	X	X	X	X	X	X	X	X	X		
M20									X	X	X	X	X	X	X	X	X	X	X	X	X	X	X	X	X	X	X	
M24										X	X	X	X	X	X	X	X	X	X	X	X	X	X	X	X	X	X	X

PREFERRED ISO METRIC SCREW LENGTHS AND DIAMETERS (B.S. 3692:1963 APPENDIX)

Nominal Size D	Nominal Length																																
	4	5	6	8	10	12	14	16	20	25	30	35	40	45	50	55	60	65	70	75	80	90	100	110	120	130	140	150	160	170	180	190	200
Coarse Thread																																	
M3	X	X	X	X	X	X	X	X	X	X																							
M4		X	X	X	X	X	X	X	X	X	X	X	X																				
M5			X	X	X	X	X	X	X	X	X	X	X	X	X	X	X	X	X														
M6			X	X	X	X	X	X	X	X	X	X	X	X	X	X	X	X	X	X	X	X											
M8				X	X	X	X	X	X	X	X	X	X	X	X	X	X	X	X	X	X	X	X										
M10				X	X	X	X	X	X	X	X	X	X	X	X	X	X	X	X	X	X	X	X	X	X	X	X	X					
M12						X	X	X	X	X	X	X	X	X	X	X	X	X	X	X	X	X	X	X	X	X	X	X	X				
M16							X	X	X	X	X	X	X	X	X	X	X	X	X	X	X	X	X	X	X	X	X	X	X	X			
M20								X	X	X	X	X	X	X	X	X	X	X	X	X	X	X	X	X	X	X	X	X	X	X	X	X	X
M24									X	X		X	X	X	X	X	X	X	X	X	X	X	X	X	X	X	X	X	X	X	X	X	X
M30												X	X	X	X	X	X	X	X	X	X	X	X	X	X	X	X	X	X	X	X	X	X
M36													X	X	X	X	X	X	X	X	X	X	X	X	X	X	X	X	X	X	X	X	X
Fine Thread																																	
M8				X	X	X	X	X	X	X	X	X	X	X	X	X	X	X	X	X	X	X	X										
M10				X	X	X	X	X	X	X	X	X	X	X	X	X	X	X	X	X	X	X	X	X	X	X	X	X	X				
M12				X	X	X	X	X	X	X	X	X	X	X	X	X	X	X	X	X	X	X	X	X	X	X	X	X	X	X	X		
M16						X	X	X	X	X	X	X	X	X	X	X	X	X	X	X	X	X	X	X	X	X	X	X	X	X	X	X	X
M20								X	X	X	X	X	X	X	X	X	X	X	X	X	X	X	X	X	X	X	X	X	X	X	X	X	X
M24								X	X	X	X	X	X	X	X	X	X	X	X	X	X	X	X	X	X	X	X	X	X	X	X	X	X

BASIC FORM OF THE ISO METRIC THREAD

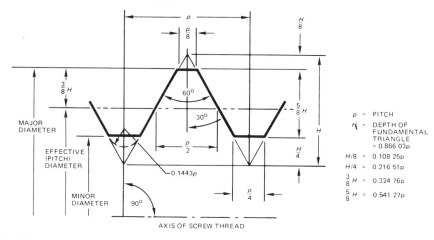

ISO METRIC THREADS

Nominal Size mm	Coarse Series			Fine Series	
	Pitch mm	Tap Drill mm	Clearance Drill mm	Pitch mm	Tap Drill mm
	Approximately 75% Thread				
1.4	0.3	1.1	1.55	. . .	. . .
1.6	0.35	1.25	1.8	. . .	. . .
2	0.4	1.6	2.2	. . .	. . .
2.5	0.45	2.05	2.6	. . .	. . .
3	0.5	2.5	3.2	. . .	. . .
4	0.7	3.3	4.2	. . .	. . .
5	0.8	4.2	5.2	. . .	. . .
6	1.0	5.0	6.2	. . .	. . .
8	1.25	6.75	8.2	1	7.0
10	1.5	8.5	10.2	1.25	8.75
12	1.75	10.25	12.2	1.25	10.50
14	2	12.00	14.2	1.5	12.50
16	2	14.00	16.45	1.5	14.50
18	2.5	15.50	18.20	1.5	16.50
20	2.5	17.50	20.50	1.5	18.50
22	2.5	19.50	22.80	1.5	20.50
24	3	21.00	24.60	2	22.00
27	3	24.00	27.95	2	25.00

Appendix D

The need to convert from one measuring system to another is the most difficult part of going metric. Trying to use two measuring systems or two languages to communicate is always difficult.

Until America becomes fully metric it will be necessary to make some conversions. The following tables will help you with these conversions.

COMMON CONVERSION FACTORS (APPROXIMATE) BY MULTIPLICATION FACTORS

	When You Know:	You Can Find:	If You Multiply By:
Length	Inches	Millimetres	25
	Feet	Centimetres	30
	Yards	Metres	0.9
	Miles	Kilometres	1.6
	Millimetres	Inches	0.04
	Centimetres	Inches	0.4
	Metres	Yards	1.1
	Kilometres	Miles	0.6
Area	Square inches	Square centimetres	6.5
	Square feet	Square metres	0.09
	Square yards	Square metres	0.8
	Square miles	Square kilometres	2.6
	Acres	Square hectometres (hectares)	0.4
	Square centimetres	Square inches	0.16
	Square metres	Square yards	1.2
	Square kilometres	Square miles	0.4
	Square hectometres (hectares)	Acres	2.5
Mass	Ounces	Grams	28
	Pounds	Kilograms	0.45
	Short tons	Megagrams (metric tons)	0.9
	Grams	Ounces	0.035
	Kilograms	Pounds	2.2
	Megagrams (metric tons)	Short tons	1.1
Liquid Volume	Ounces	Millilitres	30
	Pints	Litres	0.47
	Quarts	Litres	0.95
	Gallons	Litres	3.8
	Millilitres	Ounces	0.034
	Litres	Pints	2.1
	Litres	Quarts	1.06
	Litres	Gallons	0.26
Temperature	Degrees Fahrenheit	Degrees Celsius	$\frac{5}{9}$ (after subtracting 32)
	Degrees Celsius	Degrees Fahrenheit	$\frac{9}{5}$ (then add 32)

COMPARISON U.S. STANDARD AND METRIC DRAWING SCALES

Style	Existing English Graduation	Equivalent Direct Ratio	Nearest Standard Metric Ratio	Difference In Drawing Size*
Primary Architects	$1/8'' = 1'$	1:96	1:100	− 4%
	$1/4'' = 1'$	1:48	1:50	− 4%
	$3/8'' = 1'$	1:32	1:30	+ 7%
	$1/2'' = 1'$	1:24	1:20	+20%
	$3/4'' = 1'$	1:16	1:20	−20%
	$1'' = 1'$	1:12	1:10	+20%
	$1 1/2'' = 1'$	1:8	1:10	−20%
	$3'' = 1'$	1:4	1:5	−20%
Secondary Architects	$2'' = 1'$	1:6	1:5	+20%
	$4'' = 1'$	1:3	1:3	0%
	$3/16'' = 1'$	1:64	1:50	+28%
	$3/32'' = 1'$	1:128	1:100	+28%
	$1/16'' = 1'$	1:192	1:200	− 4%
	$1/32'' = 1'$	1:384	1:300	+28%
Primary Engineers	$1'' = 10'$	1:120	1:100	+20%
	$1'' = 20'$	1:240	1:200	+20%
	$1'' = 30'$	1:360	1:300	+20%
	$1'' = 40'$	1:480	1:500	− 4%
	$1'' = 50'$	1:600	1:500	+20%
	$1'' = 60'$	1:720	1:1000	−28%
Secondary Engineers	$1'' = 80'$	1:960	1:1000	− 4%
	$1'' = 100'$	1:1200	1:1000	−28%
Mechanical	Half size	1:2	1:2	0%
	Quarter size	1:4	1:5	−20%
	Eighth size	1:8	1:10	−20%
	$3/8$ size	$1:2\,2/3$	1:3	−11%
	2x size	2:1	2x	0%
	3x size	3:1	$3\,1/3$ x	+11%
	5x size	5:1	5x	0%

PREFIXES

Multiples and Submultiples of Base and Special Named Units   Example—Base Unit = Metre

Name of Factor	Multiplication Factors	Prefix	Symbol	Pronunciation	Multi. Factor		Quantity	Symbol
One quintillion	$1\,000\,000\,000\,000\,000\,000 = 10^{18}$	Exa	E	exa	10^{18}	metres =	exametre	= Em
One quadrillion	$1\,000\,000\,000\,000\,000 = 10^{15}$	Peta	P	pe'ta	10^{15}	metres =	petametre	= Pm
One trillion	$1\,000\,000\,000\,000 = 10^{12}$	Tera	T	ter'a	10^{12}	metres =	terametre	= Tm
One billion	$1\,000\,000\,000 = 10^{9}$	Giga	G	ji'ga	10^{9}	metres =	gigametre	= Gm
One million*	$1\,000\,000 = 10^{6}$	Mega	M	meg'a	10^{6}	metres =	megametre	= Mm
One thousand*	$1\,000 = 10^{3}$	kilo	k	kil'o	10^{3}	metres =	kilometre	= km
One hundred	$100 = 10^{2}$	hecto	h	hec'to	10^{2}	metres =	hectometre	= hm
Ten	$10 = 10^{1}$	deka	da	dek'a	10^{1}	metres =	dekametre	= dam
One (base unit)**	Base unit $1 = 10^{0}$				10^{0}	metres =	metre	= m
One-tenth	$0.1 = 10^{-1}$	deci	d	des'i	10^{-1}	metre =	decimetre	= dm
One-hundredth	$0.01 = 10^{-2}$	centi	c	sen'ti	10^{-2}	metre =	centimetre	= cm
One-thousandth*	$0.001 = 10^{-3}$	milli	m	mil'i	10^{-3}	metre =	millimetre	= mm
One-millionth*	$0.000\,001 = 10^{-6}$	micro	μ	mi'kro	10^{-6}	metre =	micrometre	= μm
One-billionth	$0.000\,000\,001 = 10^{-9}$	nano	n	nin'o	10^{-9}	metre =	nanometre	= nm
One-trillionth	$0.000\,000\,000\,001 = 10^{-12}$	pico	p	pe'ko	10^{-12}	metre =	picometre	= pm
One-quadrillionth	$0.000\,000\,000\,000\,001 = 10^{-15}$	femto	f	fem'to	10^{-15}	metre =	femtometre	= fm
One-quintillionth	$0.000\,000\,000\,000\,000\,001 = 10^{-18}$	atto	a	at'to	10^{-18}	metre =	attometre	= am

*Preferred multiplication factors
**Base unit to which the prefixes are assigned

METRIC-INCH EQUIVALENTS

Inches (Fractions)	Inches (Decimals)	Milli-metres	Inches (Fractions)	Inches (Decimals)	Milli-metres
	.004	.10		.472	12.00
	.008	.20	31/64	.484	12.30
	.012	.30	1/2	.500	12.70
1/64	.016	.39		.512	13.00
	.017	.40	33/64	.516	13.10
	.020	.50	17/32	.531	13.49
	.024	.60	35/64	.547	13.90
	.028	.70		.551	14.00
1/32	.031	.79	9/16	.563	14.29
	.032	.80	37/64	.578	14.69
	.035	.90		.591	15.00
	.039	1.00	19/32	.594	15.08
3/64	.047	1.19	39/64	.609	15.48
1/16	.063	1.59	5/8	.625	15.88
5/64	.078	1.98		.630	16.00
	.079	2.00	41/64	.641	16.27
3/32	.094	2.38	21/32	.656	16.67
7/64	.109	2.78		.669	17.00
	.118	3.00	43/64	.672	17.07
1/8	.125	3.18	11/16	.688	17.46
9/64	.141	3.57	45/64	.703	17.86
5/32	.156	3.97		.709	18.00
	.157	4.00	23/32	.719	18.26
11/64	.172	4.37	47/64	.734	18.65
3/16	.188	4.76		.748	19.00
	.197	5.00	3/4	.750	19.05
13/64	.203	5.16	49/64	.766	19.45
7/32	.219	5.56	25/32	.781	19.84
15/64	.234	5.95		.787	20.00
	.236	6.00	51/64	.797	20.24
1/4	.250	6.35	13/16	.813	20.64
17/64	.266	6.75		.827	21.00
	.276	7.00	53/64	.828	21.03
9/32	.281	7.14	27/32	.844	21.43
19/64	.297	7.54	55/64	.859	21.83
5/16	.313	7.94		.866	22.00
	.315	8.00	7/8	.875	22.23
21/64	.328	8.33	57/64	.891	22.62
11/32	.344	8.73		.905	23.00
	.354	9.00	29/32	.906	23.01
23/64	.359	9.13	59/64	.922	23.42
3/8	.375	9.53	15/16	.938	23.81
25/64	.391	9.92		.945	24.00
	.394	10.00	61/64	.953	24.21
13/32	.406	10.32	31/32	.969	24.61
27/64	.422	10.72		.984	25.00
	.433	11.00	63/64	.985	25.01
7/16	.438	11.11	1	1.000	25.40
29/64	.453	11.51	1 1/64	1.016	25.45
15/32	.469	11.91			

GENERAL AREA MEASUREMENTS WITH MULTIPLICATION FACTORS

To Convert From	To	Multiply By
	metre^2 (m^2)	$4.046\ 856 \times 10^3$
	metre^2 (m^2)	$1.000\ 000 \times 10^{-28}$
	metre^2 (m^2)	$5.067\ 000 \times 10^{-10}$
	metre^2 (m^2)	$9.290\ 304 \times 10^{-2}$
	metre^2 (m^2)	$6.541\ 600 \times 10^{-4}$
	metre^2 (m^2)	$2.589\ 988 \times 10^6$
	metre^2 (m^2)	$2.589\ 988 \times 10^6$
	metre^2 (m^2)	$9.323\ 957 \times 10^7$
	metre^2 (m^2)	$8.361\ 274 \times 10^{-1}$

	5	6	7	8	9	10	11
02	127	152	178	203	229	254	279
06	432	457	483	508	533	559	584
11	737	762	787	813	838	864	889
16	1.041	1.067	1.092	1.118	1.143	1.168	1.194
21	1.346	1.372	1.397	1.422	1.448	1.473	1.499
26	1.651	1.676	1.702	1.727	1.753	1.778	1.803
30	1.956	1.981	2.007	2.032	2.057	2.083	2.108
35	2.261	2.286	2.311	2.337	2.362	2.388	2.413
40	2.565	2.591	2.616	2.642	2.667	2.692	2.718
45	2.870	2.896	2.921	2.946	2.972	2.997	3.023
50	3.175	3.200	3.226	3.251	3.277	3.302	3.327
54	3.480	3.505	3.531	3.556	3.581	3.607	3.632
59	3.785	3.810	3.835	3.861	3.886	3.912	3.937
64	4.089	4.115	4.140	4.166	4.191	4.216	4.242
69	4.394	4.420	4.445	4.470	4.496	4.521	4.547

ADD

SUMMER FALL SPRING

UNITS

INSTRUCTOR

HOURS

DAYS

COURSE DESCRIPTION

NUMBER – SECTION

DEPT

NAME   LAST   FIRST   MI

SOC. SEC. NO.

STUDENT'S COPY — PINK COPY
INSTRUCTOR — CANARY COPY
ADMISSIONS & RECORDS - DATA PROC. — WHITE COPY

DIVISION CHAIRMAN'S SIGNATURE

COMMENTS:

ADMISSIONS & RECORDS

OR INSTRUCTOR'S SIGNATURE   DATE

STUDENT'S SIGNATURE   DATE

THIS FORM MUST BE PROCESSED BY ADMISSIONS WITHIN 24 HRS. OF THE ABOVE DATE

Feet	Inches											
	0	1	2	3	4	5	6	7	8	9	10	11
	Metres and Millimetres											
15	4.572	4.597	4.623	4.648	4.674	4.699	4.724	4.750	4.775	4.801	4.826	4.851
16	4.877	4.902	4.928	4.953	4.978	5.004	5.029	5.055	5.080	5.105	5.131	5.156
17	5.182	5.207	5.232	5.258	5.283	5.309	5.334	5.359	5.385	5.410	5.436	5.461
18	5.486	5.512	5.537	5.563	5.588	5.613	5.639	5.664	5.690	5.715	5.740	5.766
19	5.791	5.817	5.842	5.867	5.893	5.918	5.944	5.969	5.994	6.020	6.045	6.071
20	6.096	6.121	6.147	6.172	6.198	6.223	6.248	6.274	6.299	6.325	6.350	6.375
21	6.401	6.426	6.452	6.477	6.502	6.528	6.553	6.579	6.604	6.629	6.655	6.680
22	6.706	6.731	6.756	6.782	6.807	6.833	6.858	6.883	6.909	6.934	6.960	6.985
23	7.010	7.036	7.061	7.087	7.112	7.137	7.163	7.188	7.214	7.239	7.264	7.290
24	7.315	7.341	7.366	7.391	7.417	7.442	7.468	7.493	7.518	7.544	7.569	7.595
25	7.620	7.645	7.671	7.696	7.722	7.747	7.772	7.798	7.823	7.849	7.874	7.899
26	7.925	7.950	7.976	8.001	8.026	8.052	8.077	8.103	8.128	8.153	8.179	8.204
27	8.230	8.255	8.280	8.306	8.331	8.357	8.382	8.407	8.433	8.458	8.484	8.509
28	8.534	8.560	8.585	8.611	8.636	8.661	8.687	8.712	8.738	8.763	8.788	8.814
29	8.839	8.865	8.890	8.915	8.941	8.966	8.992	9.017	9.042	9.068	9.093	9.119
30	9.144	9.169	9.195	9.220	9.246	9.271	9.296	9.322	9.347	9.373	9.398	9.423
31	9.449	9.474	9.500	9.525	9.550	9.576	9.601	9.627	9.652	9.677	9.703	9.728
32	9.754	9.779	9.804	9.830	9.855	9.881	9.906	9.931	9.957	9.982	10.008	10.033
33	10.058	10.084	10.109	10.135	10.160	10.185	10.211	10.236	10.262	10.287	10.312	10.338
34	10.363	10.389	10.414	10.439	10.465	10.490	10.516	10.541	10.566	10.592	10.617	10.643
35	10.668	10.693	10.719	10.744	10.770	10.795	10.820	10.846	10.871	10.897	10.922	10.947
36	10.973	10.998	11.024	11.049	11.074	11.100	11.125	11.151	11.176	11.201	11.227	11.252
37	11.278	11.303	11.328	11.354	11.379	11.405	11.430	11.455	11.481	11.506	11.532	11.557
38	11.582	11.608	11.633	11.659	11.684	11.709	11.735	11.760	11.786	11.811	11.836	11.862
39	11.887	11.913	11.938	11.963	11.989	12.014	12.040	12.065	12.090	12.116	12.141	12.167
40	12.192	12.217	12.243	12.268	12.294	12.319	12.344	12.370	12.395	12.421	12.446	12.471
41	12.497	12.522	12.548	12.573	12.598	12.624	12.649	12.675	12.700	12.725	12.751	12.776
42	12.802	12.827	12.852	12.878	12.903	12.929	12.954	12.979	13.005	13.030	13.058	13.081
43	13.106	13.132	13.157	13.183	13.208	13.233	13.259	13.284	13.310	13.335	13.360	13.386
44	13.411	13.437	13.462	13.487	13.513	13.538	13.564	13.589	13.614	13.640	13.665	13.691
45	13.716	13.741	13.767	13.792	13.818	13.843	13.868	13.894	13.919	13.945	13.970	13.995
46	14.021	14.046	14.072	14.097	14.122	14.148	14.173	14.199	14.224	14.249	14.275	14.300
47	14.326	14.351	14.376	14.402	14.427	14.453	14.478	14.503	14.529	14.554	14.580	14.605
48	14.630	14.656	14.681	14.707	14.732	14.757	14.783	14.808	14.834	14.859	14.884	14.910
49	14.935	14.961	14.986	15.011	15.037	15.062	15.088	15.113	15.138	15.164	15.189	15.215
50	15.240	15.265	15.291	15.316	15.342	15.367	15.392	15.418	15.443	15.469	15.494	15.519
51	15.545	15.570	15.596	15.621	15.646	15.672	15.697	15.723	15.748	15.773	15.799	15.824
52	15.850	15.875	15.900	15.926	15.951	15.977	16.002	16.027	16.053	16.078	16.104	16.129
53	16.154	16.180	16.205	16.231	16.256	16.281	16.307	16.332	16.358	16.383	16.408	16.434
54	16.459	16.485	16.510	16.535	16.561	16.586	16.612	16.637	16.662	16.688	16.713	16.739
55	16.764	16.789	16.815	16.840	16.866	16.891	16.916	16.942	16.967	16.993	17.018	17.043
56	17.069	17.094	17.120	17.145	17.170	17.196	17.221	17.247	17.272	17.297	17.323	17.348
57	17.374	17.399	17.424	17.450	17.475	17.501	17.526	17.551	17.577	17.602	17.628	17.653

0′ TO 100′ BY ONE INCH INCREMENTS
FEET AND INCHES TO METRES (CONT.)

Feet	Inches											
	0	1	2	3	4	5	6	7	8	9	10	11
	Metres and Millimetres											
58	17.678	17.704	17.729	17.755	17.780	17.805	17.830	17.856	17.882	17.907	17.932	17.958
59	17.983	18.009	18.034	18.059	18.085	18.110	18.136	18.161	18.186	18.212	18.237	18.263
60	18.288	18.313	18.339	18.364	18.390	18.415	18.440	18.466	18.491	18.517	18.542	18.567
61	18.593	18.618	18.644	18.669	18.694	18.720	18.745	18.771	18.796	18.821	18.847	18.872
62	18.898	18.923	18.948	18.974	18.999	19.025	19.050	19.075	19.101	19.126	19.152	19.177
63	19.202	19.228	19.253	19.279	19.304	19.329	19.355	19.380	19.406	19.431	19.456	19.482
64	19.507	19.533	19.558	19.583	19.609	19.634	19.660	19.685	19.710	19.736	19.761	19.787
65	19.812	19.837	19.863	19.888	19.914	19.939	19.964	19.990	20.015	20.041	20.066	20.091
66	20.117	20.142	20.168	20.193	20.218	20.244	20.269	20.295	20.320	20.345	20.371	20.396
67	20.422	20.447	20.472	20.498	20.523	20.549	20.574	20.599	20.625	20.650	20.676	20.701
68	20.726	20.752	20.777	20.803	20.828	20.853	20.879	20.904	20.930	20.955	20.980	21.006
69	21.031	21.057	21.082	21.107	21.133	21.158	21.184	21.209	21.234	21.260	21.285	21.311
70	21.336	21.361	21.387	21.412	21.438	21.463	21.488	21.514	21.539	21.565	21.590	21.615
71	21.641	21.666	21.692	21.717	21.742	21.768	21.793	21.819	21.844	21.869	21.895	21.920
72	21.946	21.971	21.996	22.022	22.047	22.073	22.098	22.123	22.149	22.174	22.200	22.225
73	22.250	22.276	22.301	22.327	22.352	22.377	22.403	22.428	22.454	22.479	22.504	22.530
74	22.555	22.581	22.606	22.631	22.657	22.682	22.708	22.733	22.758	22.784	22.809	22.835
75	22.860	22.885	22.911	22.936	22.962	22.987	23.012	23.038	23.063	23.089	23.114	23.139
76	23.165	23.190	23.216	23.241	23.266	23.292	23.317	23.343	23.368	23.393	23.419	23.444
77	23.470	23.495	23.520	23.546	23.571	23.597	23.622	23.647	23.673	23.698	23.724	23.749
78	23.774	23.800	23.825	23.851	23.876	23.901	23.927	23.952	23.978	24.003	24.028	24.054
79	24.079	24.105	24.130	24.155	24.181	24.206	24.232	24.257	24.282	24.308	24.333	24.359
80	24.384	24.409	24.435	24.460	24.486	24.511	24.536	24.562	24.587	24.613	24.638	24.663
81	24.689	24.714	24.740	24.765	24.790	24.816	24.841	24.867	24.892	24.917	24.943	24.968
82	24.994	25.019	25.044	25.070	25.095	25.121	25.146	25.171	25.197	25.222	25.248	25.273
83	25.298	25.324	25.349	25.375	25.400	25.425	25.451	25.476	25.502	25.527	25.552	25.578
84	25.603	25.629	25.654	25.679	25.705	25.730	25.756	25.781	25.806	25.832	25.857	25.883
85	25.908	25.933	25.959	25.984	26.010	26.035	26.060	26.086	26.111	26.137	26.162	26.187
86	26.213	26.238	26.264	26.289	26.314	26.340	26.365	26.391	26.416	26.441	26.467	26.492
87	26.518	26.543	26.568	26.594	26.619	26.645	26.670	26.695	26.721	26.746	26.772	26.797
88	26.822	26.848	26.873	26.899	26.924	26.949	26.975	27.000	27.026	27.051	27.076	27.102
89	27.127	27.153	27.178	27.203	27.229	27.254	27.280	27.305	27.330	27.356	27.381	27.407
90	27.432	27.457	27.483	27.508	27.534	27.559	27.584	27.610	27.635	27.661	27.686	27.711
91	27.737	27.762	27.788	27.813	27.838	27.864	27.889	27.915	27.940	27.965	27.991	28.016
92	28.042	28.067	28.092	28.118	28.143	28.169	28.194	28.219	28.245	28.270	28.296	28.321
93	28.346	28.372	28.397	28.423	28.448	28.473	28.499	28.524	28.550	28.575	28.600	28.626
94	28.651	28.677	28.702	28.727	28.753	28.778	28.804	28.829	28.854	28.880	28.905	28.931
95	28.956	28.981	29.007	29.032	29.058	29.083	29.108	29.134	29.159	29.185	29.210	29.235
96	29.261	29.286	29.312	29.337	29.362	29.388	29.413	29.439	29.464	29.489	29.515	29.540
97	29.566	29.591	29.616	29.642	29.667	29.693	29.718	29.743	29.769	29.794	29.820	29.845
98	29.870	29.896	29.921	29.947	29.972	29.997	30.023	30.048	30.074	30.099	30.124	30.150
99	30.175	30.201	30.226	30.251	30.277	30.302	30.328	30.353	30.378	30.404	30.429	30.455
100	30.480	· · ·	· · ·	· · ·	· · ·	· · ·	· · ·	· · ·	· · ·	· · ·	· · ·	· · ·

0 TO 500 SQUARE FEET BY ONE SQ. FT. INCREMENTS
SQUARE FEET TO SQUARE METRES

Square Feet	0	1	2	3	4	5	6	7	8	9
	Square Metres (m²)									
0	· · ·	0.09	0.19	0.28	0.37	0.46	0.56	0.65	0.74	0.84
10	0.93	1.02	1.11	1.21	1.30	1.39	1.49	1.58	1.67	1.77
20	1.86	1.95	2.04	2.14	2.23	2.32	2.42	2.51	2.60	2.69
30	2.79	2.88	2.97	3.07	3.16	3.25	3.34	3.44	3.53	3.62
40	3.72	3.81	3.90	3.99	4.09	4.18	4.27	4.37	4.46	4.55
50	4.65	4.74	4.83	4.92	5.02	5.11	5.20	5.30	5.39	5.48
60	5.57	5.67	5.76	5.85	5.95	6.04	6.13	6.22	6.32	6.41
70	6.50	6.60	6.69	6.78	6.87	6.97	7.06	7.15	7.25	7.34
80	7.43	7.53	7.62	7.71	7.80	7.90	7.99	8.08	8.18	8.27
90	8.36	8.45	8.55	8.64	8.73	8.83	8.92	9.01	9.10	9.20
100	9.29	9.38	9.48	9.57	9.66	9.75	9.85	9.94	10.03	10.13
110	10.22	10.31	10.41	10.50	10.59	10.68	10.78	10.87	10.96	11.06
120	11.15	11.24	11.33	11.43	11.52	11.61	11.71	11.80	11.89	11.98
130	12.08	12.17	12.26	12.36	12.45	12.54	12.63	12.73	12.82	12.91
140	13.01	13.10	13.19	13.29	13.38	13.47	13.56	13.66	13.75	13.84
150	13.94	14.03	14.12	14.21	14.31	14.40	14.49	14.59	14.68	14.77
160	14.86	14.96	15.05	15.14	15.24	15.33	15.42	15.51	15.61	15.70
170	15.79	15.89	15.98	16.07	16.17	16.26	16.35	16.44	16.54	16.63
180	16.72	16.82	16.91	17.00	17.09	17.19	17.28	17.37	17.47	17.56
190	17.65	17.74	17.84	17.93	18.02	18.12	18.21	18.30	18.39	18.49
200	18.58	18.67	18.77	18.86	18.95	19.05	19.14	19.23	19.32	19.42
210	19.51	19.60	19.70	19.79	19.88	19.97	20.07	20.16	20.25	20.35
220	20.44	20.53	20.62	20.72	20.81	20.90	21.00	21.09	21.18	21.27
230	21.37	21.46	21.55	21.65	21.74	21.83	21.93	22.02	22.11	22.20
240	22.30	22.39	22.48	22.58	22.67	22.76	22.85	22.95	23.04	23.13
250	23.23	23.32	23.41	23.50	23.60	23.69	23.78	23.88	23.97	24.06
260	24.15	24.25	24.34	24.43	24.53	24.62	24.71	24.81	24.90	24.99
270	25.08	25.18	25.27	25.36	25.46	25.55	25.64	25.73	25.83	25.92
280	26.01	26.11	26.20	26.29	26.38	26.48	26.57	26.66	26.76	26.85
290	26.94	27.03	27.13	27.22	27.31	27.41	27.50	27.59	27.69	27.78
300	27.87	27.96	28.06	28.15	28.24	28.34	28.43	28.52	28.61	28.71
310	28.80	28.89	28.99	29.08	29.17	29.26	29.36	29.45	29.54	29.64
320	29.73	29.82	29.91	30.01	30.10	30.19	30.29	30.38	30.47	30.57
330	30.66	30.75	30.84	30.94	31.03	31.12	31.22	31.31	31.40	31.49
340	31.59	31.68	31.77	31.87	31.96	32.05	32.14	32.24	32.33	32.42
350	32.52	32.61	32.70	32.79	32.89	32.98	33.07	33.17	33.26	33.35
360	33.45	33.54	33.63	33.72	33.82	33.91	34.00	34.10	34.19	34.28
370	34.37	34.47	34.56	34.65	34.75	34.84	34.93	35.02	35.12	35.21
380	35.30	35.40	35.49	35.58	35.67	35.77	35.86	35.95	36.05	36.14
390	36.23	36.33	36.42	36.51	36.60	36.70	36.79	36.88	36.98	37.07
400	37.16	37.25	37.35	37.44	37.53	37.63	37.72	37.81	37.90	38.00
410	38.09	38.18	38.28	38.37	38.46	38.55	38.65	38.74	38.83	38.93
420	39.02	39.11	39.21	39.30	39.39	39.48	39.58	39.67	39.76	39.86
430	39.95	40.04	40.13	40.23	40.32	40.41	40.51	40.60	40.69	40.78

0 TO 500 SQUARE FEET BY ONE SQ. FT. INCREMENTS
SQUARE FEET TO SQUARE METRES (CONT.)

Square Feet	0	1	2	3	4	5	6	7	8	9
	Square Metres (m²)									
440	40.88	40.97	41.06	41.16	41.25	41.34	41.43	41.53	41.62	41.71
450	41.81	41.90	41.99	42.09	42.18	42.27	42.36	42.46	42.55	42.64
460	42.74	42.83	42.92	43.01	43.11	43.20	43.29	43.39	43.48	43.57
470	43.66	43.76	43.85	43.94	44.04	44.13	44.22	44.31	44.41	44.50
480	44.59	44.69	44.78	44.87	44.97	45.06	45.15	45.24	45.34	45.43
490	45.52	45.62	45.71	45.80	45.89	45.99	46.08	46.17	46.27	46.36
500	46.45	· · ·	· · ·	· · ·	· · ·	· · ·	· · ·	· · ·	· · ·	· · ·

0 TO 100 SQUARE INCHES BY ONE SQ. IN. INCREMENTS
SQUARE INCHES TO SQUARE MILLIMETRES

Square Inches	0	1	2	3	4	5	6	7	8	9
	Square Millimetres (mm²)									
0	· · ·	645.2	1290.3	1935.5	2580.6	3225.8	3871.0	4516.1	5161.3	5806.4
10	6451.6	7096.8	7741.9	8387.1	9032.2	9677.4	10322.6	10967.7	11612.9	12258.0
20	12903.2	13548.4	14193.5	14838.7	15483.8	16129.0	16774.2	17419.3	18064.5	18709.6
30	19354.8	20000.0	20645.1	21290.3	21935.4	22580.6	23225.8	23870.9	24516.1	25161.2
40	25806.4	26451.6	27096.7	27741.9	28387.0	29032.2	29677.4	30322.5	30967.7	31612.8
50	32258.0	32903.2	33548.3	34193.5	34838.6	35483.8	36129.0	36774.1	37419.3	38064.4
60	38709.6	39354.8	39999.9	40645.1	41290.2	41935.4	42580.6	43225.7	43870.9	44516.0
70	45161.2	45806.4	46451.5	47096.7	47741.8	48387.0	49032.2	49677.3	50322.5	50967.6
80	51612.8	52258.0	52903.1	53548.3	54193.4	54838.6	55483.8	56128.9	56774.1	57419.2
90	58064.4	58709.6	59354.7	59999.9	60645.0	61290.2	61935.4	62580.5	63225.7	63870.8
100	64516.0	· · ·	· · ·	· · ·	· · ·	· · ·	· · ·	· · ·	· · ·	· · ·

0 TO 500 U.S. GALLONS BY ONE-GALLON INCREMENTS

U.S. Gallons to Litres Conversions

U.S. Gallons	U.S. Gallons									
	0	1	2	3	4	5	6	7	8	9
	Litres									
0	· · ·	3.785	7.571	11.36	15.14	18.93	22.71	26.50	30.28	34.07
10	37.85	41.64	45.42	49.21	53.00	56.78	60.57	64.35	68.14	71.92
20	75.71	79.49	83.28	87.06	90.85	94.64	98.42	102.2	106.0	109.8
30	113.6	117.3	121.1	124.9	128.7	132.5	136.3	140.1	143.8	147.6
40	151.4	155.2	159.0	162.8	166.6	170.3	174.1	177.9	181.7	185.5
50	189.3	193.1	196.8	200.6	204.4	208.2	212.0	215.8	219.6	223.3
60	227.1	230.9	234.7	238.5	242.3	246.1	249.8	253.6	257.4	261.2
70	265.0	268.8	272.5	276.3	280.1	283.9	287.7	291.5	295.3	299.0
80	302.8	306.6	310.4	314.2	318.0	321.8	325.5	329.3	333.1	336.9
90	340.7	344.5	348.3	352.0	355.8	359.6	363.4	367.2	371.0	374.8
100	378.5	382.3	386.1	389.9	393.7	397.5	401.3	405.0	408.8	412.6
110	416.4	420.2	424.0	427.8	431.5	435.3	439.1	442.9	446.7	450.5
120	454.2	458.0	461.8	465.6	469.4	473.2	477.0	480.7	484.5	488.3
130	492.1	495.9	499.7	503.5	507.2	511.0	514.8	518.6	522.4	526.2
140	530.0	533.7	537.5	541.3	545.1	548.9	552.7	556.5	560.2	564.0
150	567.8	571.6	575.4	579.2	583.0	586.7	590.5	594.3	598.1	601.9
160	605.7	609.5	613.2	617.0	620.8	624.6	628.4	632.2	635.9	639.7
170	643.5	647.3	651.1	654.9	658.7	662.4	666.2	670.0	673.8	677.6
180	681.4	685.2	688.9	692.7	696.5	700.3	704.1	707.9	711.7	715.4
190	719.2	723.0	726.8	730.6	734.4	738.2	741.9	745.7	749.5	753.3
200	757.1	760.9	764.7	768.4	772.2	776.0	779.8	783.6	787.4	791.2
210	794.9	798.7	802.5	806.3	810.1	813.9	817.6	821.4	825.2	829.0
220	832.8	836.6	840.4	844.1	847.9	851.7	855.5	859.3	863.1	866.9
230	870.6	874.4	878.2	882.0	885.8	889.6	893.4	897.1	900.9	904.7
240	908.5	912.3	916.1	919.9	923.6	927.4	931.2	935.0	938.8	942.6
250	946.4	950.1	953.9	957.7	961.5	965.3	969.1	972.9	976.6	980.4
260	984.2	988.0	991.8	995.6	999.3	1003	1007	1011	1014	1018
270	1022	1026	1030	1033	1037	1041	1045	1049	1052	1056
280	1060	1064	1067	1071	1075	1079	1083	1086	1090	1094
290	1098	1102	1105	1109	1113	1117	1120	1124	1128	1132
300	1136	1139	1143	1147	1151	1155	1158	1162	1166	1170
310	1173	1177	1181	1185	1189	1192	1196	1200	1204	1208
320	1211	1215	1219	1223	1226	1230	1234	1238	1242	1245
330	1249	1253	1257	1261	1264	1268	1272	1276	1279	1283
340	1287	1291	1295	1298	1302	1306	1310	1314	1317	1321
350	1325	1329	1332	1336.	1340	1344	1348	1351	1355	1359

0 TO 500 U.S. GALLONS BY ONE-GALLON INCREMENTS (CONT.)

U.S. Gallons to Litres Conversions

U.S. Gallons	U.S. Gallons									
	0	1	2	3	4	5	6	7	8	9
	Litres									
360	1363	1367	1370	1374	1378	1382	1385	1389	1393	1397
370	1401	1404	1408	1412	1416	1420	1423	1427	1431	1435
380	1438	1442	1446	1450	1454	1457	1461	1465	1469	1473
390	1476	1480	1484	1488	1491	1495	1499	1503	1507	1510
400	1514	1518	1522	1526	1529	1533	1537	1541	1544	1548
410	1552	1556	1560	1563	1567	1571	1575	1579	1582	1586
420	1590	1594	1597	1601	1605	1609	1613	1616	1620	1624
430	1628	1632	1635	1639	1643	1647	1650	1654	1658	1662
440	1666	1669	1673	1677	1681	1685	1688	1692	1696	1700
450	1703	1707	1711	1715	1719	1722	1726	1730	1734	1738
460	1741	1745	1749	1753	1756	1760	1764	1768	1772	1775
470	1779	1783	1787	1790	1794	1798	1802	1806	1809	1813
480	1817	1821	1825	1828	1832	1836	1840	1843	1847	1851
490	1855	1859	1862	1866	1870	1874	1878	1881	1885	1889
500	1893	1896	1900	1904	1908	1912	1915	1919	1923	1927

0 TO 100 CUBIC FEET BY ONE CUBIC FOOT INCREMENTS
CUBIC FEET TO CUBIC METRES

Cubic Feet	0	1	2	3	4	5	6	7	8	9
	Cubic Metres (m³)									
0	· · ·	0.03	0.06	0.08	0.11	0.14	0.17	0.20	0.23	0.25
10	0.28	0.31	0.34	0.37	0.40	0.42	0.45	0.48	0.51	0.54
20	0.57	0.59	0.62	0.65	0.68	0.71	0.73	0.76	0.79	0.82
30	0.85	0.88	0.91	0.93	0.96	0.99	1.02	1.05	1.08	1.10
40	1.13	1.16	1.19	1.22	1.25	1.27	1.30	1.33	1.36	1.39
50	1.42	1.44	1.47	1.50	1.53	1.56	1.59	1.61	1.64	1.67
60	1.70	1.73	1.76	1.78	1.81	1.84	1.87	1.90	1.93	1.95
70	1.98	2.01	2.04	2.07	2.10	2.12	2.15	2.18	2.21	2.24
80	2.27	2.29	2.32	2.35	2.38	2.41	2.44	2.46	2.49	2.52
90	2.55	2.58	2.61	2.63	2.66	2.69	2.72	2.75	2.78	2.80
100	2.83	· · ·	· · ·	· · ·	· · ·	· · ·	· · ·	· · ·	· · ·	· · ·

Index

ٲ